Mathematics Study Resources

Volume 20

This series comprises direct translations of successful foreign language titles, especially from the German language.

Powered by advances in automated translation, these books draw on global teaching excellence to provide students and lecturers with diverse materials for teaching and study.

Karl Mosler · Rainer Dyckerhoff ·
Christoph Scheicher

Mathematical Methods for Business and Economics

From Calculus to Optimization and Dynamic Systems

Springer

Karl Mosler
Department of Econometrics and Statistics
University of Cologne
Cologne, Germany

Rainer Dyckerhoff
Department of Econometrics and Statistics
University of Cologne
Cologne, Germany

Christoph Scheicher
Department of Econometrics and Statistics
University of Cologne
Cologne, Germany

ISSN 2731-3824 ISSN 2731-3832 (electronic)
Mathematics Study Resources
ISBN 978-3-662-73830-6 ISBN 978-3-662-73831-3 (eBook)
https://doi.org/10.1007/978-3-662-73831-3

This book is a translation of the original German edition "Mathematische Methoden für Ökonomen," 3rd edition, by Karl Mosler et al., published by Springer-Verlag GmbH, DE in 2018. The translation was done with the help of an artificial intelligence machine translation tool. A subsequent human revision was done primarily in terms of content, so that the book will read stylistically differently from a conventional translation. Springer Nature works continuously to further the development of tools for the production of books and on the related technologies to support the authors.

Responsible Editor: Iris Ruhmann
This Springer imprint is published by the registered company Springer-Verlag GmbH, DE, part of Springer Nature
The registered company address is: Heidelberger Platz 3, 14197 Berlin, Germany

Preface

In the business world, calculations are made. It's about quantities and prices, revenues and profits and their relationships to each other. Many economic problems can be solved by adding and multiplying numbers and calculating percentages, but not all! Modern economics uses mathematical symbols and terms to represent economic facts in a general and concise way. It mainly uses differential and integral calculus as well as linear models and systems of equations. The economic literature is therefore full of formulas and mathematical derivations. Those studying business administration or economics may initially find this terrible or at least need some getting used to. However, a deep understanding of mathematics is not necessary for the basic concepts, and the benefits are easy to see. In principle, everything that can be described by a mathematical formula can also be expressed in plain words, but this usually requires much more space. And: Who wants to read all this – like this preface – and also remember it? The mathematical notation is largely used as shorthand to formulate facts concisely and memorably. In addition, the mathematical language is more precise than ordinary language. Finally, mathematical formulation allows calculations and computational methods, for example with the help of differential calculus or linear programming to calculate optimal solutions for an economic problem. This book aims to introduce these concepts. It should enable students to read economic texts, understand economic models and apply optimization methods.

A special feature of the book is the close link between mathematical concepts and those of economics and operations research. Production and utility functions are introduced early on, as are models of interest and growth of an economy. This is intended to clarify the role of mathematical methods as part of the basic training in economics and facilitate the transfer to other economic disciplines.

The book is aimed at students in bachelor's and master's programs in economic subjects. It is intended to accompany students throughout their studies and serve as a reference book alongside courses in micro- and macroeconomics, econometrics, operations research, probability theory and special topics in economics and business administration. Numerous learning aids, worked examples and exercises with short solutions enable self-study.

The prerequisite for this textbook is high school mathematics, including introductory courses on real functions and differential calculus as well as algebraic operations, trigonometric functions and analytical geometry. Strictly speaking, we only assume some experience and skill in dealing with mathematical concept formation and formulas. Appendices on set operations, combinatorics, propositional logic, mathematical proofs, and complex numbers provide summaries of specific material for short reference. The bibliography refers to textbooks of different academic levels, for repetition of elementary mathematics as well as for further reading.

The book is based on lectures that the authors have regularly and for many years given to students of economics and business administration in Hamburg and Cologne and continue to give. The material of chapters 1 to 6 is structured in such a way that one can reach optimization in several variables as early as possible in an introductory course, as required in a possibly parallel course "Microeconomics". If the "Microeconomics" falls into a later semester, it may be advisable to bring forward the linear algebra presented in chapters 8 and 9.

In addition to introductory courses in the bachelor's and master's programs in economics, the textbook is also suitable for more specialized courses, such as a course "Optimization of Linear Systems", which consists of chapters 2 and 8 to 11, or a course "Dynamic Systems in Continuous Variables" from chapters 1, 3, 4, 7 and 12. Similarly, chapters 2, 8, 9, 10 and 13 (which is independent of chapter 12) can form the basis for a course "Dynamic Systems in Discrete Variables", possibly followed by a course on the econometric analysis of time series.

Two areas of practical mathematics are not covered in this book: statistics and probability on the one hand, and game theory on the other. Both are important in modern economics, but our focus is on the mathematical methods that lie at the core of economics. It is also worth noting that several Nobel Prize winners in Economics, among them Leonid Kantorovich, John F. Nash Jr., Robert J. Aumann, and Lloyd S. Shapley, have been awarded the prize for their fundamental contributions to the mathematics that underlie economics.

The present edition is based on the third German edition of the textbook, which has been completely reviewed, corrected, and supplemented. In the editing and correction of the various editions of the textbook, we have always been actively supported by the research assistants and student assistants of the Institute of Econometrics and Statistics at the University of Cologne.

Iris Ruhmann and Verena Nörthen from Springer-Verlag have accompanied the creation of this edition with great commitment and patience. We would like to express our heartfelt thanks to all of them.

Cologne, March 2026

Rainer Dyckerhoff
Karl Mosler
Christoph Scheicher

Contents

Chapter 1

Functions

Economic models establish quantitative relationships between quantities and prices, and other variables. We will start with an example from production.

Example 1.1 (Production). A company produces a milk drink from two ingredients (inputs): whole milk and milk powder. From x_1 liters of whole milk and x_2 kilograms of milk powder, we get

$$q = 0.7x_1 + 0.1x_2$$

liters of the drink. The proportions of whole milk and milk powder can be freely chosen. An **isoquant** is the set of all input combinations in the (x_1, x_2)-plane that produce a certain quantity, q. Figure 1.1 shows the isoquants of production for $q = 100, 200, 300$.

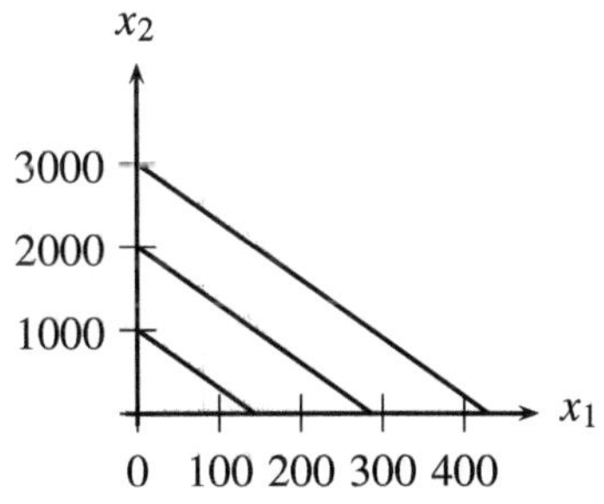

Figure 1.1: Isoquants of a production function.

More generally, we consider a good that is produced from one or more factors of production. Given the technical conditions of production, the quantity produced

K. Mosler et al., *Mathematical Methods for Business and Economics*, Mathematics Study Resources 20, https://doi.org/10.1007/978-3-662-73831-3_1

(output) depends only on the quantities of the factors used (inputs). Typical questions are: By how much does the output change if all inputs are increased by one percent? How can one production factor be replaced by another to achieve the same output? How can a given output be produced at minimal cost?

1.1 Basic concepts

To handle economic variables in a quantitative way, we need the mathematical concept of a function.

Definition 1.1: Function

Let A and B be non-empty sets. To each element $x \in A$ exactly one element $y = f(x) \in B$ is assigned. This is referred to as a **function** f from A to B, symbolically

$$\begin{aligned} f : A &\to B\,, \\ x &\mapsto f(x)\,. \end{aligned}$$

A is called **domain (of definition)**, B is called **codomain** of the function f, x denotes the variable, and $f(x)$ the function value at the point x.

A function thus consists of three things: domain, codomain, and unique assignment. Instead of domain and codomain, one also says **source** and **target**. For most of the functions that interest us, the **function value** $f(x)$ is a real number, and the **argument** x consists of one or more real numbers; such functions are briefly called **real-valued functions** of one or more variables (all variables being real numbers unless otherwise stated). A real-valued function of one variable is simply called a **real function**.

In the following, $\mathbb{R}$ denotes the set of **real numbers**; they correspond to the points on the real line. $\mathbb{R}^2$ stands for the set of all pairs (x_1, x_2) of real numbers, i.e., the points on the number plane. $\mathbb{R}_+$ denotes the set of all non-negative real numbers and $\mathbb{R}_+^2$ the set of all pairs of such numbers. $\mathbb{N}$ stands for the **natural numbers**, $\mathbb{N} = \{1, 2, 3, \ldots\}$, and $\mathbb{Z}$ for the **integers**, $\mathbb{Z} = \{\ldots, -2, -1, 0, 1, 2, \ldots\}$. A closed interval on the real line is denoted by $[a, b]$, an open one by (a, b); half-open intervals are $(a, b]$ and $[a, b)$. Here, a, b are numbers from $\mathbb{R}$ with $a < b$.[1]

Example 1.2 (Square root function). The function

$$\begin{aligned} f_1 : \mathbb{R}_+ &\to \mathbb{R} \\ x &\mapsto y = \sqrt{x} \end{aligned}$$

is referred to as the **square root function**. The square root function is depicted by its graph in the (x, y)-plane (Figure 1.2):

[1] More about intervals and other special subsets of $\mathbb{R}$ can be found in Appendix B.

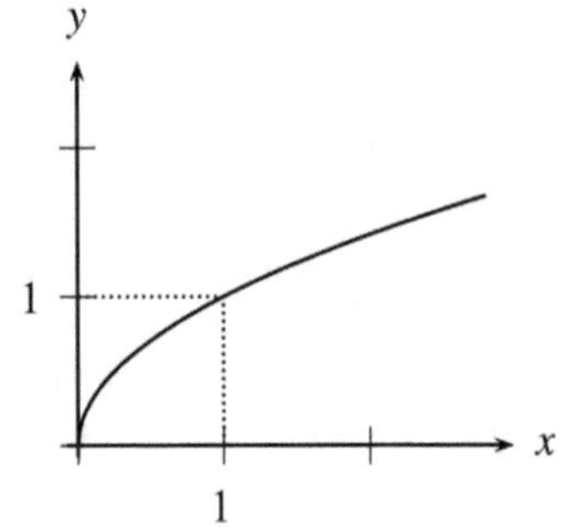

Figure 1.2: Square root function $f : \mathbb{R}_+ \to \mathbb{R}, x \mapsto \sqrt{x}$.

Example 1.3. The function

$$\begin{aligned} f_2 : \mathbb{R}_+^2 &\to \mathbb{R} \\ (x_1, x_2) &\mapsto y = x_1 \sqrt{x_2} \end{aligned}$$

is an example of a function of two arguments. The graph of such a function can be visualized as a "mountain range" over the (x_1, x_2)-plane; see Figure 1.3.

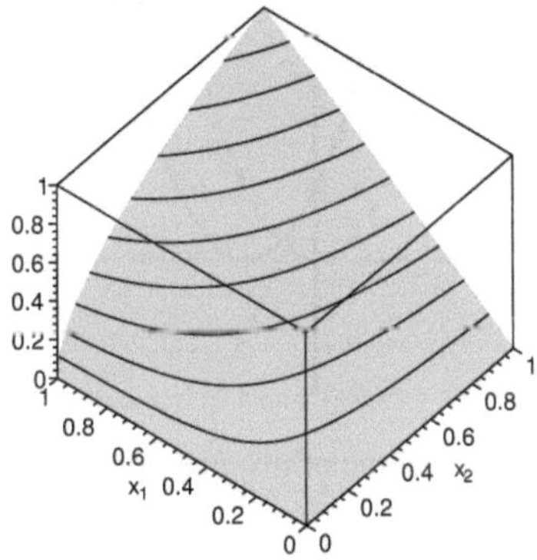

Figure 1.3: Graph of the function $f(x_1, x_2) = x_1 \sqrt{x_2}$.

Example 1.4 (Cobb-Douglas function). A function that plays a major role in economics is the so-called **Cobb-Douglas function**[2]. It is often used as a production or utility function.

$$\begin{aligned} f_3 : \mathbb{R}_+^2 &\to \mathbb{R} \\ (x_1, x_2) &\mapsto y = c\, x_1^{\alpha} x_2^{\beta} \end{aligned}$$

with parameters $\alpha, \beta, c > 0$.

If, as in these three examples, the codomain of a function is equal to the real numbers $\mathbb{R}$, the function is also written in the form $f(x)$, $x \in A$. Then the example functions

[2]Charles Wiggins Cobb (1875–1949) and Paul Howard Douglas (1892–1976)

are written as

$$f_1(x) = \sqrt{x}, \qquad x \in \mathbb{R}_+,$$
$$f_2(x_1, x_2) = x_1\sqrt{x_2}, \quad (x_1, x_2) \in \mathbb{R}^2_+,$$
$$f_3(x_1, x_2) = c x_1^{\alpha} x_2^{\beta}, \quad (x_1, x_2) \in \mathbb{R}^2_+.$$

While Example 1.2 represents a real function, i.e., a real-valued function of one variable, the other two examples are real-valued functions of two variables. A function can, as in Example 1.4, also depend on parameters. If we specify them as $\alpha = 1, \beta = \frac{1}{2}$ and $c = 1$, Example 1.3 is obtained. If in Example 1.3 the first variable is fixed at $x_1 = 1$ and only x_2 varies, we get Example 1.2.

Real-valued **functions**, whose domains coincide, can be **added** by adding their values. Given two functions $f : A \to \mathbb{R}$ and $g : A \to \mathbb{R}$, one defines their sum $f + g$ as

$$f + g : A \to \mathbb{R}, \quad x \mapsto f(x) + g(x).$$

Example 1.5. The function $h : x \mapsto x^2 - 2x + 1$ is the sum of the functions $f : x \mapsto x^2$ and $g : x \mapsto -2x + 1$; see Figure 1.4.

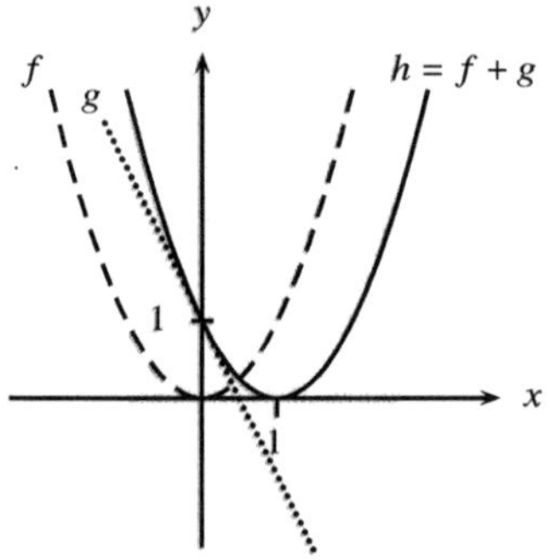

Figure 1.4: Sum of two functions, $h = f + g$.

Similarly, two functions having a common domain A can be **subtracted**, **multiplied** and **divided**:

$$f - g : A \to \mathbb{R}, \quad x \mapsto f(x) - g(x),$$
$$f \cdot g : A \to \mathbb{R}, \quad x \mapsto f(x) \cdot g(x),$$
$$\frac{f}{g} : A \to \mathbb{R}, \quad x \mapsto \frac{f(x)}{g(x)}.$$

Of course the quotient $\frac{f}{g}$ is defined only at arguments x for which the denominator does not vanish. If $g(x) = 0$ at some $x \in A$, the domain must be accordingly restricted.

Similarly, the **minimum** and the **maximum** of f and g are defined,

$$\min\{f, g\} : A \to \mathbb{R}, \quad x \mapsto \min\{f(x), g(x)\},$$
$$\max\{f, g\} : A \to \mathbb{R}, \quad x \mapsto \max\{f(x), g(x)\},$$

i.e., at each x the smaller resp. larger of the two function values $f(x)$ and $g(x)$ is taken.

Example 1.6. We consider the functions $f(x) = x$ and $g(x) = -x$, each for $x \in \mathbb{R}$; see Figure 1.5. Their maximum is the **absolute value function** $|\cdot| : \mathbb{R} \to \mathbb{R}$,

$$x \mapsto |x| = \max\{x, -x\} = \begin{cases} x & \text{if } x \geq 0\,, \\ -x & \text{if } x < 0\,. \end{cases} \tag{1.1}$$

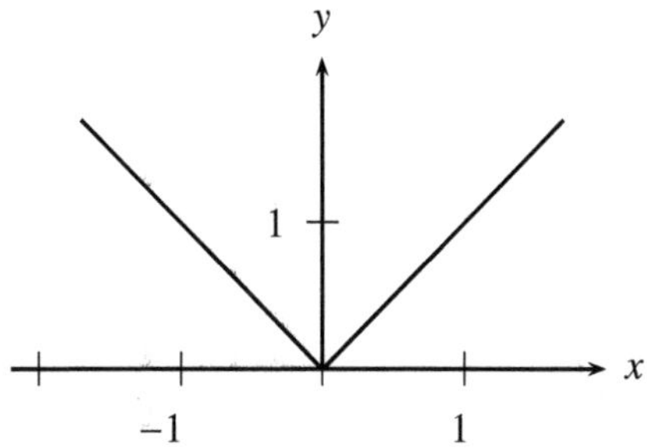

Figure 1.5: Absolute value function $|\cdot| : \mathbb{R} \to \mathbb{R}$, $x \mapsto |x|$.

The following are further examples of real functions. Their domain is either all of $\mathbb{R}$ or a proper subset of $\mathbb{R}$, their codomain, unless otherwise stated, is $\mathbb{R}$.

Example 1.7 (Polynomial of degree r). A function of the form $p : \mathbb{R} \to \mathbb{R}$,

$$p(x) = a_n x^n + a_{n-1} x^{n-1} + \ldots + a_1 x + a_0\,,$$

with coefficients $a_n, a_{n-1}, \ldots, a_1, a_0 \in \mathbb{R}$ and $n \in \mathbb{N}$ is called a **polynomial**, more precisely a polynomial of degree n, if $a_n \neq 0$. For example, the function in Example 1.5 is a polynomial of degree two. The function $p(x) = -5x^4 + x$ is a polynomial of degree four.

Definition 1.2: Graph, image of a function

For a given function $f : A \to B$, the set

$$G_f = \{(x, y) \;:\; y = f(x),\, x \in A\}$$

is called the **graph** of f, the set

$$f(A) = \{y \in B \;:\; y = f(x),\, x \in A\}$$

is called the **image** of A under f.

The image of a function is thus a subset of its codomain. The graph of a real function is a subset of $\mathbb{R}^2$. It can be visualized in the plane, see, e.g., Figures 1.2 and 1.4.

We now come to terms that describe the growth behavior of a real function.

Definition 1.3: Monotonically increasing and decreasing functions

Let $f : A \to B$ with $A, B \subset \mathbb{R}$. The function f **increases monotonically**, if the function value increases or remains unchanged with each increase of its argument, formally: if for all $x, x' \in A$ the implication[3]

$$x < x' \quad \Longrightarrow \quad f(x) \leq f(x')$$

holds.
f **increases strictly monotonically**, if the function value increases with each increase of its argument, formally: if

$$x < x' \quad \Longrightarrow \quad f(x) < f(x')\,.$$

f **decreases monotonically** or **decreases strictly monotonically**, if from $x < x'$ the inequality $f(x) \geq f(x')$ or $f(x) > f(x')$ follows, respectively.

Example 1.8 (Floor function, ceiling function). The floor function assigns to each $x \in \mathbb{R}$ the largest integer $\lfloor x \rfloor$ which is less than or equal to x. Similarly, the ceiling function assigns to each $x \in \mathbb{R}$ the smallest integer $\lceil x \rceil$, which is larger than or equal to x. Figure 1.6 shows the floor function; it is monotonically increasing, but not strictly monotonically increasing.

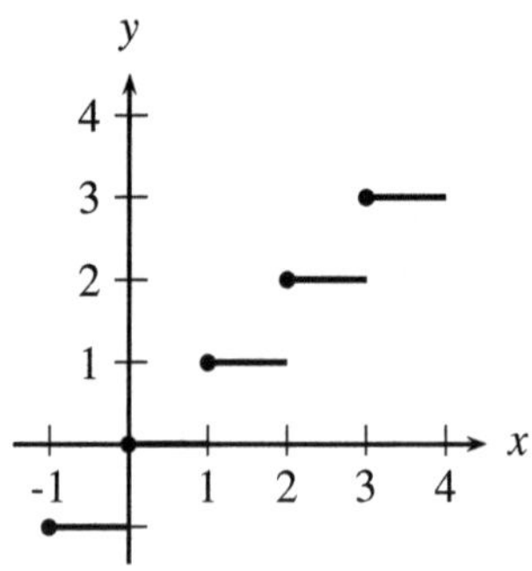

Figure 1.6: Floor function $\lfloor \cdot \rfloor : \mathbb{R} \to \mathbb{R}, x \mapsto \lfloor x \rfloor$.

The **floor function** (as well as the ceiling function) is a special **step function**. The jumps of a step function do not have to be of equal height. Similarly, the "steps", i.e., the intervals on which the function is constant, can be of different lengths.

Whether a certain function is monotonically increasing or decreasing can often be seen directly from its graph. For example, the square root function (Figure 1.2) is strictly monotonically increasing. The function h in Figure 1.4 is neither monotonically increasing nor monotonically decreasing. However, if we restrict its domain to

[3]An **implication** is an if-then statement of the form: If statement P holds, then statement Q also holds, in symbols $P \Longrightarrow Q$. In this case P is called a **sufficient condition** for Q, and Q is called a **necessary condition** for P. We write $P \Longleftrightarrow Q$, if both $P \Longrightarrow Q$ and $Q \Longrightarrow P$ hold and then say, the statements P and Q are **equivalent** (see Appendix F).

the interval $[1, \infty)$, a monotonically increasing function,

$$\widetilde{h} : [1, \infty) \to \mathbb{R}, \quad x \mapsto x^2 - 2x + 1,$$

is obtained. Correspondingly, its **restriction** to the domain $(-\infty, 1]$ yields a monotonically decreasing function.

The graphs of Examples 1.2 and 1.5 show further qualitative properties: The straight line connecting two points of the graph of the square root function (Figure 1.2) always lies below the graph. Such a function is called **concave**. For the graph of the quadratic function (Figure 1.4), it is the opposite: The connecting line between any two points lies above the graph. A function with this property is called **convex**. However, to be able to work with the notions of concavity and convexity of functions, we need a formal definition: Let f be a function defined on an interval A. We consider two arbitrary points of the graph of f, $(x, f(x))$ and $(z, f(z))$, and the line that connects them; see Figure 1.7. Every point of the connecting line has the form

$$(\alpha x + (1 - \alpha)z, \, \alpha f(x) + (1 - \alpha) f(z))$$

with $\alpha \in [0, 1]$.

Definition 1.4: Concave and convex functions

The function $f : A \to \mathbb{R}$ is called **concave**, if for all $x, z \in A$ and for all $\alpha \in (0, 1)$ it holds

$$f(\alpha x + (1 - \alpha)z) \geq \alpha f(x) + (1 - \alpha) f(z).$$

f is called **convex**, if for all $x, z \in A$ and for all $\alpha \in (0, 1)$ it holds:

$$f(\alpha x + (1 - \alpha)z) \leq \alpha f(x) + (1 - \alpha) f(z).$$

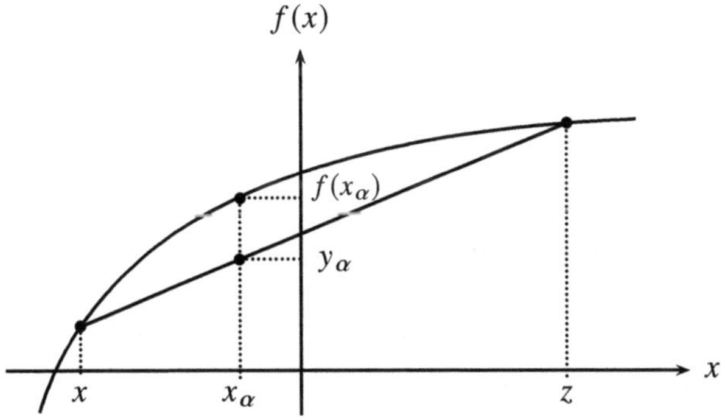

Figure 1.7: Illustration of Definition 1.4 for a concave function.

Figure 1.7 illustrates the definition of concavity: At every point $x_\alpha = \alpha x + (1 - \alpha)z$ between x and z the function value $f(\alpha x + (1 - \alpha)z)$ lies above the corresponding value $y_\alpha = \alpha f(x) + (1 - \alpha) f(z)$ on the connecting line.

Another qualitative property of functions is **boundedness**. Some functions, such as the square root function and the logarithm function take arbitrarily large and/or small values. Others, like the following function, do not.

Example 1.9 (Negative-exponential function). The function

$$f : \mathbb{R}_+ \to \mathbb{R}, \quad x \mapsto 1 - \exp(-\lambda x),$$

is called **negative-exponential function**[4]. Its function values lie between zero and one for any choice of the parameter $\lambda > 0$; see Figure 1.8.

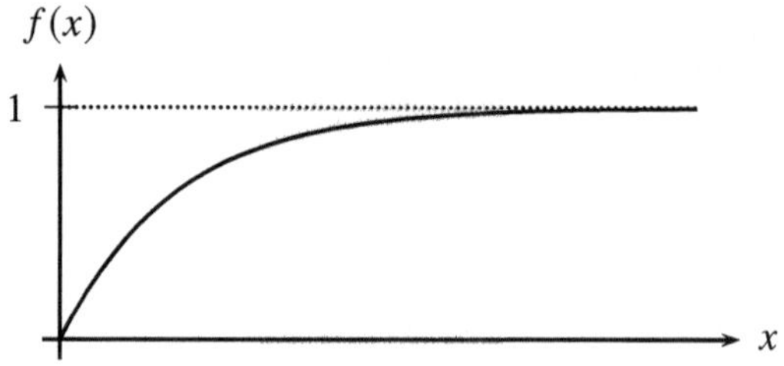

Figure 1.8: Negative-exponential function, $f(x) = 1 - \exp(-\lambda x), x \geq 0$, with $\lambda = 2$.

A function $f : A \to \mathbb{R}$ is called **bounded**, if there is a number M such that

$$|f(x)| \leq M \quad \text{for all } x \in A\,.$$

Obviously, this is satisfied for the negative-exponential function with $M = 1$, but also with any number $M > 1$. The square root function $x \mapsto \sqrt{x}$, $x \in \mathbb{R}$, and the logarithm function $x \mapsto \ln(x)$, $x \in \mathbb{R}$, are not bounded. If you consider an arbitrary bound $M > 0$ for the square root function or the logarithm function, the function takes values greater than $\sqrt{M^2} = M$ or $\ln(e^M) = M$ at all points $x > M^2$ or $x > e^M$.

Example 1.10. The restriction of the square root function to a finite interval, for example

$$f : [0, 100] \to \mathbb{R}, \quad x \mapsto \sqrt{x},$$

is bounded. Here all function values lie between $\sqrt{0} = 0$ and $\sqrt{100} = 10$; see left panel in Figure 1.9.

Example 1.11. The **hyperbolic function** $f(x) = \frac{1}{x}$ is defined for all $x \neq 0$, i.e., $x \in \mathbb{R} \setminus \{0\}$. It is not bounded, as it takes arbitrarily large values for $x > 0$ and arbitrarily small values for $x < 0$; see right panel in Figure 1.9.

Below you will find examples how functions are used in economic problems.

Example 1.12 (Interest on capital). Let K be a capital that earns interest over time. The capital that is available at time t is denoted by $f(t)$. Depending on whether the interest is credited periodically (for example at the end of each year)

[4] Here $\exp(z) = e^z$, $z \in \mathbb{R}$, denotes the exponential function. For its definition by an infinite series see Definition 3.7.

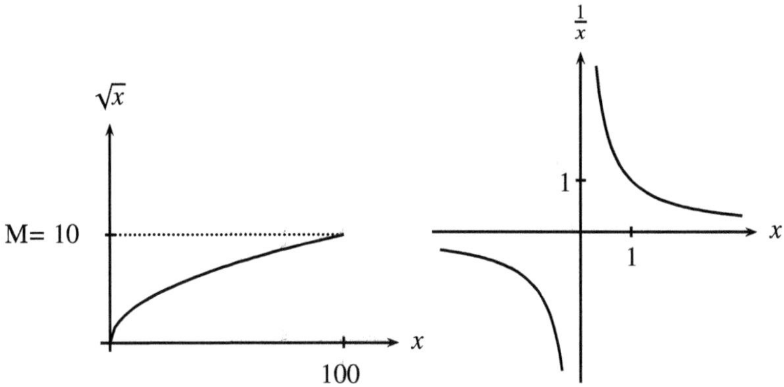

Figure 1.9: Restricted root function and hyperbolic function.

or continuously, we distinguish two cases:

a) **Periodic interest**:

$$f(t) = K(1+p)^t\,, \quad t \in \mathbb{N}\,.$$

Time t is measured in whole numbers (= periods), while p denotes the interest rate per period. The interest is credited at the end of each period t to the capital.

b) **Continuous interest** (= exponential growth):

$$f(t) = Ke^{pt}\,, \quad t \in (0,\infty)\,.$$

The interest is continuously credited to the capital at each point in time. p is the continuous interest rate.

Periodic interest is the subject of Section 3.6.

Example 1.13 (Growth with saturation). In some growth processes, a given saturation limit cannot be exceeded, such as the equipping of households in a country with a certain technology (e.g. a DSL connection). Such growth can be described by the **logistic function**,

$$f(x) = \frac{G}{1+\alpha e^{-\lambda G x}}\,, \quad x \geq 0\,,$$

see Figure 1.10, which has the parameters $\alpha > 0$, $G > 0$ and $\lambda > 0$. The saturation level is G. Typical questions are:

- How high is the maximum increase, in what way does it depend on λ and when does it occur?
- From what point in time on does the growth decrease again?

Questions like these are answered in Chapter 4 with the help of differential calculus. In Example 12.16 we will return to the logistic function.

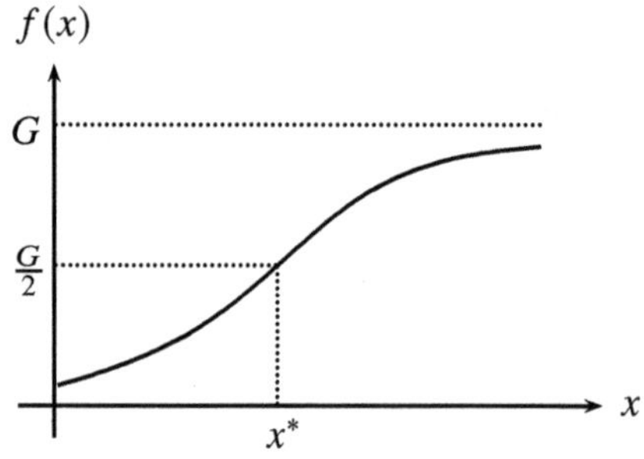

Figure 1.10: Logistic function, $x^* = \frac{\ln(a)}{\lambda G}$.

1.2 Inverse function, composition

Of particular interest is whether a given function f has an "inverse", i.e., whether for each element y of the codomain there is exactly one element x of the domain for which $y = f(x)$ holds. This is the case when each element of the codomain is reached as a function value and no two arguments have the same function value. In particular, the latter is satisfied when the function is strictly increasing (or strictly decreasing) in its entire domain. For functions on arbitrary sets, we define:

Definition 1.5: Inverse function

Let $f : A \to B$ be a function. f is called **invertible**, if for each $y \in B$ there is exactly one $x \in A$ for which $f(x) = y$ holds. Then the function

$$f^{-1} : B \to A, \quad y = f(x) \mapsto x,$$

is called the **inverse function** of f (or short **inverse** of f).

If there is an inverse function f^{-1}, it is in turn invertible, and its inverse function is the original function f.

For real functions, the question of invertibility can usually be easily examined on the image of the graph: For each y in the codomain of f, the graph G_f must intersect the horizontal line at height y exactly once. The graph of the inverse function is then obtained by reflecting G_f on the main diagonal, in other words, by swapping the x-with the y-axis. This is illustrated in the following examples.

Example 1.14. The square function $f(x) = x^2$, $x \in \mathbb{R}$, is invertible if we restrict the domain A and the codomain B to the non-negative numbers,

$$f : \mathbb{R}_+ \to \mathbb{R}_+ , \quad x \mapsto y = x^2 .$$

Its inverse function is then the square root function (Figure 1.11),

$$f^{-1} : \mathbb{R}_+ \to \mathbb{R}_+ , \quad y \mapsto x = \sqrt{y} .$$

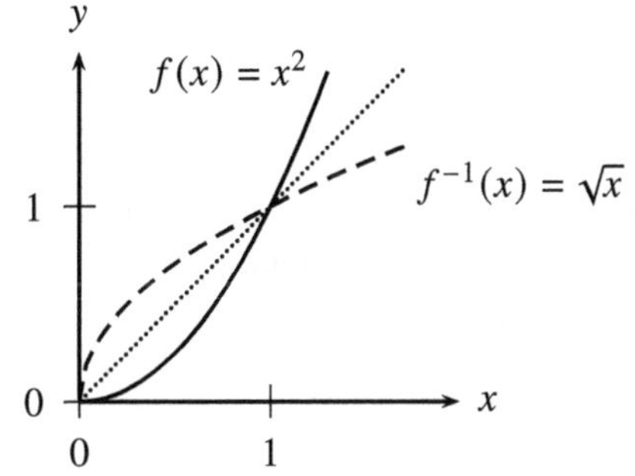

Figure 1.11: Square function and square root function.

Example 1.15. The **natural logarithm** $\ln(x)$ is the number to which the Euler's constant ($e \approx 2.71828$) must be raised to obtain x; it is defined for $x > 0$. The logarithm function

$$\ln : (0, \infty) \to \mathbb{R} , \quad x \mapsto y = \ln(x) ,$$

is (see Figure 1.12) strictly monotonically increasing and invertible. Its inverse function is the **exponential function**, $\ln^{-1} = \exp$,

$$\exp : \mathbb{R} \to (0, \infty) , \quad y \mapsto x = \exp(y) = e^y .$$

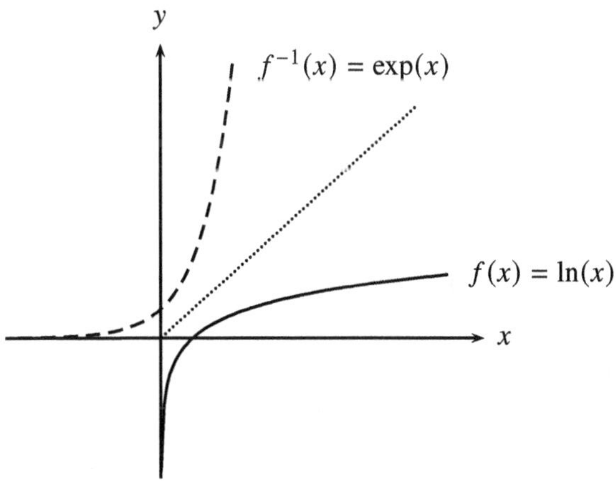

Figure 1.12: Logarithm function and exponential function.

The following example shows how to calculate the inverse function of a given function.

Example 1.16. Given is the function $f(x) = \frac{1}{x^2+1}$, $x \in \mathbb{R}_+$. To determine the inverse function of f, we solve the equation

$$y = \frac{1}{x^2 + 1}$$

for x:

$$x^2 + 1 = \frac{1}{y} \quad \Longleftrightarrow \quad x^2 = \frac{1}{y} - 1 = \frac{1-y}{y} \quad \overset{x \geq 0}{\Longleftrightarrow} \quad x = \sqrt{\frac{1-y}{y}}$$

The domain of f is given by $x \geq 0$; it corresponds to the image of the inverse function f^{-1}. The image of f, i.e., the domain of f^{-1}, includes all values y with $0 < y \leq 1$. Thus, the correspondingly restricted function f,

$$f : \mathbb{R}_+ \to (0, 1] \,, \quad x \mapsto \frac{1}{x^2 + 1} \,,$$

has the inverse function

$$f^{-1} \, : \, (0, 1] \to \mathbb{R}_+ \,, \quad y \mapsto \sqrt{\frac{1-y}{y}} \,.$$

Often the argument of a function is again a function of another variable. Two functions are **composed** by substituting the function values of one into the other. The prerequisite is that the image of the first function is contained in the domain of the second function. A new function is then obtained, which is called **composition** of the two original functions, as illustrated in Figure 1.13.

Definition 1.6: Composition of functions

Let $f : A \to B^*$, and $g : B \to C$ with $B^* \subset B$. By defining[5]

$$(g \circ f)(x) := g(f(x)) \quad \text{for } x \in A$$

we obtain the function

$$g \circ f : A \to C \,,$$

in words, “g of f” or “g after f”.

Example 1.17.

a) Let $f : \mathbb{R} \to \mathbb{R}_+, x \mapsto y = x^2 + 1$, and $g : \mathbb{R}_+ \to \mathbb{R}_+, y \mapsto \sqrt{y}$. Then

$$g \circ f : \mathbb{R} \to \mathbb{R}_+ \,, \quad x \mapsto \sqrt{x^2 + 1} \,.$$

[5] The symbol := means “is defined to be equal to”.

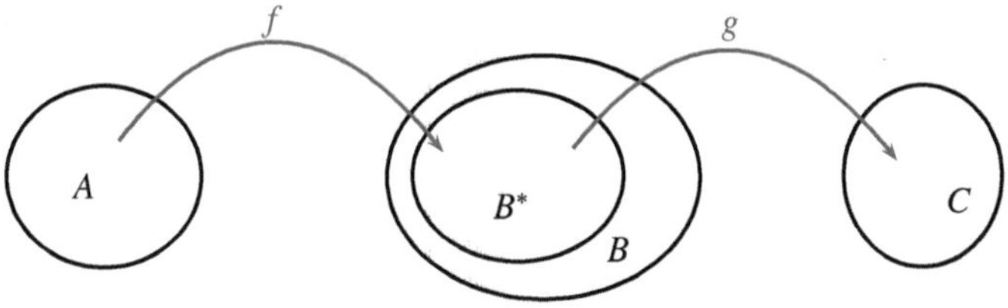

Figure 1.13: Illustration of a composite function $g \circ f$.

b) Let $h : \mathbb{R}_+ \to \mathbb{R}_+, x \mapsto x^2$, and g as before. Then

$$\begin{aligned} g \circ h : \mathbb{R}_+ \to \mathbb{R}_+, &\quad x \mapsto \sqrt{x^2} = x\,, \\ h \circ g : \mathbb{R}_+ \to \mathbb{R}_+, &\quad y \mapsto (\sqrt{y})^2 = y\,. \end{aligned}$$

If, as in Example 1.17 b) an invertible function $f : A \to B$ is composed with its inverse, one obtains $f^{-1}(f(x)) = x$ for every $x \in A$. Similarly, with the composition in reverse order, it holds $f(f^{-1}(y)) = y$ for every $y \in B$. The composition of an invertible function with its inverse is thus the **identity function** $x \mapsto x$, $x \in A$, or conversely $y \mapsto y$, $y \in B$.

Example 1.18. The **general exponential function** with base $a > 0$ is defined by

$$a^x = e^{x \cdot \ln(a)} = \exp\big(x \cdot \ln(a)\big)\,, \quad x > 0\,.$$

It is the composition of the function $x \mapsto x \cdot \ln(a)$, $x > 0$, with the exponential function.

1.3 Bivariate functions

Most of the previous examples referred to **univariate functions**, i.e., real-valued functions of one variable, and their graphical representation. A **bivariate function** is a real-valued function of two variables, $f : A \to \mathbb{R}$ with $A \subset \mathbb{R}^2$, It can be represented as well by its graph,

$$G_f = \{(x_1, x_2, f(x_1, x_2)) \; : \; (x_1, x_2) \in A\}\,.$$

G_f is a subset of $\mathbb{R}^3$, which is the set of all triples of real numbers. Intuitively, the set $\mathbb{R}^3$ corresponds to three-dimensional physical space. G_f can be described – similar to the earth's surface on a physical map (Figure 1.14) – by **contour lines**. The **contour line** $H_f(c)$ of f is the location of all points (x_1, x_2) that have the same "height", i.e., the same function value c,

$$H_f(c) = \{(x_1, x_2) \in \mathbb{R}^2 \; : \; f(x_1, x_2) = c\}\,,$$

where c runs through all values in the image of f. In Figure 1.16 selected contour lines of a bivariate function are shown. The contour lines of a production function are referred to as **isoquants**; see Example 1.1 .

The contour line of height zero is the set of **zeros** of a function. From the contour lines, one can recognize the slope behavior of a function and determine where maxima and minima are approximately located.

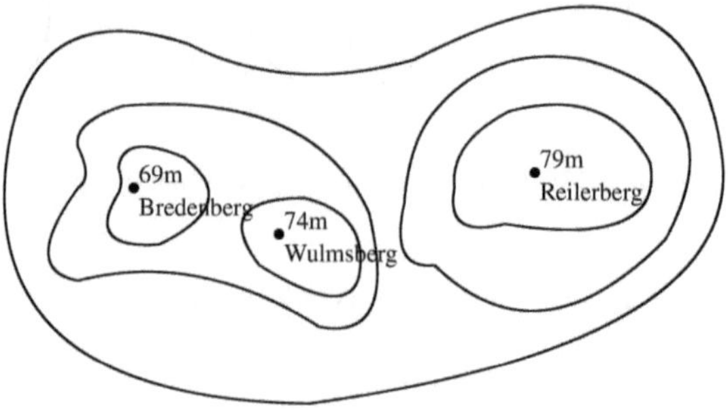

Figure 1.14: Contour lines of a physical map.

Example 1.19. The following bivariate function is defined on the whole $\mathbb{R}^2$,

$$\begin{aligned} f : \mathbb{R}^2 &\longrightarrow \mathbb{R}\,, \\ (x_1, x_2) &\longmapsto \tfrac{(x_1+2)^2}{4} + \tfrac{(x_2+1)^2}{9} - 1\,. \end{aligned}$$

Figure 1.15 depicts the graph of f. Some contour lines are shown in Figure 1.16. The set of zeros of f,

$$\left\{(x_1, x_2) \;:\; \frac{(x_1+2)^2}{4} + \frac{(x_2+1)^2}{9} = 1\right\},$$

is an ellipse around $(-2, -1)$ with major axes of length 2 and 3 that parallel the coordinate axes.

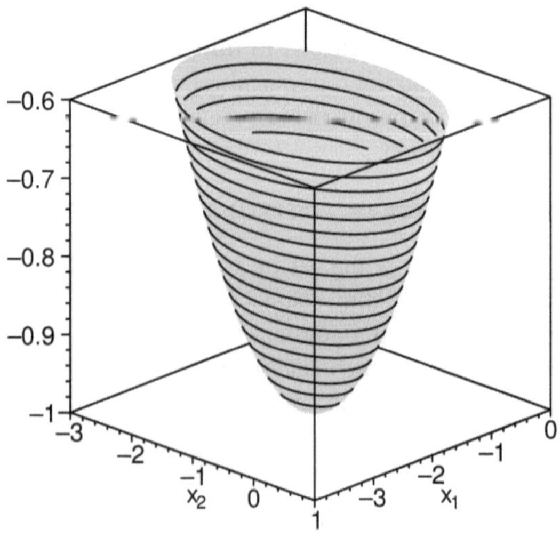

Figure 1.15: Graph of the bivariate function from Example 1.19.

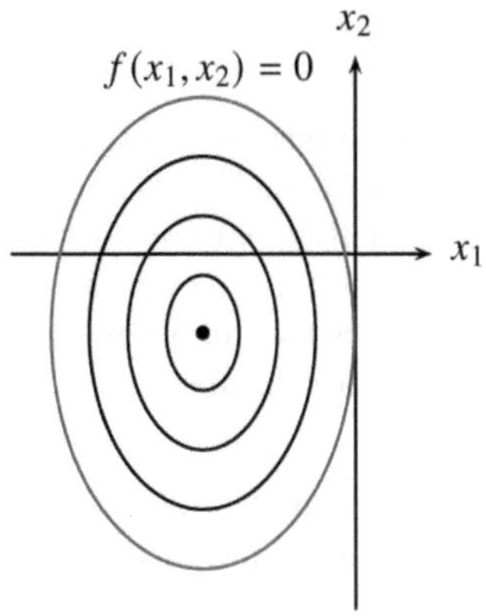

Figure 1.16: Contour lines of the function f from Example 1.19.

Example 1.20. We consider the three functions $f, g, h : \mathbb{R}^2_+ \to \mathbb{R}$,

$$f(x_1,x_2) = x_1 \cdot x_2\,, \qquad g(x_1,x_2) = x_1 + x_2\,, \qquad h(x_1,x_2) = \min\{x_1,x_2\}\,.$$

Selected contour lines are exhibited in Figure 1.17. By inspecting the graphs in Figure 1.18 we see immediately that all three functions are minimum at $(x_1,x_2) = (0,0)$.

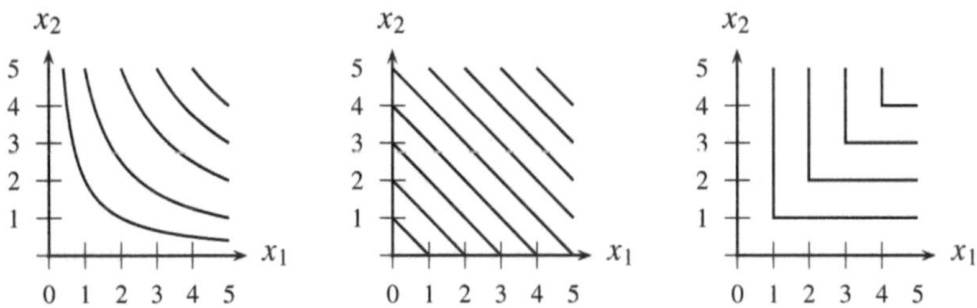

Figure 1.17: Contour lines of the functions f, g and h from Example 1.20.

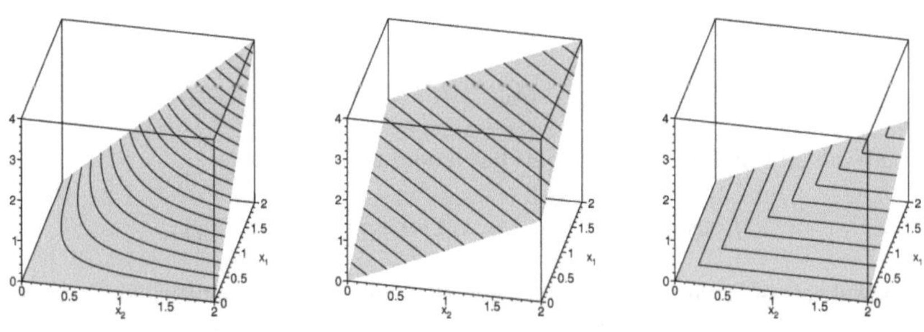

Figure 1.18: Graphs of the functions f, g and h from Example 1.20.

Definition 1.7: Monotonically increasing and decreasing functions in $\mathbb{R}^2$

A bivariate real-valued function **increases monotonically** if it increases monotonically with respect to each of the two variables (with the other variable held constant). It **increases strictly monotonically** if it increases strictly monotonically with respect to each of the two variables. Correspondingly, we say that the function **decreases (strictly) monotonically** if it decreases (strictly) monotonically with respect to each variable.

Thus, the functions f and h in Example 1.20 are monotonically increasing. The function g is even strictly monotonically increasing. This is illustrated by the contour lines in Figure 1.17.

Definition 1.8: Concave and convex functions in $\mathbb{R}^2$

A bivariate real-valued function is called **concave**, if the straight line connecting two points of its graph always lies below the graph. The function is called **convex**, if this connecting line always lies above the graph.

For example, the function in Example 1.19 (Figure 1.15) is a convex function.

1.4 Multivariate functions

The functions that appear in economic models are often functions of more than two variables: The production of a good requires the use of n factors, the utility of an individual is derived from the consumption of n goods, where n is any natural number. Such a real-valued function of several variables is referred to as a **multivariate function**.

Analogous to the points (x_1, x_2) in $\mathbb{R}^2$ and (x_1, x_2, x_3) in $\mathbb{R}^3$, we group n real numbers together as a "point" and refer to the set of all such "points" as the n-dimensional **Euclidean[6] space** $\mathbb{R}^n$,

$$\mathbb{R}^n = \{(x_1, x_2, \dots, x_n) \; : \; x_i \in \mathbb{R}, \; i = 1, 2, \dots, n\} .$$

A real-valued function in n variables has a domain $A \subset \mathbb{R}^n$,

$$f : A \to \mathbb{R}, \quad (x_1, \dots, x_n) \mapsto y = f(x_1, \dots, x_n) .$$

Some subsets of $\mathbb{R}^n$ are of particular interest: As many economic variables are either positive or zero (quantities, prices, etc.), the **non-negative orthant** of $\mathbb{R}^n$,

$$\mathbb{R}^n_+ = \{(x_1, x_2, \dots, x_n) \in \mathbb{R}^n \; : \; x_i \geq 0, \; i = 1, 2, \dots, n\} .$$

[6]Euclid of Alexandria (ca. 365 BC – ca. 300 BC)

serves as the domain of numerous n-variate economic functions. Often, each variable x_i is limited to an open interval (a_i, b_i); then the corresponding points form an open **n-dimensional interval** of the form

$$(\boldsymbol{a}, \boldsymbol{b}) = (a_1, b_1) \times \ldots \times (a_n, b_n).$$

Analogously, we form n-dimensional closed and half-open intervals. Such intervals can be finite or infinite. Thus, the whole $\mathbb{R}^n$ is an interval, and also the $\mathbb{R}^n_+$. A finite two-dimensional interval is a rectangle (see Figure 1.19), a finite three-dimensional interval is a cuboid.

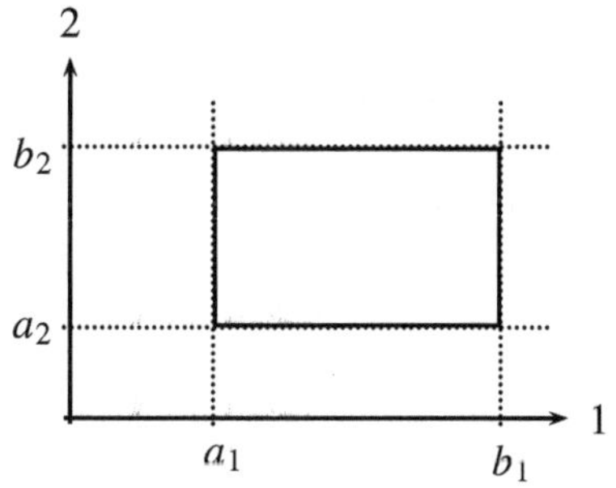

Figure 1.19: Two-dimensional closed interval.

Example 1.21 (Cobb-Douglas production function with n inputs). The general form of the Cobb-Douglas function, which we already got to know in Example 1.4 for two inputs, is

$$f(x_1, x_2, \ldots, x_n) = c\, x_1^{\alpha_1} x_2^{\alpha_2} \cdot \ldots \cdot x_n^{\alpha_n}, \qquad (x_1, x_2, \ldots, x_n) \in \mathbb{R}^n_+,$$

with parameters $c > 0, \alpha_1, \ldots, \alpha_n > 0$.

Example 1.22 (Utility of a bundle of goods). Let $f(x_1, \ldots, x_n)$ be the utility that a consumer derives from consuming the quantities x_1 of good 1, etc. up to x_n of good n. A special utility function is the **additive-logarithmic function,**

$$f(x_1, \ldots, x_n) = b_1 \ln(x_1) + \ldots + b_n \ln(x_n), \quad x_1, \ldots, x_n > 0,$$

with parameters $b_1, \ldots, b_n > 0$. It holds $f(x_1, \ldots, x_n) = \ln(x_1^{b_1} \cdot \ldots \cdot x_n^{b_n})$, so this utility function is a logarithmized Cobb-Douglas function.

A real-valued function of several variables **increases** or **decreases (strictly) monotonically**, if it increases or decreases (strictly) monotonically with respect to each individual variable while the other variables are held constant.

Clearly, the Cobb-Douglas function in Example 1.21 is strictly monotonically increasing. It attains its minimum when $(x_1, x_2, \dots, x_n) = (0, 0, \dots, 0)$ while a maximum does not exist. The isoquant for a given output q consists of all input bundles $(x_1, x_2, \dots, x_n)$, with which the output q is achieved. Each input bundle is associated with certain costs. A typical question is: How can a given output be produced at minimum cost?

Since the utility function in Example 1.22 is a logarithmized Cobb-Douglas function, it is also strictly monotonically increasing. Here a similar question arises: How can a given utility level be achieved at minimum cost? Questions like these can be answered with the help of calculus in several variables. This is the subject of Chapters 5 and 6.

1.5 Further properties of multivariate functions

Let A be an n-dimensional interval.

Definition 1.9: Concave and convex functions in $\mathbb{R}^n$

A real-valued function in n variables $f : A \to \mathbb{R}$ is called **concave**, if for all $\boldsymbol{x}, \boldsymbol{z} \in A$ and for all $\lambda \in (0, 1)$ it holds

$$f(\lambda \boldsymbol{x} + (1 - \lambda)\boldsymbol{z}) \geq \lambda f(\boldsymbol{x}) + (1 - \lambda) f(\boldsymbol{z}).$$

f is called **convex**, if for all $\boldsymbol{x}, \boldsymbol{z} \in A$ and for all $\lambda \in (0, 1)$ it holds

$$f(\lambda \boldsymbol{x} + (1 - \lambda)\boldsymbol{z}) \leq \lambda f(\boldsymbol{x}) + (1 - \lambda) f(\boldsymbol{z}).$$

The Cobb-Douglas production function in Example 1.21 and the utility function in Example 1.22 are concave functions; see Figure 1.20 .

Example 1.23. Every **linear function** $f : \mathbb{R}^n \to \mathbb{R}$,

$$f(x_1, x_2, \dots, x_n) = \alpha_1 x_1 + \alpha_2 x_2 + \dots + \alpha_n x_n ,$$

with coefficients $\alpha_1, \alpha_2, \dots, \alpha_n \in \mathbb{R}$ is both concave and convex.

Definition 1.10: Quasiconcave function in $\mathbb{R}^n$

A real-valued function in n variables $f : A \to \mathbb{R}$ is called **quasiconcave**, if for every $y \in \mathbb{R}$ the set

$$\{\boldsymbol{x} \in A : f(\boldsymbol{x}) \geq y\}$$

is convex[7].

In general, every concave function is quasiconcave, but not vice versa. For example, every linear function is not only concave, but also quasiconcave.

[7] For the definition of convexity of a set, see Section 2.7, page 50.

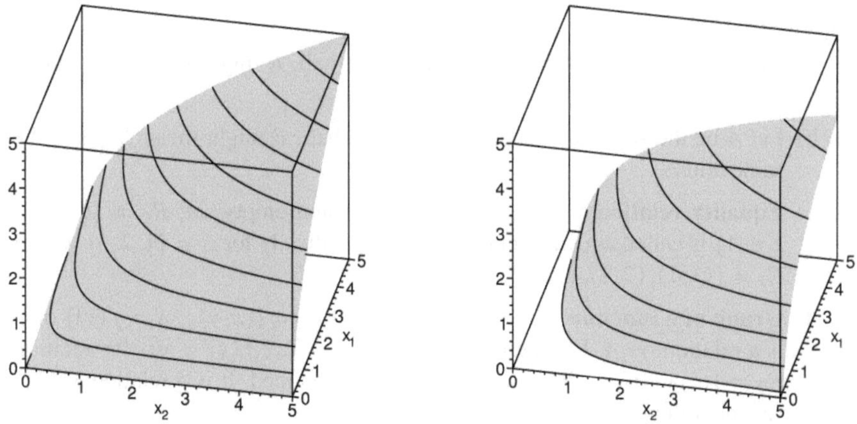

Figure 1.20: Graph of the Cobb-Douglas function $f(x_1,x_2) = x_1^{\frac{1}{2}} x_2^{\frac{1}{2}}$ and the utility function $f(x_1,x_2) = \ln(x_1) + \ln(x_2)$.

A monotonically increasing function of two variables is quasiconcave, if its contour lines are convex. This is the case for all three functions in example 1.20; see Figure 1.17. The function $f(x_1,x_2) = x_1 x_2$ is quasiconcave, but not concave. To see that f is not concave, consider the two points $\boldsymbol{x} = (0,0)$ and $\boldsymbol{x} = (1,1)$. The functions g and h are concave; see Figure 1.18.

1.6 Orders and equivalence relations

A basic task of mathematical economic analysis is to order objects to formally characterize situations such as the following: A consumer prefers the consumption of one good over another good; an investor considers an investment, described by certain indicators, to be riskier than another. The equivalence of objects should also be characterized in a general way, such as the situation that two goods are equally desirable or that two investments are considered equally risky.

Order or equivalence of objects are formally described by so-called relations on a set.

Definition 1.11: Relation

Consider a non-empty set A. A **relation** R on A is a set of pairs from A.

If $(x, y) \in R$, it is said that the element $x \in A$ is in relation R to the element $y \in A$. This is also expressed by the notation xRy.

The concept of relation is very general. We illustrate its range through different examples.

Example 1.24.

a) Let A be the set of all students in a lecture hall. R might mean "is younger than".

b) Let A be the set of all students in a lecture hall. R might mean "lives in the same house".

c) **Equality relation** on a set A: Let A be a non-empty set. $R_1 = \{(x, x) : x \in A\}$ is called equality relation on A. Specifically for $A = \{1, 2, 3\}$ we get $R_1 = \{(1, 1), (2, 2), (3, 3)\}$.

d) **Graph of a function** $f : A \to A$. The graph $R_2 = \{(x, y) : y = f(x)\}$ of f is a relation on A. For the function $f : \mathbb{R}_+ \to \mathbb{R}_+$, $f(x) = \sqrt{x}$, the relation $R_2 = \{(x, y) : x \in \mathbb{R}_+, y \in \mathbb{R}, y = \sqrt{x}\}$ is obtained. It is a relation on $\mathbb{R}_+$; see the following figure.

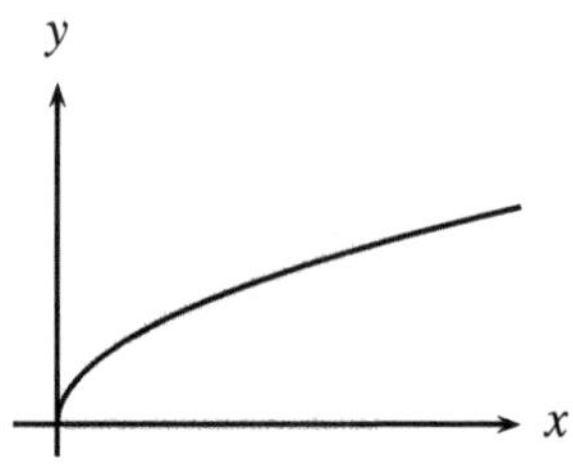

e) **Absolute value relation**. $R_3 = \{(x, y) \in \mathbb{R}^2 : |x| = |y|\}$ is a relation on $\mathbb{R}$. As a subset of the (x, y)-plane, it consists of the two diagonals; see the following figure.

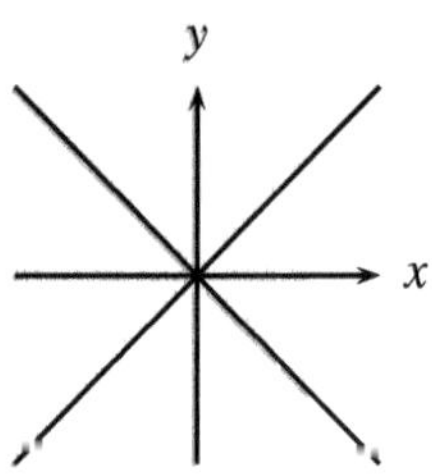

f) **Less-than-or-equal-to relation**. For $x, y \in \mathbb{R}$ we define xR_4y, if $x \leq y$. R_4 is a relation on $\mathbb{R}$. Graphically, it is the northwest half-plane, bounded by the main diagonal:

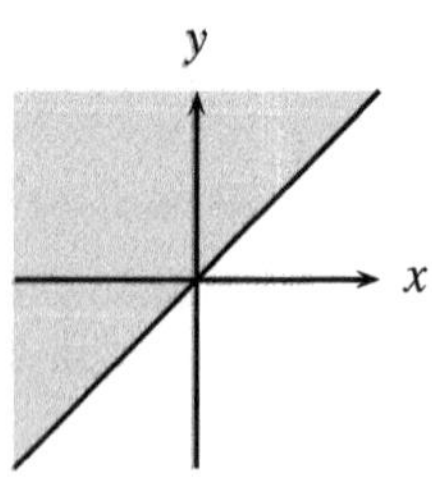

g) **Preference relation**. A student owns a bar that contains four bottles; two labels of rum (r_1, r_2), one whiskey (w) and one liqueur (l). A visitor expresses his preferences on $A = \{r_1, r_2, w, l\}$ as follows:

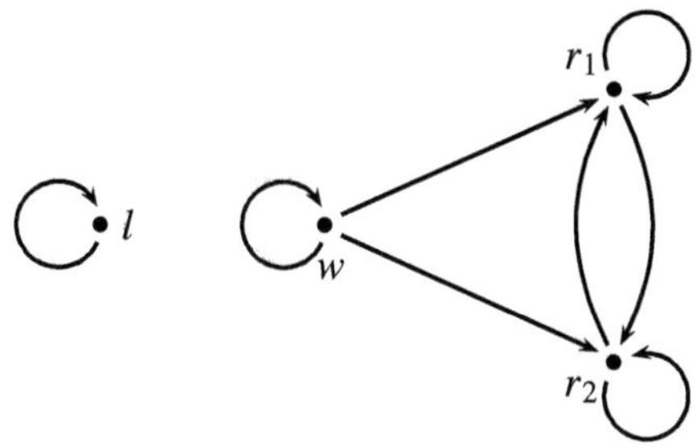

In this way, a relation P on A is defined. The arrow from w to r_1 means, e.g., that the first rum is at least as strongly preferred as the whiskey, symbolically wPr_1. Of course, each bottle is at least as strongly preferred as itself, hence the circular arrows from each bottle to itself. Such a graph is called a **relation graph**.

Remark: Every **relation on a finite set** can be graphically represented by points and arrows. Similarly, it can be described by a **relation table**. The relation table for P in the last example reads:

	r_1	r_2	w	l
r_1	1	1		
r_2	1	1		
w	1	1	1	
l				1

In general, the rows of a relation table, as well as the columns, correspond to the elements of the set A. An entry 1 means that the element x associated with the row is in relation to the element y associated with the column, xRy. Relation tables are particularly suited for representing relations when the relation graphs become too large or confusing.

We now consider general properties of relations and relate them to the above examples.

Definition 1.12: Reflexivity, symmetry, transitivity

(i) A relation R on A is **reflexive**, if

$$xRx \quad \text{for all } x \in A$$

holds. That is, every x is in relation to itself.

(ii) R is **symmetric**, if for all $x, y \in A$ it holds

$$xRy \quad \Longrightarrow \quad yRx\,.$$

That is, if x is in relation to y, then y is also in relation to x.

(iii) R is **transitive**, if for all $x, y, z \in A$ it holds

$$xRy \text{ and } yRz \quad \Longrightarrow \quad xRz\,.$$

That is, if x is in relation to y and y is in relation to z then x is also in relation to z.

Example 1.25. We consider the relations from Example 1.24.

- "is younger than" is transitive.
- "lives in the same house" is reflexive, transitive and symmetric.
- R_1 is reflexive, symmetric and transitive.
- R_2 is not reflexive (e.g. $2 \neq \sqrt{2}$), not symmetric (e.g. $2 = \sqrt{4} \;\not\Longrightarrow\; 4 = \sqrt{2}$), and also not transitive (e.g. $2 = \sqrt{4}, 4 = \sqrt{16} \;\not\Longrightarrow\; 2 = \sqrt{16}$).
- R_3 is reflexive ($|x| = |x|$ for all x), symmetric ($|x| = |y| \Longrightarrow |y| = |x|$) and transitive ($|x| = |y|, |y| = |z| \Longrightarrow |x| = |z|$).
- R_4 is reflexive ($x \leq x$ holds for all x), transitive ($x \leq y, y \leq z \Longrightarrow x \leq z$), but not symmetric ($3 \leq 5, 5 \nleq 3$).
- The preference relation P is transitive, reflexive, but not symmetric.

The relations in Examples 1.24 b), c) and e) express a certain equivalence or interchangeability of the objects. They represent so-called equivalence relations, whose general definition is as follows:

Definition 1.13: Equivalence relation

A relation R on A is called **equivalence relation**, if it is symmetric, reflexive and transitive.

As a symbol of an equivalence relation, we usually use $\sim$ instead of R and say, x is **equivalent** to y, if $x \sim y$ holds. For a given $x \in A$, we call

$$B_x = \{y \in A \,:\, x \sim y\}$$

equivalence class of x with respect to $\sim$. For example, for the above equivalence relation R_3, $B_5 = \{-5, 5\}$ is the equivalence class of 5 and in general $B_x = \{-x, x\}$ is the equivalence class of $x \in \mathbb{R}$.

An equivalence relation divides the set A into subsets consisting of equivalent elements.

A set $\mathcal{Z}$ of subsets of A is called a **partition** of A, if A is equal to the union of these sets and the subsets are all non-empty and pairwise disjoint[8].

[8] See Appendix B.

Theorem 1.1: Partition into equivalence classes

Let A be a non-empty set.

(i) Every partition $\mathcal{Z}$ of A defines an equivalence relation through $x \sim y$ if x and y are elements of the same set of the partition.

(ii) Every equivalence relation on A defines a partition $\mathcal{Z}$, namely the partition into equivalence classes related to R,

$$\mathcal{Z} = \left\{ B_x \ : \ B_x = \{y \in A \ : \ xRy\},\ x \in A \right\}.$$

Example 1.26.

- $\mathcal{Z} = \left\{\{a,b\},\{c\}\right\}$ is a partition of the set $A = \{a,b,c\}$. The corresponding equivalence relation is $R = \{(a,a),(b,b),(a,b),(b,a),(c,c)\}$.
- Let $A = \{1,2,\ldots,9\}$ and xRy hold if and only if $x + y$ is an even number. The corresponding partition is $\mathcal{Z} = \left\{\{1,3,5,7,9\},\{2,4,6,8\}\right\}$.
- Let A be any non-empty set. $\mathcal{Z}_1 = \{\{x\} \ : \ x \in A\}$ is a partition of A. It belongs to the equivalence relation R_1.
- Let $A = \mathbb{R}$. $\mathcal{Z}_3 = \{\{x,-x\} \ : \ x \in \mathbb{R}\}$ is the partition of $\mathbb{R}$ which belongs to the relation R_3.

We now turn to relations that express an order between elements of a set.

Definition 1.14: Preorder, order

A relation R on A is called **preorder**, if it is reflexive and transitive. Instead of xRy we then write $x \preceq y$.

R is called **order** if, in addition, R is **antisymmetric**, i.e., if for all $x, y \in A$ it holds

$$x \preceq y \text{ and } y \preceq x \quad \Longrightarrow \quad x = y.$$

R is called **complete**, if for all $x, y \in A$ it holds

$$x \preceq y \text{ or } y \preceq x.$$

Asymmetry states that no elements are mutually related unless they are equal. An order that is complete is also called **total order** or **linear order**. In a complete order, all elements of A can be compared with each other.

Example 1.27.

- P and R_4 are preorders.
- Every equivalence relation is a preorder.
- R_4 is antisymmetric ($x \leq y, y \leq x \implies x = y$), so it is an order.

- R_4 is complete (for all $x, y \in \mathbb{R}$ it holds $x \leq y$ or $y \leq x$).
- P is neither antisymmetric nor complete.

Apparently, the usual order $\leq$ of the real numbers is a complete order of the set $\mathbb{R}$. The **usual order of** $\mathbb{R}^2$ is defined componentwise:

$$(x_1, x_2) \leq (y_1, y_2), \quad \text{if } x_1 \leq y_1 \text{ and } x_2 \leq y_2 .$$

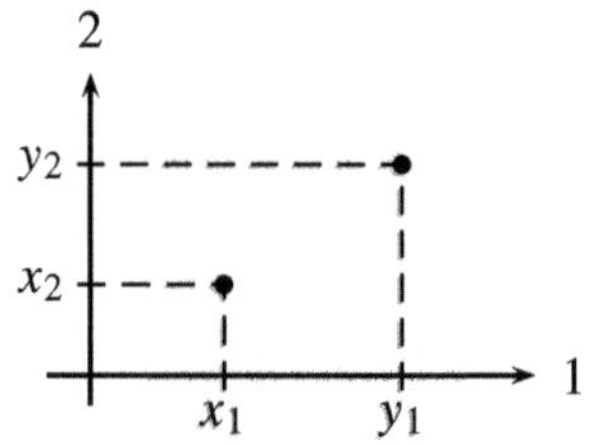

It is easy to see that the relation $\leq$ in $\mathbb{R}^2$ is reflexive, antisymmetric and transitive, thus an order. However, it is not complete, since e.g. the points $(1, 2)$ and $(3, 1)$ cannot be compared with each other. The same applies for the **usual order of** $\mathbb{R}^n$, $n \geq 2$:

$$(x_1, x_2, \ldots, x_n) \leq (y_1, y_2, \ldots, y_n), \quad \text{if } x_i \leq y_i \quad \text{for all } i = 1, 2, \ldots, n .$$

$\leq$ is an order on $\mathbb{R}^n$, but not complete for $n \geq 2$.

A complete order on $\mathbb{R}^n$ is the **lexicographic order** $\leq_L$:

$$\begin{aligned}(x_1, \ldots, x_n) \leq_L (y_1, \ldots, y_n), \quad & \text{if } x_1 < y_1 \\ & \text{or } (x_1 = y_1 \text{ and } x_2 < y_2) \\ & \text{or } (x_1 = y_1, x_2 = y_2 \text{ and } x_3 < y_3) \\ & \vdots \\ & \text{or } (x_1 = y_1, \ldots, x_{n-1} = y_{n-1}, x_n \leq y_n)\end{aligned}$$

For example, $(0, 0) \leq_L (0, 3)$, $(0, 3) \leq_L (\frac{1}{1000}, 0)$. The lexicographic order $\leq_L$ is a complete order. It includes the ordinary order, since $x \leq y \implies x \leq_L y$. The name "lexicographic order" comes about as follows: If you consider the points $(x_1, ..., x_n)$ as words of n letters and arrange these words alphabetically as it is done in dictionaries, you obtain the lexicographic order $\leq_L$ in $\mathbb{R}^n$.

In many applications, one searches for largest or smallest elements of a given preorder: Is there a good that a consumer values most, or an investment whose risk is minimal? If a preorder on a finite set A is complete, i.e., an order, such elements obviously exist and can be determined by pairwise checks. As we also consider incomplete preorders, special distinctions have to be made for this search.

Definition 1.15: Smallest, largest, maximum, minimum element

Let "$\preceq$" be a preorder on A. Let $B \subset A$ and $a, b \in B$. Then we call

(i) a **smallest element** of B, if $a \preceq y$ for all $y \in B$.

(ii) b **largest element** of B, if $y \preceq b$ for all $y \in B$.

(iii) a **minimal element** of B, if for all $y \in B$ $(y \preceq a \implies a \preceq y)$ holds, which means that for any $y \in B$ the relation $y \preceq a$ cannot hold unless y and a are equivalent.

(iv) b **maximal element** of B, if for all $y \in B$ $(b \preceq y \implies y \preceq b)$ holds. I.e., any $y \in B$ with $b \preceq y$ must be equivalent to b.

Note that a greatest element must necessarily be comparable with all other elements. The same applies to a smallest element.

Example 1.28. We consider the set B according to the following figure and the two orders "$\leq$" and "$\leq_L$" in $\mathbb{R}^2$.

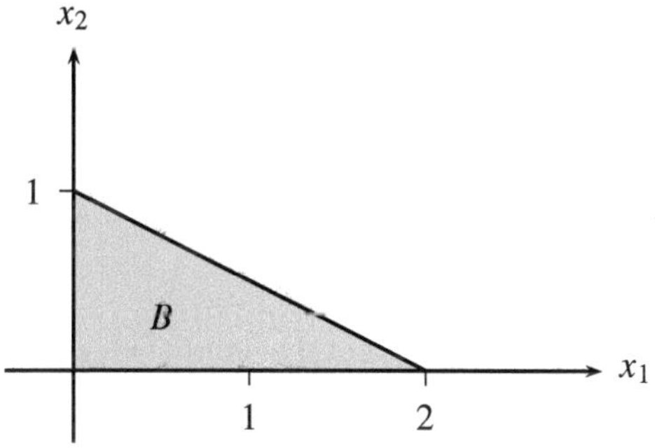

Smallest elements: It holds $(0,0) \leq (x_1, x_2)$ for all $(x_1, x_2) \in B$, so $(0,0)$ is the smallest element of B with respect to "$\leq$". $(0,0)$ is also the smallest element with respect to "$\leq_L$".

Largest elements: If $(x_1, x_2) \leq (b_1, b_2)$ for all $(x_1, x_2) \in B$, then $2 \leq b_1, 1 \leq b_2$ must hold. There is no such b in B, so there is no greatest element with respect to "$\leq$". $(2,0)$ is the greatest element with respect to "$\leq_L$".

Minimal elements: $(0,0)$ is minimal with respect to "$\leq$". $(0,0)$ is also the minimal element with respect to "$\leq_L$".

Maximal elements: Every point on the connecting line of the points $(2,0)$ and $(0,1)$ is maximal with respect to "$\leq$", as it is not smaller than any other point from B with respect to "$\leq$". $(2,0)$ is the maximal element with respect to "$\leq_L$".

Important terms for review

After reading this chapter, the following terms should be familiar:

- Function, domain, codomain
- Step function, bounded function
- Monotonically increasing or decreasing function
- Convex function, concave function
- Inverse function, composition of functions
- Exponential function, logarithm
- Bivariate function, multivariate function
- Contour lines of a bivariate function, isoquants
- Cobb-Douglas function
- Order, relation, equivalence relation

Self-test

You can check your understanding of this chapter using the following true-or-false questions. For each statement, indicate whether it is true (T) or false (F). Short solutions to these questions can be found in the Appendix H.

I. Given are two functions $f : \mathbb{R} \to \mathbb{R}$ and $g : \mathbb{R} \to \mathbb{R}$. Then for all $x \in \mathbb{R}$ ...

	T	F
... $f(x) + g(x) = \max\{f(x), g(x)\} + \min\{f(x), g(x)\}$.	☐	☐
... $\lvert f(x)\rvert = f(\lvert x\rvert)$.	☐	☐
... $f^{-1}(x) = \frac{1}{f(x)}$, if f is invertible.	☐	☐
... $f(g(x)) = (g \circ f)(x)$.	☐	☐

II. Are the following statements true or false?

	T	F
The function $f : \mathbb{R} \to \mathbb{R}$ with $f(x) = x^3$ is both convex and concave.	☐	☐
There is no function $f : \mathbb{R} \to \mathbb{R}$ that is both convex and concave.	☐	☐
For $x, y \in \mathbb{R}$ it holds: $e^x = e^y \implies x = y$.	☐	☐
For $x, y > 0$ it holds: $\ln(x) = \ln(y) \implies x = y$.	☐	☐
For $x, y \in \mathbb{R}$ it holds: $x^2 = y^2 \implies x = y$.	☐	☐
For $x, y \in \mathbb{R}$ it holds: $x^3 = y^3 \implies x = y$.	☐	☐

III. Are the following statements about an arbitrary function $\mathbb{R}^2 \to \mathbb{R}$ true or false?

	T	F
With some "contour lines", the function can be roughly represented in a two-dimensional mapping.	☐	☐
A "contour line" cannot be equal to the whole $\mathbb{R}^2$.	☐	☐
A "contour line" can be equal to the empty set.	☐	☐

IV. Are the following statements true or false?

	T	F
The order relation "$\leq$" in the real numbers is transitive.	☐	☐
The relation "wins against" in the single encounter of three football teams is always transitive.	☐	☐
The relation "$\geq$" on the integers is symmetric.	☐	☐

Exercises

Exercise 1
Given are the functions

a) $f_1(x) = e^x, \quad x \in \mathbb{R},$

b) $f_2(x) = 4x^4 - 4, \quad x \in \mathbb{R}_+,$

c) $f_3(x) = 4x^2 - 4, \quad x \in [-1, 1],$

d) $f_4(x) = \ln(x) + \sqrt{2x}, \quad x \in [1, \infty),$

e) $f_5(x) = \sqrt[3]{x^2} + x, \quad x \in [1, 4].$

Draw the graphs and determine the image of each function. Using the graphs, decide whether the functions are monotonic increasing, convex, and/or bounded.

Exercise 2
Check whether the following functions are invertible and provide the inverse function if possible:

a) $f_1(x) = \frac{1}{4}x^3 - 2, \quad x \in \mathbb{R},$

b) $f_2(x) = \ln(x + 1)^2, \quad x \in \mathbb{R}_+,$

c) $f_3(x) = x^3 - x^2 + 4, \quad x \in \mathbb{R}.$

Exercise 3
Determine for the following production functions the isoquants (contour lines of the production function) and sketch them for outputs of 1, 2 and 3 units:

a) $f_1(x, y) = 2x^2y, \quad x, y \geq 0,$

b) $f_2(x, y) = 2x + \frac{1}{2}y, \quad x, y \geq 0,$

c) $f_3(x, y) = \min\{2x, 4y\}, \quad x, y \geq 0.$

Exercise 4
Assume a brewer would use the following ingredients to produce one hectoliter [hl] of Kölsch[9]:

- 18 kg of Pilsner malt,
- 4 kg of wheat malt,
- 0.2 kg of hops.

Set up the production function and sketch the factor combinations with which an output of 5 [hl] can be produced. The mixing ratio must be strictly adhered to. Other production factors not mentioned here are to be neglected.

[9]Kölsch is a special brand of top-fermented beer that is exclusively brewed in Cologne, Germany.

Chapter 2

Matrices and vectors

A matrix is a rectangular scheme of numbers. Matrices are an essential part of mathematical notation and have numerous applications, such as in the description of production processes or in solving systems of linear equations. A vector is a special matrix, viz. a column or row of numbers. With the help of matrix-vector notation, complex economic relationships can often be represented in a clear and transparent way.

We begin this chapter with two examples that illustrate the usefulness of the matrix-vector notation.

Example 2.1 (Linear production function). A company produces a good. For the production of this good, n different factors are needed. The output, y, of the good, given inputs $x_1, \ldots, x_n$, can be described by a **linear production function**,

$$y = f(x_1, \ldots, x_n) = a_1 x_1 + a_2 x_2 + \ldots + a_n x_n \,, \quad x_1, \ldots, x_n \geq 0 \,.$$

Example 2.2 (Leontief's[1] open input-output model). We consider an economy that consists of n sectors $1, \ldots, n$ (agriculture, energy, construction, banking, etc.). In each sector, a single good is produced as **output**. The sectors are interconnected, so that the production of a good requires **inputs** from the other sectors. Let x_i denote the output of the i-th sector and d_i the final demand for the i-th good, the so-called **exogenous demand**. Let a_{ij} denote the amount of the i-th good required to produce one unit of the j-th good. The amount of good i that is used as input for the sectors $1, \ldots, n$, the so-called **endogenous demand**, is therefore given by

$$a_{i1} x_1 + a_{i2} x_2 + \ldots + a_{in} x_n \,.$$

[1] Wassily Leontief (1906–1999); Nobel Prize in Economics 1973.

K. Mosler et al., *Mathematical Methods for Business and Economics*, Mathematics Study Resources 20, https://doi.org/10.1007/978-3-662-73831-3_2

In order for the demand d_i to be met, the output of the i-th sector must equal the sum of exogenous and endogenous demand, i.e.,

$$x_i = (a_{i1}x_1 + a_{i2}x_2 + \ldots + a_{in}x_n) + d_i \,.$$

The question now arises whether a given exogenous demand can be satisfied, and how large the outputs x_i of the individual sectors must be. Formally, this leads to the problem of whether the following system of equations has a solution, and, if so, for which values $x_1, \ldots, x_n$ it is satisfied:

$$\begin{array}{ccccccccccc} x_1 & = & a_{11}x_1 & + & a_{12}x_2 & + & \ldots & + & a_{1n}x_n & + & d_1 \\ x_2 & = & a_{21}x_1 & + & a_{22}x_2 & + & \ldots & + & a_{2n}x_n & + & d_2 \\ \vdots & & \vdots & & \vdots & & & & \vdots & & \vdots \\ x_n & = & a_{n1}x_1 & + & a_{n2}x_2 & + & \ldots & + & a_{nn}x_n & + & d_n \end{array}$$

To write and solve systems of equations like this in a concise form, matrices and vectors are used. In the following, we give a short introduction into the principal notions of matrices and vectors and the calculus by which they can be added and multiplied.

2.1 Matrices

Definition 2.1: Matrix, transpose

Let $m, n \in \mathbb{N}$. A rectangular scheme

$$\boldsymbol{A} = \begin{bmatrix} a_{11} & a_{12} & \cdots & a_{1n} \\ a_{21} & a_{22} & \cdots & a_{2n} \\ \vdots & \vdots & & \vdots \\ a_{m1} & a_{m2} & \cdots & a_{mn} \end{bmatrix}$$

of numbers $a_{ij} \in \mathbb{R}$, $i = 1, \ldots, m$, $j = 1, \ldots, n$, is called a **matrix** with m **rows** and n **columns** or a matrix of **format** $m \times n$, in short $m \times n$ matrix. We also write $\boldsymbol{A} = (a_{ij})_{\substack{i=1,\ldots,m \\ j=1,\ldots,n}}$ or $\boldsymbol{A} = (a_{ij})$ or $\boldsymbol{A} = (a_{ij})_{m \times n}$.

We indicate matrices with large Latin letters in boldface, so $\boldsymbol{A}$, $\boldsymbol{B}$, $\boldsymbol{A}_1$, $\boldsymbol{A}_2$, etc. The set of all $m \times n$ matrices is denoted by $\mathbb{R}^{m \times n}$.

Given some $\boldsymbol{A} = (a_{ij})_{m \times n}$, the $n \times m$ matrix

$$\boldsymbol{A}^T = \begin{bmatrix} a_{11} & a_{21} & \cdots & a_{m1} \\ a_{12} & a_{22} & \cdots & a_{m2} \\ \vdots & \vdots & & \vdots \\ a_{1n} & a_{2n} & \cdots & a_{mn} \end{bmatrix}$$

is called the **transpose** of $\boldsymbol{A}$.

When forming the transpose, rows and columns are swapped: The columns of the transpose are the rows of the original matrix and vice versa.

Example 2.3.

$$\boldsymbol{A} = \begin{bmatrix} 9 & 10 \\ 6 & 4 \\ 2 & 0 \end{bmatrix}, \qquad \boldsymbol{A}^T = \begin{bmatrix} 9 & 6 & 2 \\ 10 & 4 & 0 \end{bmatrix}.$$

Transposing a matrix that has already been transposed gives the original matrix: $(\boldsymbol{A}^T)^T = \boldsymbol{A}$. The elements $a_{11}, a_{22}, \ldots$ of a matrix form the **main diagonal**. The main diagonals of a matrix and its transpose are identical.

A matrix, whose components are all zero, is called **zero matrix**. We denote the zero matrix of format $m \times n$ with $\mathbf{0}_{m\times n}$,

$$\mathbf{0}_{m\times n} = \begin{bmatrix} 0 & \cdots & 0 \\ \vdots & & \vdots \\ 0 & \cdots & 0 \end{bmatrix}.$$

A matrix $\boldsymbol{A}$ with an equal number of rows and columns, i.e., with $m = n$, is called a **square matrix**. It is called **symmetric**, if $\boldsymbol{A} = \boldsymbol{A}^T$, i.e., if $a_{ij} = a_{ji}$ for all indices i, j. For example,

$$\boldsymbol{A} = \begin{bmatrix} 1 & 2 & 5 \\ 2 & 3 & 8 \\ 5 & 8 & 4 \end{bmatrix}.$$

is a symmetric matrix. A square matrix, whose main diagonal elements are all equal to one and whose remaining elements are all zero, is called **identity matrix**. We denote the identity matrix of format $n \times n$ with $\boldsymbol{I}_n$,

$$\boldsymbol{I}_n = \begin{bmatrix} 1 & 0 & \cdots & 0 \\ 0 & 1 & \cdots & 0 \\ \vdots & \vdots & \ddots & \vdots \\ 0 & 0 & \cdots & 1 \end{bmatrix}.$$

A square matrix is called **diagonal matrix**, if all the elements outside the main diagonal are zero, i.e., $a_{ij} = 0$, if $i \neq j$, while the numbers on the main diagonal can be arbitrary numbers. For such a matrix we briefly write $\mathrm{diag}(a_{11}, a_{22}, \ldots, a_{nn})$.

A square matrix is called **upper triangular**, if the elements below the main diagonal are all zero, i.e., $a_{ij} = 0$ for $i > j$. It is called a **lower triangular**, if the elements above the main diagonal are all zero, i.e., $a_{ij} = 0$ for $i < j$.

Example 2.4.

$$D = \begin{bmatrix} 2 & 0 & 0 \\ 0 & 5 & 0 \\ 0 & 0 & 1 \end{bmatrix}, \quad U = \begin{bmatrix} 2 & 6 & 4 \\ 0 & 5 & 3 \\ 0 & 0 & 1 \end{bmatrix}, \quad L = \begin{bmatrix} 2 & 0 & 0 \\ 3 & 5 & 0 \\ 4 & 6 & 1 \end{bmatrix}.$$

$\boldsymbol{D}$ is a diagonal matrix, $\boldsymbol{U}$ is an upper and $\boldsymbol{L}$ is a lower triangular matrix.

Definition 2.2: Column vector, row vector

A matrix of format $n \times 1$ is referred to as a **column vector** with n components,

$$\boldsymbol{x} = \begin{bmatrix} x_1 \\ x_2 \\ \vdots \\ x_n \end{bmatrix} = [x_1, x_2, \dots x_n]^T.$$

x_i is called the ***i*-th component** of $\boldsymbol{x}$, $i = 1, \dots, n$. Correspondingly, a matrix $\boldsymbol{y}$ of format $1 \times n$ is referred to as a **row vector** with n components,

$$\boldsymbol{y} = [y_1, y_2, \dots, y_n].$$

When the difference between row and column form is not important, we simply say **vector**. Vectors are indicated with small Latin letters in boldface, $\boldsymbol{x}, \boldsymbol{y}, \boldsymbol{z}, \boldsymbol{u}, \boldsymbol{v}, \boldsymbol{a}, \boldsymbol{b}$, etc. Then the same non-boldfaced letter with an index i denotes the i-th component of the vector. For example, u_4 is the fourth component of the vector $\boldsymbol{u}$.

Example 2.5. Examples of vectors are

$$\boldsymbol{x} = \begin{bmatrix} 2 \\ 1 \\ 3 \end{bmatrix} \in \mathbb{R}^3, \quad \boldsymbol{a} = \begin{bmatrix} 2 \\ 1 \\ 3 \\ 0 \end{bmatrix} \in \mathbb{R}^4, \quad \boldsymbol{a}^T = \begin{bmatrix} 2, 1, 3, 0 \end{bmatrix} = (2, 1, 3, 0) \in \mathbb{R}^4.$$

A column vector having n components, $\boldsymbol{x} = [x_1, x_2, \dots, x_n]^T$, corresponds to a point $(x_1, x_2, \dots, x_n) \in \mathbb{R}^n$. The same holds for a row vector $[x_1, x_2, \dots, x_n]$. In the following, we identify the points of n-dimensional Euclidean space $\mathbb{R}^n$ with the column vectors of format $n \times 1$. Thus, the space $\mathbb{R}^n$ becomes the set of all column vectors having n components.

The origin of $\mathbb{R}^n$ corresponds to the **zero vector** $[0, 0, \dots, 0]^T$. The vector whose i-th component is equal to one and whose other components are all zero is called the ***i*-th canonical unit vector** and denoted by $\boldsymbol{e}_i$. $\mathbb{R}^n$ contains n such canonical unit vectors. For example, the three canonical unit vectors of $\mathbb{R}^3$ are

$$e_1 = \begin{bmatrix} 1 \\ 0 \\ 0 \end{bmatrix}, \; e_2 = \begin{bmatrix} 0 \\ 1 \\ 0 \end{bmatrix}, \; e_3 = \begin{bmatrix} 0 \\ 0 \\ 1 \end{bmatrix}.$$

2.2 Addition and scalar multiplication of matrices

You can calculate with matrices in a way similar to numbers. Basic operations are the addition and subtraction of two matrices as well as the multiplication of a matrix with a real number.

The addition of two matrices is done in a straightforward way. Two matrices are added by adding their corresponding components.

Definition 2.3: Addition of matrices

Let $\boldsymbol{A}$ and $\boldsymbol{B}$ be $m \times n$ matrices, $\boldsymbol{A} = (a_{ij})$, $\boldsymbol{B} = (b_{ij})$. The sum of $\boldsymbol{A}$ and $\boldsymbol{B}$ is defined as follows:

$$\boldsymbol{A} + \boldsymbol{B} = (a_{ij} + b_{ij}) = \begin{bmatrix} a_{11} + b_{11} & a_{12} + b_{12} & \cdots & a_{1n} + b_{1n} \\ a_{21} + b_{21} & a_{22} + b_{22} & \cdots & a_{2n} + b_{2n} \\ \vdots & \vdots & & \vdots \\ a_{m1} + b_{m1} & a_{m2} + b_{m2} & \cdots & a_{mn} + b_{mn} \end{bmatrix}.$$

Note: Two matrices can be only added if they have *the same format*, i.e., the same number of rows and the same number of columns!

Example 2.6.

$$\boldsymbol{A} = \begin{bmatrix} 2 & 0 \\ 1 & 5 \\ 3 & 2 \end{bmatrix}, \quad \boldsymbol{B} = \begin{bmatrix} -2 & 0 \\ 7 & 1 \\ -5 & 2 \end{bmatrix}, \quad \boldsymbol{A} + \boldsymbol{B} = \begin{bmatrix} 0 & 0 \\ 8 & 6 \\ -2 & 4 \end{bmatrix}.$$

Obviously, the addition of matrices is commutative, it always holds $\boldsymbol{A}+\boldsymbol{B} = \boldsymbol{B}+\boldsymbol{A}$. If you add a matrix to a zero matrix of the same format, you get the original matrix: $\boldsymbol{0}_{m\times n} + \boldsymbol{A} = \boldsymbol{A} + \boldsymbol{0}_{m\times n} = \boldsymbol{A}$.

As the addition, the subtraction of two matrices is done component-wise.

Definition 2.4: Subtraction of matrices

Let $\boldsymbol{A}$ and $\boldsymbol{B}$ be $m \times n$ matrices, $\boldsymbol{A} = (a_{ij})$, $\boldsymbol{B} = (b_{ij})$. The difference between

A and B is defined as follows:

$$A - B = (a_{ij} - b_{ij}) = \begin{bmatrix} a_{11} - b_{11} & a_{12} - b_{12} & \cdots & a_{1n} - b_{1n} \\ a_{21} - b_{21} & a_{22} - b_{22} & \cdots & a_{2n} - b_{2n} \\ \vdots & \vdots & & \vdots \\ a_{m1} - b_{m1} & a_{m2} - b_{m2} & \cdots & a_{mn} - b_{mn} \end{bmatrix}.$$

If you subtract a matrix from itself, you get the zero matrix of the corresponding format: $A - A = 0_{m \times n}$.

Example 2.7 (Continuation of Example 2.6). With A and B from Example 2.6 we get Then

$$A - B = \begin{bmatrix} 4 & 0 \\ -6 & 4 \\ 8 & 0 \end{bmatrix}.$$

Another basic operation is the multiplication of a matrix with a real number. As a real number is also mentioned as a **scalar**, this multiplication is referred to as **scalar multiplication**. The scalar multiplication of a matrix is done by multiplying each component of the matrix with the scalar, i.e., the real number.

Definition 2.5: Scalar multiplication

Let A be a $m \times n$ matrix and $\lambda \in \mathbb{R}$. (To distinguish the scalar from the elements of the matrix, it is usually notated by a Greek letter, say λ.) The multiplication of A with the scalar λ is defined as follows:

$$\lambda A = (\lambda a_{ij}) = \begin{bmatrix} \lambda a_{11} & \lambda a_{12} & \cdots & \lambda a_{1n} \\ \lambda a_{21} & \lambda a_{22} & \cdots & \lambda a_{2n} \\ \vdots & \vdots & & \vdots \\ \lambda a_{m1} & \lambda a_{m2} & \cdots & \lambda a_{mn} \end{bmatrix}.$$

Example 2.8 (Continuation of Example 2.6).

$$\left(-\frac{1}{2}\right) \cdot A = \begin{bmatrix} -1 & 0 \\ -\frac{1}{2} & -\frac{5}{2} \\ -\frac{3}{2} & -1 \end{bmatrix}$$

The matrix $-A$ is understood to be the matrix $(-1) \cdot A$.

Remark: Obviously, $A - B = A + (-1) \cdot B$. So, the difference of two matrices could also have been defined using scalar multiplication and addition.

We summarize the most important rules for addition and scalar multiplication of matrices in the following theorem.

Theorem 2.1: Rules for matrices

Let $\boldsymbol{A}, \boldsymbol{B}$ and $\boldsymbol{C}$ be $m \times n$ matrices, $\lambda, \mu \in \mathbb{R}$. Then the following statements hold:

(i)	$\boldsymbol{A} + \boldsymbol{B} = \boldsymbol{B} + \boldsymbol{A}$,	commutative (for "+")
(ii)	$(\boldsymbol{A} + \boldsymbol{B}) + \boldsymbol{C} = \boldsymbol{A} + (\boldsymbol{B} + \boldsymbol{C})$,	associative (for "+")
(iii)	$\boldsymbol{A} + \boldsymbol{0}_{m\times n} = \boldsymbol{A}$,	neutral element (for "+")
(iv)	$\boldsymbol{A} + (-\boldsymbol{A}) = \boldsymbol{0}_{m\times n}$,	negative element (for "+")
(v)	$\lambda(\boldsymbol{A} + \boldsymbol{B}) = \lambda\boldsymbol{A} + \lambda\boldsymbol{B}$,	distributive
(vi)	$(\lambda + \mu)\boldsymbol{A} = \lambda\boldsymbol{A} + \mu\boldsymbol{A}$,	distributive
(vii)	$\lambda(\mu\boldsymbol{A}) = (\lambda\mu)\boldsymbol{A}$,	associative (for "·")
(viii)	$1 \cdot \boldsymbol{A} = \boldsymbol{A}$.	neutral element (for "·")

Addition and scalar multiplication of points of $\mathbb{R}^n$ Since vectors are special matrices (namely those with only one column or row), addition, subtraction and scalar multiplication of matrices immediately carry over to vectors, viz. are component-wise done. The above rules apply as well to vectors, and in the same way, to points of $\mathbb{R}^n$. In particular, two points $\boldsymbol{x} = (x_1, \ldots, x_n)$ and $\boldsymbol{y} = (y_1, \ldots, y_n)$ are added component-wise:

$$\boldsymbol{x} + \boldsymbol{y} = (x_1, \ldots, x_n) + (y_1, \ldots, y_n) = (x_1 + y_1, \ldots, x_n + y_n)\,.$$

Geometrically, the addition of $\boldsymbol{x}$ and $\boldsymbol{y}$ corresponds to the concatenation of the two arrows pointing from $\boldsymbol{0}$ to $\boldsymbol{x}$ resp. $\boldsymbol{y}$; see left panel in Figure 2.1.

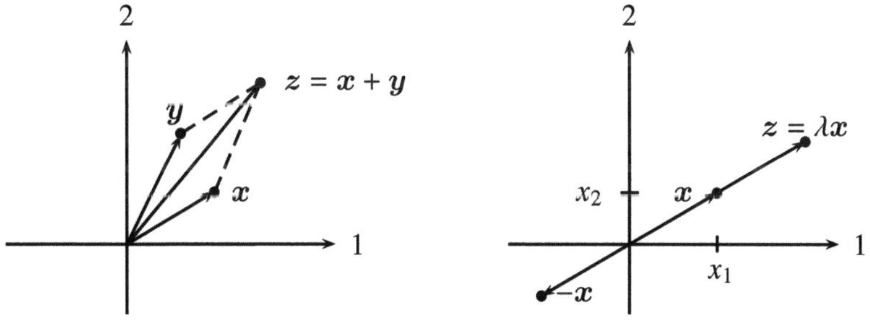

Figure 2.1: Sum of vectors, $\boldsymbol{z} = \boldsymbol{x} + \boldsymbol{y}$, and multiples of a vector, $\boldsymbol{z} = \lambda\boldsymbol{x}$.

Subtracting a point $\boldsymbol{y}$ from a point $\boldsymbol{x}$ and multiplying a point $\boldsymbol{x}$ with a scalar $\lambda \in \mathbb{R}$, is again done component-wise:

$$\begin{aligned}\boldsymbol{x} - \boldsymbol{y} &= (x_1, \ldots, x_n) - (y_1, \ldots, y_n) = (x_1 - y_1, \ldots, x_n - y_n)\,,\\ \lambda\boldsymbol{x} &= (\lambda x_1, \ldots, \lambda x_n)\,.\end{aligned}$$

Scalar multiplication of a vector corresponds to stretching or shrinking the vector, possibly combined with a reflection at the origin. This is illustrated in the right panel in Figure 2.1.

Example 2.9 (Aggregation of product bundles). A company produces four products that are stored at three different locations. At a certain point in time, there were $[100, 2000, 300, 0]$ units in warehouse A, $[40, 0, 0, 700]$ units in warehouse B and $[400, 200, 0, 0]$ units in warehouse C. Here, for each warehouse, the quantities of the four products were summarized by a row vector in $\mathbb{R}^4$. So in total there were

$$[100, 2000, 300, 0] + [40, 0, 0, 700] + [400, 200, 0, 0] = [540, 2200, 300, 700]$$

units available. Due to a storm, 80% of the products in warehouse A were destroyed, which are

$$0.8 \cdot [100, 2000, 300, 0] = [80, 1600, 240, 0]$$

units. Afterwards, there are still

$$[540, 2200, 300, 700] - [80, 1600, 240, 0] = [460, 600, 60, 700]$$

units of the four products in stock.

Example 2.10. In the following figure three points of $\mathbb{R}^2$ are depicted: $\boldsymbol{x} = (3, 1)$, $\boldsymbol{y} = (2, 2)$, $\boldsymbol{z} = (-2, 6)$. Thus, $\boldsymbol{x} + \boldsymbol{y} = (5, 3)$ and $\boldsymbol{x} - \boldsymbol{z} = (5, -5)$ hold. For arbitrary numbers λ we consider the multiples $\lambda\boldsymbol{x}$. They form the set

$$U = \{(3\lambda, \lambda) : \lambda \in \mathbb{R}\} = \left\{(x_1, x_2) : x_2 = \frac{1}{3}x_1\right\}.$$

This is the line through the origin and the point $(3, 1)$ in $\mathbb{R}^2$.

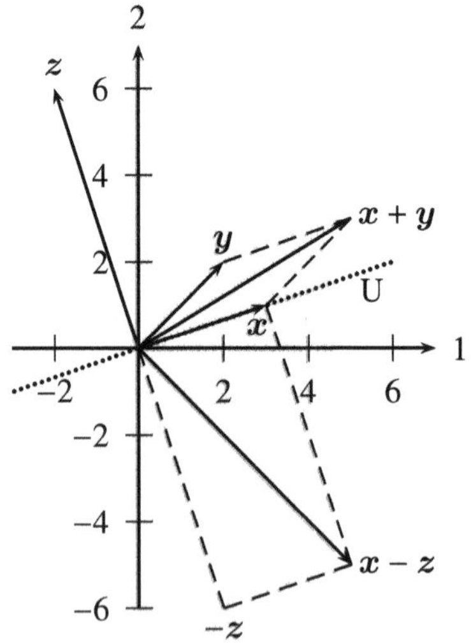

2.3 Multiplication of matrices

Matrices can also be multiplied with each other. The prerequisite is that they be compatible in format: The left-hand matrix in a product must have as many columns as the right-hand matrix has rows.

Definition 2.6: Matrix product

Let $\boldsymbol{A} \in \mathbb{R}^{m\times k}$ and $\boldsymbol{B} \in \mathbb{R}^{k\times n}$ with $\boldsymbol{A} = (a_{il})$ and $\boldsymbol{B} = (b_{lj})$. The **product** of $\boldsymbol{A}$ and $\boldsymbol{B}$ is the $m \times n$ matrix $\boldsymbol{AB} = (c_{ij})$ with the elements

$$c_{ij} = a_{i1}b_{1j} + a_{i2}b_{2j} + \ldots + a_{ik}b_{kj} = \sum_{l=1}^{k} a_{il}b_{lj}.$$

The element c_{ij} is thus obtained by multiplying the i-th row of the matrix $\boldsymbol{A}$ with the j-th column of the matrix $\boldsymbol{B}$.

Example 2.11. Given are the matrices

$$\boldsymbol{A} = \begin{bmatrix} 1 & 0 \\ 0 & 1 \\ 2 & 1 \end{bmatrix} \quad \text{and} \quad \boldsymbol{B} = \begin{bmatrix} -1 & 2 \\ -3 & 0 \end{bmatrix}.$$

As $\boldsymbol{A}$ has two columns and $\boldsymbol{B}$ two rows, the product $\boldsymbol{AB}$ can be formed.

$$\begin{array}{cc|cc} & & -1 & 2 \\ & & -3 & 0 \\ \hline 1 & 0 & -1 & 2 \\ 0 & 1 & -3 & 0 \\ 2 & 1 & -5 & 4 \end{array} \; \begin{array}{c} \boldsymbol{B} \\ \\ \boldsymbol{AB} \end{array}, \quad \text{thus} \quad \boldsymbol{AB} = \begin{bmatrix} -1 & 2 \\ -3 & 0 \\ -5 & 4 \end{bmatrix}.$$

(with $\boldsymbol{A}$ written to the left of its rows)

For example, the element in the first row and the first column of $\boldsymbol{AB}$ is calculated as

$$1 \cdot (-1) + 0 \cdot (-3) = -1\,.$$

The other elements of $\boldsymbol{AB}$ are accordingly determined. Furthermore, let $\boldsymbol{x} = [3, 7]^T$. We obtain

$$\boldsymbol{Ax} = \begin{bmatrix} 3 \\ 7 \\ 13 \end{bmatrix} \in \mathbb{R}^3, \;\; \boldsymbol{Bx} = \begin{bmatrix} 11 \\ -9 \end{bmatrix} \in \mathbb{R}^2, \;\; \boldsymbol{x}^T\boldsymbol{x} = 3^2 + 7^2 = 58\,.$$

The product $\boldsymbol{BA}$ is not defined, since the number of columns of $\boldsymbol{B}$ does not match the number of rows of $\boldsymbol{A}$. Next, we calculate $\boldsymbol{AA}^T$ and $\boldsymbol{A}^T\boldsymbol{A}$. Both products are defined, but generally different, as we see from our example:

$$\begin{array}{ccc|ccc} & & & 1 & 0 & 2 & \boldsymbol{A}^T \\ & & & 0 & 1 & 1 & \\ \hline & 1 & 0 & 1 & 0 & 2 & \\ \boldsymbol{A} & 0 & 1 & 0 & 1 & 1 & \boldsymbol{A}\boldsymbol{A}^T \\ & 2 & 1 & 2 & 1 & 5 & \end{array} , \quad \text{but} \quad \begin{array}{cccc|cc} & & & & 1 & 0 & \\ & & & & 0 & 1 & \boldsymbol{A} \\ & & & & 2 & 1 & \\ \hline \boldsymbol{A}^T & 1 & 0 & 2 & 5 & 2 & \boldsymbol{A}^T\boldsymbol{A} \\ & 0 & 1 & 1 & 2 & 2 & \end{array} .$$

Theorem 2.2: Rules for matrix multiplication

Let $\boldsymbol{A} \in \mathbb{R}^{m\times n}$, $\boldsymbol{B}$ and $\boldsymbol{C} \in \mathbb{R}^{n\times k}$, $\boldsymbol{D} \in \mathbb{R}^{k\times r}$, $\lambda \in \mathbb{R}$. Then the following statements hold:

$$\begin{array}{rrl} \text{(i)} & \boldsymbol{A}(\boldsymbol{B}\boldsymbol{D}) = (\boldsymbol{A}\boldsymbol{B})\boldsymbol{D}\,, & \text{associative} \\ \text{(ii)} & \boldsymbol{A}\boldsymbol{I}_{n\times n} = \boldsymbol{A}\,, & \rbrace\ \text{neutral element} \\ \text{(iii)} & \boldsymbol{I}_{m\times m}\boldsymbol{A} = \boldsymbol{A}\,, & \\ \text{(iv)} & \boldsymbol{A}(\boldsymbol{B}+\boldsymbol{C}) = \boldsymbol{A}\boldsymbol{B}+\boldsymbol{A}\boldsymbol{C}\,, & \rbrace\ \text{distributive} \\ \text{(v)} & (\boldsymbol{B}+\boldsymbol{C})\boldsymbol{D} = \boldsymbol{B}\boldsymbol{D}+\boldsymbol{C}\boldsymbol{D}\,, & \\ \text{(vi)} & \boldsymbol{A}(\lambda\boldsymbol{B}) = \lambda(\boldsymbol{A}\boldsymbol{B})\,, & \\ \text{(vii)} & (\boldsymbol{A}\boldsymbol{B})^T = \boldsymbol{B}^T\boldsymbol{A}^T\,. & \end{array}$$

Remark: While addition and subtraction of matrices and vectors essentially have the same properties as addition and subtraction of real numbers, this is not the case with the multiplication of matrices. We want to highlight two points where matrix multiplication fundamentally differs from the multiplication of real numbers.

- Matrix multiplication is **not commutative**, i.e., the order of the factors does matter. If the product $\boldsymbol{A}\boldsymbol{B}$ of two matrices is defined, this does not mean that the product $\boldsymbol{B}\boldsymbol{A}$ is defined as well. Even if the product $\boldsymbol{B}\boldsymbol{A}$ is defined, the matrices $\boldsymbol{A}\boldsymbol{B}$ and $\boldsymbol{B}\boldsymbol{A}$ need not be the same, which is illustrated by the following example:

$$\boldsymbol{A}\boldsymbol{B} = \begin{bmatrix} 1 & 2 \\ 3 & 4 \end{bmatrix}\begin{bmatrix} 1 & 1 \\ 1 & 2 \end{bmatrix} = \begin{bmatrix} 3 & 5 \\ 7 & 11 \end{bmatrix},$$
$$\boldsymbol{B}\boldsymbol{A} = \begin{bmatrix} 1 & 1 \\ 1 & 2 \end{bmatrix}\begin{bmatrix} 1 & 2 \\ 3 & 4 \end{bmatrix} = \begin{bmatrix} 4 & 6 \\ 7 & 10 \end{bmatrix}.$$

- If the product of two real numbers is zero, at least one of the two factors must be zero. However, the product of two matrices can be a zero matrix without any of the factors being zero. For example,

$$\boldsymbol{A}\boldsymbol{B} = \begin{bmatrix} 1 & 2 \\ 2 & 4 \end{bmatrix}\begin{bmatrix} 2 & -4 \\ -1 & 2 \end{bmatrix} = \begin{bmatrix} 0 & 0 \\ 0 & 0 \end{bmatrix} = \boldsymbol{0}_{2\times 2}\,,$$

although neither $\boldsymbol{A}$ nor $\boldsymbol{B}$ is the zero matrix.

2.4 Inverse matrices

A real number x which is not zero has a reciprocal, namely the number $1/x = x^{-1}$. If you multiply some $x \neq 0$ from the right or left with its reciprocal, the result always equals one: $x \cdot x^{-1} = x^{-1} \cdot x = 1$.

The question is whether there is something comparable for matrices. Does every matrix which is not the zero matrix have a "matrix reciprocal", such that the multiplication of the matrix with its "reciprocal" gives the identity matrix? Indeed, if a matrix $\boldsymbol{A}$ is not square, no matrix $\boldsymbol{B}$ can be found for which both $\boldsymbol{AB}$ and $\boldsymbol{BA}$ are identity matrices. However, in a large class of square matrices, there exists an analogue to the reciprocal, the so-called **inverse matrix**.

Definition 2.7: Inverse matrix

Let $\boldsymbol{A}$ be an $n \times n$ matrix. If there is an $n \times n$ matrix $\boldsymbol{B}$, such that $\boldsymbol{AB} = \boldsymbol{I}_n$ is, then $\boldsymbol{B}$ is called **inverse matrix** to $\boldsymbol{A}$ or **inverse** of $\boldsymbol{A}$. We write $\boldsymbol{A}^{-1} = \boldsymbol{B}$. Then it holds

$$\boldsymbol{A}\boldsymbol{A}^{-1} = \boldsymbol{A}^{-1}\boldsymbol{A} = \boldsymbol{I}_n .$$

If the inverse $\boldsymbol{A}^{-1}$ exists for a matrix $\boldsymbol{A}$, then $\boldsymbol{A}$ is called **invertible** or **regular**. If there is no inverse for $\boldsymbol{A}$, then $\boldsymbol{A}$ is called **non-invertible** or **singular**.

Example 2.12.

- The matrix $\boldsymbol{A} = \left[\begin{smallmatrix} 4 & 7 \\ 1 & 2 \end{smallmatrix}\right]$ is invertible with

$$\boldsymbol{A}^{-1} = \begin{bmatrix} 2 & -7 \\ -1 & 4 \end{bmatrix} .$$

  This is checked by multiplying $\boldsymbol{A}$ with $\boldsymbol{A}^{-1}$:

$$\boldsymbol{A}\boldsymbol{A}^{-1} = \begin{bmatrix} 4 & 7 \\ 1 & 2 \end{bmatrix} \cdot \begin{bmatrix} 2 & -7 \\ -1 & 4 \end{bmatrix} = \begin{bmatrix} 1 & 0 \\ 0 & 1 \end{bmatrix} .$$

- The identity matrix $\boldsymbol{I}_n$ is regular. It holds $\boldsymbol{I}_n^{-1} = \boldsymbol{I}_n$.
- A zero matrix $\boldsymbol{0}_{n \times n}$ is not invertible, as its product with any other matrix is again a zero matrix (provided the product exists).

A matrix $\boldsymbol{A} = [a_{11}] \in \mathbb{R}^{1 \times 1}$, which corresponds to the real number a_{11}, is obviously invertible if and only if $a_{11} \neq 0$. The inverse is then $\boldsymbol{A}^{-1} = [a_{11}^{-1}]$. Also for 2×2 matrices one can easily decide, whether the matrix is invertible, and determine the inverse matrix in a simple way.

Theorem 2.3: Inverse, $n = 2$

For $\boldsymbol{A} = \left[\begin{smallmatrix} a & b \\ c & d \end{smallmatrix}\right]$, the inverse $\boldsymbol{A}^{-1}$ exists if and only if $ad - bc \neq 0$. Then,

$$\boldsymbol{A}^{-1} = \frac{1}{ad - bc}\begin{bmatrix} d & -b \\ -c & a \end{bmatrix}.$$

E.g. the matrix $\boldsymbol{B} = \left[\begin{smallmatrix} 1 & 2 \\ 2 & 4 \end{smallmatrix}\right]$ is not invertible, since $1 \cdot 4 - 2 \cdot 2 = 0$. The quantity $ad - bc$ is the so-called determinant[2] of the matrix $\boldsymbol{A}$. Thus, the mnemonic for inverting a 2×2 matrix is:

Mnemonic 2.1: Inverse, $n = 2$

Swap main diagonal elements, change the sign of the off-diagonal elements, and divide the matrix by its determinant.

Example 2.13. The inverse of the matrix $\boldsymbol{A} = \left[\begin{smallmatrix} 4 & 7 \\ 1 & 2 \end{smallmatrix}\right]$ from example 2.12 is easily calculated as follows:

$$\boldsymbol{A}^{-1} = \frac{1}{4 \cdot 2 - 7 \cdot 1}\begin{bmatrix} 2 & -7 \\ -1 & 4 \end{bmatrix} = 1 \cdot \begin{bmatrix} 2 & -7 \\ -1 & 4 \end{bmatrix} = \begin{bmatrix} 2 & -7 \\ -1 & 4 \end{bmatrix}.$$

Also for larger matrices formulas for the inverse exist. For example, the inverse of a 3×3 matrix is given by

$$\begin{bmatrix} a & b & c \\ d & e & f \\ g & h & i \end{bmatrix}^{-1} = \frac{1}{\lambda}\begin{bmatrix} ei - fh & ch - bi & bf - ce \\ fg - di & ai - cg & cd - af \\ dh - eg & bg - ah & ae - bd \end{bmatrix},$$

where $\lambda = aei + bfg + cdh - gec - hfa - idb$, which equals the corresponding determinant. The inverse exists if and only if $\lambda \neq 0$. Obviously, this formula is hard to remember. In section 8.4 we will learn a simple method that allows not only to decide whether a matrix is invertible (or not) but also to directly calculate the inverse in case it exists.

Further properties of the inverse of a matrix are listed below.

Theorem 2.4: Inverse of a product

Let $\boldsymbol{A}$ and $\boldsymbol{B}$ be invertible $n \times n$ matrices. Then, $\boldsymbol{AB}$ is invertible and it holds

$$(\boldsymbol{AB})^{-1} = \boldsymbol{B}^{-1}\boldsymbol{A}^{-1}.$$

[2] For determinants see Chapter 10.

Note that for $\boldsymbol{B}^{-1}\boldsymbol{A}^{-1}$ being the inverse of $\boldsymbol{AB}$ it must hold $(\boldsymbol{AB})(\boldsymbol{B}^{-1}\boldsymbol{A}^{-1}) = \boldsymbol{I}_n$, which is seen from the following calculation:

$$(\boldsymbol{AB})(\boldsymbol{B}^{-1}\boldsymbol{A}^{-1}) = \boldsymbol{A}(\boldsymbol{BB}^{-1})\boldsymbol{A}^{-1} = \boldsymbol{AI}_n\boldsymbol{A}^{-1} = \boldsymbol{AA}^{-1} = \boldsymbol{I}_n\,.$$

This theorem can be easily remembered with the **socks and shoes principle**: When getting dressed, you first put on the sock ($\boldsymbol{A}$) and then the shoe ($\boldsymbol{B}$). The inverse of getting dressed ($\boldsymbol{AB}$) is undressing ($(\boldsymbol{AB})^{-1}$). Here, you must first take off the shoe ($\boldsymbol{B}^{-1}$) and then take off the sock ($\boldsymbol{A}^{-1}$). So $(\boldsymbol{AB})^{-1} = \boldsymbol{B}^{-1}\boldsymbol{A}^{-1}$, see Figure 2.2.

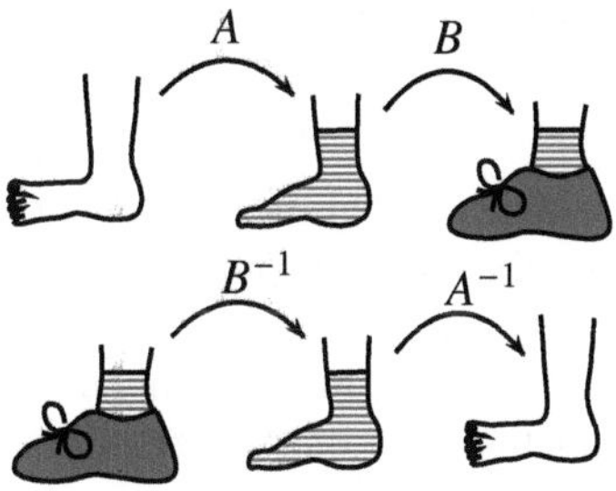

Figure 2.2: Illustration of $(AB)^{-1} = B^{-1}A^{-1}$.

Theorem 2.5: Rules for inversion

If $\boldsymbol{A}$ is invertible, then:

(i) $(\boldsymbol{A}^{-1})^{-1} = \boldsymbol{A}$,

(ii) $(\boldsymbol{A}^{T})^{-1} = (\boldsymbol{A}^{-1})^{T}$,

(iii) If $\boldsymbol{A}$ is symmetric, then $\boldsymbol{A}^{-1}$ is also symmetric,

(iv) For every real number $\lambda \neq 0$, $(\lambda\boldsymbol{A})^{-1} = \frac{1}{\lambda}\boldsymbol{A}^{-1}$.

2.5 Linear transformations

We now know what matrices or vectors are about and how to do calculations with them. Vectors specifically are used to describe points in multidimensional space. But what do matrices describe? For this, consider again the examples 2.1 and 2.2.

Example 2.14 (Linear production function). If we summarize the coefficients $a_1, \ldots, a_n$ of the linear production function in a row vector, $\boldsymbol{a} = [a_1, \ldots, a_n]$, and the inputs $x_1, \ldots, x_n$ in a column vector, $\boldsymbol{x} = [x_1, \ldots, x_n]^T$, we can write the linear production function in the form

$$y = f(\boldsymbol{x}) = \boldsymbol{ax}$$

This clearly shows the analogy to linear functions from $\mathbb{R}$ to $\mathbb{R}$, which are known to have the form $y = ax$.

Example 2.15 (Leontief's open input-output model). Analogously to the previous example, we summarize the production coefficients in a matrix $\boldsymbol{A} = (a_{ij})$, which is called **production matrix**. Correspondingly, we summarize the exogenous demands $d_1, \ldots, d_n$ and the outputs $x_1, \ldots, x_n$ in column vectors, $\boldsymbol{d} = [d_1, \ldots, d_n]^T$ and $\boldsymbol{x} = [x_1, \ldots, x_n]^T$. In this notation, the system of equations of the Leontief model becomes an equation between vectors of $\mathbb{R}^n$,

$$\boldsymbol{x} = \boldsymbol{A}\boldsymbol{x} + \boldsymbol{d}.$$

Substituting $\boldsymbol{I}\boldsymbol{x} = \boldsymbol{x}$, the above equation becomes $\boldsymbol{I}\boldsymbol{x} = \boldsymbol{A}\boldsymbol{x} + \boldsymbol{d}$, which simplifies to

$$(\boldsymbol{I} - \boldsymbol{A})\boldsymbol{x} = \boldsymbol{d}$$

by the rules of matrix arithmetic. This is the well-known equation of the open input-output model in matrix-vector notation. It is a **linear equation** between vectors of $\mathbb{R}^n$. To check whether a given exogenous demand $\boldsymbol{d}$ can be met and what total output $\boldsymbol{x}$ is necessary for this, you only need to solve this equation for $\boldsymbol{x}$. How to solve linear equations like this will be discussed in detail in Chapter 8.

The importance of matrices lies in the fact that they describe linear functions from $\mathbb{R}^n$ to $\mathbb{R}^m$.

Definition 2.8: Linear function

A function $f : \mathbb{R}^n \to \mathbb{R}^m$ is called **linear**, if for all $\boldsymbol{x}, \boldsymbol{y} \in \mathbb{R}^n$ and $\lambda \in \mathbb{R}$ the following two conditions hold:

Lin1: $f(\boldsymbol{x} + \boldsymbol{y}) = f(\boldsymbol{x}) + f(\boldsymbol{y})$,

Lin2: $f(\lambda \boldsymbol{x}) = \lambda f(\boldsymbol{x})$.

A linear function is also mentioned as a **linear transformation** or **linear map**.

Any given $m \times n$ matrix $\boldsymbol{A}$ defines a function $f_{\boldsymbol{A}} : \mathbb{R}^n \to \mathbb{R}^m$,

$$\boldsymbol{x} \mapsto f_{\boldsymbol{A}}(\boldsymbol{x}) = \boldsymbol{A}\boldsymbol{x}.$$

From the rules of matrix calculation, it immediately follows that $f_{\boldsymbol{A}}$ is a linear transformation. The fundamental importance of matrices comes from the fact that *every* linear transformation has the form $f_{\boldsymbol{A}}$ with a suitably chosen matrix $\boldsymbol{A}$. Thus, any $m \times n$ matrix corresponds to a linear transformation from $\mathbb{R}^n$ to $\mathbb{R}^m$.

However, given a linear function f, how can we find a corresponding matrix $\boldsymbol{A}$? If $\boldsymbol{A}$ is the matrix corresponding to f and $\boldsymbol{a}_1, \boldsymbol{a}_2, \ldots, \boldsymbol{a}_n$ the columns of $\boldsymbol{A}$, then the j-th canonical unit vector $\boldsymbol{e}_j$ has function value

$$f(\boldsymbol{e}_j) = \boldsymbol{A}\boldsymbol{e}_j = [\boldsymbol{a}_1, \ldots, \boldsymbol{a}_j, \ldots, \boldsymbol{a}_m]\, \boldsymbol{e}_j = \boldsymbol{a}_j.$$

In other words: The j-th column of the matrix $\boldsymbol{A}$ is the image of the j-th canonical unit vector. We formulate this as a mnemonic.

Mnemonic 2.2: Matrix and corresponding linear transformation

Let $f : \mathbb{R}^n \to \mathbb{R}^m$ be a linear transformation and $\boldsymbol{A}$ the corresponding $m \times n$ matrix. Then the columns $\boldsymbol{a}_j$ of $\boldsymbol{A}$ are the images of the canonical unit vectors, $\boldsymbol{a}_j = f(\boldsymbol{e}_j), j = 1, \ldots, n$.

From this it follows directly that a linear function $f : \mathbb{R}^n \to \mathbb{R}^m$ is uniquely determined by its values at the n canonical unit vectors of $\mathbb{R}^n$.

Example 2.16. Let $f : \mathbb{R}^3 \to \mathbb{R}^4$ be a linear transformation and let

$$f(\boldsymbol{e}_1) = [1,\ 3,\ 4,\ 2]^T, \quad f(\boldsymbol{e}_2) = [2,\ 0,\ 5,\ 1]^T, \quad f(\boldsymbol{e}_3) = [0,\ 3,\ 2,\ 4]^T.$$

Then $f(\boldsymbol{x}) = \boldsymbol{A}\boldsymbol{x}$ with

$$\boldsymbol{A} = \begin{bmatrix} 1 & 2 & 0 \\ 3 & 0 & 3 \\ 4 & 5 & 2 \\ 2 & 1 & 4 \end{bmatrix}.$$

Linear functions from $\mathbb{R}$ to $\mathbb{R}$ have the form $f(x) = ax$. Unless $a = 0$, they have an inverse, which is given by $f^{-1}(x) = \frac{1}{a}x$. Under what conditions does a linear function $f_{\boldsymbol{A}} : \mathbb{R}^n \to \mathbb{R}^m$, $\boldsymbol{x} \mapsto \boldsymbol{A}\boldsymbol{x}$, have an inverse function and what does it look like? If $m = n$ and $\boldsymbol{A}$ is invertible, then the linear function $f_{\boldsymbol{A}^{-1}}$ is the inverse function of $f_{\boldsymbol{A}}$, which is immediately seen from the following two equations:

$$f_{\boldsymbol{A}^{-1}}\big(f_{\boldsymbol{A}}(\boldsymbol{x})\big) = \boldsymbol{A}^{-1}\boldsymbol{A}\boldsymbol{x} = \boldsymbol{I}_n\boldsymbol{x} = \boldsymbol{x},$$
$$f_{\boldsymbol{A}}\big(f_{\boldsymbol{A}^{-1}}(\boldsymbol{x})\big) = \boldsymbol{A}\boldsymbol{A}^{-1}\boldsymbol{x} = \boldsymbol{I}_n\boldsymbol{x} = \boldsymbol{x}.$$

The following theorem tells that this is the only case where $f_{\boldsymbol{A}}$ has an inverse. It further says that the composition of linear functions amounts to the multiplication of the corresponding matrices. This, in fact, is the main reason why matrix multiplication is defined the way it is.

Theorem 2.6: Composition and inverse function of a linear function

Let $f_{\boldsymbol{A}} : \mathbb{R}^n \to \mathbb{R}^m$, $\boldsymbol{x} \mapsto \boldsymbol{A}\boldsymbol{x}$, and $f_{\boldsymbol{B}} : \mathbb{R}^m \to \mathbb{R}^k$, $\boldsymbol{x} \mapsto \boldsymbol{B}\boldsymbol{x}$ be linear functions, then the composition $f_{\boldsymbol{B}} \circ f_{\boldsymbol{A}}$ is again a linear function. $f_{\boldsymbol{B}} \circ f_{\boldsymbol{A}}$ is given by

$$f_{\boldsymbol{B}} \circ f_{\boldsymbol{A}} : \mathbb{R}^n \to \mathbb{R}^k, \quad \boldsymbol{x} \mapsto (\boldsymbol{B}\boldsymbol{A})\boldsymbol{x}.$$

If $f_{\boldsymbol{A}} : \mathbb{R}^n \to \mathbb{R}^m$, $\boldsymbol{x} \mapsto \boldsymbol{A}\boldsymbol{x}$ is a linear function, $f_{\boldsymbol{A}}$ has an inverse if and only if $m = n$ and $\boldsymbol{A}$ is invertible. In that case, the inverse function is given by

$$f_{\boldsymbol{A}^{-1}} : \mathbb{R}^n \to \mathbb{R}^n, \quad \boldsymbol{x} \mapsto \boldsymbol{A}^{-1}\boldsymbol{x}.$$

Note that a linear transformation $f_{\boldsymbol{A}}$ always maps the origin of $\mathbb{R}^n$ to the origin of $\mathbb{R}^m$, $f_{\boldsymbol{A}}(\boldsymbol{0}_m) = \boldsymbol{A}\boldsymbol{0}_n = \boldsymbol{0}_m$. Transformations that allow for an additional shift are called **affine-linear** functions $\mathbb{R}^n \longrightarrow \mathbb{R}^m$,

$$\boldsymbol{x} \longmapsto \boldsymbol{A}\boldsymbol{x} + \boldsymbol{b},$$

with a suitable choice of an $m \times n$ matrix $\boldsymbol{A}$ and a column vector $\boldsymbol{b} \in \mathbb{R}^m$.

2.6 Geometry of $\mathbb{R}^n$

The $\mathbb{R}^2$ can be intuitively thought of as a two-dimensional drawing plane, the $\mathbb{R}^3$ as the three-dimensional space surrounding us. For the plane and the three-dimensional space, we know what is meant by terms like "length", "distance", "angle" or "perpendicular". The $\mathbb{R}^n$ is the natural generalization of $\mathbb{R}^2$ and $\mathbb{R}^3$, but unlike these two, it lacks direct intuitiveness when $n > 3$. However, some geometric notions that have a visual meaning in $\mathbb{R}^2$ and $\mathbb{R}^3$ can be transferred to the general $\mathbb{R}^n$. This will be done in the present section.

First, the **scalar product** of two vectors is defined. As we will show, it is fundamental for the generalization of the notions.

Definition 2.9: Scalar product

For $\boldsymbol{x}, \boldsymbol{y} \in \mathbb{R}^n$, $\boldsymbol{x}^T\boldsymbol{y}$ is called **scalar product** or **inner product**. It is given by

$$\boldsymbol{x}^T\boldsymbol{y} = \boldsymbol{y}^T\boldsymbol{x} = x_1y_1 + x_2y_2 + \ldots + x_ny_n = \sum_{i=1}^{n} x_iy_i \in \mathbb{R}.$$

The scalar product of $\boldsymbol{x}$ and $\boldsymbol{y}$ is equal to the matrix product of the row vector $\boldsymbol{x}^T$ and the column vector $\boldsymbol{y}$.

Example 2.17. Let

$$\boldsymbol{a} = \begin{bmatrix} 1 \\ 2 \\ 3 \end{bmatrix}, \quad \boldsymbol{b} = \begin{bmatrix} -2 \\ 0 \\ 2 \end{bmatrix}, \quad \boldsymbol{c} = \begin{bmatrix} \frac{1}{\sqrt{2}} \\ 0 \\ \frac{1}{\sqrt{2}} \end{bmatrix}.$$

Then $\boldsymbol{a}^T\boldsymbol{b} = 1 \cdot (-2) + 2 \cdot 0 + 3 \cdot 2 = 4$, and $\boldsymbol{b}^T\boldsymbol{c} = 0$.

Example 2.18 (Continuation of Example 2.9). The four products of the company have a value of 500, 700, 200 and 100 Euro per unit, respectively. The total value of the products destroyed by the storm amounts to

$$\begin{aligned} \begin{bmatrix} 80, & 1600, & 240, & 0 \end{bmatrix} \begin{bmatrix} 500 \\ 700 \\ 200 \\ 100 \end{bmatrix} &= 80 \cdot 500 + 1600 \cdot 700 + 240 \cdot 200 + 0 \cdot 100 \\ &= 1\,208\,000 \text{ Euro}. \end{aligned}$$

The inventory value after the storm is only

$$[460,\ 600,\ 60,\ 700]\,[500,\ 700,\ 200,\ 100]^T = 732\,000\,\text{Euro}.$$

If $\boldsymbol{x} = (x_1, x_2)$ is a point in $\mathbb{R}^2$, then the distance from $\boldsymbol{x}$ to the origin $\mathbf{0}$ can be calculated with the Pythagorean[3] theorem. It is $\sqrt{x_1^2 + x_2^2}$. Similarly, for a point $\boldsymbol{x} \in \mathbb{R}^3$, the distance of the point from $\mathbf{0}$ is given by $\sqrt{x_1^2 + x_2^2 + x_3^2}$. The distance of a point or vector $\boldsymbol{x}$ from the origin is also called the **(Euclidean) norm** of $\boldsymbol{x}$. Even though we no longer have a visual representation of $\mathbb{R}^4$ and higher dimensional spaces, the concept of the norm (and thus of the distance to the origin) can be generalized by the corresponding formula:

Definition 2.10: (Euclidean) norm, unit vector

For $\boldsymbol{x} \in \mathbb{R}^n$ we call

$$\|\boldsymbol{x}\| := \sqrt{\sum_{i=1}^{n} x_i^2} = \sqrt{x_1^2 + x_2^2 + \ldots + x_n^2} = \sqrt{\boldsymbol{x}^T \boldsymbol{x}}$$

the **Euclidean norm**, or simply the **norm**, of $\boldsymbol{x}$.

A vector $\boldsymbol{x}$ with norm one is called a **unit vector**.

In the case $n = 1$, i.e., when x is a real number, we have $\|x\| = \sqrt{x^2} = |x|$. So the norm can be seen as a generalization of the absolute value. If you consider a vector in $\mathbb{R}^2$ or $\mathbb{R}^3$ as an arrow pointing from the origin $\mathbf{0}$ to the point $\boldsymbol{x}$, then $\|\boldsymbol{x}\|$ corresponds to the length of the arrow; see left panel in Figure 2.3.

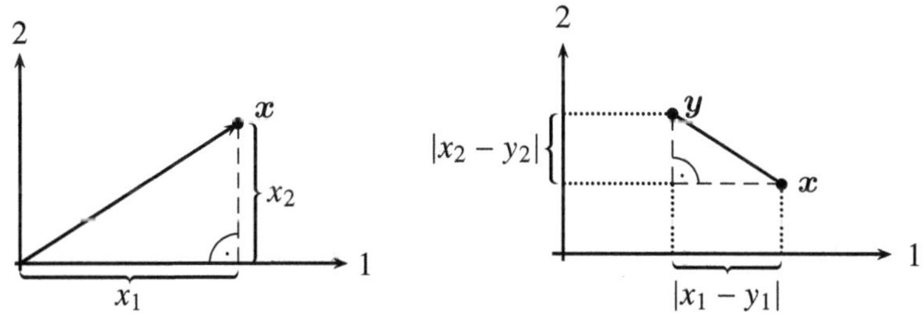

Figure 2.3: Norm and distance of vectors in $\mathbb{R}^2$.

Theorem 2.7: Properties of the norm

For all $\boldsymbol{x}, \boldsymbol{y} \in \mathbb{R}^n$ the following holds:

[3] Pythagoras of Samos (ca. 570 BC – ca. 500 BC)

(i) $||\boldsymbol{x}|| \geq 0$,

(ii) $||\boldsymbol{x}|| = 0$ if and only if $\boldsymbol{x} = \mathbf{0}$,

(iii) $||\lambda\boldsymbol{x}|| = |\lambda|||\boldsymbol{x}||$ for every real number λ,

(iv) $||\boldsymbol{x} + \boldsymbol{y}|| \leq ||\boldsymbol{x}|| + ||\boldsymbol{y}||$.

Property (iv) is referred to as **triangle inequality**. Geometrically, it means that in a triangle one side is at most as long as the sum of the other two sides, see the following figure.

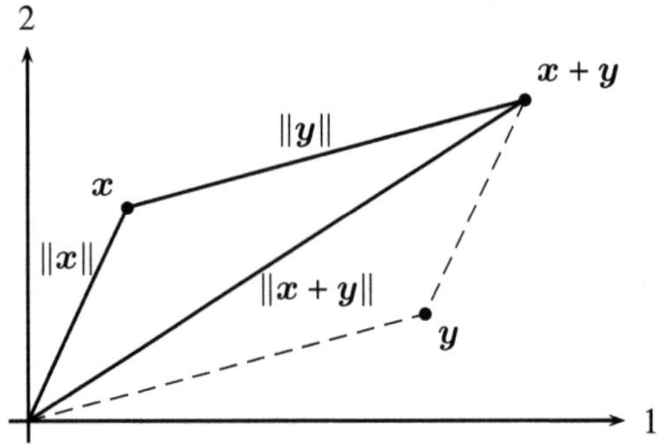

With the concept of norm, we have defined what we can interpret as the distance of a point from the origin in $\mathbb{R}^n$. How should we define the distance between two points $\boldsymbol{x}, \boldsymbol{y} \in \mathbb{R}^n$? The right panel in Figure 2.3 shows that in $\mathbb{R}^2$ the distance between two points $\boldsymbol{x}$ and $\boldsymbol{y}$ is given by $\sqrt{(x_1 - y_1)^2 + (x_2 - y_2)^2} = ||\boldsymbol{x} - \boldsymbol{y}||$. Similarly, the distance between two points in $\mathbb{R}^3$ amounts to $||\boldsymbol{x} - \boldsymbol{y}||$.

In general, we therefore define the distance between two points $\boldsymbol{x}, \boldsymbol{y} \in \mathbb{R}^n$ as the norm of their difference,

$$||\boldsymbol{x} - \boldsymbol{y}|| = \sqrt{(x_1 - y_1)^2 + (x_2 - y_2)^2 + \ldots + (x_n - y_n)^2}\,.$$

We denote $||\boldsymbol{x} - \boldsymbol{y}||$ as **Euclidean distance**. Obviously, this distance is symmetric; it holds $||\boldsymbol{x} - \boldsymbol{y}|| = ||\boldsymbol{y} - \boldsymbol{x}||$. For $n = 1, 2, 3$, the concept of Euclidean distance agrees with our usual intuition.

We now come to the concept of the **angle** between two vectors, which is based on the following theorem.

Theorem 2.8: Cauchy-Schwarz inequality[4]

For $x, y \in \mathbb{R}^n$ it holds

$$|\boldsymbol{x}^T\boldsymbol{y}| \leq ||\boldsymbol{x}|| \cdot ||\boldsymbol{y}||\,.$$

[4] Augustin Louis Baron Cauchy (1789–1857) and Hermann Amandus Schwarz (1843–1921)

From the Cauchy-Schwarz inequality it follows that for vectors $\boldsymbol{x}, \boldsymbol{y} \neq \boldsymbol{0}$ always

$$-1 \leq \frac{\boldsymbol{x}^T \boldsymbol{y}}{||\boldsymbol{x}|| \cdot ||\boldsymbol{y}||} \leq 1 .$$

This quotient is used to define the **angle** $\alpha(\boldsymbol{x}, \boldsymbol{y})$ of two vectors $\boldsymbol{x}, \boldsymbol{y} \neq \boldsymbol{0}$ by

$$\cos(\alpha(\boldsymbol{x}, \boldsymbol{y})) = \frac{\boldsymbol{x}^T \boldsymbol{y}}{||\boldsymbol{x}|| \cdot ||\boldsymbol{y}||} .$$

This definition corresponds in $\mathbb{R}^2$ to the classical definition of the cosine (adjacent leg divided by hypotenuse) at the right triangle. In Figure 2.4, the hypotenuse has length $||\boldsymbol{y}||$, the adjacent leg length $\boldsymbol{x}^T \boldsymbol{y} / ||\boldsymbol{x}||$.

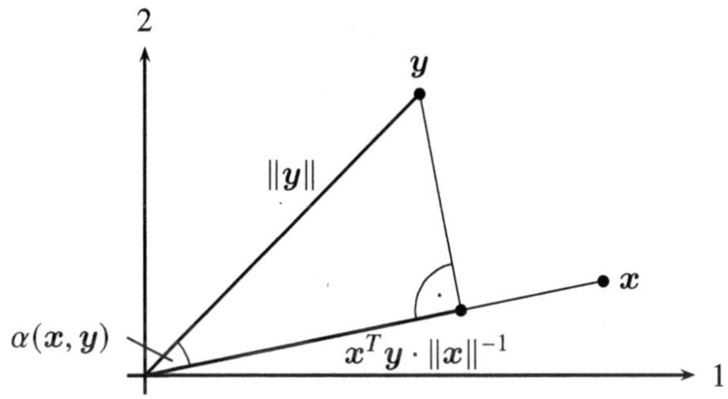

Figure 2.4: $\cos(\alpha(\boldsymbol{x}, \boldsymbol{y})) = \frac{\text{adjacent leg}}{\text{hypotenuse}} = \frac{\boldsymbol{x}^T \boldsymbol{y} \cdot ||\boldsymbol{x}||^{-1}}{||\boldsymbol{y}||}$.

The definition of the length of a vector as well as the angle between two vectors allows us to interpret the scalar product of two vectors geometrically. It holds

$$\boldsymbol{x}^T \boldsymbol{y} = ||\boldsymbol{x}|| \cdot ||\boldsymbol{y}|| \cdot \cos(\alpha(\boldsymbol{x}, \boldsymbol{y})) .$$

The scalar product is therefore the product of the lengths (norms) of the vectors with the cosine of the angle between them.

If the angle formed by two vectors $\boldsymbol{x}$ and $\boldsymbol{y}$ is 90°, we say that the vectors are orthogonal (or perpendicular) to each other. The scalar product $\boldsymbol{x}^T \boldsymbol{y}$ and thus the cosine are zero in this case. This fact is used for a formal definition of orthogonality.

Definition 2.11: Orthogonal vectors

If $\boldsymbol{x}^T \boldsymbol{y} = 0$, then $\boldsymbol{x}$ and $\boldsymbol{y}$ are called **orthogonal** (or **perpendicular**) to each other, in short: $\boldsymbol{x} \perp \boldsymbol{y}$.

Example 2.19. In Example 2.17, $\boldsymbol{b} \perp \boldsymbol{c}$.

Let $\boldsymbol{d} = [2, 1]^T$ and $\boldsymbol{e} = [2, -4]^T$, then $\boldsymbol{d} \perp \boldsymbol{e}$. The following figure shows $\boldsymbol{d}$ and $\boldsymbol{e}$ as well as the two sets $G = \{\boldsymbol{x} \in \mathbb{R}^2 : \boldsymbol{d}^T\boldsymbol{x} \geq 0\}$ (light gray) and $H = \{\boldsymbol{x} \in \mathbb{R}^2 : \boldsymbol{d}^T\boldsymbol{x} \leq 0\}$ (dark gray), which are the halfplanes in the direction of $\boldsymbol{d}$ resp. $-\boldsymbol{d}$. The set $U = \{\boldsymbol{x} \in \mathbb{R}^2 : \boldsymbol{x} = \lambda\boldsymbol{e}, \lambda \in \mathbb{R}\}$ is the dashed line, on which all vectors that are orthogonal to $\boldsymbol{d}$ are located.

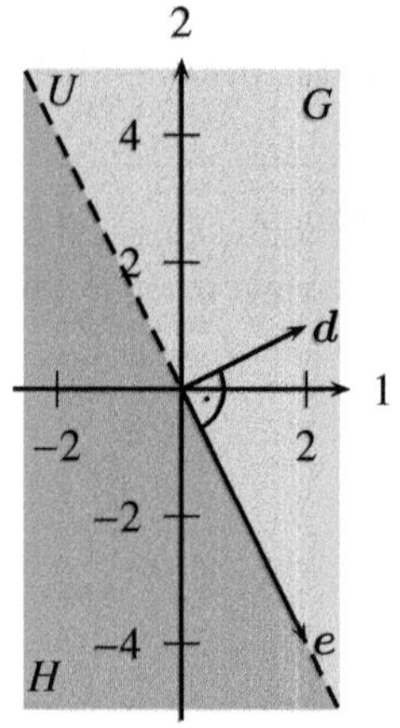

The following table summarizes the geometric concepts developed in this section and their formalization in $\mathbb{R}^n$.

Geometric concept	Formalization
Length of a vector $\boldsymbol{x}$	$\|\|\boldsymbol{x}\|\|$,
Distance between $\boldsymbol{x}$ and $\boldsymbol{0}$	$\|\|\boldsymbol{x}\|\|$,
Distance between $\boldsymbol{x}$ and $\boldsymbol{y}$	$\|\|\boldsymbol{x} - \boldsymbol{y}\|\|$,
Angle between $\boldsymbol{x}$ and $\boldsymbol{y}$	$\cos(\alpha(\boldsymbol{x}, \boldsymbol{y})) = \frac{\boldsymbol{x}^T\boldsymbol{y}}{\|\|\boldsymbol{x}\|\| \, \|\|\boldsymbol{y}\|\|}$,
$\boldsymbol{x}$ and $\boldsymbol{y}$ are orthogonal	$\boldsymbol{x}^T\boldsymbol{y} = 0$.

2.7 Further properties of sets in $\mathbb{R}^n$

The set of all points whose norm is at most one forms the **closed unit ball** in $\mathbb{R}^n$, which is the "ball" with "radius" one around the origin,

$$B(\boldsymbol{0}, 1) = \{\boldsymbol{x} = (x_1, x_2, \ldots, x_n) \in \mathbb{R}^n : ||\boldsymbol{x}|| \leq 1\}.$$

For $n = 2$ we obtain the closed disk with radius one,

$$\left\{(x_1, x_2) \in \mathbb{R} : \sqrt{x_1^2 + x_2^2} \leq 1\right\}.$$

For $n = 3$ it is the ball with radius one about the origin, and for $n = 1$ the closed interval $[-1, 1]$.

More general, one considers the **ball** $B(\boldsymbol{a}, \varepsilon)$ with center $\boldsymbol{a} \in \mathbb{R}^n$ and radius $\varepsilon > 0$,

$$B(\boldsymbol{a}, \varepsilon) = \{\boldsymbol{x} = (x_1, x_2, \ldots, x_n) \in \mathbb{R}^n \, : \, ||\boldsymbol{x} - \boldsymbol{a}|| \leq \varepsilon\} \, ,$$

which is the set of all points in $\mathbb{R}^n$ that have a distance no more than ε from the point $\boldsymbol{a}$; see Figure 2.5.

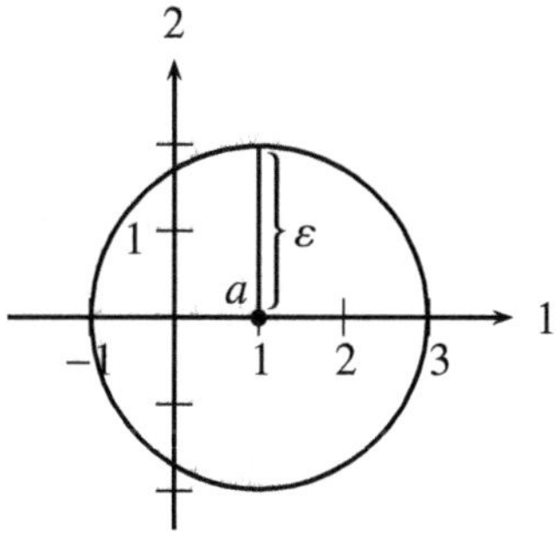

Figure 2.5: Ball $B(\boldsymbol{a}, \varepsilon)$ in $\mathbb{R}^2$ (= disk) with $\boldsymbol{a} = (0, 1), \varepsilon = 2$.

A set $A \subset \mathbb{R}^n$ is called **bounded**, if it is contained in a sufficiently large ball around $\boldsymbol{0}$, i.e., if there is a number M such that

$$A \subset B(\boldsymbol{0}, M) \, .$$

For example, every ball $B(\boldsymbol{a}, \varepsilon)$ having center $\boldsymbol{a} \in \mathbb{R}^n$ and radius $\varepsilon > 0$ is bounded. For every $\boldsymbol{x} \in B(\boldsymbol{a}, \varepsilon)$ it holds that $||\boldsymbol{x} - \boldsymbol{a}|| \leq \varepsilon$, so

$$||\boldsymbol{x}|| = ||\boldsymbol{x} - \boldsymbol{a} + \boldsymbol{a}|| \leq ||\boldsymbol{x} - \boldsymbol{a}|| + ||\boldsymbol{a}|| \leq \varepsilon + ||\boldsymbol{a}|| \, .$$

Therefore, $B(\boldsymbol{a}, \varepsilon)$ lies in the ball around $\boldsymbol{0}$ with radius $M = \varepsilon + ||\boldsymbol{a}||$.

Also, every finite n-dimensional interval $[\boldsymbol{a}, \boldsymbol{b}]$ with $\boldsymbol{a}, \boldsymbol{b} \in \mathbb{R}^n$ is bounded.

Next, we consider the boundary points of a set $A \subset \mathbb{R}^n$. A point $\boldsymbol{z}$ is called **boundary point** of A, if every small ball around $\boldsymbol{z}$ contains both points from A and points from the complement of A; see Figure 2.6. The set of all boundary points is referred to as the **boundary** $\mathrm{bd}(A)$ of the set A.

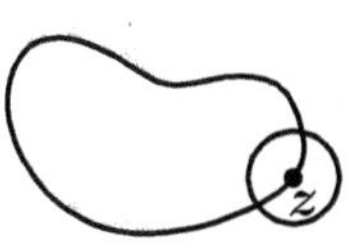

Figure 2.6: Boundary point of a set.

For example, the unit ball $B(\boldsymbol{0}, 1)$ in $\mathbb{R}^2$ has the boundary

$$\{(x_1, x_2) \, : \, x_1^2 + x_2^2 = 1\} \, ,$$

which is the circle around $\mathbf{0}$ with radius 1. For $a, b \in \mathbb{R}$ the ordinary closed interval $[a, b] \subset \mathbb{R}$ has the boundary points a and b. Similarly, the open interval (a, b) and the half-open intervals $[a, b)$ and $(a, b]$ have the boundary points a and b. Obviously, boundary points can belong to the set or not.

A set $A \subset \mathbb{R}^n$ is called **closed**, if it contains all its boundary points. The set is called **open**, if none of the boundary points are elements of A. $A \setminus \mathrm{bd}(A)$ is referred to as **interior** of A, $A \cup \mathrm{bd}(A)$ as **closure** of A. A set $A \subset \mathbb{R}^n$, which is closed and bounded, is called **compact**.

For example, the ordinary closed interval $[a, b] \subset \mathbb{R}$ is closed, the corresponding open interval (a, b) is open. The half-open intervals $[a, b)$ and $(a, b]$ are neither closed nor open. The unit ball $B(\mathbf{0}, 1)$ in $\mathbb{R}^2$ is closed. More generally, every ball $B(\boldsymbol{a}, \varepsilon)$ in $\mathbb{R}^n$ is closed, hence compact. The nonnegative orthant $\mathbb{R}^n_+$ is closed, but not bounded.

A set $A \subset \mathbb{R}^n$ is called **convex** if, for any two points in A, the line segment connecting them also lies in A, i.e., if for all $\boldsymbol{x}, \boldsymbol{y} \in A$ and all $\lambda \in [0, 1]$ it holds

$$\lambda \boldsymbol{x} + (1 - \lambda)\boldsymbol{y} \in A\,.$$

Examples of convex sets are the intervals in $\mathbb{R}$, but also all n-dimensional intervals $[\boldsymbol{a}, \boldsymbol{b}]$. Furthermore, every ball $B(\boldsymbol{a}, \varepsilon)$ in $\mathbb{R}^n$ is convex. Also, $\mathbb{R}^n_+$ and $\mathbb{R}^n$ are convex. Figure 2.7 shows a convex set (left) and a non-convex set (right).

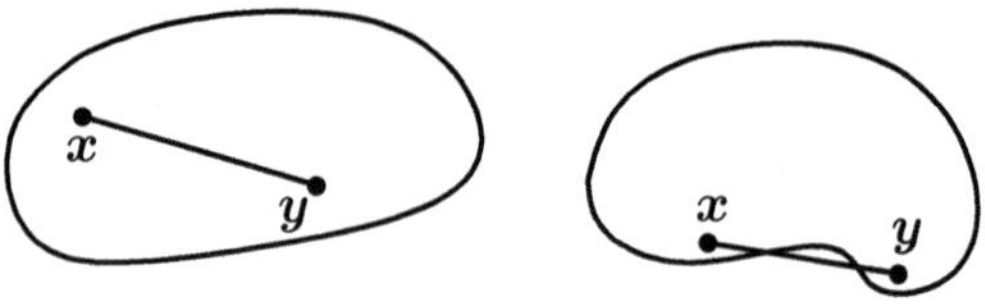

Figure 2.7: A convex and a non-convex set.

2.8 Orthogonal matrices and transformations

An important class of matrices are the so-called orthogonal matrices.

Definition 2.12: Orthogonal matrix

An $n \times n$ matrix $\boldsymbol{Q}$ is called **orthogonal** if all columns of $\boldsymbol{Q}$ have norm one and any two different columns are orthogonal.

If $\boldsymbol{q}_i$ denotes the i-th column of $\boldsymbol{Q}$, then

$$\boldsymbol{Q} \text{ orthogonal} \quad \Longleftrightarrow \quad \boldsymbol{q}_i^T \boldsymbol{q}_j = \begin{cases} 1, & \text{if } i = j, \\ 0, & \text{if } i \neq j. \end{cases}$$

Orthogonal matrices have some very nice properties. First, consider the product $Q^T Q$. Since the element at position (i, j) of the product is the scalar product of the i-th row of Q^T and the j-th column of Q, this is equal to $q_i^T q_j$. (The columns of Q are the rows of Q^T!) From the definition it follows that $Q^T Q$ contains ones on the main diagonal and zeros at all other positions. In other words: $Q^T Q$ is the identity matrix. This means that Q^T is the inverse of Q. Then QQ^T is also the identity matrix. From this it follows that Q^T is an orthogonal matrix as well. So the columns of Q^T are orthogonal to each other and have norm one. Since the columns of Q^T are the rows of Q, the same applies for the rows of Q. We summarize these considerations in a theorem:

Theorem 2.9: Properties of orthogonal matrices

Let Q be an $n \times n$ matrix. Then the following statements are equivalent:

(i) Q is orthogonal,

(ii) Q^T is orthogonal,

(iii) $Q^T Q = I_n$,

(iv) $QQ^T = I_n$,

(v) Q is invertible and $Q^{-1} = Q^T$,

(vi) the columns of Q are orthogonal and have norm one,

(vii) the rows of Q are orthogonal and have norm one.

We have seen in Section 2.5 that the linear transformations from $\mathbb{R}^n$ to $\mathbb{R}^n$ are precisely the transformations $f_A : x \mapsto Ax$ where A is an $n \times n$ matrix. Linear transformations of the form $f_Q : x \mapsto Qx$, where Q is an **orthogonal** matrix, are called **orthogonal transformations** . The importance of orthogonal transformations lies in the fact that they are **length-** and **angle-preserving**.

Theorem 2.10: Length and angle preservation of orthogonal functions

Let Q be an orthogonal matrix. Then the function f_Q

(i) is **length-preserving**, i.e., for arbitrary vectors x, y holds

$$\|Qx - Qy\| = \|x - y\|,$$

(ii) is **angle-preserving**, i.e., for arbitrary vectors $x, y \neq 0$ holds

$$\alpha(Qx, Qy) = \alpha(x, y).$$

Proof: Property (i) follows from

$$\|Qx - Qy\| = \|Q(x - y)\|$$

$$
\begin{aligned}
&= \sqrt{[Q(x-y)]^T[Q(x-y)]} \\
&= \sqrt{(x-y)^T Q^T Q(x-y)} \\
&= \sqrt{(x-y)^T(x-y)} \\
&= \|x-y\| .
\end{aligned}
$$

For (ii), first note that

$$(Qx)^T(Qy) = x^T Q^T Q y = x^T y .$$

Due to (i), however, $\|Qx\| = \|x\|$ and $\|Qy\| = \|y\|$. Property (ii) then follows directly from the formula for the angle between two vectors

$$\cos\bigl(\alpha(Qx, Qy)\bigr) = \frac{(Qx)^T(Qy)}{\|Qx\|\,\|Qy\|} = \frac{x^T y}{\|x\|\,\|y\|} = \cos\bigl(\alpha(x, y)\bigr) .$$

□

In particular, length preservation implies that $\|Qx\| = \|x\|$.

Thus, orthogonal transformations leave distances as well as angles between two points unchanged. The orthogonal transformations from $\mathbb{R}^2$ to $\mathbb{R}^2$ are precisely the **rotations** around the origin and the **reflections** at straight lines through the origin.

Geometric interpretation of orthogonal matrices

What are the orthogonal transformations of $\mathbb{R}^2$? Since an orthogonal transformation corresponds to an orthogonal matrix Q, we only need to find the orthogonal 2×2 matrices. Since orthogonal transformations preserve length, the image of the first unit vector must have a distance of one from the origin, i.e., it must lie somewhere on the unit circle. If α is the angle between e_1 and the image Qe_1, then $Qe_1 = [\cos(\alpha), \sin(\alpha)]^T$.

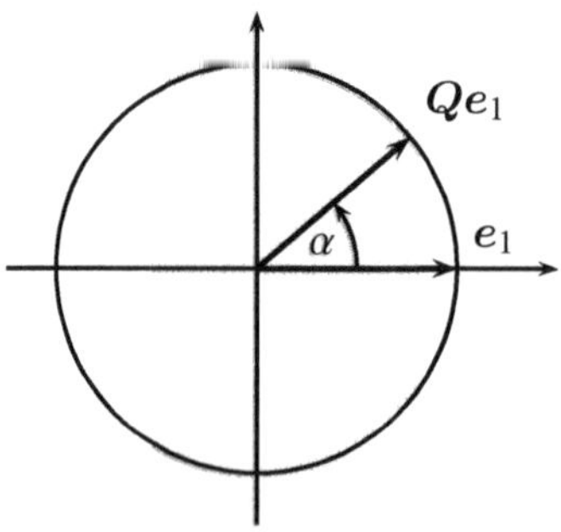

Now, what is the image of the second unit vector? Since orthogonal transformations preserve angles and the unit vectors e_1 and e_2 form a right angle, the images of the two unit vectors must also form a right angle. Since the image of the second unit vector must also lie on the unit circle due to the preservation of length, there are only the following two possibilities:

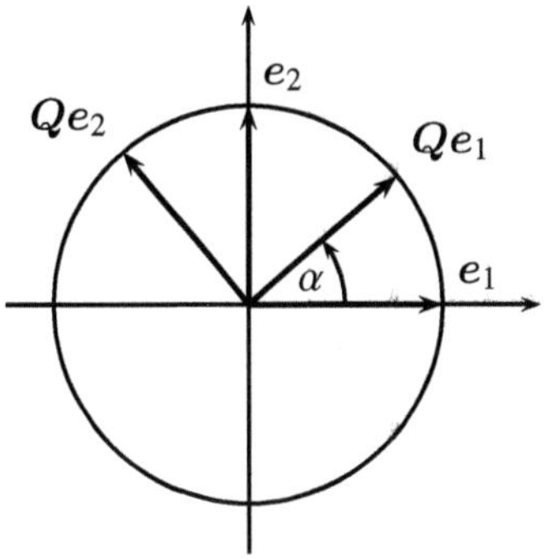

In the first case, $Qe_2 = [-\sin(\alpha), \cos(\alpha)]^T$ and in the second case correspondingly $Qe_2 = [\sin(\alpha), -\cos(\alpha)]^T$. Since the images of the unit vectors are the columns of the matrix, an orthogonal 2×2 matrix Q must therefore have one of the following two forms:

$$Q = \begin{bmatrix} \cos(\alpha) & -\sin(\alpha) \\ \sin(\alpha) & \cos(\alpha) \end{bmatrix} \quad \text{or} \quad Q = \begin{bmatrix} \cos(\alpha) & \sin(\alpha) \\ \sin(\alpha) & -\cos(\alpha) \end{bmatrix}.$$

Geometrically, a matrix of the first form describes a rotation by an angle α around the origin. A matrix of the second form yields a reflection on the straight line through the origin, forming an angle $\alpha/2$ with the x-axis.

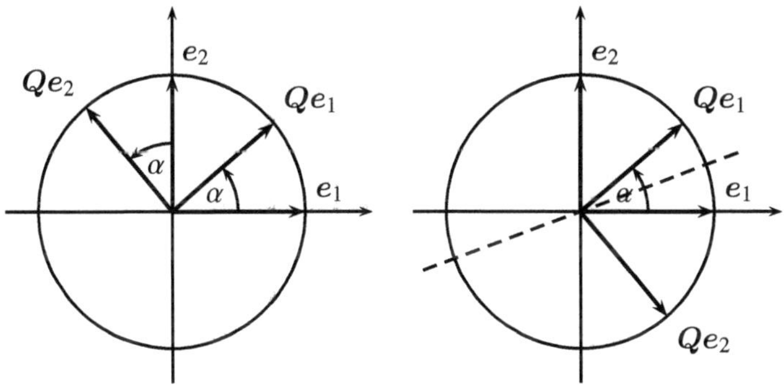

Hence, we have completely classified the orthogonal 2×2 matrices. We summarize the results in the following theorem.

Theorem 2.11: Classification of orthogonal 2×2 matrices

If Q is an orthogonal 2×2 matrix, then there is an angle $\alpha \in (-\pi, \pi]$, such that Q has one of the following forms:

$$Q = \begin{bmatrix} \cos(\alpha) & -\sin(\alpha) \\ \sin(\alpha) & \cos(\alpha) \end{bmatrix} \quad \text{or} \quad Q = \begin{bmatrix} \cos(\alpha) & \sin(\alpha) \\ \sin(\alpha) & -\cos(\alpha) \end{bmatrix}$$

In the first case, Q describes a **rotation** by the angle α, in the second case a **reflection** on the x-axis rotated by $\alpha/2$.

In higher dimensions, the geometric interpretation of orthogonal matrices is significantly more complicated. However, it generally holds that every orthogonal matrix corresponds to a sequence of rotations and reflections.

Important terms for review

After reading this chapter, the following terms should be familiar:

- Matrix
- Identity matrix, zero matrix
- Transpose
- Symmetric matrix
- Scalar multiplication of a matrix
- Addition and multiplication of two matrices
- Row vector, column vector
- Scalar product of vectors, orthogonal, orthogonal matrix
- Norm, unit vector
- Inverse matrix
- Linear transformation, linear function

Self-test

You can check your understanding of this chapter using the following true-or-false questions. For each statement, indicate whether it is true (T) or false (F). Short solutions to these questions can be found in Appendix H.

I. The multiplication of . . .

	T	F
. . . two diagonal matrices yields a diagonal matrix.	☐	☐
. . . a diagonal matrix with a lower triangular matrix yields a diagonal matrix.	☐	☐
. . . two upper triangular matrices yields an upper triangular matrix.	☐	☐

II. Given are a number $\lambda \in \mathbb{R}$ and matrices $\boldsymbol{A}$, $\boldsymbol{B}$ and $\boldsymbol{C}$ of suitable format. Are the following statements true or false?

	T	F
$\boldsymbol{A}(\boldsymbol{B}+\boldsymbol{C}) = \boldsymbol{A}\boldsymbol{B}+\boldsymbol{C}\boldsymbol{A}$	☐	☐
$\boldsymbol{A}(\boldsymbol{B}+\lambda\boldsymbol{C}) = \boldsymbol{A}\boldsymbol{B}+\lambda\boldsymbol{A}\boldsymbol{C}$	☐	☐

$((BA)^{-1})^T = (B^{-1})^T(A^T)^{-1}$ ☐ ☐

$(A + B)(C + D) = AC + BC + AD + BD$ ☐ ☐

III. Given are numbers $\lambda, \mu \in \mathbb{R}$ and vectors $x, y, z \in \mathbb{R}^3$. Are the following statements true or false?

	T	F
$\|\|\lambda x + \mu y\|\| \leq \|\lambda\| \cdot \|\|x\|\| + \|\mu\| \cdot \|\|y\|\|$	☐	☐
From $x \perp y$ and $y \perp z$ follows that $x \perp z$.	☐	☐

IV. Are the following statements true or false?

	T	F
The union of two convex sets is again convex.	☐	☐
The identity matrix is the only diagonal matrix that is orthogonal.	☐	☐
The matrix $\begin{bmatrix} -1 & 0 \\ 0 & -1 \end{bmatrix}$ is orthogonal.	☐	☐

Exercises

Exercise 1

Given are the following vectors in $\mathbb{R}^4$: $x = [3, 4, 5, 1]^T$, $y = [-1, 4, 3, -2]^T$ and $z = [1, 1, -1, -2]^T$. Calculate

a) $x + 3y - z$,

b) $x^T \cdot (\lambda z), \quad \lambda \in \mathbb{R}^+ \setminus \{0\}$,

c) the distance between the point y and the origin.

Exercise 2

Given are the vectors $a = [1, 2, 3, 4, 5]^T$, $b = [0, 1, 0, 1, 0]^T$, $c = [0, -1, -2, 3]^T$ and $d = [0, 2, 4, -1]^T$. Calculate, if possible,

a) $\frac{1}{2}a + b$,

b) $a + c$,

c) $(a^T b)c$,

d) $a(b^T c)$,

e) $a^T b + c^T d$.

Exercise 3
Given are the matrices and vectors

$$A = \begin{bmatrix} 1 & 6 \\ 2 & 5 \\ 3 & 4 \end{bmatrix}, \quad B = \begin{bmatrix} 4 & 1 & 2 \\ 5 & -1 & 3 \end{bmatrix}, \quad C = \begin{bmatrix} 1 & -2 & 3 \\ -2 & 1 & -2 \\ 3 & -2 & 1 \end{bmatrix},$$

$$d = \begin{bmatrix} -2, & 0, & 1 \end{bmatrix}, \quad e = \begin{bmatrix} 5 \\ -1 \\ -3 \end{bmatrix}.$$

Determine if possible

a) AB, b) BA, c) ABC, d) $3CAB$, e) $5de$, f) $e^T A$,

g) dC, h) eC, i) $A + B^T$, j) $AB + C - ed$, k) $C(d^T + e)$.

Exercise 4
Assume that all matrices considered below have a compatible format. Compute

a) $(AB)^T(B^{-1}A^{-1})^T$,

b) $(A(A^{-1} + B^{-1})B)(B + A)^{-1}$,

c) $(I^{-1})^T A - (I^T)^{-1} A + 0 \cdot I$.

Exercise 5

Given are the matrices $A = \begin{bmatrix} 1 & 2 \\ 3 & 5 \end{bmatrix}$ and $B = \begin{bmatrix} 1 & 2 \\ 1 & 4 \end{bmatrix}$. Calculate

a) $A^{-1} + B^{-1}$,

b) $(A^{-1}B)^{-1}$,

c) $((A^{-1}B)^T)^{-1}$.

Exercise 6
Given are the vectors $x^T = \begin{bmatrix} 1, 2, 3 \end{bmatrix}$, $y^T = \begin{bmatrix} 0, -3, 2 \end{bmatrix}$, $z^T = \begin{bmatrix} -1, 1, 1 \end{bmatrix}$.

a) Find lower bounds for the expressions $||x|| + ||y||$ and $||x|| \cdot ||z||$. Further, calculate the exact values of these expressions.

b) What can you say about the relative positions of the vectors x and y?

Exercise 7
Check whether the following matrices are orthogonal:

$$\text{a)} \begin{bmatrix} \frac{\sqrt{3}}{2} & \frac{1}{2} \\ \frac{1}{2} & -\frac{\sqrt{3}}{2} \end{bmatrix}, \qquad \text{b)} \begin{bmatrix} 1 & 0 & 0 \\ 0 & \frac{3}{5} & -\frac{4}{5} \\ 0 & \frac{4}{5} & \frac{3}{5} \end{bmatrix}.$$

Chapter 3

Sequences, series and continuous functions

In this chapter, we first consider sequences of real numbers and provide criteria for their convergence. Then we introduce the notion of an infinite series and study their properties. Using the geometric series, formulas of financial mathematics are derived. Further, based on the convergence of sequences, the continuity of functions in one and several variables is introduced. Finally, fixed points of a function are discussed.

Example 3.1 (Cobweb model). The so-called cobweb model describes the mutual adjustment of supply and demand in a market. We consider a dynamic period model for the price and quantity of a certain good sold: The price of the good in period n is denoted by p_n. Assume that the supply s_n in period n linearly depends on the **price of the previous period**, $s_n = a + bp_{n-1}$, where a and b are parameters with $b > 0$, so that a higher price p_{n-1} leads to a larger supply s_n. Further assume that the **demand** d_n in period n depends linearly on the **current price**, $d_n = c + dp_n$, with parameter $d < 0$, implying that a higher price p_n creates a lower demand d_n. The cobweb model describes the typical situation that the suppliers react to the demand only with a time delay, which is one period here. Such a delay is generally mentioned as a **(time) lag**. Finally assume that in each period the price is set so that the whole supply is sold, i.e., $d_n = s_n$. From this it follows that the price in period n is related to the price p_{n-1} in the previous period as follows:

$$p_n = \frac{b}{d} p_{n-1} - \frac{c-a}{d}. \tag{3.1}$$

K. Mosler et al., *Mathematical Methods for Business and Economics*,
Mathematics Study Resources 20, https://doi.org/10.1007/978-3-662-73831-3_3

The development of the price and the supply is shown in Figure 3.1. Let q_n, $n = 1, 2, \ldots$, denote the quantity supplied (and demanded) in period n. We obtain

$$q_n = c + dp_n = c + d\left(\frac{b}{d}p_{n-1} - \frac{c-a}{d}\right) = bp_{n-1} + a$$
$$= \frac{b}{d}(c + dp_{n-1}) + \frac{ad - bc}{d} = \frac{b}{d}q_{n-1} + \frac{ad - bc}{d}.$$

The name of the model is derived from its graphical illustration (Figure 3.1), which resembles a spider's web.

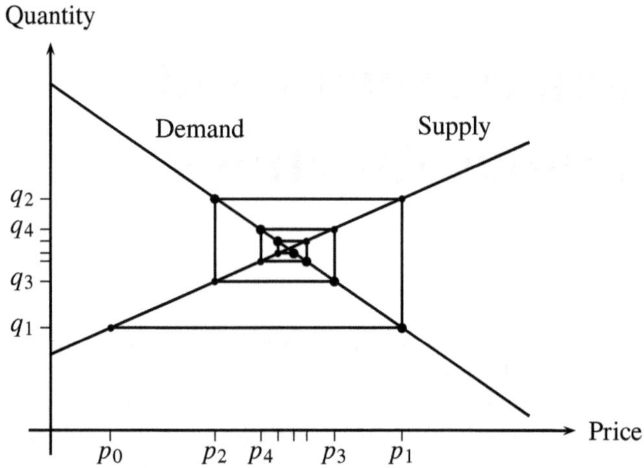

Figure 3.1: The cobweb model.

In the period $n = 0$, the good is sold at a price p_0. The suppliers use this price to determine their supply quantity q_1 for period $n = 1$. The buyers are willing to buy the offered quantity q_1 at a price of p_1 each. The suppliers use this price again to plan the production quantity q_2 for the next period, and so on. The price-quantity combinations that are actually realized are those lying on the demand curve.

In the situation shown in Figure 3.1, the sequence of prices seems to approach an equilibrium price, which results from the intersection of the supply and demand curves. The equilibrium price $\overline{p}$, at which supply equals demand, is obtained from the equation $a + b\overline{p} = c + d\overline{p}$ as

$$\overline{p} = \frac{c-a}{b-d}.$$

An interesting question is, under which conditions the convergence to an equilibrium price takes place.

3.1 Sequences of real numbers

A sequence of real numbers is mathematically regarded as a function from the natural numbers to the real numbers, i.e., to each natural number n a real number a_n is assigned.

Definition 3.1: Sequence in $\mathbb{R}$

A function

$$\mathbb{N} \to \mathbb{R}, \quad n \mapsto a_n,$$

is called a **sequence of real numbers, sequence in $\mathbb{R}$** or **real-valued sequence**. We notate a sequence by $(a_n)_{n\in\mathbb{N}}$ or simply (a_n). The number a_n is referred to as the n-th **term of the sequence**.

Example 3.2 (Sequences in $\mathbb{R}$).

a) $(1, 4, 9, 16, \ldots), a_n = n^2$,

b) $(-1, 1, -1, 1, -1, 1, \ldots), a_n = (-1)^n$,

c) $(1, \frac{1}{4}, \frac{1}{9}, \frac{1}{16}, \ldots), a_n = \frac{1}{n^2}$,

d) $a_1 = b, a_n = c\, a_{n-1} + d$ for $n \geq 2$, $(b, c, d \in \mathbb{R})$,

e) $a_1 = 2, a_n = 1 - \frac{1}{a_{n-1}}$ for $n \geq 2$.

When considering the sequence $(1/n)_{n\in\mathbb{N}} = (1, \frac{1}{2}, \frac{1}{3}, \frac{1}{4}, \ldots)$, it becomes apparent that the terms of the sequence tend toward zero. The same holds for the sequence $(2^{-n})_{n\in\mathbb{N}} = (\frac{1}{2}, \frac{1}{4}, \frac{1}{8}, \frac{1}{16}, \ldots)$. So one may call zero the "limit" of these sequences. Although one has an intuitive understanding of what it means for a sequence to "tend towards" a certain value or to have a "limit", it is not obvious how to precisely define this notion.

Two points are crucial. Firstly a number must exist that is finally approached by the sequence (a_n). I.e., the distance between this number and the elements of the sequence must become arbitrarily small. Secondly, in the long run, a once achieved small distance to this number must not be exceeded by any of the subsequent terms of the sequence. The distance of the n-th term a_n from a fixed value a is given by the absolute value $|a_n - a|$. For a to be called the "limit" of a sequence $(a_n)_{n\in\mathbb{N}}$, we require that beyond a certain term, *all subsequent terms* have a distance less than a given small value, say $\varepsilon = 0.1$ from a. However, with $\varepsilon = 0.1$ these distances may be still too large. Therefore one requires that, no matter how small $\varepsilon > 0$ is chosen, beyond a certain term all subsequent terms of the sequence have a distance less than ε from a. Formally, this means:

Definition 3.2: Convergence of a sequence

A sequence $(a_n)_{n\in\mathbb{N}}$ of real numbers **converges** to a real number a, if for every (arbitrarily small) $\varepsilon > 0$ there is a number $N_\varepsilon \in \mathbb{N}$ such that

$$|a_n - a| < \varepsilon \quad \text{for all } n \geq N_\varepsilon.$$

The number a is called the **limit** of the sequence. We write $\lim_{n\to\infty} a_n = a$ or briefly $\lim a_n = a$. Other notations are $a_n \xrightarrow{n\to\infty} a$ or simply $a_n \longrightarrow a$.

A sequence $(a_n)_{n\in\mathbb{N}}$ in $\mathbb{R}$ is called **convergent**, if it converges to a real number a; otherwise, it is called **divergent**.

A sequence having limit zero is mentioned as a **null sequence**.

Whether a sequence is convergent or not does not depend on the values of its initial terms, but only on the behavior of its later terms. So if you change the first ten sequence elements of a convergent sequence with limit a, the resulting sequence is still convergent towards a. In general, one can change a finite number of terms without changing the convergence behavior. A convergent sequence is shown in Figure 3.2.

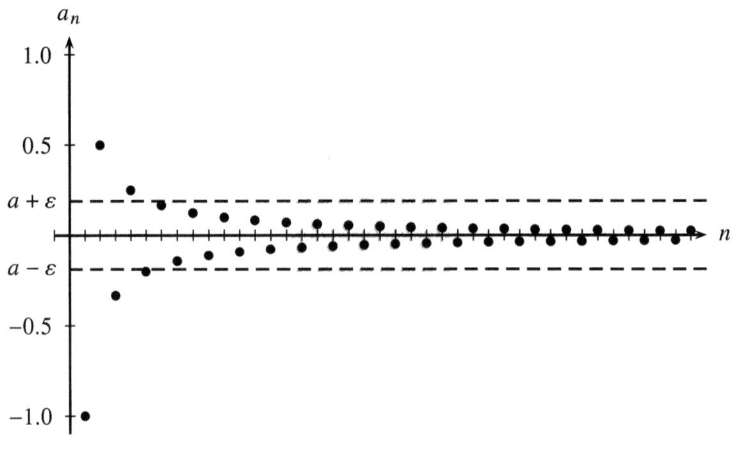

Figure 3.2: Illustration of a convergent sequence.

In order to decide whether a given sequence converges or diverges, we may try to directly check Definition 3.2. However, in this definition the value of the limit has to be known (or at least guessed) beforehand, which works for rather simple sequences only. For more complicated sequences, the limit is determined using rules that will be presented below.

Example 3.3. For the sequence $(\frac{1}{n})_{n\in\mathbb{N}} = (1, \frac{1}{2}, \frac{1}{3}, \ldots)$ one easily guesses the limit to be zero. Thus, we need to check whether we can find a number N_ε for every (no matter how small) ε such that

$$\left|\frac{1}{n} - 0\right| < \varepsilon \quad \text{for all } n \geq N_\varepsilon$$

is true. We have

$$\left|\frac{1}{n} - 0\right| = \left|\frac{1}{n}\right| = \frac{1}{n}$$

and $\frac{1}{n} < \varepsilon$ if and only if $n > \frac{1}{\varepsilon}$. So our requirement is met, if we choose for N_ε some natural number greater than $\frac{1}{\varepsilon}$.

For example, for $\varepsilon = 0.1$ we may choose $N_\varepsilon = 11$. Starting from the eleventh term, all terms of the sequence have a distance from zero less than 0.1. For $\varepsilon = 0.000001$, $N_\varepsilon = 1000001$ will suit. Starting from this term, all terms have a distance from zero less than 0.000001.

Example 3.4.

a) The sequence (a_n) with $a_n = \frac{1}{n^\alpha}$ and some $\alpha \in \mathbb{Q}$, $\alpha > 0$, is a null sequence. One chooses N_ε as the smallest n with $\frac{1}{n^\alpha} < \varepsilon$, i.e., $n > \varepsilon^{-\frac{1}{\alpha}}$.

b) The sequence (a_n) with $a_n = (-1)^n$ is not a null sequence. For $\varepsilon = \frac{1}{2}$, $|a_n| = 1 > \varepsilon$ holds for all $n \in \mathbb{N}$.

The sequence $(n)_{n\in\mathbb{N}} = (1, 2, 3, \ldots)$ does not converge. However, this sequence shows a clear limit behavior. The elements of the sequence become larger and larger and eventually exceed any given bound. Such a sequence is called **divergent to infinity**.

Definition 3.3: Divergence to Infinity

A sequence $(a_n)_{n\in\mathbb{N}}$ is called **divergent to infinity**, if for every real number M there is a number $N_M \in \mathbb{N}$ such that

$$a_n > M \quad \text{for all } n \geq N_M.$$

It is called **divergent to minus infinity**, if for every real number M there is a number N_M such that

$$a_n < M \quad \text{for all } n \geq N_M.$$

If $(a_n)_{n\in\mathbb{N}}$ is divergent to ∞, we write $\lim_{n\to\infty} a_n = \infty$ or briefly $\lim a_n = \infty$. Other notations are $a_n \xrightarrow{n\to\infty} \infty$ or simply $a_n \longrightarrow \infty$. Corresponding notations are used for sequences divergent to minus infinity.

Even though the notation $\lim_{n\to\infty} a_n = \infty$ is used for sequences divergent to ∞, such sequences are neither convergent, nor can ∞ be properly referred to as the limit of the sequence. Note that ∞ and $-\infty$ are no real numbers and we cannot calculate with them as we do with real numbers.

In Chapter 1, we have already got to know the notions of bounded and monotone functions. Since sequences are functions from $\mathbb{N}$ to $\mathbb{R}$, these notions apply immediately to sequences.

A sequence $(a_n)_{n\in\mathbb{N}}$ is called **bounded**, if there is a real number M such that $|a_n| \leq M$ for all $n \in \mathbb{N}$, i.e., all elements of the sequence lie between $-M$ and M.

A sequence $(a_n)_{n\in\mathbb{N}}$ is called **monotonically increasing**, if $a_n \leq a_{n+1}$ for all $n \in \mathbb{N}$. In the opposite case, if $a_n \geq a_{n+1}$ for all $n \in \mathbb{N}$, the sequence is called **monotonically decreasing**. A sequence is called **monotonic**, if it is monotonically increasing or monotonically decreasing.

Theorem 3.1: Convergence and boundedness I

Every convergent sequence of real numbers is bounded.

On the other hand, a bounded sequence is not necessarily convergent. This becomes clear when considering, for example, the sequence $((-1)^n)_{n\in\mathbb{N}} = (-1, 1, -1, 1, \ldots)$. This sequence jumps between the points -1 and $+1$. Obviously, the sequence is bounded, but it does not have a limit. However, if a bounded sequence is monotonic, say increasing, the terms continuously increase (or at least do not decrease). In a sense, there is insufficient 'space' above, so the terms must eventually accumulate at a value $a \leq M$, see Figure 3.3.

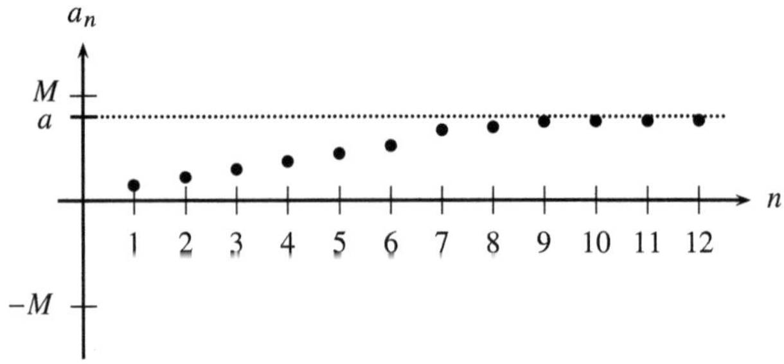

Figure 3.3: A bounded, monotonically increasing sequence that converges to a.

Theorem 3.2: Convergence and boundedness II

Every bounded and monotonic sequence of real numbers is convergent.

The preceding theorem merely states that a monotonic and bounded sequence is convergent. It does not make any statement about the limit of the sequence.

An important sequence in applications is the following:

Example 3.5. For any real number q, the sequence $(q^n)_{n\in\mathbb{N}}$ is called a **geometric sequence**. For example, with $q = 1/2$ we get the sequence $(\frac{1}{2}, \frac{1}{4}, \frac{1}{8}, \ldots)$. The convergence behavior of the geometric sequence depends on the value of q. Five cases have to be distinguished:

Case 1 ($|q| < 1$)**:** In this case, the sequence (q^n) is convergent and $\lim_{n\to\infty} q^n = 0$.

Case 2 ($q = 1$)**:** Then $q^n = 1$ for all n. The sequence is convergent with limit one.

Case 3 ($q = -1$)**:** The sequence is divergent, alternating between the values 1 and -1.

Case 4 ($q > 1$)**:** The sequence is divergent to $+\infty$.

Case 5 ($q < -1$)**:** The sequence is divergent; it is unbounded with alternating sign.

How can we find the limit of a sequence? A simple possibility is the following **squeeze theorem**. If we know two sequences that converge to a common limit, it is intuitively clear that any sequence that lies between these two sequences must converge to the same limit.

Theorem 3.3: Squeeze theorem

Let $(a_n)_{n\in\mathbb{N}}$, $(b_n)_{n\in\mathbb{N}}$ and $(c_n)_{n\in\mathbb{N}}$ be sequences in $\mathbb{R}$. If there is a number N_0 such that

$$b_n \leq a_n \leq c_n \quad \text{for all } n \geq N_0,$$

and if $\lim_{n\to\infty} b_n = \lim_{n\to\infty} c_n = a$, then also $\lim_{n\to\infty} a_n = a$.

With this theorem, we can prove the convergence of further sequences and determine their limit.

Example 3.6.

a) Consider the sequence $(a_n)_{n\in\mathbb{N}}$ with $a_n = \frac{1}{2^n}$. To show that $(a_n)_{n\in\mathbb{N}}$ is a null sequence, we aim at enclosing a_n between two known null sequences. Since

$$0 \leq \frac{1}{2^n} \leq \frac{1}{n} \quad \text{for all } n \in \mathbb{N}.$$

we choose $b_n = 0$ and $c_n = \frac{1}{n}$. Then the squeeze theorem implies that $\lim_{n\to\infty} \frac{1}{2^n} = 0$ as well.

b) As a second example, consider $(a_n)_{n\in\mathbb{N}}$ with $a_n = \frac{1}{n!}$.[1] Here, again we have

$$0 \leq \frac{1}{n!} \leq \frac{1}{n} \quad \text{for all } n \in \mathbb{N}.$$

According to the squeeze theorem, $\lim_{n\to\infty} \frac{1}{n!} = 0$.

[1]The symbol $n!$ ("n factorial") denotes the product of the first n natural numbers, $n! = 1 \cdot 2 \cdot \ldots \cdot n$.

Theorem 3.4: Rules for limits of sequences

Let $(a_n)_{n\in\mathbb{N}}$ and $(b_n)_{n\in\mathbb{N}}$ be two convergent sequences with $\lim_{n\to\infty} a_n = a$ and $\lim_{n\to\infty} b_n = b$. Then:

(i) The sequence $(a_n+b_n)_{n\in\mathbb{N}}$ is convergent with $\lim_{n\to\infty}(a_n+b_n) = a+b$.

(ii) The sequence $(a_n-b_n)_{n\in\mathbb{N}}$ is convergent with $\lim_{n\to\infty}(a_n-b_n) = a-b$.

(iii) The sequence $(a_n \cdot b_n)_{n\in\mathbb{N}}$ is convergent with $\lim_{n\to\infty}(a_n \cdot b_n) = a \cdot b$.

(iv) If $b \neq 0$, then there exists an $N_0 \in \mathbb{N}$ such that $b_n \neq 0$ for all $n \geq N_0$ and the sequence $\left(\frac{a_n}{b_n}\right)_{n\geq N_0}$ converges to $\lim_{n\to\infty}\left(\frac{a_n}{b_n}\right) = \frac{a}{b}$.

Remark: Why is part (iv) of the previous theorem formulated so strangely? We cannot exclude that certain terms b_n are equal to zero, in which case the quotient a_n/b_n is not defined. However, as the limit is $b \neq 0$, starting from a certain number N_0, all subsequent terms b_n, $n \geq N_0$, are close enough to b so that $b_n \neq 0$ holds. Thus, for $n \geq N_0$ all terms (a_n/b_n) are defined. For this reason we restrict the sequence to the quotients (a_n/b_n) with $n \geq N_0$.

Slightly less precise, but short and memorable, the previous statements are written as:

$$\lim_{n\to\infty}(a_n + b_n) = \lim_{n\to\infty} a_n + \lim_{n\to\infty} b_n\,,$$
$$\lim_{n\to\infty}(a_n - b_n) = \lim_{n\to\infty} a_n - \lim_{n\to\infty} b_n\,,$$
$$\lim_{n\to\infty}(a_n \cdot b_n) = \lim_{n\to\infty} a_n \cdot \lim_{n\to\infty} b_n\,,$$
$$\lim_{n\to\infty}\left(\frac{a_n}{b_n}\right) = \frac{\lim_{n\to\infty} a_n}{\lim_{n\to\infty} b_n}\,.$$

It is crucial here that the sequences $(a_n)_{n\in\mathbb{N}}$ and $(b_n)_{n\in\mathbb{N}}$ are convergent. The theorem cannot be applied to sequences that are divergent to plus or minus infinity and, in particular, one is not allowed to calculate with the symbols $+\infty$ or $-\infty$ as if they were real numbers. If one or both sequences are divergent to $+\infty$ or $-\infty$, the sequence $(a_n + b_n)_{n\in\mathbb{N}}$ may diverge (possibly to $+\infty$ or $-\infty$) or converge.

In the following examples, limits will be determined by the above rules.

Example 3.7.

a) Consider the sequence $(a_n)_{n\in\mathbb{N}}$,

$$a_n = \frac{5}{n \cdot 2^n} - \left(10 + \frac{7}{\sqrt{n}}\right) = 5 \cdot \frac{1}{n} \cdot \frac{1}{2^n} - 10 - 7 \cdot \frac{1}{\sqrt{n}}\,.$$

Because

$$\lim_{n\to\infty}\frac{1}{n} = 0\,, \qquad \lim_{n\to\infty}\frac{1}{2^n} = 0 \quad \text{and} \quad \lim_{n\to\infty}\frac{1}{\sqrt{n}} = 0\,,$$

we get

$$\lim_{n\to\infty} a_n = 5 \cdot 0 \cdot 0 - 10 - 7 \cdot 0 = -10 \,.$$

b) Consider the sequence $(b_n)_{n\in\mathbb{N}}$,

$$b_n = \frac{6n^5 - 3n}{\sqrt{n} + 1 - 2n^5} = \frac{6 - \frac{3}{n^4}}{\frac{1}{n^{9/2}} + \frac{1}{n^5} - 2} \,.$$

Here, the quotient was canceled by n^5. Now, all the sequences in the numerator and denominator are convergent and we get

$$\lim_{n\to\infty} b_n = \frac{6-0}{0+0-2} = -3 \,.$$

In general, a product of two sequences does not converge if one of its factor sequences diverges. However, if one factor is bounded and the other factor is a null sequence, the product converges to zero as well:

Theorem 3.5: Null sequence times bounded sequence

Let $(a_n)_{n\in\mathbb{N}}$ and $(b_n)_{n\in\mathbb{N}}$ be two sequences. If $(a_n)_{n\in\mathbb{N}}$ is a null sequence and $(b_n)_{n\in\mathbb{N}}$ is bounded, then $(a_n \cdot b_n)_{n\in\mathbb{N}}$ is also a null sequence.

If $(a_n)_{n\in\mathbb{N}}$ is a null sequence and $(b_n)_{n\in\mathbb{N}}$ is not bounded, no general statement can be made about the convergence of the sequence $(a_n \cdot b_n)_{n\in\mathbb{N}}$. Whether this sequence converges or not, depends, roughly speaking, on whether $(a_n)_{n\in\mathbb{N}}$ converges to zero "faster" than $(b_n)_{n\in\mathbb{N}}$ goes to infinity. This can be easily seen from a simple example:

Example 3.8. For the sequence $(a_n)_{n\in\mathbb{N}}$ we choose the null sequence $\left(\frac{1}{n^2}\right)_{n\in\mathbb{N}}$. For the second sequence we first choose $b_n = n$. Then $a_n \cdot b_n = \frac{1}{n}$ and $\lim_{n\to\infty}(a_n \cdot b_n) = 0$. If, on the other hand, we choose $b_n = n^3$, then $a_n \cdot b_n = n$ and $\lim_{n\to\infty}(a_n \cdot b_n) = \infty$.

To conclude this section, we will once again examine the cobweb model.

Example 3.9 (Cobweb model; continued). Instead of the price, we now consider the deviation z_n of the price from the equilibrium price $\overline{p} = \frac{c-a}{b-d}$, i.e., $z_n = p_n - \overline{p}$. By substituting $p_n = z_n + \overline{p}$ and $p_{n-1} = z_{n-1} + \overline{p}$ in Equation (3.1) we get

$$z_n + \overline{p} = \frac{b}{d}(z_{n-1} + \overline{p}) - \frac{c-a}{d} \,.$$

Further,

$$\begin{aligned} z_n &= \frac{b}{d} z_{n-1} + \frac{b}{d}\overline{p} - \overline{p} - \frac{c-a}{d} = \frac{b}{d} z_{n-1} + \frac{b-d}{d}\overline{p} - \frac{c-a}{d} \\ &= \frac{b}{d} z_{n-1} + \frac{c-a}{d} - \frac{c-a}{d} = \frac{b}{d} z_{n-1} \,, \end{aligned}$$

and finally

$$z_n = \left(\frac{b}{d}\right) z_{n-1} = \left(\frac{b}{d}\right)^2 z_{n-2} = \ldots = \left(\frac{b}{d}\right)^n z_0 .$$

Three cases have to be distinguished:

Case 1 ($|b| < |d|$): In this case, $|b/d| < 1$ and according to Example 3.5 we have $\lim_{n\to\infty} z_n = \lim_{n\to\infty}(b/d)^n z_0 = 0$ and thus $\lim_{n\to\infty} p_n = \lim_{n\to\infty} z_n + \overline{p} = \overline{p}$, i.e., the sequence of prices converges to the equilibrium price.

Case 2 ($|b| = |d|$): Then (due to $b > 0$ and $d < 0$) $b/d = -1$ and the prices lie alternately above and below the equilibrium price, i.e., they oscillate between $\overline{p} + z_0$ and $\overline{p} - z_0$. In this case, the sequence of prices converges only if $z_0 = 0$, that is the initial price p_0 already equals the equilibrium price $\overline{p}$; otherwise it diverges.

Case 3 ($|b| > |d|$): In this case, $|b/d| > 1$ and according to Example 3.5 the sequence (z_n) diverges.

3.2 Multidimensional sequences

In the cobweb model, prices and quantities depend on each other. Besides considering the two sequences separately, it makes sense to examine their joint development over time. In each period, price and quantity form a pair of real numbers, which corresponds to a point in $\mathbb{R}^2$. This yields a sequence of points in $\mathbb{R}^2$.

In this section we consider sequences of points in $\mathbb{R}^d$ for any dimension $d \in \mathbb{N}$. (Since the letter n is already used as an index of the sequence, here and in Section 3.11, we denote the Euclidean space with $\mathbb{R}^d$ instead of $\mathbb{R}^n$.)

Definition 3.4: Sequence in $\mathbb{R}^d$

A function

$$\mathbb{N} \to \mathbb{R}^d, \quad n \mapsto \boldsymbol{a}_n,$$

is called a sequence in $\mathbb{R}^d$. For the sequence we write $(\boldsymbol{a}_n)_{n\in\mathbb{N}}$ or simply $(\boldsymbol{a}_n)$.

For $d = 1$ we get a sequence in $\mathbb{R}$. The case $d = 2$ is illustrated with two examples.

Example 3.10. Let $\boldsymbol{a}_n = \left(\frac{3}{n}, 2\cdot(-1)^n\right)$. Then $(\boldsymbol{a}_n)_{n\in\mathbb{N}}$ is a sequence in $\mathbb{R}^2$. Its first four elements are $(3, -2)$, $(1.5, 2)$, $(1, -2)$, $(0.75, 2)$ and shown in the following figure.

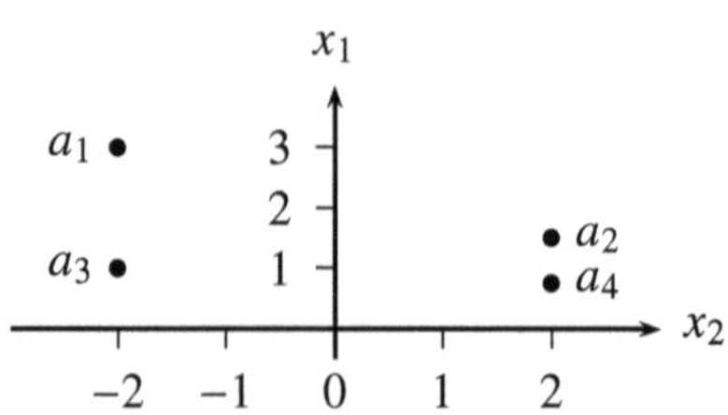

Example 3.11 (Cobweb model; continued). In the cobweb model, the prices p_n and quantities q_n in period n form a sequence $(p_n, q_n)_{n\in\mathbb{N}}$ in $\mathbb{R}^2$. With the notation of Example 3.1, we have

$$\begin{aligned} p_n &= \frac{b}{d} p_{n-1} - \frac{c-a}{d}, \\ q_n &= \frac{b}{d} q_{n-1} + \frac{ad-bc}{d}. \end{aligned}$$

The convergence of a sequence in $\mathbb{R}^d$ can be defined exactly as in Definition 3.2. The distance between points in $\mathbb{R}^d$ has been already defined in Section 2.6. Between two points $\boldsymbol{x}$ and $\boldsymbol{z}$ it is given by the norm $\|\boldsymbol{x} - \boldsymbol{z}\|$. Thus, we obtain the following definition of convergence of a sequence in $\mathbb{R}^d$.

Definition 3.5: Convergence of a sequence in $\mathbb{R}^d$

A sequence $(\boldsymbol{a}_n)_{n\in\mathbb{N}}$ in $\mathbb{R}^d$ **converges** to a point $\boldsymbol{a} \in \mathbb{R}^d$, if for every (arbitrarily small) $\varepsilon > 0$ there is a number $N_\varepsilon \in \mathbb{N}$ such that

$$\|\boldsymbol{a}_n - \boldsymbol{a}\| < \varepsilon \quad \text{for all } n \geq N_\varepsilon .$$

The point $\boldsymbol{a}$ is called the **limit** of the sequence. We write $\lim_{n\to\infty} \boldsymbol{a}_n = \boldsymbol{a}$ or briefly $\lim \boldsymbol{a}_n = \boldsymbol{a}$. Other notations are $\boldsymbol{a}_n \xrightarrow{n\to\infty} \boldsymbol{a}$ or simply $\boldsymbol{a}_n \longrightarrow \boldsymbol{a}$.

A sequence in $\mathbb{R}^d$ is called **convergent**, if it converges to a point $a \in \mathbb{R}^d$. A sequence $(\boldsymbol{a}_n)_{n\in\mathbb{N}}$ in $\mathbb{R}^d$ is called **divergent**, if it is not convergent.

Definition 3.5 generalizes the definition of convergence of a real-valued sequence; by setting $d = 1$, Definition 3.2 for number sequences is obtained.

Theorem 3.6: Convergence in $\mathbb{R}^d$

Let $(\boldsymbol{a}_n)_{n\in\mathbb{N}} = (a_{n1}, a_{n2}, \ldots, a_{nd})_{n\in\mathbb{N}}$ be a sequence in $\mathbb{R}^d$. The sequence $(\boldsymbol{a}_n)_{n\in\mathbb{N}}$ converges to a point $\boldsymbol{a} = (a_1, a_2, \cdots, a_d)$ if and only if for each $k = 1, 2, \ldots, d$ it holds that

$$\lim_{n\to\infty} a_{nk} = a_k .$$

This is an extremely useful criterion. It states that we can separately check each component sequence (which is a real-valued sequence) for convergence. If and only if each component sequence is convergent, the sequence $(\boldsymbol{a}_n)$ is convergent in $\mathbb{R}^d$. In that case the sequence $(\boldsymbol{a}_n)$ converges to the point whose components are the limits of the individual component sequences. We illustrate the application of the criterion with two examples.

Example 3.12.

a) Let $\boldsymbol{a}_n = \left(\frac{3}{n}, (-2)^n\right)$. Then, since $\lim_{n\to\infty} \frac{3}{n} = 0$, the first component sequence is convergent, but the second component sequence $\left((-2)^n\right)_{n\in\mathbb{N}} = (-2, 2, -2, 2, \ldots)$ is divergent. The sequence $\left(\frac{3}{n}, (-2)^n\right)_{n\in\mathbb{N}}$ is therefore divergent, so it does not have a limit.

b) Let $\boldsymbol{a}_n = \left(1-\frac{1}{n}, \frac{1}{n^2}\right)$. Then for the first component sequence $\lim_{n\to\infty} 1-\frac{1}{n} = 1$ and for the second component sequence $\lim_{n\to\infty} \frac{1}{n^2} = 0$. The sequence $\left(1-\frac{1}{n}, \frac{1}{n^2}\right)_{n\in\mathbb{N}}$ is therefore convergent to the limit $(1, 0)$,

$$\lim_{n\to\infty} \left(1-\frac{1}{n}, \frac{1}{n^2}\right) = (1, 0)\,.$$

3.3 Further properties of sequences

A disadvantage of the above definition of convergence is that, when using it to establish the convergence of a given sequence, one has to know the limit in advance. An important criterion, by which one can decide between convergence and divergence without knowing the limit, is the so-called **Cauchy criterion**, which applies to number sequences as well as to sequences in $\mathbb{R}^d$, $d \geq 2$.

Theorem 3.7: Cauchy criterion for sequences in $\mathbb{R}^d$

A sequence $(\boldsymbol{a}_n)_{n\in\mathbb{N}}$ converges if and only if for every $\varepsilon > 0$ there is a natural number N_ε such that

$$||\boldsymbol{a}_m - \boldsymbol{a}_n|| < \varepsilon \quad \text{for all } m, n \geq N_\varepsilon .$$

This theorem states that a sequence converges if and only if, regardless of how small ε, there exists a number such that from this number on, any two terms of the sequence have a distance smaller than ε.

We now turn our attention to which of the results of Section 3.1 also apply to sequences in $\mathbb{R}^d$. To do this, we first define, when a sequence in $\mathbb{R}^d$ is bounded: In the corresponding definition for real-valued sequences, the absolute value is simply replaced by the norm. Thus, a sequence $(\boldsymbol{a}_n)_{n\in\mathbb{N}}$ in $\mathbb{R}^d$ is called **bounded**, if there is a real number M such that $||\boldsymbol{a}_n|| \leq M$ for all terms of the sequence. Next, for two points $\boldsymbol{a}$ and $\boldsymbol{b}$ in $\mathbb{R}^d$, $\boldsymbol{a} \leq \boldsymbol{b}$ means that the points are component-wise ordered, i.e., $a_k \leq b_k$ holds for each of their components ($k = 1, \ldots, d$). A sequence $(\boldsymbol{a}_n)_{n\in\mathbb{N}}$ in $\mathbb{R}^d$ is called **monotonically increasing** if $\boldsymbol{a}_n \leq \boldsymbol{a}_{n+1}$ for all $n \in \mathbb{N}$. In the reverse case, i.e., if $\boldsymbol{a}_n \geq \boldsymbol{a}_{n+1}$ for all $n \in \mathbb{N}$, the sequence is called **monotonically decreasing**. A sequence is mentioned as **monotonic**, if it is either monotonically increasing or monotonically decreasing. With these definitions, the Theorems 3.1 and 3.2 as well as the squeeze theorem 3.3 immediately extend to sequences in $\mathbb{R}^d$.

Theorem 3.8: Convergence and boundedness in $\mathbb{R}^d$ I

Every convergent sequence in $\mathbb{R}^d$ is bounded.

Theorem 3.9: Convergence and boundedness in $\mathbb{R}^d$ II

Every bounded and monotonic sequence in $\mathbb{R}^d$ is convergent.

Now, we formulate an extension of Theorem 3.4 to sequences in $\mathbb{R}^d$. Remember that points in $\mathbb{R}^d$ are added and subtracted by adding resp. subtracting their components. Thus, the convergence rules for sums and differences of sequences remain valid in $\mathbb{R}^d$. Moreover, similar rules apply to scalar products and inner products of such sequences.

Theorem 3.10: Rules for limits of sequences in $\mathbb{R}^d$

Let $(\boldsymbol{a}_n)_{n\in\mathbb{N}}$ and $(\boldsymbol{b}_n)_{n\in\mathbb{N}}$ be two convergent sequences in $\mathbb{R}^d$ with $\lim_{n\to\infty} \boldsymbol{a}_n = \boldsymbol{a}$ and $\lim_{n\to\infty} \boldsymbol{b}_n = \boldsymbol{b}$. Furthermore, let $(\lambda_n)_{n\in\mathbb{N}}$ be a sequence in $\mathbb{R}$ with $\lim_{n\to\infty} \lambda_n = \lambda$.

(i) The sequence $(\boldsymbol{a}_n+\boldsymbol{b}_n)_{n\in\mathbb{N}}$ is convergent with $\lim_{n\to\infty}(\boldsymbol{a}_n+\boldsymbol{b}_n) = \boldsymbol{a}+\boldsymbol{b}$.

(ii) The sequence $(\boldsymbol{a}_n-\boldsymbol{b}_n)_{n\in\mathbb{N}}$ is convergent with $\lim_{n\to\infty}(\boldsymbol{a}_n-\boldsymbol{b}_n) = \boldsymbol{a}-\boldsymbol{b}$.

(iii) The sequence $(\lambda_n\boldsymbol{a}_n)_{n\in\mathbb{N}}$ is convergent with $\lim_{n\to\infty}(\lambda_n\boldsymbol{a}_n) = \lambda\boldsymbol{a}$.

(iv) The sequence $(\boldsymbol{a}_n^T\boldsymbol{b}_n)_{n\in\mathbb{N}}$ is convergent with $\lim_{n\to\infty}(\boldsymbol{a}_n^T\boldsymbol{b}_n) = \boldsymbol{a}^T\boldsymbol{b}$.

Obviously, part (iv) of Theorem 3.4 cannot be generalized to the case $d \geq 2$, since the quotient of two points in $\mathbb{R}^d$ is not defined.

3.4 Series

Example 3.13 (Investment and national income). We want to investigate the impact of an investment on national income in a closed economy, i.e., an economy without foreign trade. To do this, we assume that a one-time investment of amount K is made. How much does this investment increase the national income? The investment increases the income of its recipients by a total of K. We assume that all recipients use a share q of their additional income for consumption purposes. q is called the **marginal propensity to consume** and lies between zero and one. Thus, the recipients of the initial investment additionally spend a total of qK, which increases the national income by qK. But since the recipients of this amount also use a share q of the additional income for consumption purposes, the national income increases further by an additional amount, $q \cdot qK = q^2K$. This process continues indefinitely, so that through the initial investment K the national income is increased by a total amount of

$$K + qK + q^2K + q^3K + \dots$$

Here we face the task of summing up an infinite number of values. It seems reasonable to sum up the terms $a_n = q^n K$ one after the other and thus form the sequence of **partial sums** $s_n = \sum_{i=0}^{n} q^i K$. As n increases, these partial sums form increasingly better approximations of the sum to be determined, the "infinite sum". Therefore, one refers to the limit of the sequence of partial sums – if it exists – as the value of the "infinite sum".

Definition 3.6: Infinite series

Let $(a_n)_{n\in\mathbb{N}}$ be a real-valued sequence. We denote the sum of the first n terms with s_n. Hence,

$$s_n = a_1 + a_2 + \dots + a_n = \sum_{i=1}^{n} a_i .$$

The sum s_n is called the n-th **partial sum** of the sequence $(a_n)_{n\in\mathbb{N}}$. The sequence $(s_n)_{n\in\mathbb{N}}$ of partial sums is called an **infinite series**. It is also denoted by the symbol $\sum_{i=1}^{\infty} a_i$.

If the sequence of partial sums converges to a value $s \in \mathbb{R}$, i.e., if $\lim_{n\to\infty} s_n = s$, then we say that the **series converges**. We call s the **value of the infinite series** and write

$$\sum_{i=1}^{\infty} a_i = s .$$

Note that the symbol $\sum_{i=1}^{\infty} a_i$ represents no number, unless the series converges. The lower limit of summation is not necessarily equal to one. In many cases – as with the geometric series, which will be considered in the following section – the lower limit of summation is zero. In general, any integer can serve as the lower limit of summation.

3.5 Geometric series

If we set $K = 1$ in the previous example, the task is to determine the value of the infinite series $\sum_{i=0}^{\infty} q^i$. This infinite series is of particular importance and occurs in many contexts. It is called **geometric series**. To determine its convergence behavior, we examine the partial sums. A partial sum

$$\sum_{i=0}^{n} q^i = 1 + q + q^2 + \dots + q^n$$

is also called **finite geometric series**. It holds

$$\begin{aligned}(1-q)\sum_{i=0}^{n} q^i &= 1 + q + q^2 + \ldots + q^{n-1} + q^n \\ &\quad - q - q^2 - \ldots - q^{n-1} - q^n - q^{n+1} \\ &= 1 - q^{n+1}\,.\end{aligned}$$

Therefore, provided $q \neq 1$, we get

$$\boxed{\sum_{i=0}^{n} q^i = \frac{1-q^{n+1}}{1-q}\,.} \tag{3.2}$$

This equation is referred to as the **summation formula for the finite geometric series**. With the help of this formula, we can easily examine the convergence behavior of the geometric series. If $|q| < 1$, we get

$$\lim_{n\to\infty}\sum_{i=0}^{n} q^i = \lim_{n\to\infty}\frac{1-q^{n+1}}{1-q} = \frac{1-\lim_{n\to\infty} q^{n+1}}{1-q} = \frac{1}{1-q}\,,$$

i.e., the geometric series is convergent for $|q| < 1$ and its value is

$$\boxed{\sum_{i=0}^{\infty} q^i = \frac{1}{1-q} \quad \text{if } |q| < 1\,.}$$

It is easy to see that for $|q| \geq 1$ the geometric series diverges.

Example 3.14. For $q = 1/5$ it holds

$$\sum_{i=0}^{\infty}\frac{1}{5^i} = \sum_{i=0}^{\infty}\left(\frac{1}{5}\right)^i = \frac{1}{1-\frac{1}{5}} = \frac{5}{4} = 1.25\,.$$

Remark: Occasionally, the lower summation limit of the geometric series is not zero but one. We then obtain the formulae

$$\sum_{i=1}^{n} q^i = \frac{q-q^{n+1}}{1-q} \tag{3.3}$$

and

$$\sum_{i=1}^{\infty} q^i = \frac{q}{1-q} \quad \text{if } |q| < 1.$$

Example 3.15 (Investment and national income). As already mentioned above, we assume that the marginal propensity to consume q lies between zero and one,

$0 < q < 1$. Therefore, the geometric series is convergent. From the summation formula it follows that an initial investment of amount K increases the national income by

$$\sum_{i=0}^{\infty} q^i K = K \sum_{i=0}^{\infty} q^i = K \frac{1}{1-q}$$

Thus, the increase in national income results from the amount of the investment multiplied by the **multiplier** $1/(1-q)$. For a value $q = 0.84$ we get a multiplier of 6.25. An investment of 1 000 Euro thus increases the national income by 6 250 Euro. The higher the marginal consumption rate, the larger is the multiplier.

3.6 Application: Financial mathematics

In this section, we will derive several fundamental formulas of financial mathematics. They relate to the interest earned on an account and are based on the finite geometric series.

We consider the following general model. An account has an initial balance K_0. The account is compounded over n interest periods of equal length. These can be years, quarters, months, or days, etc. For simplicity assume that no deposits or withdrawals are made within an interest period, but only at the end of a period. Let Z_i denote the deposit or withdrawal (depending on whether Z_i is greater or less than zero) booked to the account at the end of the i-th period, and K_i the balance of the account including all deposits, withdrawals and interest payments at the end of period i. The interest rate is p, i.e., an interest rate of $p = 0.08$ corresponds to an interest of 8%. We assume that the interest rate remains the same in all periods. Overall, we use the following notations:

K_0	initial balance,
n	number of interest periods,
Z_i	deposit or withdrawal at the end of the i-th period,
K_i	balance at the end of the i-th period,
p	interest rate.

In the case of **ordinary interest**, interest is credited to the account at the end of each period, based on the balance at the beginning of the period. The interest is compounded with the account in the following periods. Besides ordinary interest, also **advance interest**[2] occurs, which is less common and will not be discussed further here. With ordinary interest, the balance at the end of the first period is composed of the initial balance K_0, the interest payment pK_0 and the deposit Z_1:

$$K_1 = K_0 + pK_0 + Z_1 = (1+p)K_0 + Z_1 .$$

Correspondingly, the balance at the end of period i is composed of the balance K_{i-1}

[2] Advance interest is credited to the account at the beginning of a period. The resulting formulas differ from those of ordinary interest, but can be derived in a similar way.

at the beginning of period i, the interest payment pK_{i-1} and the deposit Z_i:

$$\boxed{K_i = (1+p)K_{i-1} + Z_i\,.} \tag{3.4}$$

Theorem 3.11: Compound interest formula

For all $n \in \mathbb{N}$ it holds

$$K_n = (1+p)^n K_0 + \sum_{i=1}^{n} (1+p)^{n-i} Z_i\,. \tag{3.5}$$

Proof: We prove this formula by mathematical induction[3]. For $n = 1$, the formula is obviously correct. We now assume that the balance after $n-1$ periods is given by Equation (3.5). The balance after n periods is then obtained with Equation (3.4) as

$$\begin{aligned} K_n &= (1+p)K_{n-1} + Z_n \\ &= (1+p)\left[(1+p)^{n-1}K_0 + \sum_{i=1}^{n-1}(1+p)^{n-1-i}Z_i\right] + Z_n \\ &= \left[(1+p)^n K_0 + \sum_{i=1}^{n-1}(1+p)^{n-i}Z_i\right] + Z_n \\ &= (1+p)^n K_0 + \sum_{i=1}^{n}(1+p)^{n-i}Z_i\,. \end{aligned}$$

By mathematical induction, the statement is proven. □

The factor $1 + p$ is called **capitalization factor** (or **future value factor**). From the compound interest formula (3.5) and the summation formula for the finite geometric series (3.2) or (3.3) we will now derive special formulas for financial economics.

Repayment of a debt Assume that the account has an initial debt W. The debt is to be fully repaid by n payments, each being put into effect at the end of n successive periods. Let $Z_1, \ldots, Z_n$ denote the payments at periods $1, \ldots, n$, and p the interest rate per period. We ask for the value of the initial debt, given the payments and the interest rate.

By setting $K_0 = -W$ and $K_n = 0$ in (3.5), we get

$$0 = -(1+p)^n W + \sum_{i=1}^{n}(1+p)^{n-i}Z_i\,.$$

[3] See Appendix G on proof techniques.

Solving this equation for W then gives

$$W = (1+p)^{-n} \sum_{i=1}^{n} (1+p)^{n-i} Z_i = \sum_{i=1}^{n} (1+p)^{-i} Z_i \,.$$

The initial debt that is repaid by the payment series is therefore

$$W = \sum_{i=1}^{n} \left(\frac{1}{1+p} \right)^i Z_i \,. \tag{3.6}$$

The factor $1/(1+p)$ is called **discount factor**. The value W in (3.6) is referred to as the **present value of the payment series** $Z_1, \ldots, Z_n$. The present value of a payment series can be interpreted differently depending on the perspective. From the perspective of a creditor, W is the loan that is repaid by the payments $Z_1, \ldots, Z_n$. From the perspective of an investor, W is the discounted value of the investments $Z_1, \ldots, Z_n$ over time, and from the perspective of a pension recipient, W is the present value of future pension payments $Z_1, \ldots, Z_n$, i.e., the sum to be paid now for a pension plan.

Present value of a capital A capital that we own today is more valuable than a capital that we will own in the future, as interest has to be taken into account. Today's value of a capital that will be disposable in the future after n periods is called its **present value**. To determine the present value of a capital Z_n, we use Equation (3.6), specifically inserting $Z_1 = \ldots = Z_{n-1} = 0$. So we get

$$W = \frac{Z_n}{(1+p)^n} \,. \tag{3.7}$$

To determine the present value of 10 000 Euro disposable after five years, we assume an interest rate of 6% per year. By inserting $n = 5$, $Z_5 = 10\,000$ and $p = 0.06$ we get

$$W = \frac{10\,000 \text{ Euro}}{1.06^5} = 7472.58 \text{ Euro} \,.$$

Present value of an ordinary annuity When there are repeated payments of an amount R at the times $1, \ldots, n$, this is referred to as an ordinary annuity having term n. The present value of such an annuity is obtained by specifically choosing $Z_1 = \ldots = Z_n = R$ in Equation (3.6). With the summation formula (3.3) we then get

$$W = R \sum_{i=1}^{n} \left(\frac{1}{1+p} \right)^i = R \frac{\frac{1}{1+p} - \left(\frac{1}{1+p} \right)^{n+1}}{1 - \frac{1}{1+p}} \,.$$

Expanding with $1 + p$ yields

$$\boxed{W = \frac{R}{p}\left[1 - \left(\frac{1}{1+p}\right)^n\right].}$$

Since the expression in brackets is less than one, it follows that the present value amounts to $W < R/p$.

As a numerical example, consider a monthly annuity of 1 000 Euro with a term of six years at a monthly interest rate of 0.5%. Then $R = 1\,000$ Euro, $p = 0.005$, and $n = 6 \cdot 12 = 72$, and we get as present value

$$W = \frac{1\,000 \text{ Euro}}{0.005}\left[1 - \frac{1}{1.005^{72}}\right] = 60\,339.51 \text{ Euro}.$$

Annuity for debt repayment To fully repay a debt W within n periods, constant payments of the amount A are to be made at the end of each interest period. A is then also referred to as **annuity**. To determine the annuity needed to fully repay a debt of W in n periods, set $Z_1 = \ldots = Z_n = A$ in (3.6) and get analogously to the previous calculation of present value

$$W = \frac{A}{p}\left[1 - \left(\frac{1}{1+p}\right)^n\right].$$

Solving for A gives

$$\boxed{A = \frac{pW}{1 - \left(\frac{1}{1+p}\right)^n}.} \tag{3.8}$$

If a debt of 100 000 Euro at an interest rate of 7% is to be repaid in 25 years, the annuity is

$$A = \frac{0.07 \cdot 100\,000 \text{ Euro}}{1 - \left(\frac{1}{1.07}\right)^{25}} = 8\,581.05 \text{ Euro}.$$

Term of an annuity loan Given a debt W, an interest rate p and a constant annuity A, the duration of the repayment, i.e., the term of the annuity loan, can be calculated by solving Equation (3.8) for n. (3.8) is equivalent to

$$1 - \left(\frac{1}{1+p}\right)^n = \frac{pW}{A} \quad \text{or} \quad \left(\frac{1}{1+p}\right)^n = 1 - \frac{pW}{A}.$$

Taking logarithms gives

$$n \ln\left(\frac{1}{1+p}\right) = \ln\left(1 - \frac{pW}{A}\right)$$

and finally

$$n = \frac{\ln\left(1 - \frac{pW}{A}\right)}{\ln\left(\frac{1}{1+p}\right)}. \tag{3.9}$$

Let $\overline{n}$ denote the smallest integer greater or equal n. Then the loan will be terminated during the $\overline{n}$-th period.

Example 3.16. A loan of 100 000 Euro is to be repaid at an interest rate of 7% through annuities. What is the term of the loan if the initial repayment is 1%? Since the interest rate is 7%, 7 000 Euro in interest accrue in the first year. If 1% of the debt is to be repaid in the first year, this corresponds to an amount of 1 000 Euro. The annuity in the first year – and thus in all subsequent years – is therefore $A = 8\,000$ Euro. If we further set $W = 100\,000$ Euro and $p = 0.07$ in Equation (3.9), we get

$$n = \frac{\ln\left(1 - \frac{7\,000\,€}{8\,000\,€}\right)}{\ln\left(\frac{1}{1.07}\right)} = \frac{\ln(0.125)}{\ln\left(\frac{1}{1.07}\right)} = 30.7343\,.$$

Hence, the loan is repaid within 31 years.

As the remaining debt decreases from year to year due to the repayment, the interest portion of the annuity becomes smaller and the repayment portion correspondingly larger. This can be nicely seen in the following repayment plan. The remaining debt at the *beginning* of the year as well as the interest and repayment portion of the annuity are shown. At the end of the 31st year, the sum of the remaining debt and interest is less than the annuity of 8 000 Euro, so the remaining payment is correspondingly smaller.

Year	Debt	Interest	Repayment	Annuity
1	100 000.00	7 000.00	1 000.00	8 000.00
2	99 000.00	6 930.00	1 070.00	8 000.00
3	97 930.00	6 855.10	1 144.90	8 000.00
4	96 785.10	6 774.96	1 225.04	8 000.00
5	95 560.06	6 689.20	1 310.80	8 000.00
⋮	⋮	⋮	⋮	⋮
27	31 323.53	2 192.65	5 807.35	8 000.00
28	25 516.18	1 786.13	6 213.87	8 000.00
29	19 302.31	1 351.16	6 648.84	8 000.00
30	12 653.47	885.74	7 114.26	8 000.00
31	5 539.21	387.74	5 539.21	5 926.96

3.7 Convergence criteria for series

Sometimes, it can be difficult to prove whether a given series converges or not. Only in rare cases, it is possible to find closed expressions for the partial sums of a series, as it was the case with the geometric series. Therefore, it is important to have simple criteria by which one can decide whether a series diverges or not, without having to explicitly form the partial sums of the series. A necessary condition for the convergence of a series is given by the following theorem.

Theorem 3.12: Necessary condition for the convergence of series

If $\sum_{n=1}^{\infty} a_n$ is convergent, then $(a_n)_{n\in\mathbb{N}}$ is a null sequence.

How does this theorem help us to check the convergence of a series? The theorem provides a **necessary condition** for the convergence of a series, i.e., we can use it only to establish non-convergence: If the sequence $(a_n)_{n\in\mathbb{N}}$ is not a null sequence, the series diverges. It is important that the reverse conclusion is not valid. If you have an infinite series with $(a_n)_{n\in\mathbb{N}}$ being a null sequence, you cannot conclude that the series converges. An example of a series that diverges though $(a_n)_{n\in\mathbb{N}}$ is a null sequence is the so-called **harmonic series** (see Example 3.17 below). Before we can show that the harmonic series diverges, we need the so-called **Cauchy criterion for series**. This is simply the Cauchy criterion for sequences (Theorem 3.7) being applied to the sequence of partial sums of a series.

Theorem 3.13: Cauchy criterion for series

The series $\sum_{n=1}^{\infty} a_n$ is convergent if and only if for every $\varepsilon > 0$ there exists a number N_ε such that for all i, j with $j \geq i \geq N_\varepsilon$ holds

$$\left| \sum_{n=i+1}^{j} a_n \right| < \varepsilon \,.$$

Example 3.17 (Harmonic series). The series

$$\sum_{n=1}^{\infty} \frac{1}{n}$$

is called the **harmonic series**. The harmonic series is divergent to ∞, although the sequence $(1/n)_{n\in\mathbb{N}}$ is a null sequence.

The divergence of the harmonic series is easily shown with the Cauchy criterion. It is

$$\sum_{n=2}^{2} \frac{1}{n} = \frac{1}{2},$$

$$\sum_{n=3}^{4} \frac{1}{n} = \frac{1}{3} + \frac{1}{4} \geq \frac{1}{2},$$

$$\sum_{n=5}^{8} \frac{1}{n} = \frac{1}{5} + \frac{1}{6} + \frac{1}{7} + \frac{1}{8} \geq \frac{1}{2},$$

$$\sum_{n=9}^{16} \frac{1}{n} = \frac{1}{9} + \frac{1}{10} + \frac{1}{11} + \frac{1}{12} + \frac{1}{13} + \frac{1}{14} + \frac{1}{15} + \frac{1}{16} \geq \frac{1}{2}.$$

In general, consider the sum that runs from $n = 2^{k-1} + 1$ to $n = 2^k$. This sum consists of 2^{k-1} terms, each of which has at least the value $1/2^k$. Therefore,

$$\sum_{n=2^{k-1}+1}^{2^k} \frac{1}{n} \geq 2^{k-1} \frac{1}{2^k} = \frac{1}{2}.$$

For any $\varepsilon \leq 1/2$, there is no number N_ε that satisfies the condition of the Cauchy criterion. According to the Cauchy criterion for series, the harmonic series is therefore divergent.

Since series are just a special type of sequences, we can apply all results obtained for sequences to the case of infinite series. By Theorem 3.4 the sum and difference of convergent series are again convergent. Also, when we multiply a convergent series with a constant number, a convergent series is obtained:

Theorem 3.14: Rules for series

Let $\sum_{n=1}^{\infty} a_n$ and $\sum_{n=1}^{\infty} b_n$ be convergent series and let λ be a real number. Then the series

$$\sum_{n=1}^{\infty}(a_n + b_n), \quad \sum_{n=1}^{\infty}(a_n - b_n) \quad \text{and} \quad \sum_{n=1}^{\infty}(\lambda a_n)$$

are also convergent, and it holds

$$\sum_{n=1}^{\infty}(a_n + b_n) = \sum_{n=1}^{\infty} a_n + \sum_{n=1}^{\infty} b_n,$$

$$\sum_{n=1}^{\infty}(a_n - b_n) = \sum_{n=1}^{\infty} a_n - \sum_{n=1}^{\infty} b_n,$$

$$\sum_{n=1}^{\infty}(\lambda a_n) = \lambda \sum_{n=1}^{\infty} a_n.$$

Two convergence criteria for series follow: the comparison test and the ratio test.

Theorem 3.15: Direct comparison test

Let $\sum_{k=1}^{\infty} b_k$ be convergent, with all terms b_k being nonnegative. If there exists a number N_0 such that

$$|a_k| \le b_k \quad \text{for all } k \ge N_0\,,$$

then $\sum_{k=1}^{\infty} |a_k|$ and thus also $\sum_{k=1}^{\infty} a_k$ are convergent.

Theorem 3.16: Ratio test

Let $a_k \neq 0$ for all k. If there exists a number $N_0 \in \mathbb{N}$ and a real number q with $0 < q < 1$, such that

$$\left|\frac{a_{k+1}}{a_k}\right| \le q \quad \text{for all } k \ge N_0$$

then the series $\sum_{k=1}^{\infty} a_k$ is convergent.

We illustrate the ratio test with an example.

Example 3.18. Consider the series

$$\sum_{k=0}^{\infty} \frac{x^k}{k!}$$

for a given $x \in \mathbb{R}$. Here $a_k = x^k/k!$. For the ratios a_{k+1}/a_k we get

$$\left|\frac{a_{k+1}}{a_k}\right| = \left|\frac{x^{k+1}/(k+1)!}{x^k/k!}\right| = \left|\frac{x^{k+1}\cdot k!}{x^k\cdot(k+1)!}\right| = \frac{|x|}{k+1} < \frac{1}{2} \quad \text{for } k > 2|x| - 1\,.$$

Consequently, $\sum_{k=0}^{\infty} \frac{x^k}{k!}$ converges according to Theorem 3.16.

Since the above series converges for every $x \in \mathbb{R}$, it defines a function $\mathbb{R} \to \mathbb{R}$.

Definition 3.7: Exponential function

The function $\exp : \mathbb{R} \to \mathbb{R}$,

$$\exp(x) = \sum_{k=0}^{\infty} \frac{x^k}{k!}, \quad x \in \mathbb{R},$$

is called **exponential function**.

Example 3.19. The following list contains some convergent series and their values.

a) $\sum_{i=0}^{\infty} \frac{1}{i!} = e$, where $e = 2.718281828\ldots$ is Euler's number

b) $\sum_{i=0}^{\infty} (-1)^i \frac{1}{i!} = \frac{1}{e}$

c) $\sum_{i=1}^{\infty} (-1)^{i-1} \frac{1}{i} = \ln(2)$ (alternating harmonic series)

d) $\sum_{i=0}^{\infty} \frac{1}{2^i} = 2$ (geometric series, $q = \frac{1}{2}$)

e) $\sum_{i=0}^{\infty} (-1)^i \frac{1}{2^i} = \frac{2}{3}$ (geometric series, $q = -\frac{1}{2}$)

f) $\sum_{i=1}^{\infty} (-1)^{i-1} \frac{1}{2i-1} = \frac{\pi}{4}$

g) $\sum_{i=1}^{\infty} \frac{1}{i(i+1)} = 1$

h) $\sum_{i=1}^{\infty} \frac{1}{(2i-1)(2i+1)} = \frac{1}{2}$

i) $\sum_{i=2}^{\infty} \frac{1}{(i-1)(i+1)} = \frac{3}{4}$

j) $\sum_{i=1}^{\infty} \frac{1}{i^2} = \frac{\pi^2}{6}$

k) $\sum_{i=1}^{\infty} (-1)^i \frac{1}{i^2} = \frac{\pi^2}{12}$

3.8 Continuity of functions

In this section, we formulate the concept of continuity for functions of one variable in a mathematically precise way and present properties of continuous functions. The intuitive notion of a continuous function is that of a function whose graph can be drawn "without lifting the pen". This implies in particular that a continuous function must not have "jumps". However, there are other types of so-called discontinuity points and conversely, some functions that one would intuitively describe as discontinuous are continuous in a mathematical sense. The core of the concept of "continuity" can be vaguely formulated as that small changes in the arguments of a function should only result in small changes in the function value. A function is not continuous if small changes in the arguments can result in abrupt changes in the function value.

Before the definition of continuity is discussed, we must first define what is meant by the limit of a function.

The concept of continuity essentially concerns the question of whether the values of a function f tend to a certain value when the argument x tends to a certain value a. So far, however, we have only learned about the concept of convergence of a sequence. The idea is to choose a sequence $(x_n)_{n\in\mathbb{N}}$ that converges to a and define the limit $\lim_{x\to a} f(x)$ as the limit $\lim_{n\to\infty} f(x_n)$. The problem with this "definition" is that this limit may depend on the choice of the sequence $(x_n)_{n\in\mathbb{N}}$. Therefore, we will only say that the function f has a limit at the point a if for **every** sequence $(x_n)_{n\in\mathbb{N}}$ that converges to a, the limit of function values $\lim_{n\to\infty} f(x_n)$ is the same.

Example 3.20. Consider the following piecewise defined function

$$f(x) = \begin{cases} 0 & \text{for } x \leq a, \\ 1 & \text{for } x > a. \end{cases}$$

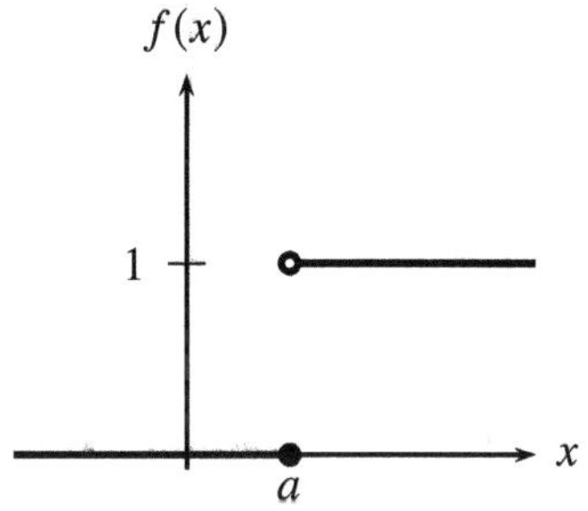

If we consider a sequence that converges to a, consisting only of values less than a, the limit is 0. However, if we consider a sequence that converges to a, consisting only of values greater than a, the limit is 1. Thus, f has no limit at a.

We define the limit of a function in the following definition.

Definition 3.8: Limit of a function

Let $D \subset \mathbb{R}$ be a subset of the real numbers and $f : D \to \mathbb{R}$ a real function. We write

$$\lim_{x\to a} f(x) = c\,,$$

if for every sequence $(x_n)_{n\in\mathbb{N}}$ with $x_n \in D$ and $\lim_{n\to\infty} x_n = a$ it holds:

$$\lim_{n\to\infty} f(x_n) = c\,.$$

For a, the values $+\infty$ and $-\infty$ are allowed.

Therefore, a theorem analogous to Theorem 3.10 applies:

Theorem 3.17: Rules for limits of functions

Let $D \subset \mathbb{R}$ be a subset of the real numbers, and let $f, g : D \to \mathbb{R}$ be real functions. If the limits $\lim\limits_{x\to a} f(x)$ and $\lim\limits_{x\to a} g(x)$ exist, the following holds:

(i) $\lim\limits_{x\to a} (f(x) \pm g(x)) = \lim\limits_{x\to a} f(x) \pm \lim\limits_{x\to a} g(x)$,

(ii) $\lim\limits_{x\to a} (f(x) \cdot g(x)) = \lim\limits_{x\to a} f(x) \cdot \lim\limits_{x\to a} g(x)$,

(iii) $\lim\limits_{x\to a} \left(\dfrac{f(x)}{g(x)}\right) = \dfrac{\lim\limits_{x\to a} f(x)}{\lim\limits_{x\to a} g(x)}$, if $\lim\limits_{x\to a} g(x) \neq 0$.

Remarks

1. The value a does not have to be in the domain D of the function f. It is only required that at least one sequence in D exists that converges to a.
2. If a is in the domain of f, one can choose the constant sequence, i.e., $x_n = a$ for all n. For this sequence, $\lim_{n\to\infty} f(x_n) = \lim_{n\to\infty} f(a) = f(a)$ is obtained. So if a is in the domain of f, either f does not have a limit at the point a or the limit is $f(a)$. In the second case, f is called continuous at a. We will return to this in Definition 3.9 below.
3. The definition of the limit of a function is not consistent in the literature. Occasionally, the limit of a function is defined in such a way that only sequences are considered in which all elements x_n of the sequence are different from a. In this definition, the limit $\lim_{x\to a} f(x)$ can also be different from $f(a)$.

We can now also define **one-sided limits** of a function. We write

$$\lim_{x\nearrow a} f(x) = c\,,$$

if for every sequence $(x_n)_{n\in\mathbb{N}}$ with $x_n \in D$, $x_n < a$ and $\lim_{n\to\infty} x_n = a$ it holds:

$$\lim_{n\to\infty} f(x_n) = c\,.$$

The limit $\lim_{x\nearrow a} f(x) = c$ is called **left-sided limit** of f at the point a. Instead of $\lim_{x\nearrow a} f(x) = c$, we also write more briefly $f(a-) = c$.

Accordingly, we write

$$\lim_{x\searrow a} f(x) = c\,,$$

if for every sequence $(x_n)_{n\in\mathbb{N}}$ with $x_n \in D$, $x_n > a$ and $\lim_{n\to\infty} x_n = a$ holds:

$$\lim_{n\to\infty} f(x_n) = c\,.$$

In this case, the limit is called the **right-sided limit** of f at the point a, and we write more briefly $f(a+) = c$. Here too, the values $+\infty$ and $-\infty$ are allowed for a.

We now define the concept of continuity of a function.

Definition 3.9: Continuity

Let $D \subset \mathbb{R}$ be a subset of the real numbers, $f : D \to \mathbb{R}$ a real function and $a \in D$. The function f is called **continuous at** a, if it holds:

$$\lim_{x \to a} f(x) = f(a)\,.$$

f is called **continuous** if f is continuous at every point of its domain.

The so-called distribution functions that occur in probability theory are generally not continuous, but have the weaker property that the right-sided limit exists at every point a and is equal to the function value at the point a. Such a function is called **right-continuous**:

Definition 3.10: Right-continuous (left-continuous) function

Let $D \subset \mathbb{R}$ be a subset of the real numbers, $f : D \to \mathbb{R}$ a real function and $a \in D$. The function f is called **right-continuous at** a, if it holds:

$$f(a+) = f(a)\,.$$

If

$$f(a-) = f(a)\,,$$

f is called **left-continuous at** a.

f is called **right-continuous (left-continuous)** if f is right-continuous (left-continuous) at every point of its domain.

Example 3.21. The function

$$f(x) = \begin{cases} 0 & \text{for } x \leq 4\,, \\ 1 & \text{for } x > 4\,, \end{cases}$$

is not continuous at $x = 4$, but only left-continuous. The function

$$g(x) = \begin{cases} 0 & \text{for } x < 4\,, \\ 1 & \text{for } x \geq 4\,, \end{cases}$$

is not continuous at $x = 4$, but only right-continuous. At all other points, f and g are continuous.

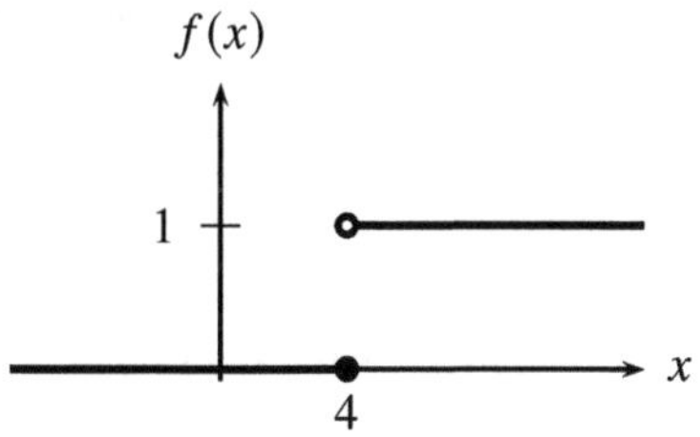

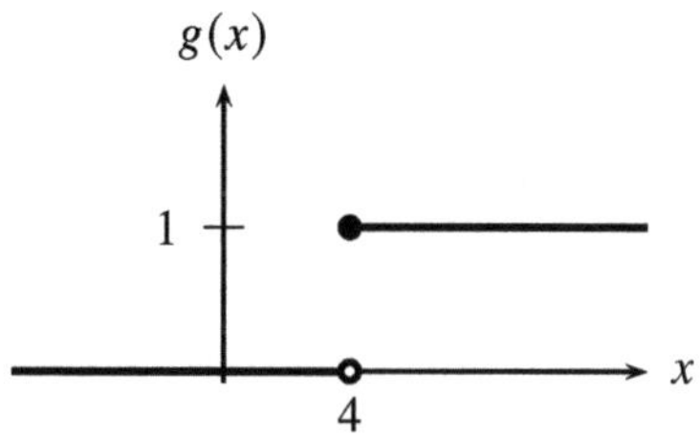

For points in the interior of the domain, continuity is equivalent to the function being both left- and right-continuous there.

Theorem 3.18: Continuity of a real function

Let $D \subset \mathbb{R}$ be an interval and $a \in D$ not a boundary point of D. The real function $f : D \to \mathbb{R}$ is continuous at a if and only if

$$f(a-) = f(a) = f(a+) .$$

Example 3.22. The following functions are continuous:

- Every **constant function** $f : \mathbb{R} \to \mathbb{R}, x \mapsto c$,
- the **identity** $\mathrm{id} : \mathbb{R} \to \mathbb{R}, x \mapsto x$.

Theorem 3.19: Preservation properties of continuous functions

Let $f, g : D \to \mathbb{R}$ be functions that are continuous at $a \in D$, and let $\lambda \in \mathbb{R}$. Then the functions $f \pm g, \lambda f, fg, \max\{f, g\}$ and $\min\{f, g\}$ are also continuous at a. The function f/g is continuous at a, provided $g(a) \neq 0$.

When you add, subtract, multiply by a number or multiply continuous functions, you get continuous functions again. The same applies for the minimum and maximum of continuous functions. You just have to be careful with division since f/g is not defined at a if $g(a) = 0$. However, if $g(a) \neq 0$, then the division of continuous functions yields a function that is continuous at a.

Theorem 3.20: Continuity of the composition

Let $A, B \subset \mathbb{R}$ and $f : A \to B$, $g : B \to \mathbb{R}$. If f is continuous at the point $a \in A$ and g is continuous at the point $f(a)$, then the function $g \circ f : A \to \mathbb{R}$ is also continuous at the point a.

With theorems 3.19 and 3.20 you can now prove continuity for many functions. According to Example 3.22, every constant function and the identity are continuous. Then the function $x \mapsto x^n$ is continuous for every n, since it is the n-fold product of the identity with itself. Likewise, of course, $x \mapsto a_n x^n$ is continuous for every $a_n \in \mathbb{R}$. Since the sum of continuous functions is again continuous, we finally get that every polynomial $x \mapsto a_0 + a_1 x + a_2 x^2 + \ldots + a_k x^k$ is continuous.

Example 3.23. The following functions are continuous:

- Every polynomial $f : \mathbb{R} \to \mathbb{R}, x \mapsto \sum_{i=0}^{k} a_i x^i = a_0 + a_1 x + a_2 x^2 + \ldots + a_k x^k$,
- every rational function $f : A \to \mathbb{R}, x \mapsto \frac{p(x)}{q(x)}$, where p and q are polynomials and $A = \mathbb{R} \setminus \{x : q(x) = 0\}$.

It is more difficult to show the continuity of the following functions:

Example 3.24. The following functions are continuous:

- The trigonometric functions $\sin(x)$, $\cos(x)$, $\tan(x)$, $\cot(x)$,
- the inverse functions of the trigonometric functions $\arcsin(x)$, $\arccos(x)$, $\arctan(x)$, $\text{arccot}(x)$,
- the exponential function $\exp(x)$,
- the natural logarithm $\ln(x)$.

Example 3.25. The general exponential function $x \mapsto a^x = e^{x\ln(a)}$, where $a > 0$, is continuous, because it is the composition of two continuous functions:

$$x \mapsto x\ln(a) = y \mapsto e^y = e^{x\ln(a)} .$$

3.9 Further properties of continuous functions

Often, the continuity of a function is not defined as in Definition 3.9, but, in an equivalent way, by the so-called ε-δ definition.

Theorem 3.21: ε-δ definition

Let $f : D \to \mathbb{R}$ be a function and $a \in D$. f is continuous at a if and only if, for every $\varepsilon > 0$, there exists a $\delta > 0$ such that

$$|f(x) - f(a)| < \varepsilon \quad \text{for all } x \in D \text{ with } |x - a| < \delta.$$

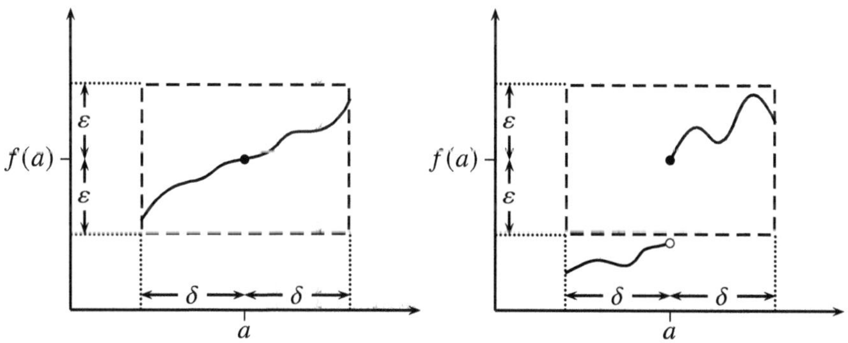

Figure 3.4: Illustration of the ε-δ definition of continuity.

The ε-δ definition of continuity can be best illustrated with a graph. In Figure 3.4, a rectangle with the center point $(a, f(a))$, the height 2ε and the width 2δ is drawn. f is continuous at x if and only if, no matter how small the height of the rectangle is, the rectangle can always be made so narrow that in the interval $(a-\delta, a+\delta)$ the graph of the function f lies completely inside the rectangle. In the left panel, this is obviously always possible. In the right panel, the function has a jump at a. If ε is smaller than the jump height, then the graph of f will be outside the rectangle immediately left of a. Thus, δ cannot be chosen so small that in the interval $(a - \delta, a + \delta)$ the graph lies within the rectangle .

Let us now assume that at a point a, the function value $f(a)$ is greater than zero. If ε is chosen so small that the ε-δ rectangle does not intersect with the x-axis, then for a continuous function, δ can always be chosen so small that in the interval $(a-\delta, a+\delta)$ the graph of f lies completely within the ε-δ rectangle. In particular, $f(x) > 0$ for all $x \in (a - \delta, a + \delta)$. Thus, if f is continuous and $f(a) > 0$, then for x sufficiently close to a, the function value $f(x)$ is also positive.

Theorem 3.22: Non-zero in a neighborhood

Let $f : D \to \mathbb{R}$ be continuous and let $a \in D$ with $f(a) \neq 0$. Then there is a δ-neighborhood of a such that $f(x) \neq 0$ for all $x \in (a - \delta, a + \delta)$.

A graphical illustration of this theorem is given in Figure 3.5 .

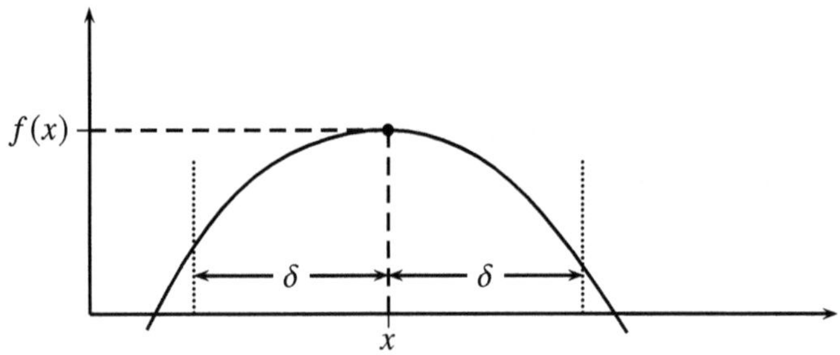

Figure 3.5: Illustration of Theorem 3.22 .

In simple terms, a continuous function (within the domain) has no jumps. If for two values $a < b$ it holds $f(a) \neq f(b)$, one would expect that a continuous function cannot "skip" any value between $f(a)$ and $f(b)$, i.e., every value between $f(a)$ and $f(b)$ should be taken at (at least) one point between a and b. This is the statement of the so-called intermediate value theorem.

Theorem 3.23: Intermediate value theorem

Let $f : [a, b] \to \mathbb{R}$ be a continuous function and v a value between $f(a)$ and $f(b)$. Then there is a value $u \in [a, b]$ such that $f(u) = v$.

The statement of the intermediate value theorem can be illustrated by Figure 3.6.

Another important property of continuous functions is that a continuous function on a closed and bounded interval has both a maximum and a minimum.

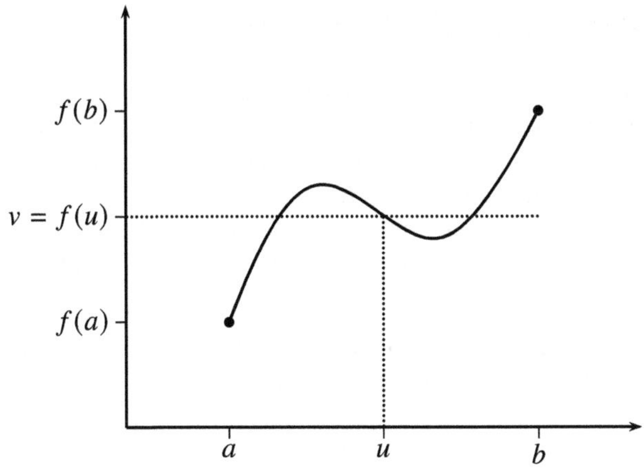

Figure 3.6: Illustration of the intermediate value theorem.

Theorem 3.24: Maxima and minima of a continuous function

Let $f : [a, b] \to \mathbb{R}$ be a continuous function. Then the function f is bounded and attains its maximum and its minimum[4] on $[a, b]$.

The following examples show that f generally does not attain a minimum and maximum if one of the three conditions (f continuous, closed domain, bounded domain) is violated.

Example 3.26.

a) Let $f : [1, \infty) \to \mathbb{R}, x \mapsto \frac{1}{x}$. Then f is continuous; the domain is closed, but not bounded. $f(x)$ is positive everywhere, asymptotically approaches zero, but does not attain a minimum.

b) Let $f : (0, 1] \to \mathbb{R}, x \mapsto \frac{1}{x}$. Then f is continuous; the domain is bounded, but not closed. $f(x)$ becomes arbitrarily large for x close to zero, but does not attain a maximum.

c) Let $f : [0, 1] \to \mathbb{R}$ with

$$f(x) = \begin{cases} x & \text{if } 0 \le x \le \frac{1}{2}, \\ \frac{1}{x} & \text{if } \frac{1}{2} < x \le 1. \end{cases}$$

The domain of f is closed and bounded; f is not continuous. For $x > 1/2$ close to $1/2$, $f(x)$ approaches the value 2 arbitrarily closely but does not attain a maximum.

[4] More precisely, one or more global maxima and one or more global minima are attained. Moreover, if there are local maxima or minima, they are attained as well on $[a, b]$.

3.10 Fixed points of a function

In many economic applications, the **fixed points** of a function f are of particular interest. A fixed point is a value x such that $f(x) = x$. A typical situation arises when the variable x_n describes the state of an economic system at time n. The state of the system at time $n+1$ is derived from the state x_n at time n according to $x_{n+1} = f(x_n)$. If the state of the system at some time n is equal to a fixed point x, then the system remains in state x forever. The fixed point x thus describes an **equilibrium state** of the system.

Example 3.27 (Cobweb model; continued). In the cobweb model the price at time n is derived from the previous price according to

$$p_n = f(p_{n-1}) = \frac{b}{d}\,p_{n-1} - \frac{c-a}{d}\,.$$

The function f that maps the price of one period to the price of the following period is thus given by $f(x) = \frac{b}{d}x - \frac{c-a}{d}$. For the price $\overline{p} = \frac{c-a}{b-d}$, which results from the intersection of the supply and demand curves, it holds

$$f(\overline{p}) = \frac{b}{d}\overline{p} - \frac{c-a}{d} = \frac{b}{d}\frac{c-a}{b-d} - \frac{c-a}{d} = \frac{c-a}{d}\left[\frac{b}{b-d} - 1\right] = \frac{c-a}{b-d} = \overline{p}\,.$$

$\overline{p}$ is therefore a fixed point of f and is referred to as the equilibrium price.

In the context of fixed points, two questions are of particular interest:

1. Does a given function f have a fixed point at all?
2. Does the sequence of values given by $x_n = f(x_{n-1})$ converge to a fixed point?

The following fixed-point theorem provides an answer to the first question.

Theorem 3.25: Brouwer[5] fixed-point theorem

A continuous function $f : [a,b] \to [a,b]$, which maps a bounded closed interval to itself, has at least one fixed point.

Proof: The Brouwer fixed-point theorem follows directly from the intermediate value theorem. If $f : [a,b] \to [a,b]$ is continuous, then also $g(x) = x - f(x)$. If $f(a) = a$ or $f(b) = b$, then a or b is a fixed point. Otherwise, $f(a) > a$ and $f(b) < b$. Then, however, $g(a) = a - f(a) < 0$ and $g(b) = b - f(b) > 0$. According to the intermediate value theorem, g has a zero z. Because $g(z) = z - f(z) = 0$, z is a fixed point of f. □

A key requirement of the Brouwer fixed-point theorem is that the domain and the codomain of the function coincide. Such a function is also called a **self-map**.

[5]Luitzen Egbertus Jan Brouwer (1881–1966)

The statement of the Brouwer fixed-point theorem can be illustrated by Figure 3.7. The fixed-point theorem states that every continuous function whose graph lies inside the square must intersect the diagonal at least once.

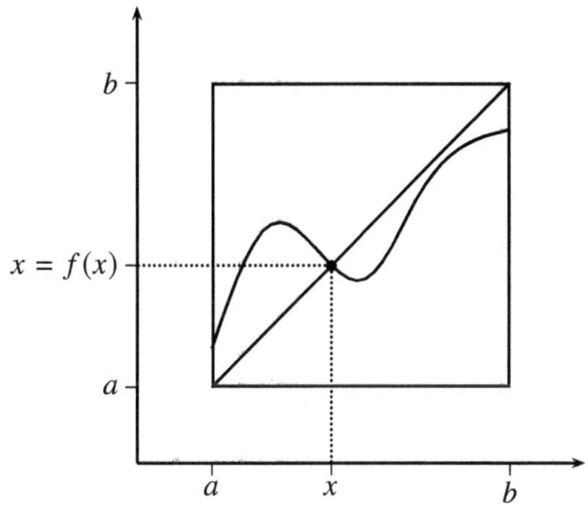

Figure 3.7: Illustration of Brouwer's fixed-point theorem.

An answer to the second question is provided by the so-called Banach fixed-point theorem. Fundamental to this theorem is the concept of a **contracting map**. This refers to a function for which there is a constant $q < 1$, such that for arbitrary values x, y it holds

$$|f(x) - f(y)| < q|x - y| \qquad \text{with } q < 1.$$

Intuitively, this means that the distance between two function values is always less than the distance of the corresponding arguments. Overall, the domain is contracted to a smaller range. From the ε-δ condition of continuity (Theorem 3.21) it follows with $\delta = \varepsilon$, that every contracting map is continuous.

Theorem 3.26: Banach[6] fixed-point theorem

Every contracting self-map $f : [a, b] \to [a, b]$ has exactly one fixed point z. For any initial value x_0, the sequence defined by the iteration $x_n = f(x_{n-1})$ converges to z,

$$\lim_{n\to\infty} x_n = z\,.$$

Figure 3.8 provides a graphical illustration of Banach's fixed-point theorem.

Example 3.28 (Cobweb model; continued). We have already seen in Example 3.9 that under the condition $|b| < |d|$ the sequence of prices in the cobweb model converges to an equilibrium price. With the Banach fixed-point theo-

[6] Stefan Banach (1892–1945)

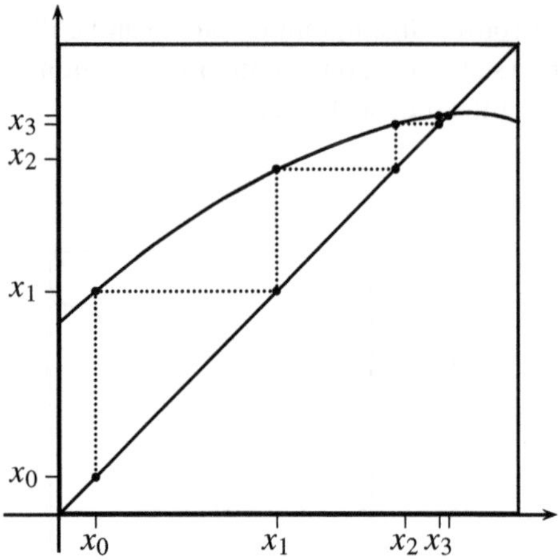

Figure 3.8: Illustration of Banach's fixed-point theorem

rem, the convergence of prices to an equilibrium can be examined for more general supply and demand functions than the linear ones considered so far. If $s_n = s(p_{n-1})$ and $d_n = d(p_n)$, then from the requirement $s_n = d_n$ we get the equation $s(p_{n-1}) = d(p_n)$ or

$$p_n = d^{-1}(s(p_{n-1}))\,.$$

We now only need to examine under what conditions the map $p \mapsto d^{-1}(s(p))$ is a contracting self-map. Then, the convergence of prizes to an equilibrium follows with the Banach fixed-point theorem.

3.11 Continuity of functions of several variables

We have defined in Section 3.8 what is meant by the continuity of a function of a real variable and have discussed the properties of continuous functions. In this section, the concept of continuity is to be generalized to functions of several real variables. More precisely, we will consider functions whose domain is a subset of $\mathbb{R}^d$ and whose codomain is (a subset of) $\mathbb{R}^m$. We will also examine which of the results in Section 3.9 extend to this more general context.

The basic idea of generalizing the concept of continuity is the same as in Section 3.2, where we generalized the concept of convergence of real-valued sequences to the multidimensional case by generalizing the concept of distance. Roughly speaking, we simply replaced absolute values with the norm. This general principle will appear repeatedly throughout this section.

The definition of the limit of a function can be almost literally adopted.

Definition 3.11: Limit of a function

Let $D \subset \mathbb{R}^d$ be a subset of $\mathbb{R}^d$ and $f : D \to \mathbb{R}^m$ a function. We write

$$\lim_{\boldsymbol{x}\to\boldsymbol{a}} f(\boldsymbol{x}) = \boldsymbol{c},$$

if for every sequence $(\boldsymbol{x}_n)_{n\in\mathbb{N}}$ with $\boldsymbol{x}_n \in D$, $\boldsymbol{x}_n \neq \boldsymbol{a}$ and $\lim_{n\to\infty} \boldsymbol{x}_n = \boldsymbol{a}$ it holds

$$\lim_{n\to\infty} f(\boldsymbol{x}_n) = \boldsymbol{c}.$$

With Definition 3.11 one can define the continuity of functions of several variables analogously to Definition 3.9.

Definition 3.12: Continuity

Let $D \subset \mathbb{R}^d$ be a subset of $\mathbb{R}^d$, $f : D \to \mathbb{R}^m$ a function and $\boldsymbol{a} \in D$. The function f is called **continuous at** $\boldsymbol{a}$, if it holds:

$$\lim_{\boldsymbol{x}\to\boldsymbol{a}} f(\boldsymbol{x}) = f(\boldsymbol{a}).$$

f is called **continuous** if f is continuous at every point of its domain.

One-sided limits (and thus left and right continuity), as we have defined them in Section 3.8, cannot be meaningfully defined for functions of several variables. This is because in $\mathbb{R}^d$ there are many different ways a sequence can converge to a point. While every real number x is either greater or less than a given number $a \neq x$, this is not the case for vectors in $\mathbb{R}^d$. A criterion like Theorem 3.18 does not exist for functions of several variables. However, the ε-δ definition remains valid if one replaces absolute values by the norm.

Theorem 3.27: ε-δ definition

Let $D \subset \mathbb{R}^d$, $f : D \to \mathbb{R}^m$ be a function and $\boldsymbol{a} \in D$. f is continuous at a if and only if for every $\varepsilon > 0$ there is a $\delta > 0$ such that

$$\|f(\boldsymbol{x}) - f(\boldsymbol{a})\| < \varepsilon \quad \text{for all } \boldsymbol{x} \in D \text{ with } \|\boldsymbol{x} - \boldsymbol{a}\| < \delta.$$

A function $f : \mathbb{R}^d \to \mathbb{R}^m$ can be viewed as a collection of m functions $f_1, f_2, \ldots, f_m : \mathbb{R}^d \to \mathbb{R}$, such that

$$f(x_1, \ldots, x_d) = (f_1(x_1, \ldots, x_d), \ldots, f_m(x_1, \ldots, x_d)).$$

The functions $f_1, \ldots, f_m$ are referred to as **component functions** of f. An important criterion asserts that testing the continuity of a function $f : \mathbb{R}^d \to \mathbb{R}^m$ can be reduced to testing the continuity of its component functions.

Theorem 3.28: Continuity of a function $\mathbb{R}^d \to \mathbb{R}^m$

A function $f : \mathbb{R}^d \to \mathbb{R}^m$ is continuous if and only if all its component functions $f_1, \ldots, f_m$ are continuous.

Theorem 3.20 about the continuity of the composition now applies more generally.

Theorem 3.29: Continuity of the composition

Let $A \subset \mathbb{R}^d$, $B \subset \mathbb{R}^m$, and let the functions $f : A \to B$ and $g : B \to \mathbb{R}^n$ be continuous, then the function $g \circ f : A \to \mathbb{R}^n$ is also continuous.

Likewise, Theorem 3.19 now applies in a more general context. We formulate this theorem only for functions with codomain $\mathbb{R}$.

Theorem 3.30: Preservation properties of continuous functions

Let $D \subset \mathbb{R}^d$ and $\lambda \in \mathbb{R}$. If $f, g : D \to \mathbb{R}$ are functions that are continuous at $\boldsymbol{a} \in D$, then the functions $f \pm g$, λf, fg, $\max\{f, g\}$, and $\min\{f, g\}$ are also continuous at $\boldsymbol{a}$. The function f/g is continuous at $\boldsymbol{a}$, provided $g(\boldsymbol{a}) \neq 0$.

Also Theorem 3.22 is still valid. We formulate it for functions with codomain $\mathbb{R}$.

Theorem 3.31: Non-zero in a neighborhood

Let $D \subset \mathbb{R}^d$, $f : D \to \mathbb{R}$ be continuous, and let $\boldsymbol{a} \in D$ with $f(\boldsymbol{a}) \neq 0$. Then there is a δ-neighborhood of $\boldsymbol{a}$, such that $f(\boldsymbol{x}) \neq 0$ for all $\boldsymbol{x}$ with $\|\boldsymbol{x} - \boldsymbol{a}\| < \delta$.

Also, Theorem 3.24 about the boundedness of a continuous function on a closed interval $[a, b]$ applies with minor modifications. In place of $[a, b]$ we use a domain in $\mathbb{R}^d$ that is compact, i.e., bounded and closed (see the definitions in Section 2.7).

Theorem 3.32: Maxima and minima of a continuous function

Let $D \subset \mathbb{R}^d$ be compact. If $f : D \to \mathbb{R}$ is continuous, then f is bounded and attains a maximum and a minimum on D.

To establish an intermediate value theorem for functions of several variables, we need a further restriction on their domain. In addition to compactness of the domain, it must be assumed that the domain has no "gaps". Formally, this means that the domain must be **connected**[7]; see Figure 3.9.

[7] A set $D \subset \mathbb{R}^d$ is called **connected** if D cannot be divided into two (non-empty) subsets A and B that have no common boundary point.

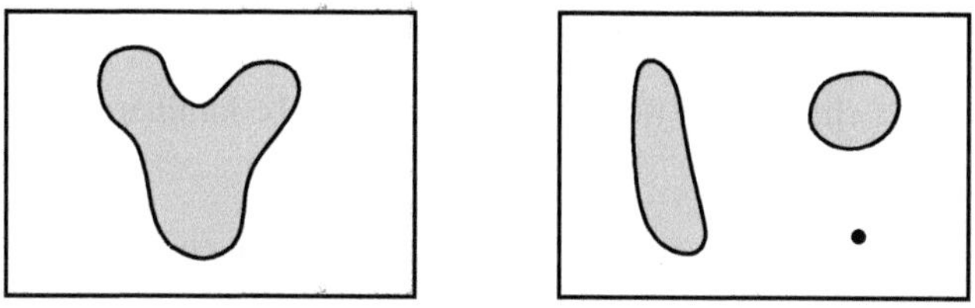

Figure 3.9: A connected set (left panel) and a non-connected set (right panel).

Theorem 3.33: Intermediate value theorem

Let $D \subset \mathbb{R}^d$ be compact and connected, and let the function $f : D \to \mathbb{R}$ be continuous. Furthermore, let a be the minimum and b the maximum of f on D. Then for every value $y \in (a, b)$ there exists a point $\boldsymbol{x} \in D$, such that $f(\boldsymbol{x}) = y$.

To conclude this section, we show how the fixed-point theorems 3.25 and 3.26 can be generalized. As already explained, a self-map is a function whose domain and codomain coincide. The fixed-point theorem 3.25 remains valid if one replaces the closed interval $[a, b]$ with a set that is convex and compact.

Theorem 3.34: Brouwer fixed-point theorem

Every continuous self-map of a convex and compact subset of $\mathbb{R}^d$ has at least one fixed point.

The Brouwer fixed-point theorem is best illustrated with a cup of coffee. If you stir the coffee, at least one particle will be in the same place after stirring as before.

To generalize the Banach fixed-point theorem, we define: A map f is called **contracting**, if there is a constant $q < 1$ such that for arbitrary values $\boldsymbol{x}$, $\boldsymbol{y}$ it holds

$$\|f(\boldsymbol{x}) - f(\boldsymbol{y})\| < q\|\boldsymbol{x} - \boldsymbol{y}\| .$$

This corresponds to the definition from Section 3.10 with the difference that the absolute values have been replaced by the norm.

The contraction property is so strong that it already implies Banach's fixed-point theorem if the domain is closed:

Theorem 3.35: Banach fixed-point theorem

Every contracting self-map $f : D \to D$ of a closed subset $D \subset \mathbb{R}^d$ has exactly one fixed point $\boldsymbol{z}$. For any initial value $\boldsymbol{x}_0$, the sequence defined by the iteration $\boldsymbol{x}_n = f(\boldsymbol{x}_{n-1})$ converges to $\boldsymbol{z}$,

$$\lim_{n\to\infty} \boldsymbol{x}_n = \boldsymbol{z} .$$

Important terms for review

After reading this chapter, the following terms should be familiar:

- Number sequence
- Limit of a sequence, convergence, divergence of a sequence
- Divergence to ∞ or $-\infty$
- Bounded sequence
- Monotone sequence
- Geometric sequence
- Multidimensional sequence
- Infinite series, partial sum
- Geometric series
- Compounding and discounting, present value, future value
- Limit of a function
- Right- and left-sided limit
- Continuous function
- Fixed-point theorem

Self-test

You can check your understanding of this chapter using the following true-or-false questions. For each statement, indicate whether it is true (T) or false (F). Short solutions to these questions can be found in Appendix H.

I. Assess these statements about sequences:

	T	F
The sequence $a_n = (-\frac{1}{2})^n$ is a null sequence.	☐	☐
If $(a_n)_{n\in\mathbb{N}}$ and $(b_n)_{n\in\mathbb{N}}$ are null sequences, then $(a_n + b_n)_{n\in\mathbb{N}}$ is also a null sequence.	☐	☐
The sequence $a_n = (-1)^n \frac{1}{n} + \ln(n)$ is convergent.	☐	☐
Every null sequence is convergent.	☐	☐
There are unbounded null sequences.	☐	☐
The sequence $a_n = (-1)^n$ has the limits -1 and 1.	☐	☐

II. Assess the following statements about series:

	T	F
Every infinite series is convergent.	☐	☐

If $(a_n)_{n\in\mathbb{N}}$ is a null sequence, then $\sum_{n=0}^{\infty} a_n$ is convergent.	☐	☐
If $(a_n)_{n\in\mathbb{N}}$ is not a null sequence, then $\sum_{n=0}^{\infty} a_n$ is divergent.	☐	☐
The geometric series converges for $q = 1$.	☐	☐

III. Determine the number of fixed points in each case.

	T	F
The function $f : [0, 1] \to [0, 1]$, $f(x) = x^{\alpha}$, with $\alpha > 0$ has at least one fixed point.	☐	☐
The function $f : [0, 5] \to [0, 5]$, $f(x) = \lfloor x \rfloor$, has exactly one fixed point.	☐	☐
The function $f : [0, 1] \to [0, 1]$, $f(x) = x$, has infinitely many fixed points.	☐	☐

Exercises

Exercise 1

Check whether the sequence $(a_n)_{n\in\mathbb{N}}$ is convergent and find its limit if it exists.

a) $a_n = \frac{1+e^n}{e^n}$

b) $a_n = \frac{4n^3+\sqrt[3]{-8n}}{\sqrt{n^6}+5n^2}$

c) $a_n = -\frac{\sqrt{n^3}}{n}$

d) $a_n = 2^{10000}\frac{1}{(n+1)!}\frac{1}{2^n}$

e) $a_n = \frac{1}{n^2}\sin(n)$

Exercise 2

An initial capital of 700 Euro is invested for 60 years. Calculate the final amount in each of the following cases:

a) The nominal annual interest rate is 3%, and interest is credited at the end of each month (ordinary interest).

b) Everything as in a), but interest is credited at the beginning of each month (advance interest).

c) What annual interest rate, with interest credited once per year at the end of each year, would yield the same final amount as in a) or b) (i.e., what is the "effective interest rate"?

Exercise 3
A loan of 40 000 Euro is interest-free and repayment-free for the first two years. From the third year onwards, fixed installments are to be paid at the end of each quarter. These consist of 6% interest, 2% repayment and 0.4% fees per year, where repayment and fees refer to the initial loan amount. The quarterly installment is therefore 840 Euro. When is the loan repaid?

Exercise 4
Consider example 3.28 (Cobweb model; continued). Suppose the demand function is $d_n = d(p_n) = 100 - p_n$, for $p_n \in [0, 100]$ and the supply function is $s_n(p_{n-1}) = 0.01p_{n-1}$. Does the sequence of prices converge to an equilibrium price?

Exercise 5
The functions $f_1, f_2 : \mathbb{R}^2 \to \mathbb{R}$ are given by

$$f_1(x, y) = \begin{cases} \dfrac{|x| + |y|}{\sqrt{x^2 + y^2}}, & \text{if } (x, y) \neq (0, 0), \\ 1, & \text{if } (x, y) = (0, 0), \end{cases}$$

$$f_2(x, y) = \begin{cases} \dfrac{x^2y^2}{x^2 + y^2}, & \text{if } (x, y) \neq (0, 0), \\ 0, & \text{if } (x, y) = (0, 0). \end{cases}$$

Check whether the functions f_1 and f_2 are continuous.

Chapter 4

Differentiable functions of one variable

This chapter deals with differential calculus for real-valued functions of a single variable.

Example 4.1 (Production). In a production process, a good is produced from one or more inputs. The produced quantity, the **output**, depends on the quantities used, the **inputs**. In this chapter, we limit ourselves to one input. For example, the production of light fuel oil is essentially based on one input, crude oil. Typical questions are:

- How does the output change at a certain level of production if the input is increased by a small unit?
- By how many percent does the output change if the input is increased by a small percentage?

The production is described by a real function $y = f(x)$, $x \in A$. We are interested in the absolute and relative change of the function f when the value of the variable x changes: By how many units does y change when x is increased by Δx units? And by how many percent does y change when x is increased by $\frac{\Delta x}{x} \cdot 100$ percent?

With the help of differential calculus, we will analyze the growth behavior of a real function. The first derivative describes the local increase of the function, the second derivative its curvature. The two derivatives characterize maxima and minima of the function as well as certain qualitative properties of it – monotonicity, convexity, and concavity.

K. Mosler et al., *Mathematical Methods for Business and Economics*, Mathematics Study Resources 20, https://doi.org/10.1007/978-3-662-73831-3_4

4.1 Derivative, differential, elasticity

First, we investigate the *local growth behavior* of a function, that is, its growth at a certain point a in its domain. To do this, we approximate the graph of the function by its tangent at a given point a. The slope of the tangent is called the **derivative** of the function; it depends on a. In this section, the derivative and concepts related to the slope such as the local approximation of a function are defined, and rules for calculating the derivative are given.

Figure 4.1 shows the graph of a real function $y = f(x), x \in \mathbb{R}$, and the tangent (solid line) to the graph at the point $x = a$. Let Δx denote an increase in the variable x. The change of y relative to that of x,

$$\frac{\Delta y}{\Delta x} = \frac{f(a + \Delta x) - f(a)}{\Delta x},$$

is called the **difference quotient**. It equals the slope of the dashed line that passes through the two points $(a, f(a))$ and $(a + \Delta x, f(a + \Delta x))$.

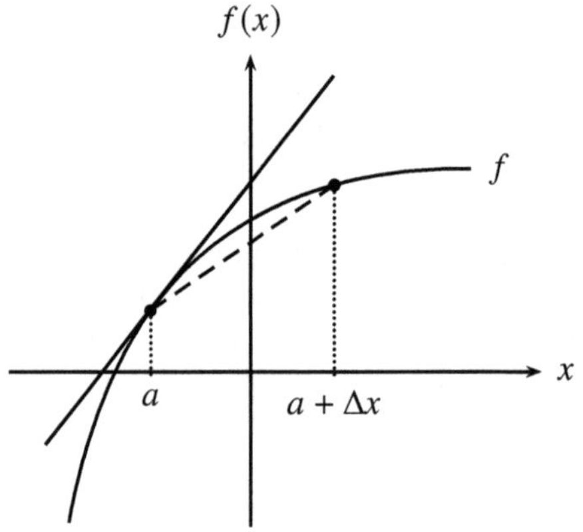

Figure 4.1: Difference quotient and differential quotient.

As Δx approaches zero, this line becomes the tangent to the graph of f at the point $(a, f(a))$, and the difference quotient approaches the differential quotient.

Definition 4.1: Differential quotient, derivative

Consider a function $f : A \to \mathbb{R}$, defined on an interval[1] A. If for $a \in A$ the limit

$$\frac{df(a)}{dx} := \lim_{\Delta x \to 0} \frac{f(a + \Delta x) - f(a)}{\Delta x} \tag{4.1}$$

exists, f is called **differentiable at** a. The limit is called the **differential quotient**

[1]The interval can be bounded or unbounded, in particular it can be equal to $\mathbb{R}$.

or **derivative** at a and denoted by $f'(a)$. Alternative notations are

$$\frac{d}{dx}f(a), \quad \frac{df}{dx}(a), \quad \text{and} \quad \frac{dy}{dx}(a).$$

The limit $\Delta x \to 0$ in (4.1) is to be understood in such a way that for every (!) sequence $(h_n)_{n\in\mathbb{N}}$ that converges to zero (with $a + h_n \in A$ and $h_n \neq 0$ for all n) the difference quotient

$$\frac{f(a+h_n) - f(a)}{h_n}$$

converges to the same limit. Among them, we have to consider also sequences with both positive and negative terms h_n. The difference quotient is thus the limit of a function in the sense of Definition 3.8. The derivative is equal to the slope of the tangent to the graph at $(a, f(a))$; see Figure 4.1. The slope of the tangent, that is the derivative, is mentioned as the **slope of the function at** a**.**

As an example, we consider the function $f(x) = |x|, x \in \mathbb{R}$; see Figure 1.5. Its graph has a tangent at every $x > 0$, namely the main diagonal of the coordinate system with slope 1. Similarly, it has a tangent at every $x < 0$ with slope -1. However, at the point $x = 0$ there is no tangent, because at $x = 0$ the slope changes from -1 to 1. The function is not differentiable at $x = 0$; the graph has a corner there.

The tangent is the graph of a function, namely the **tangent function** to f at the point a. It is given by

$$t_{f,a}(x) = f(a) + f'(a) \cdot (x - a), \quad x \in \mathbb{R}. \tag{4.2}$$

A function $f : \mathbb{R} \to \mathbb{R}$ is named **affine-linear** if its graph is a straight line. Thus, the tangent function is an affine-linear function.

Near the point a, the graph of a differentiable function f does not differ much from its tangent at a, that is, the function values of f are approximately equal to those of the tangent function. The growth of a function f at a is therefore approximately described by the growth of its tangent function: its derivative $f'(a)$. If we increase a by a small unit Δx to $a + \Delta x$, the tangent function grows **exactly** by $\Delta x \cdot f'(a)$, while the function itself grows **approximately** by $\Delta x \cdot f'(a)$. At the point a, the first derivative of the function coincides with that of its tangent. We say that in the neighbourhood of a the function is **approximated to first order** by its tangent.

This is often used in more complex scenarios by replacing f with its – simpler to handle – tangent function, particularly when f is not explicitly given or its values cannot be readily computed. However, note that these approximations are *only valid locally*. As the distance from a increases, f typically deviates more significantly from its tangent function at a.

The following two examples deal with a production function and a growth function. The derivative and tangent of the square root function and the exponential function are calculated.

Example 4.2 (Marginal output of a production). The derivative of a production function at a point $a \in A$ is referred to as the **marginal output** of production at a given level a. It indicates by how many units Δy the production approximately increases when the input is increased by Δx. This answers the first of the two questions posed at the beginning in Example 4.1.

Specifically, let a production be described by the **square root function** $y = f(x) = \sqrt{x}$, $0 \leq x \leq 100$. The current level of input be $a = 60$ units. What is the marginal output, and by approximately how many units does the output change when the input is increased by 2 units?

We expand the difference quotient, transform it by using the difference-of-two-squares formula[2], and compute the limit for $\Delta x \to 0$:

$$\begin{aligned}\frac{\sqrt{a+\Delta x}-\sqrt{a}}{\Delta x} &= \frac{(\sqrt{a+\Delta x}-\sqrt{a})(\sqrt{a+\Delta x}+\sqrt{a})}{\Delta x(\sqrt{a+\Delta x}+\sqrt{a})} \\ &= \frac{(a+\Delta x)-a}{\Delta x(\sqrt{a+\Delta x}+\sqrt{a})} = \frac{1}{\sqrt{a+\Delta x}+\sqrt{a}} \\ &\overset{\Delta x\to 0}{\longrightarrow} \frac{1}{\sqrt{a}+\sqrt{a}} = \frac{1}{2\sqrt{a}} = f'(a)\,.\end{aligned}$$

This holds for any $a > 0$. At $a = 60$ we get the marginal output

$$f'(60) = \frac{1}{2\sqrt{60}} = 0.0645\,.$$

If the input is increased by two units, the output increases approximately by twice the marginal output. So we obtain about 0.1290 additional units.

The tangent function at this point is given by

$$t_{f,60}(x) = f(60) + f'(60)\cdot(x-60) = \sqrt{60} + \frac{x-60}{2\sqrt{60}} = \frac{1}{2}\sqrt{60} + \frac{1}{2\sqrt{60}}\cdot x\,.$$

Example 4.3 (Marginal growth). An economic quantity that changes over time, such as investment capital, is considered as a function $y = f(x)$ of time x. The derivative of f at a point a then equals approximately the amount by which the quantity grows from a to $a+1$; this derivative is called **marginal growth**.

As a specific growth function, we consider the **exponential function** from Example 1.12 b). For $x \in \mathbb{R}$ it is given by the infinite series

$$f(x) = \exp(x) = e^x = \sum_{k=0}^{\infty} \frac{x^k}{k!}\,,$$

[2] Difference-of-two-squares formula: $\alpha^2 - \beta^2 = (\alpha-\beta)(\alpha+\beta)$.

see Definition 3.7. At any $a \in \mathbb{R}$ it holds[3]

$$f'(a) = \frac{d}{dx}\exp(a) = \exp(a),$$

and the tangent function is

$$t_{\exp,a}(x) = \exp(a) + \exp(a) \cdot (x - a) = \exp(a) \cdot (x + 1 - a), \quad x \in \mathbb{R}.$$

As a growth function, the exponential function has a special property: The marginal growth $\frac{d}{dx}\exp(a)$ of the quantity is equal to its achieved level $\exp(a)$ at any time a.

We now turn to the concept of the **differential**. It describes the change in an economic quantity due to a small change in an influencing variable. Equation (4.2) of the tangent function can be rewritten as

$$y - f(a) = \frac{dy}{dx}(a) \cdot (x - a). \tag{4.3}$$

The increment $y - f(a)$ in the vertical direction is proportional to the increment $x - a$ in the horizontal direction, with the derivative as the proportionality factor.

Definition 4.2: Differential

The right side of (4.3) is considered as a function of $z = x - a$,

$$df_a : z \mapsto \frac{dy}{dx}(a) \cdot z, \quad z \in \mathbb{R},$$

and is called the **differential function** of f at a. The expression

$$\frac{dy}{dx}(a) \cdot dx = f'(a)dx$$

is referred to as the **differential** of f at a.

The differential is the second term of the tangent function (4.2), where $dx = x - a$. It approximates the corresponding change in the function value; see Figure 4.2.

In the following, we will use x instead of a to denote the point at which a function is differentiated, provided that no confusion with the general argument of the function arises. For example, the differential of the square root function (Example 4.2) will be notated

$$\frac{1}{2\sqrt{x}}\,dx, \quad x > 0.$$

[3]This is shown using the series representation of the exponential function, by calculating the difference quotient $(e^{a+\Delta x} - e^a)/\Delta x$ and computing the limit for $\Delta x \to 0$.

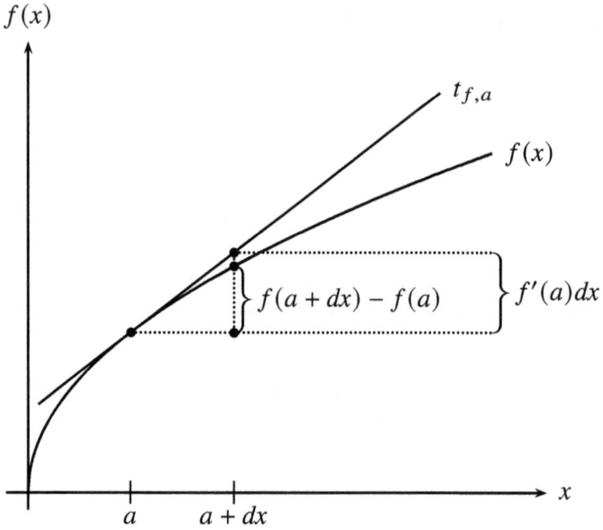

Figure 4.2: Change in the function value $f(a + dx) - f(a)$ and differential $f'(a)dx$.

Example 4.4. The differential of the function $f(x) = x^2 + x$ at any x is

$$df_x(dx) = f'(x)\,dx = (2x + 1)\,dx\,.$$

One might ask how the function value changes at $x = 2$ if the argument is increased by $dx = 0.3$ units. The approximate change in the function value can be easily determined with the differential:

$$df_2(0.3) = (2 \cdot 2 + 1) \cdot 0.3 = 1.5\,.$$

Thus, the function value increases by about 1.5 units. Compare the exact increase; it amounts to $f(2.3) - f(2) = 7.59 - 6 = 1.59$ units.

In this example, the exact change was easy to calculate. However, when the function is difficult to evaluate or not explicitly given, using the differential may help.

Elasticity Derivative and differential describe *absolute* changes in the values of a function. Often, however, one is less interested in absolute changes than in *relative changes*. The main reason for this is that relative changes do not depend on the chosen unit of measurement. In production, one could ask: by what percentage does output increase if input is increased by five percent? The answer to this question will be the same regardless of whether the input is measured in tons and the output in kilograms or in other units. Similarly to measuring absolute changes with the help of derivatives, relative changes are measured with the help of elasticities, which will be defined below:

Definition 4.3: Elasticity

Let f be differentiable at x and $f(x) \neq 0$. Then,

$$\varepsilon_f(x) = \frac{x}{f(x)} f'(x)$$

is referred to as the **elasticity** of f at the point x.

The elasticity is obtained from computing the limit of a ratio,

$$\lim_{\Delta x \to 0} \frac{\dfrac{f(x+\Delta x) - f(x)}{f(x)}}{\dfrac{(x+\Delta x) - x}{x}} = \frac{x}{f(x)} \lim_{\Delta x \to 0} \frac{f(x+\Delta x) - f(x)}{\Delta x} = \varepsilon_f(x).$$

The numerator of this fraction represents the relative change of the function value, and the denominator the relative change of the argument. The elasticity is the limit of the quotient of the relative changes. This answers the second question initially posed in Example 4.1.

Example 4.5. The elasticity of $f(x) = x^2 + x$ is

$$\varepsilon_f(x) = \frac{x}{f(x)} f'(x) = \frac{x}{x^2 + x} \cdot (2x+1) = \frac{2x+1}{x+1}.$$

For $x = 2$ we get

$$\varepsilon_f(2) = \frac{2 \cdot 2 + 1}{2 + 1} = \frac{5}{3} = 1.6667.$$

How can this value be interpreted? The elasticity, as described above, gives the approximate relative change of the function value with respect to a relative change of the argument. If we multiply the elasticity by the percentage change of the argument in the example, we get

$$\varepsilon_f(2) = 1.6667 = \frac{1.6667\%}{1\%} \quad \text{or} \quad \varepsilon_f(2) = 1.6667 = \frac{1.6667 \cdot 3\%}{3\%} = \frac{5\%}{3\%}.$$

If the variable x has been equal to 2 and is now increased by one percent, the function value changes by approximately 1.6667 percent. If, on the other hand, the argument is increased by three percent, the function value changes by approximately $1.6667 \cdot 3 = 5$ percent.

So far, when considering elasticity, we have only spoken of *change* and not of *increase* or *decrease* of the function value. This is because the sign of the elasticity of f at x does not coincide with the sign of the derivative when x and $f(x)$ have different signs. For example, the natural logarithm $f(x) = \ln(x)$, $x > 0$, is strictly monotonically increasing with derivative $f'(x) = \frac{1}{x} > 0$ for $x > 0$. So if x increases, $f(x)$ increases as well. However, the elasticity $\varepsilon_f(x) = \frac{x}{\ln(x)} \cdot \frac{1}{x} = \frac{1}{\ln(x)}$ is negative for $0 < x < 1$, since $\ln(x)$ is negative for $0 < x < 1$. Fortunately, this problem does not occur when both the arguments and the values of the function are non-negative, which is the case for many economic functions.

Example 4.6 (Linear and constant functions). We consider the **affine-linear function**

$$f : \mathbb{R} \to \mathbb{R}, \quad x \mapsto a + bx .$$

Here a and b are real constants. The difference quotient at a point $x \in \mathbb{R}$ is equal to

$$\frac{f(x + \Delta x) - f(x)}{\Delta x} = \frac{a + b(x + \Delta x) - a - bx}{\Delta x} = b$$

and hence independent of Δx and x. The limit for $\Delta x \to 0$ equals b as well. Therefore, an affine-linear function is differentiable everywhere and its derivative is $f'(x) = b$ at every $x \in \mathbb{R}$. The elasticity of the affine-linear function f is

$$\varepsilon_f(x) = \frac{x}{f(x)} f'(x) = \frac{bx}{a + bx} .$$

If $b \neq 0$, the elasticity depends on x. If $a = 0$, we get a **linear function**

$$f(x) = bx , \quad x \in \mathbb{R} .$$

It has the derivative $f'(x) = b$ and the elasticity one.

In the case $b = 0$ we get a **constant function**

$$f(x) = a , \quad x \in \mathbb{R} .$$

The constant function has derivative zero at every $x \in \mathbb{R}$. Its elasticity is also zero, provided $a \neq 0$.

The elasticity relates the relative change of the function value to the relative change of the argument. Unlike the derivative, it does not depend on the units chosen for the function value and the argument, which is illustrated by the following example.

Example 4.7. At a price of p [€] for a kilogram of a good, the demand for that good is $q_1(p) = 2000 - 200p$ [kg]. Thus the derivative at any point p is given by -200 [kg/€]. However, if we measure the price in cents and the quantity in 1000 kg, the demand function reads $q_2(p) = 2 - 0.002p$ [1000kg]. The derivative of q_2 is -0.002 [1000kg/Cent]. Therefore, the numerical value of the derivative (i.e., -200 or -0.002) can only be meaningfully interpreted if the units are taken into account. If, on the other hand, we consider the elasticity of demand with respect to price, the so-called **price elasticity of demand**, we get for the functions q_1 and q_2

$$\varepsilon_{q_1}(p) = -\frac{200p}{2000 - 200p} = -\frac{p}{10 - p}$$

and

$$\varepsilon_{q_2}(p) = -\frac{0.002p}{2 - 0.002p} = -\frac{0.01p}{10 - 0.01p}.$$

Now consider these elasticities at a price of 2 [€] resp. 200 [Cent]. In q_1 we insert $p = 2$ in Euro, while in q_2 – since the price is measured in cents – we insert $p = 200$ instead. We obtain

$$\varepsilon_{q_1}(2) = -\frac{1}{4} \quad \text{and} \quad \varepsilon_{q_2}(200) = -\frac{1}{4},$$

and hence the same elasticity. The value of the elasticity is independent of the chosen units.

In economics, a function $x \mapsto f(x)$ is referred to as

- **inelastic** at x if $|\varepsilon_f(x)| < 1$,
- **unit elastic** at x if $|\varepsilon_f(x)| = 1$,
- **elastic** at x if $|\varepsilon_f(x)| > 1$.

If a function f is inelastic at a point x, a change in x causes a less than proportional change in the function value. If f is elastic at a point x, a change in x causes a more than proportional change in the function value.

4.2 Differentiation rules

So far, we have determined derivatives as limits of difference quotients, which can result in tedious calculations. Fortunately, there are rules that allow us to calculate the derivatives of functions based on known derivatives. If the derivatives of two functions at a point x are known, the derivatives of their sum, difference, product and quotient are easily determined; similarly the derivative of a constant multiple of a function.

Theorem 4.1: Differentiation rules

Let $f, g : A \to \mathbb{R}$ be differentiable at a point x. Then the following functions are also differentiable at x:

(i) $f + g$ with derivative $(f + g)'(x) = f'(x) + g'(x)$,

(ii) $f - g$ with derivative $(f - g)'(x) = f'(x) - g'(x)$,

(iii) αf with derivative $(\alpha f)'(x) = \alpha f'(x)$ for every $\alpha \in \mathbb{R}$,

(iv) $f \cdot g$ with derivative $(f \cdot g)'(x) = f'(x)g(x) + f(x)g'(x)$,

(v) $\frac{f}{g}$ with derivative $\left(\frac{f}{g}\right)'(x) = \frac{f'(x)g(x) - f(x)g'(x)}{(g(x))^2}$ if $g(x) \neq 0$.

Mnemonic 4.1

$(f \pm g)' = f' \pm g'$	**sum and difference rule**
$(\alpha f)' = \alpha f'$	**factor rule**
$(fg)' = f'g + fg'$	**product rule**
$\left(\dfrac{f}{g}\right)' = \dfrac{f'g - fg'}{g^2}$	**quotient rule**

Example 4.8 (Square function). The function

$$h_2 : x \mapsto x^2, \qquad \mathbb{R} \to \mathbb{R},$$

is differentiable at every point $x \in \mathbb{R}$, having derivative $h_2'(x) = 2x$. To see this, we apply the product rule with $f(x) = x$ and $g(x) = x$. From Example 4.6 we know that $f'(x) = g'(x) = 1$. It follows

$$h_2'(x) = f'(x)g(x) + f(x)g'(x) = 1 \cdot x + x \cdot 1 = 2x.$$

The **power function** $h_n(x) = x^n$, $x \in \mathbb{R}$, is initially defined for exponents $n \in \{0, 1, 2, \ldots\}$. Using mathematical induction it can be shown that its derivative is

$$h_n'(x) = nx^{n-1} \quad \text{for all } x \in \mathbb{R}.$$

A **polynomial** is a sum of powers x^n multiplied by real constants; see Example 1.7. From the sum and factor rules it follows, that a polynomial is differentiable everywhere.

Example 4.9. The polynomial

$$p(x) = x^4 - 10x^3 + 35x^2 - 50x + 24, \quad x \in \mathbb{R},$$

is differentiable everywhere with derivative

$$p'(x) = 4x^3 - 30x^2 + 70x - 50, \quad x \in \mathbb{R}.$$

Example 4.10. To differentiate the function $h_{-1}(x) = \frac{1}{x}$, $x \in \mathbb{R} \setminus \{0\}$, we apply the quotient rule. It is $h_{-1} = \frac{f(x)}{g(x)}$ with $f(x) = 1$ and $g(x) = x$. The derivatives are $f'(x) = 0$ and $g'(x) = 1$, so

$$h_{-1}'(x) = \frac{f'(x)g(x) - f(x)g'(x)}{(g(x))^2} = \frac{0 \cdot x - 1 \cdot 1}{x^2} = -\frac{1}{x^2}.$$

Next, we consider the power functions $h_k(x) = x^k$, $x \in \mathbb{R} \setminus \{0\}$, whose exponents k are negative integers, $k \in \{-1, -2, -3, \ldots\}$. By mathematical induction it can be shown that their derivatives are

$$h_k'(x) = kx^{k-1} \quad \text{for all } x \in \mathbb{R} \setminus \{0\}\,.$$

Another useful rule is the **chain rule**, by which the derivative of the composition of two functions can be calculated:

Theorem 4.2: Chain rule

Let A, B be open intervals in $\mathbb{R}$. Further let

$f : A \to B$ be differentiable at $x \in A$,
$g : B \to \mathbb{R}$ be differentiable at $y = f(x) \in B$.

Then the composition $g \circ f : A \to \mathbb{R}, x \mapsto g(f(x))$, is differentiable at x, and its derivative is

$$(g \circ f)'(x) = g'(f(x))f'(x)\,.$$

Example 4.11. We want to find the derivative of $h : x \mapsto \sqrt{x^2+1}, x \in \mathbb{R}$, at the point $x = 7$. To apply the chain rule, we choose $f(x) = x^2 + 1$ and $g(y) = \sqrt{y}$. We get $h'(7) = g'(7^2+1) \cdot f'(7) = \frac{1}{2\sqrt{50}} \cdot 2 \cdot 7 = 0.9899$.

Often the problem arises to find the **inverse** $x = f^{-1}(y)$ of a given function $y = f(x)$, $x \in A$. If f is continuous and strictly monotone, then f has an inverse function, and the derivative of the inverse f^{-1} is calculated as follows.

Theorem 4.3: Existence and derivative of the inverse function

Let $A \subset \mathbb{R}$ be an open interval, and let $f : A \to \mathbb{R}$ be continuous and strictly monotonic. Then, f has an inverse $f^{-1} : B \to A$, where $B = f(A)$.

If f is differentiable at $x \in A$ and $f'(x) \neq 0$, then f^{-1} is differentiable at $y = f(x)$, and

$$\left(f^{-1}\right)'(y) = \frac{1}{f'(x)} = \frac{1}{f'(f^{-1}(y))}\,.$$

Mnemonic 4.2

$$\frac{dx}{dy} = \frac{1}{\frac{dy}{dx}}$$

Example 4.12 (Logarithm). The inverse of the exponential function $f(x) = \exp(x), x \in \mathbb{R}$, is the natural logarithm $f^{-1}(y) = \ln(y), y > 0$; see Figure 4.3.

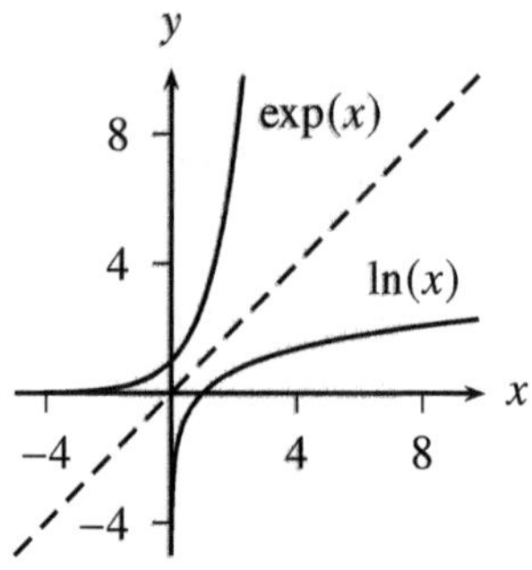

Figure 4.3: Exponential function $\exp(x)$ and logarithmic function $\ln(x)$.

With Theorem 4.3 we obtain the derivative of the logarithmic function as follows:

$$\ln'(y) = \frac{1}{\exp'(\ln(y))} = \frac{1}{\exp(\ln(y))} = \frac{1}{y}$$

In the following example the chain rule is applied.

Example 4.13 (General power function). The **general power function** is defined by

$$h_\gamma(x) = x^\gamma = \exp\big(\gamma \ln(x)\big), \quad x > 0,$$

where γ is any real number. If γ is an integer k and $x > 0$, it coincides with the power function $f(x) = x^k$ defined for $k \in \mathbb{Z}$. To find the derivative of the general power function, we write it as a composition $h_\gamma(x) = g(f(x))$ with $f(x) = \gamma \ln(x)$ and $g(x) = \exp(x)$, see above. Then the chain rule gives

$$h'_\gamma(x) = g'\big(\gamma \ln(x)\big) \cdot f'(x) = \exp\big(\gamma \ln(x)\big) \cdot \frac{\gamma}{x} = x^\gamma \cdot \frac{\gamma}{x} = \gamma x^{\gamma-1}$$

at any point $x > 0$. In particular, when $\gamma = k$ for some $k \in \mathbb{Z}$, this coincides with the previously derived rule $\frac{d}{dx}x^k = kx^{k-1}$.

4.3 First and second derivative

So far, we have examined the derivative of a function at a specific point of the domain. It describes the *local growth behavior* of the function. If a function possesses derivatives at every point of its domain, these derivatives can themselves be regarded as a function, the **derivative function**. It characterizes the *global growth behavior* of the original function, that is, its overall course of growth.

Then questions such as the following can be addressed: Is the function increasing over its entire domain? If not, at what point does it change from increasing to decreasing, or vice versa? Is the slope of the function – as measured by its derivative – continually decreasing?

Definition 4.4: Differentiable function, first derivative

Let A be an interval in $\mathbb{R}$. A function $f : A \to \mathbb{R}$ is **(continuously) differentiable**, if it is differentiable at every point $x \in A$ and the function

$$f' : A \to \mathbb{R}, \quad x \mapsto f'(x),$$

is continuous[4]. The function f' is called the **first derivative** of f.

Example 4.14. The following functions $f_i : \mathbb{R} \to \mathbb{R}$, $i = 1, 2, 3$, are continuously differentiable:

$$f_1(x) = x^2, \qquad f_1'(x) = 2x,$$
$$f_2(x) = x^3, \qquad f_2'(x) = 3x^2,$$
$$f_3(x) = \begin{cases} x^2, & x \geq 0, \\ x^3, & x < 0, \end{cases} \qquad f_3'(x) = \begin{cases} 2x, & x \geq 0, \\ 3x^2, & x < 0. \end{cases}$$

Note that f_3 is continuous at $x = 0$ and that a unique tangent exists there, as the derivatives of the left and right branches of the function tend to zero as x approaches zero.

Frequently used functions and their first derivatives are given in Table 4.1.

Next, we are interested in the growth behavior of the first derivative f'. It provides information about whether the slope of the function f is increasing or decreasing and whether there is a local minimum or maximum at a given point x.

Definition 4.5: Second derivative

If the first derivative $f' : A \to \mathbb{R}$ is differentiable at a point x, its derivative is referred to as the **second derivative** of f at x, symbolically

$$f''(x) = \frac{d}{dx} f'(x) = \frac{d^2}{dx^2} f(x).$$

If the second derivative exists at every point of the domain A, f is said to be twice differentiable, and the second derivative can be regarded as a function,

$$f'' : A \to \mathbb{R}, \quad x \mapsto f''(x).$$

The second derivative describes the growth behavior of the first derivative. If the second derivative at a point a is positive (negative), the slope increases (decreases) with increasing x. The graph of the function is **concave upward** (**concave downward**) there, that is, it lies above (below) the tangent line at a.

[4] In the following, we will only consider functions, whose derivatives are continuous. We briefly call them **differentiable** functions.

$f(x)$	defined for	$f'(x)$
$ax+b$	$x \in \mathbb{R}, a, b \in \mathbb{R}$	a
x^n	$x \in \mathbb{R}, n \in \{0, 1, 2, \ldots\}$	nx^{n-1}
x^k	$x \in \mathbb{R} \setminus \{0\}, k \in \{-1, -2, \ldots\}$	kx^{k-1}
x^α	$x > 0, \alpha \in \mathbb{R}$	$\alpha x^{\alpha-1}$
$\sum_{i=0}^{r} \alpha_i x^i$	$x \in \mathbb{R}, \alpha_0, \ldots, \alpha_r \in \mathbb{R}$	$\sum_{i=0}^{r} i\alpha_i x^{i-1}$
$\exp(x)$	$x \in \mathbb{R}$	$\exp(x)$
$\ln(x)$	$x > 0$	$\frac{1}{x}$
a^x	$x \in \mathbb{R},\ a > 0$	$a^x \ln(a)$
$\sin(x)$	$x \in \mathbb{R}$	$\cos(x)$
$\cos(x)$	$x \in \mathbb{R}$	$-\sin(x)$
$\tan(x)$	$x \neq (2n-1)\frac{\pi}{2},\ n \in \mathbb{Z}$	$\frac{1}{\cos^2(x)}$
$\cot(x)$	$x \neq n\pi,\ n \in \mathbb{Z}$	$-\frac{1}{\sin^2(x)}$

Table 4.1: Derivatives of frequently used functions

Example 4.15 (Logarithm). According to Examples 4.12 and 4.10 the second derivative of the logarithm function $f(x) = \ln(x)$ is

$$\frac{d^2}{dx^2}\ln(x) = \frac{d}{dx}\frac{1}{x} = -\frac{1}{x^2}, \quad x > 0.$$

Thus, the logarithm function is concave downward at all $x > 0$; see Figure 4.3.

4.4 Zeros and extrema

It is not only the local growth of a function that can be analyzed by differential calculus but also its global behavior. Key features of interest include zeros, maxima and minima (either global or local), as well as inflection points, saddle points, and asymptotes of a differentiable function.

We consider a real function f defined on an interval $A = [a, b]$; see Figure 4.4.

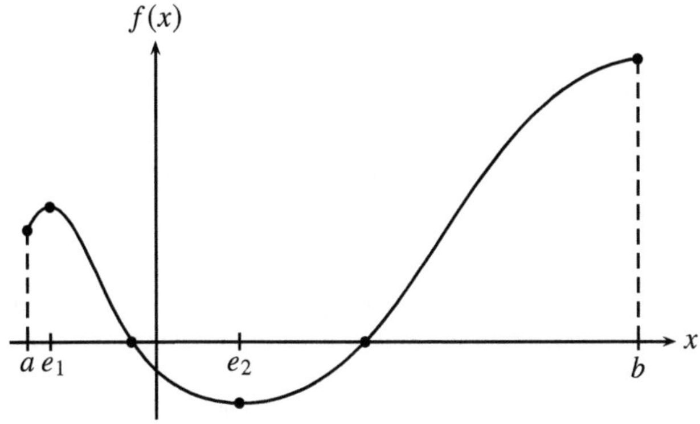

Figure 4.4: Function $[a, b] \to \mathbb{R}$ with zeros, minima and maxima.

Definition 4.6: Zero, maximum, minimum

Let $f : A \to \mathbb{R}$ and $a \in A$. We say that

- f has a **zero** (or **root**) at a, if $f(a) = 0$,
- f has a **global maximum** at a, if $f(a) \geq f(x)$ for all $x \in A$,
- f has a **global minimum** at a, if $f(a) \leq f(x)$ for all $x \in A$,
- f has a **local maximum** at a, if there exists $\varepsilon > 0$ such that $f(a) \geq f(x)$ for all $x \in A$ with $|x - a| < \varepsilon$,
- f has a **local minimum** at a, if there exists $\varepsilon > 0$ such that $f(a) \leq f(x)$ for all $x \in A$ with $|x - a| < \varepsilon$.

A local maximum is also called a **relative maximum** and a global maximum an **absolute maximum**; the same terminology applies to minima. An **extremum** is either a maximum or a minimum.

Obviously, every global maximum (minimum) is also a local maximum (minimum). The converse, however, is not true.

Example 4.16. The function f in Figure 4.4 has a global minimum in the interior of the domain at the point e_2 and a local minimum at the boundary of the domain at a. It has a global maximum at the right endpoint b of the interval and a local maximum in the interior of $[a, b]$ at the point e_1. The latter satisfies the definition of a local maximum with any $\varepsilon \leq e_1 - a$.

Theorem 4.4: Necessary condition for an extremum

Let $f : A \to \mathbb{R}$ be differentiable. If f has a local extremum at c and c is not an endpoint of A, then $f'(c) = 0$.

Theorem 4.5: Sufficient condition for an extremum

Let $f : A \to \mathbb{R}$ a be twice differentiable. If $f'(c) = 0$ and c is not an endpoint of A. Then f has a

$$\begin{aligned}&\text{local maximum at } c \text{ if } f''(c) < 0,\\&\text{local minimum at } c \text{ if } f''(c) > 0.\end{aligned}$$

Theorem 4.4 states that a local extremum of f in the interior of the domain can only occur at a point where the first derivative is zero, i.e., where a horizontal tangent exists. However, this is only a *necessary* condition for an extremum in the interior of the domain. If $f'(c) = 0$, there can also be a so-called **saddle point**. For example, the function $f(x) = (x-1)^3$ has a horizontal tangent at $c = 1$ with $f'(1) = 0$, but no extremum. Instead, it has a saddle point; see Figure 4.5.

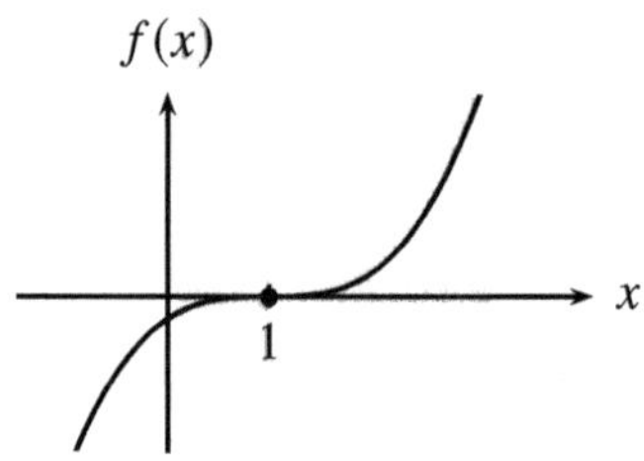

Figure 4.5: $f(x) = (x-1)^3$ has a saddle point at 1.

Theorem 4.5 complements the necessary condition with a *sufficient* one: If the slope is zero at a point in the interior and the second derivative is negative, this means that the slope of f changes from positive to negative, i.e., there is a local maximum. Similarly, if the second derivative is positive at this point, the slope of f changes from negative to positive and there is a local minimum.

For example, the function $f(x) = -4x^2 + 5x - 2$, $x \in \mathbb{R}$, has the derivatives $f'(x) = -8x + 5$ and $f''(x) = -8$ at any $x \in \mathbb{R}$. Setting $f'(x)$ to zero yields the unique solution $x = \frac{5}{8}$. So an extremum can only lie at $x = \frac{5}{8}$. Since $f''(\frac{5}{8}) < 0$, there is a local maximum there. From the graph of the function, which is a parabola opening downward, one can see that this maximum is also a global maximum.

If the domain of a function is a proper subset of $\mathbb{R}$, extrema – local and global – can also lie on the boundary of the domain. In this case, it is necessary to examine the boundary points in detail.

Example 4.17 (Unit cost function). The costs of production are described by a second-degree polynomial $C(x) = a_2x^2 + a_1x + a_0$, where $x \in [0, S]$ represents the quantity produced and a_0, a_1, a_2 are suitably chosen positive constants. The cost per unit then follows the unit cost function

$$c(x) = \frac{C(x)}{x} = \frac{a_2x^2 + a_1x + a_0}{x} = a_2x + a_1 + \frac{a_0}{x}.$$

The unit cost function is defined for $0 < x \leq S$; for x close to zero, it takes arbitrarily large values. The first two derivatives of the unit cost function on $(0, S)$ are

$$c'(x) = a_2 - \frac{a_0}{x^2}, \quad c''(x) = 2\frac{a_0}{x^3}.$$

Setting the first derivative to zero gives $x_{1,2} = \pm\sqrt{a_0/a_2}$. Since negative values of x are economically meaningless, only $x_1 = \sqrt{a_0/a_2}$ is considered.

If x_1 lies in the interior of the domain, $0 < x_1 < S$, the necessary condition for an extremum and due to $c''(x_1) > 0$ also the sufficient condition for a local minimum in x_1 is satisfied. Upon closer examination of the first derivative, we see that $c'(x) < 0$ for $0 < x < x_1$ and $c'(x) > 0$ for $x > x_1$, i.e., the function has a negative slope in the interval $(0, x_1)$ and a positive slope in the interval (x_1, S). Consequently, at x_1 the unit cost function has not only a local, but also a global minimum.

If $x_1 \geq S$, $f(x)$ is decreasing in the entire domain $[0, S]$; the global minimum of the unit cost function is therefore at the right endpoint S of the domain.

If the first derivative is zero at a, the condition that "f' changes sign at a" provides a sufficient criterion for an extremum, serving as an alternative to requiring that f is twice differentiable with $f''(a) \neq 0$. Then a local minimum occurs if the sign of f' changes from negative to positive, i.e., f decreases before a and increases after it (left panel in Figure 4.6), while a local maximum occurs if the sign changes from positive to negative, i.e., f increases before a and decreases after it (right panel in Figure 4.6).

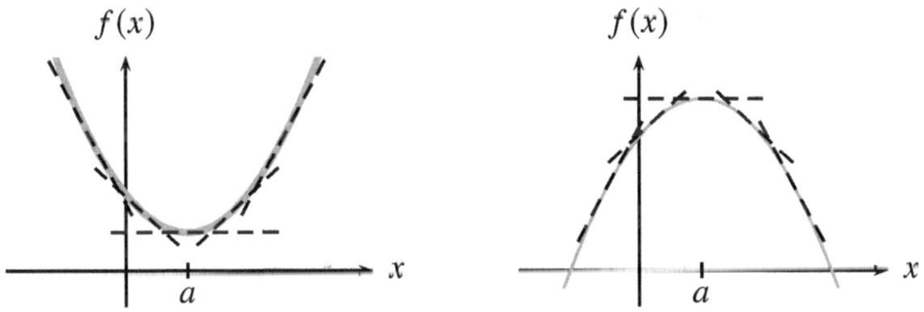

Figure 4.6: Illustration of the first derivative sign change condition for local minima (left) and local maxima (right).

If the first and second derivatives both vanish at a point a, a local extremum may or may not exist there, as demonstrated by the following two functions:

The function $f(x) = (x-1)^4, x \in \mathbb{R}$, at $a = 1$ has the derivatives $f'(1) = f''(1) = 0$. A local (and global) minimum exists at $a = 1$, as the first derivative changes its sign from negative to positive.

The function $g(x) = (x-1)^3, x \in \mathbb{R}$, at $a = 1$ has the derivatives $g'(1) = g''(1) = 0$. Unlike function f, this function has no extremum (but a saddle) at $a = 1$ since its first derivative is positive on both sides of this point.

4.5 Inflection points

To describe the behavior of a function, in addition to its zeros and extrema, other points on its graph are of particular interest.

At a zero, the function either changes its sign or has a local extremum. An extremum in the interior of the domain is characterized by a change in the sign of the first derivative. Next, we will examine what happens when the second derivative changes sign.

Definition 4.7: Inflection point, saddle point

Let f be twice differentiable, $a \in A$ and $f''(a) = 0$. If f'' changes its sign in a, a is called an **inflection point**. If in addition $f'(a) = 0$, a is a **saddle point**.

For example, the function $f : \mathbb{R} \to \mathbb{R}, x \mapsto x^3$ at $x = 0$ has an inflection point that is a saddle point.

At an inflection point, the curvature of the function's graph changes. This is nicely illustrated in Figure 4.7, which shows an inflection point that is also a saddle point. Left of the inflection point the graph is concave upward, that is, the second derivative is positive, while right of the inflection point it is concave downward, that is, the second derivative is negative.

Figure 4.7: An inflection point that is also a saddle point.

Example 4.18 (Growth with saturation). The **logistic function** $f : \mathbb{R} \to \mathbb{R}$,

$$f(x) = \frac{G}{1 + \alpha e^{-\lambda G x}}, \quad x \geq 0,$$

with $\alpha > 0$, $G > 0$ and $\lambda > 0$ (cf. Example 1.13) has the derivatives

$$\begin{aligned} f'(x) &= G^2\alpha\lambda \frac{e^{-\lambda G x}}{(1+\alpha e^{-\lambda G x})^2}, \\ f''(x) &= G^3\alpha\lambda^2 e^{-\lambda G x} \frac{\alpha e^{-\lambda G x} - 1}{(1+\alpha e^{-\lambda G x})^3}. \end{aligned}$$

Both the function and its first derivative are positive on the whole real line $\mathbb{R}$. So there are neither zeros nor local extrema. The second derivative is zero at $\alpha e^{-\lambda G x} = 1$, i.e., at $-\lambda G x = \ln(1/\alpha) = -\ln(\alpha)$. Thus, the function has an inflection point at $x = \frac{\ln(\alpha)}{\lambda G}$. The inflection point is not a saddle point.

The process of analyzing a given function for zeros, extrema, inflection points, and other geometric features of its graph is called **curve sketching**. Here is an example:

Example 4.19. We examine the function

$$f(x) = x^3 - 3x + 2, \quad x \in [-3, 2].$$

First, we determine all zeros, local and global extrema, inflection and saddle points. By trial and error we find that $x_1 = 1$ is a zero of f. The remaining zeros can be found by dividing the polynomial[5] by the factor $(x-1)$:

$$\begin{array}{llll} (x^3 & - & 3x & +2\ ) : (x-1) = x^2 + x - 2 \\ -(x^3 - x^2) & & & \\ \hline \quad x^2 & - & 3x & +2 \\ \quad -(x^2 & - & x) & \\ \hline & & -2x & +2 \\ & & -(-2x & +2) \\ \hline & & & 0 \end{array}$$

The zeros of the quadratic polynomial $x^2 + x - 2$ are calculated[6] as:

$$x_{2,3} = -\frac{1}{2} \pm \sqrt{\frac{1}{4} + 2} = -\frac{1}{2} \pm \frac{3}{2}.$$

Zeros of f are thus $x_1 = 1$, $x_2 = 1$ and $x_3 = -2$. At $x_{1,2} = 1$ a so-called **double zero** is found.

Next, we determine the first two derivatives of the function and calculate their zeros[7]:

$$f'(x) = 3x^2 - 3 \stackrel{!}{=} 0 \quad \Longleftrightarrow \quad x^2 = 1 \quad \Longleftrightarrow \quad x = \pm 1$$

[5]For polynomials the following holds: If a is a zero of a polynomial p, then the polynomial can be divided by $(x-a)$ without remainder.

[6]A quadratic polynomial $x^2 + px + q$ has the zeros $x_{1,2} = -\frac{p}{2} \pm \sqrt{\frac{p^2}{4} - q}$.

[7]If we set an expression $A(x)$ equal to zero in order to solve for x, we briefly write: $A(x) \stackrel{!}{=} 0$.

$$f''(x) = 6x \quad \stackrel{!}{=} 0 \quad \Longleftrightarrow \quad x = 0$$

At $x = 1$ a local minimum exists, since $f''(1) = 6 > 0$, while, as $f''(-1) = -6 < 0$, the function has a local maximum at $x = -1$.

At the two endpoints of the interval local extrema arise: a local minimum at $x = -3$, and a local maximum at $x = 2$. Further, f has an inflection point at $x = 0$. It is not a saddle point, as $f'(0) \neq 0$. Comparing the function values of the local minima shows that f has a global minimum at $x = -3$. Similarly, both local maxima (at $x = -1$ and $x = 3$) are global maxima. The graph of the function is shown in the figure below.

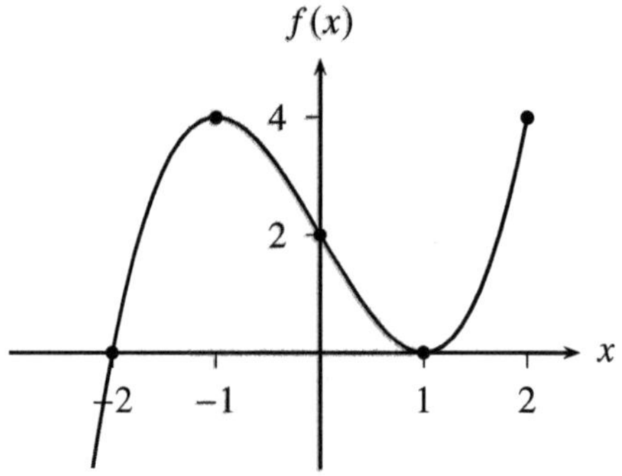

4.6 Monotonicity, concavity, convexity

The first derivative of a differentiable function is used to determine whether the function is monotonically increasing or decreasing; see Definition 1.3.

Theorem 4.6: Monotonicity of a differentiable function

Let $f : [a, b] \to \mathbb{R}$ be continuous and differentiable on the open interval (a, b).

- f is monotonically increasing if and only if $f'(x) \geq 0$ for all $x \in (a, b)$.
- f is monotonically decreasing if and only if $f'(x) \leq 0$ for all $x \in (a, b)$.

If the first derivative is always greater (smaller) than zero, then f is *strictly* monotonically increasing (decreasing). However, the converse is not true, i.e., if the function is strictly monotonic, its derivative does not have to be positive (or negative) everywhere. For example, the function $f(x) = x^3$ is strictly monotonically increasing; however, $f'(0) = 0$.

A **concave function** is concave downward in its entire domain, a **convex function** is concave upward in its entire domain; see Section 4.3. Concavity and convexity can be checked using the second derivative.

Theorem 4.7: Concavity and convexity of a twice differentiable function

Let f be twice differentiable on the open interval (a, b). Then

$$f \text{ concave} \iff f''(x) \leq 0 \text{ for all } x \in (a, b),$$
$$f \text{ convex} \iff f''(x) \geq 0 \text{ for all } x \in (a, b).$$

Example 4.20 (Utility functions). Consider the utility a person has from an income of x Euro. Two specific functions are frequently used to model the utility of money: the negative exponential function and the logarithmic function. They are shown in Figure 4.8.

a) Negative exponential utility function:

$$u(x) = 1 - e^{-\lambda x}, \quad x > 0.$$

b) Logarithmic utility function:

$$v(x) = \ln(\lambda x), \quad x > 0.$$

In both functions λ denotes a parameter that is greater than zero. The negative exponential utility is bounded, whereas the logarithmic utility is not. A typical question is: How does the utility change if x increases by one Euro?

The first derivative of a utility function is called the **marginal utility**. The marginal utilities of u and v are

$$u'(x) = \lambda e^{-\lambda x} \quad \text{and} \quad v'(x) = \frac{1}{x}, \quad x > 0.$$

They are positive everywhere, so both utility functions are strictly monotonically increasing. This corresponds to the usual requirement for utility functions, namely that an increase in income results in an increase in utility. The second derivatives,

$$u''(x) = -\lambda^2 e^{-\lambda x} \quad \text{and} \quad v''(x) = -\frac{1}{x^2}, \quad x > 0,$$

are negative everywhere. This means that the marginal utility decreases with increasing income. In other words: the two utility functions are concave.

The negative exponential utility function approaches 1 as x goes to infinity. It is said to have a **horizontal asymptote** at $y = 1$.

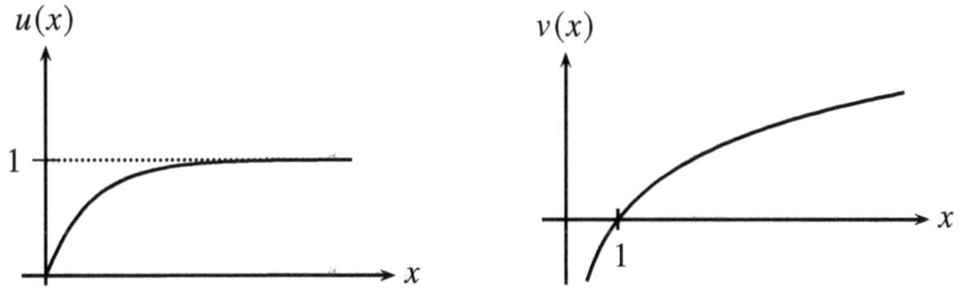

Figure 4.8: Negative exponential utility function $u(x) = 1 - e^{-x}$ (left panel) and logarithmic utility function $v(x) = \ln(\lambda x)$ (right panel).

Global extrema Concavity and convexity are useful concepts in determining global extrema. The conditions in Theorems 4.4 and 4.5 only allow the characterization of local extrema. If a function is convex, then every local minimum is also a global minimum; if it is concave, then every local maximum is also a global maximum. This not only holds for extrema in the interior of the domain but also for extrema on the boundary:

Theorem 4.8: Global extrema of concave and convex functions

Let A be an interval and $a \in A$. Then the following statements hold:

(i) If f is concave and has a **local** maximum at a, then f has a **global** maximum at a.

(ii) If f is convex and has a **local** minimum at a, then f has a **global** minimum at a.

Example 4.21. The following figure shows $f_1(x) = 9 - (x-3)^2$, $0 \le x \le 2$, and $f_2(x) = 9 - 4(x-1.5)^2$, $0 \le x \le 2$. Both functions are concave. f_1 has a local, and therefore also global maximum at the boundary of the domain at $x = 2$, f_2 has a local, and therefore also global maximum in the interior of the domain at $x = 1.5$. On the other hand, the two utility functions of Example 4.20 are concave, but have no local, and therefore also no global maximum.

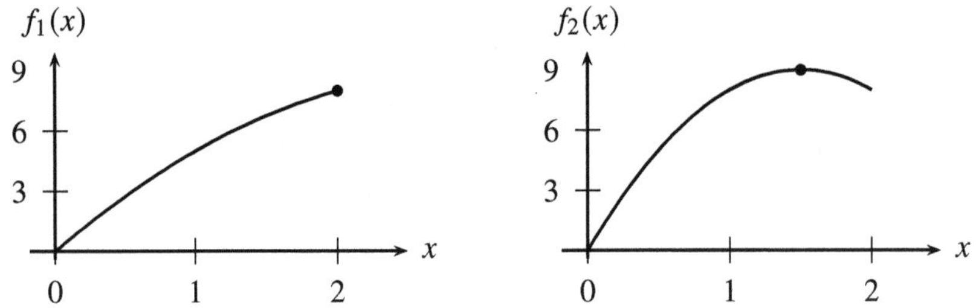

4.7 Higher order derivatives and Taylor polynomial

Often, it is desirable to approximate a complicated function by one that is much simpler to analyze. This is particularly useful when the original function is not given explicitly or when its explicit evaluation proves difficult. In Section 4.1, we saw that a differentiable function, in a neighborhood of a point a, is locally approximated by its tangent at a. As the tangent function is a first-degree polynomial, this is called an approximation to the first order.

If the function has higher derivatives, the local approximation can be improved by approximating the function with polynomials of higher degree.

Definition 4.8: Higher order derivatives

Notate $f^{(0)} = f$, and define recursively for $n \in \mathbb{N}$: If $f^{(n-1)}$ is differentiable, then the function

$$f^{(n)} = \frac{d}{dx} f^{(n-1)}$$

is called the ***n*-th derivative** of f.

Example 4.22.

a) We consider the third degree polynomial $f(x) = 7x^3 - 5x^2 + 2x + 1$. It has the derivatives $f'(x) = 21x^2 - 10x + 2$, $f''(x) = 42x - 10$, $f'''(x) = 42$, and $f^{(n)}(x) = 0$ for all $n \geq 4$.

b) The function $f(x) = \ln(x)$, $x > 0$, has the first three derivatives

$$f^{(1)}(x) = f'(x) = \frac{1}{x}, \; f^{(2)}(x) = f''(x) = -\frac{1}{x^2}, \; f^{(3)}(x) = f'''(x) = \frac{2}{x^3}.$$

Obviously, with each further derivative, the exponent in the denominator increases by 1, the numerator is multiplied by the next integer and the sign alternates. For arbitrary $n \in \mathbb{N}$ we therefore obtain

$$f^{(n)}(x) = (-1)^{n-1} \frac{(n-1)!}{x^n}.$$

Definition 4.9: Taylor[8]-Polynomial

Let a function f be n-times differentiable. Then the polynomial

$$\begin{aligned} T_{f,a,n}(x) &= \sum_{i=0}^{n} f^{(i)}(a) \frac{(x-a)^i}{i!} \\ &= f(a) + f'(a)(x-a) + f''(a)\frac{(x-a)^2}{2!} + \ldots + f^{(n)}(a)\frac{(x-a)^n}{n!} \end{aligned}$$

is called **Taylor polynomial** of degree n of f at the point a.

The tangent function of f at a is the Taylor polynomial of degree one.

The Taylor polynomial $T_{f,a,n}$ has derivatives of any order. At a its first n derivatives as well as its value coincide with those of the function f. Derivatives of higher order than n are all equal to zero.

Example 4.23. For the logarithmic function (cf. Example 4.22) we first compute the third-degree Taylor polynomial (see Figure 4.9), and then the Taylor

[8] Brook Taylor (1685–1731)

polynomial of general degree n at the point $a = 1$. It holds

$$f(1) = 0,\ f'(1) = 1,\ f''(1) = -1,\ f^{(3)}(1) = 2, \ldots, f^{(i)}(1) = (-1)^{i-1}(i-1)!$$

Therefore,

$$\begin{aligned} T_{f,1,3}(x) &= 0 + 1 \cdot (x-1) - 1 \cdot \frac{(x-1)^2}{2!} + 2\frac{(x-1)^3}{3!} \\ &= (x-1) - \frac{(x-1)^2}{2} + \frac{(x-1)^3}{3}, \end{aligned}$$

and for arbitrary $n \in \mathbb{N}$

$$\begin{aligned} T_{f,1,n}(x) &= 0 + \sum_{i=1}^{n} (-1)^{i-1}(i-1)! \frac{(x-1)^i}{i!} \\ &= \sum_{i=1}^{n} (-1)^{i-1} \frac{(x-1)^i}{i}. \end{aligned}$$

Now we let the degree of the Taylor polynomial go to infinity. It can be shown that the infinite series

$$\lim_{n\to\infty} \sum_{i=1}^{n} (-1)^{i-1} \frac{(x-1)^i}{i} = \sum_{i=1}^{\infty} (-1)^{i-1} \frac{(x-1)^i}{i}$$

converges for $0 < x \leq 2$ and has the value $\ln(x)$.

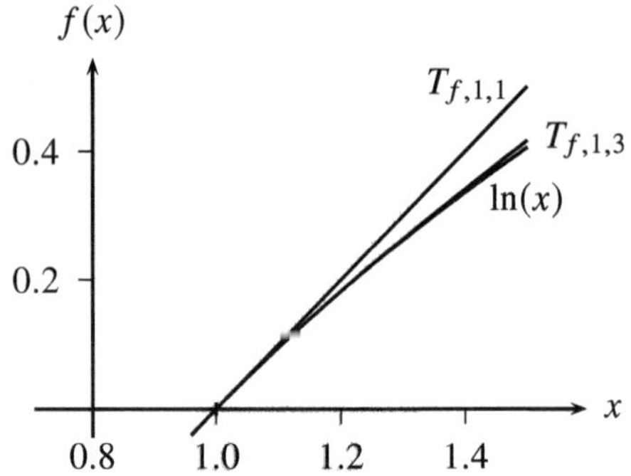

Figure 4.9: Logarithm with Taylor polynomial of first and third degree at $x = 1$.

Therefore, the **Taylor series expansion** of $\ln(x)$ at the point 1 converges for all $0 < x \leq 2$. If we substitute $x = 1 + y$, we get

$$\ln(1+y) = \sum_{i=1}^{\infty} (-1)^{i-1} \frac{y^i}{i}, \quad -1 < y \leq 1.$$

Observe that for $y > 1$ the summand does not go to null, hence the sequence diverges.

Many important functions, like the logarithmic function, have derivatives of any order and can be locally approximated by their Taylor polynomial at a point a. On some interval around a this approximation becomes arbitrarily accurate for large enough n.

4.8 L'Hôpital's rule

In this section the limit of quotients is considered, where both the numerator and the denominator approach zero. In many cases, such a limit can be determined using differential calculus. A similar situation arises when both the numerator and the denominator diverge to $+\infty$ or $-\infty$.[9]

Depending on the convergence or divergence in the numerator and denominator we distinguish three basic types,

$$\frac{0}{0}, \qquad \frac{\infty}{\infty} = \frac{-\infty}{-\infty}, \qquad \frac{-\infty}{\infty} = -\frac{\infty}{\infty} = \frac{\infty}{-\infty}.$$

The three types are also referred to as "indeterminate forms". L'Hôpital's[10] rule is used to evaluate the limits of such quotients.

Theorem 4.9: L'Hôpital's rule

Let $f, g : A \to \mathbb{R}$ be differentiable on the interval $A = (a, b)$, $a, b \in \mathbb{R} \cup \{-\infty, \infty\}$ and

$$\lim_{x \searrow a} f(x) = \lim_{x \searrow a} g(x) = 0$$

Then, if $\lim_{x \searrow a} \frac{f'(x)}{g'(x)} = c$ exists, it holds

$$\lim_{x \searrow a} \frac{f(x)}{g(x)} = c\,.$$

The same result holds at the upper endpoint b (with limits $\lim_{x \nearrow b}$),

$$\lim_{x \nearrow a} \frac{f(x)}{g(x)} = c\,.$$

These results are also true if both the numerator and the denominator diverge to ∞ or $-\infty$. Here, divergence of the quotient to $c = \infty$ or $c = -\infty$ is allowed.

In practice, proceed as follows: If taking the limits of the two functions leads to an indeterminate form of the quotient, replace the numerator and the denominator by their derivatives and compute the limit of the resulting quotient, provided it exists; otherwise you may repeat the procedure.

[9] For the definition of divergence to $+\infty$ or $-\infty$, see Definition 3.3.
[10] Guillaume François Antoine, Marquis de l'Hôpital (1661–1704)

Example 4.24. The limit

$$\lim_{x\to 1}\frac{\ln(x)}{1-x}$$

is to be calculated. Since $\lim_{x\to 1}\ln(x) = \lim_{x\to 1}(1-x) = 0$, this is an indeterminate form $\frac{0}{0}$. However, with l'Hôpital's rule we get

$$\lim_{x\to 1}\frac{\ln(x)}{1-x} = \lim_{x\to 1}\frac{\frac{1}{x}}{-1} = -1\,.$$

If necessary – as in the following example – l'Hôpital's rule may be applied more than once.

Example 4.25. It holds

$$\lim_{x\to\infty} x^2e^{-x} = \lim_{x\to\infty}\frac{x^2}{e^x} = \lim_{x\to\infty}\frac{2x}{e^x} = \lim_{x\to\infty}\frac{2}{e^x} = 0\,.$$

4.9 Mean value theorem

As we have seen, a derivative is obtained as the limit of difference quotients. A difference quotient refers to two points of the function under consideration and relates the difference of the function values to the difference of the two points at which the function is evaluated. The following theorem now gives information about the value of the difference quotient: it is equal to the value of the derivative at some point lying between the two points to which the difference quotient refers.

Theorem 4.10: Mean value theorem

Let the function $f : A \to \mathbb{R}$ be differentiable and $\alpha, \beta \in A$, $\alpha < \beta$. Then there exists $\xi \in (\alpha, \beta)$, such that

$$f'(\xi) = \frac{f(\beta) - f(\alpha)}{\beta - \alpha}\,.$$

Note that the mean value theorem guarantees *at least* one ξ, where the derivative equals the slope of the line connecting the points $(\alpha, f(\alpha))$ and $(\beta, f(\beta))$ of the function's graph. Figure 4.10 shows a function for which there are even two such points, ξ_1 and ξ_2.

A special case of the mean value theorem is **Rolle's**[11] **theorem**. If $f(\alpha) = f(\beta)$, then the difference quotient is zero and the mean value theorem guarantees at least one $\xi \in (\alpha, \beta)$, at which the derivative of f is zero. At one of these points, the function necessarily takes a minimum or a maximum. In particular, between two zeros of a differentiable function there is always an extremum.

[11] Michel Rolle (1652–1719)

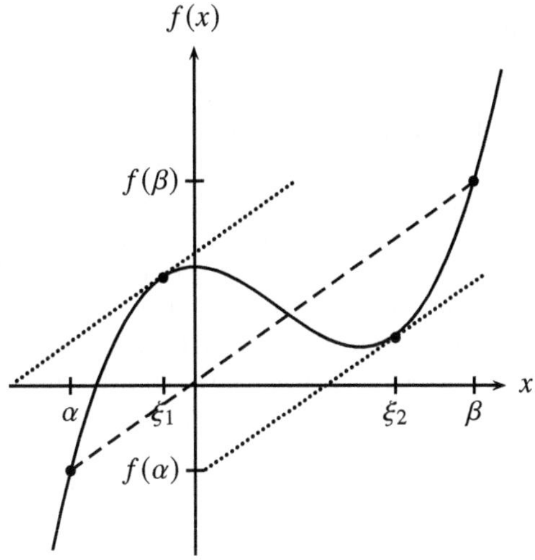

Figure 4.10: Illustration of the mean value theorem.

4.10 Numerical root-finding methods

In many applications, one must find the zeros (or roots) of a given function. This problem arises, for example, when calculating the inverse of a given function or when determining extreme values of a function. We illustrate this with two examples.

Example 4.26. The function $f : \mathbb{R} \to \mathbb{R}, x \mapsto x + e^x$, is strictly monotonically increasing and takes arbitrarily small and large values. Thus, f has an inverse f^{-1}. To calculate $f^{-1}(1)$, for example, the equation $x + e^x = 1$ must be solved for x. Since this equation cannot be explicitly solved for x, a numerical method is needed.

Example 4.27. We seek the local extrema of the function $f : \mathbb{R} \to \mathbb{R}$, $f(x) = x^2 + \sin(x)$. The first two derivatives are

$$f'(x) = 2x + \cos(x) \quad \text{and} \quad f''(x) = 2 - \sin(x)\,.$$

Since $f'(-1) < 0$ and $f'(0) > 0$, f' has a zero in the interval $[-1, 0]$. Since the second derivative is positive for all x, there is a global minimum in $[-1, 0]$. To locate it, we need to find the zero of the first derivative, i.e., the solution of the equation $2x + \cos(x) = 0$. Again, this equation is not explicitly solvable. So we need a numerical method.

In this section, we will present four methods that can be used to find the zeros of a real function.

Bisection method

The bisection method is a very simple but reliable method to determine a root of a continuous function f. Bisection can be applied when two points a_0, b_0 are known, at which f has different signs. Assume w.l.o.g. that $f(a_0) < 0$ and $f(b_0) > 0$; the reverse case can be handled analogously. According to the intermediate value theorem (Theorem 3.23), there must be at least one root in the interval (a_0, b_0). Next, we calculate the midpoint $c_0 = \frac{1}{2}(a_0 + b_0)$ of the interval. If $f(c_0) = 0$, then the method ends with the root c_0. If the function value $f(c_0) > 0$, it follows from the intermediate value theorem that there is a root in the interval (a_0, c_0). Analogously, if $f(c_0) < 0$, then there is a root in the interval (c_0, b_0). The interval (a_0, b_0) is then replaced, depending on the sign of $f(c_0)$, by (a_0, c_0) or (c_0, b_0), and the process is continued. The algorithm stops when the root has been found, or when the interval (a_n, b_n), in which the root must lie, is sufficiently small, i.e., $|a_n - b_n| < \varepsilon$ for a given "small" $\varepsilon > 0$.

Algorithm 4.1: Bisection method

The function $f : D \to \mathbb{R}$ is continuous, and for $a, b \in D$ with $a < b$, we have $f(a) < 0$ and $f(b) > 0$.

Start: Set $a_0 = a$ and $b_0 = b$.

Iteration: For $n = 0, 1, 2, \ldots$ perform the following steps:

1) Set $c_n = \frac{1}{2}(a_n + b_n)$.
2) If $|a_n - b_n| < \varepsilon$, stop the iteration.
3) If $f(c_n) > 0$, set $a_{n+1} = a_n$, $b_{n+1} = c_n$.
4) If $f(c_n) < 0$, set $a_{n+1} = c_n$, $b_{n+1} = b_n$.
5) If $f(c_n) = 0$, stop the iteration.

Output the solution c_n.

The main advantage of the bisection method is that – unlike other methods – a root is guaranteed to be found or approximated. The approximation error can be tightly estimated, since the distance from c_n to the root is always less than $\frac{1}{2}\varepsilon$. A drawback, however, is that the method converges relatively slowly.

Example 4.28. We consider the function $f(x) = x^2 - 2$. In this case, the roots $x = \pm\sqrt{2}$ can be found directly, so this example serves purely for illustration. As initial values, we choose $a_0 = 0$ with $f(a_0) = -2$ and $b_0 = 4$ with $f(b_0) = 14$. The steps of the bisection method are shown in the following table. The root is determined to eight decimal places; that is, $\frac{\varepsilon}{2} = 10^{-8}$, i.e., $\varepsilon = 2 \cdot 10^{-8}$, is chosen.

n	a_n	b_n	c_n	$f(c_n)$
0	0	4	2	2
1	0	2	1	−1
2	1	2	1.5	0.25
3	1	1.5	1.25	−0.4375
4	1.25	1.5	1.375	−0.109375
5	1.375	1.5	1.4375	0.06640625
6	1.375	1.4375	1.40625	−0.02246094
7	1.40625	1.4375	1.421875	0.02172852
8	1.40625	1.421875	1.4140625	−0.00042725
9	1.4140625	1.421875	1.41796875	0.01063538
10	1.4140625	1.41796875	1.41601563	0.00510025
11	1.4140625	1.41601563	1.41503906	0.00233555
12	1.4140625	1.41503906	1.41455078	0.00095391
13	1.4140625	1.41455078	1.41430664	0.00026327
14	1.4140625	1.41430664	1.41418457	−0.00008200
15	1.41418457	1.41430664	1.41424561	0.00009063
16	1.41418457	1.41424561	1.41421509	0.00000431
17	1.41418457	1.41421509	1.41419983	−0.00003884
18	1.41419983	1.41421509	1.41420746	−0.00001726
19	1.41420746	1.41421509	1.41421127	−0.00000647
20	1.41421127	1.41421509	1.41421318	−0.00000108
21	1.41421318	1.41421509	1.41421413	0.00000162
22	1.41421318	1.41421413	1.41421366	0.00000027
23	1.41421318	1.41421366	1.41421342	−0.00000041
24	1.41421342	1.41421366	1.41421354	−0.00000007
25	1.41421354	1.41421366	1.41421360	0.00000010
26	1.41421354	1.41421360	1.41421357	0.00000002
27	1.41421354	1.41421357	1.41421355	−0.00000003
28	1.41421355	1.41421357	1.41421356	−0.00000001
29	1.41421356	1.41421357	1.41421356	0.00000001

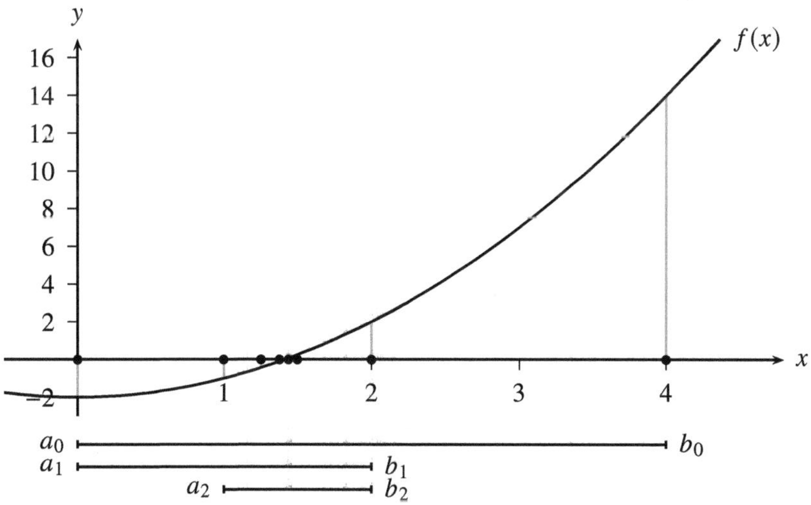

Regula falsi[12]

The regula falsi proceeds similarly to the bisection. Instead of the midpoint of the interval, however, the intersection of the secant through $(a_n, f(a_n))$ and $(b_n, f(b_n))$ with the x-axis is used as the new approximation value c_n. The equation of the secant is

$$y = f(a_n) + \frac{f(b_n) - f(a_n)}{b_n - a_n}(x - a_n)\,.$$

Substituting $y = 0$ and solving for x gives the intersection point c_n of the secant with the x-axis:

$$c_n = a_n - f(a_n)\frac{b_n - a_n}{f(b_n) - f(a_n)}\,.$$

As with the bisection method, the new interval $[a_{n+1}, b_{n+1}]$ is chosen so that it contains a root of the function. This results in the following procedure.

Algorithm 4.2: Regula falsi

The function $f : D \to \mathbb{R}$ is continuous, and for $a, b \in D$ with $a < b$, we have $f(a) < 0$ and $f(b) > 0$.

Start: Set $a_0 = a$ and $b_0 = b$.

Iteration: For $n = 0, 1, 2, \ldots$ perform the following steps:

1) Set $c_n = a_n - f(a_n)\frac{b_n - a_n}{f(b_n) - f(a_n)}$.
2) If $|c_n - c_{n-1}| < \varepsilon$, end the iteration.
3) If $f(c_n) > 0$, set $a_{n+1} = a_n$, $b_{n+1} = c_n$.
4) If $f(c_n) < 0$, set $a_{n+1} = c_n$, $b_{n+1} = b_n$.
5) If $f(c_n) = 0$, stop the iteration.

Output the solution c_n.

The regula falsi has the same advantages as the bisection method. The convergence is somewhat faster than the bisection method, but usually significantly worse than the convergence of the Newton and secant methods, which we will discuss below.

Example 4.29. We again consider the function $f(x) = x^2 - 2$ and determine its root using the regula falsi. The steps of the regula falsi (with $\varepsilon = 2 \cdot 10^{-8}$) are shown in the following table.

n	a_n	b_n	c_n	$f(c_n)$
0	0	4	0.5	-1.75
1	0.5	4	0.88888889	-1.20987654
2	0.88888889	4	1.13636364	-0.70867769

[12] Giuseppe Falsi (1762–1698)

3	1.13636364	4	1.27433628	−0.37606704
4	1.27433628	4	1.34563758	−0.18925949
5	1.34563758	4	1.38104206	−0.09272283
6	1.38104206	4	1.39827345	−0.04483136
7	1.39827345	4	1.40657821	−0.02153774
8	1.40657821	4	1.41056183	−0.01031533
9	1.41056183	4	1.41246834	−0.00493318
10	1.41246834	4	1.41337979	−0.00235757
11	1.41337979	4	1.41381530	−0.00112630
12	1.41381530	4	1.41402334	−0.00053799
13	1.41402334	4	1.41412271	−0.00025696
14	1.41412271	4	1.41417017	−0.00012273
15	1.41417017	4	1.41419284	−0.00005861
16	1.41419284	4	1.41420367	−0.00002799
17	1.41420367	4	1.41420884	−0.00001337
18	1.41420884	4	1.41421130	−0.00000639
19	1.41421130	4	1.41421248	−0.00000305
20	1.41421248	4	1.41421305	−0.00000146
21	1.41421305	4	1.41421332	−0.00000070
22	1.41421332	4	1.41421344	−0.00000033
23	1.41421344	4	1.41421351	−0.00000016
24	1.41421351	4	1.41421354	−0.00000008
25	1.41421354	4	1.41421355	−0.00000004

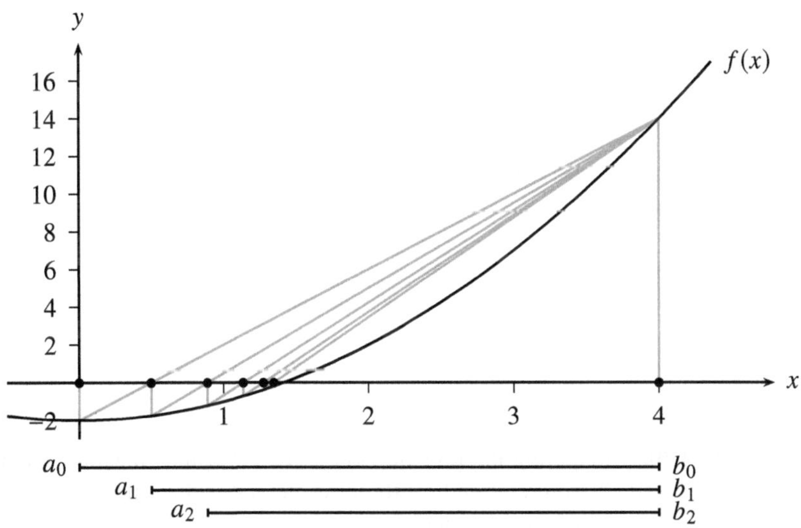

Newton's[13] method

Another method for numerical root finding is the so-called **Newton's method**. The idea behind Newton's method is to approximate the function f near a current estimate x_n by its tangent function t_{f,x_n} and to use the root of this tangent as the next approximation. Starting from an initial guess, this process is repeated: the root of the

[13] Sir Isaac Newton (1643–1727)

tangent line becomes the new (and hopefully improved) approximation. If a suitable stopping criterion is met, the iteration terminates and the last approximation is taken as the solution.

The equation of the tangent to the function f at x_n is

$$t_{f,x_n}(x) = f(x_n) + f'(x_n)(x - x_n)\,.$$

Solving the equation $t_{f,x_n}(x) \stackrel{!}{=} 0$ for x yields

$$x = x_n - \frac{f(x_n)}{f'(x_n)}\,.$$

Newton's method is outlined below. The stopping criterion used is that the distance between two successive approximations is less than a given value ε.

Algorithm 4.3: Newton's method

The function $f : D \to \mathbb{R}$ is differentiable.

Start: Choose $x_1 \in D$ arbitrarily.

Iteration: For $n = 0, 1, 2, \ldots$ perform the following steps:

1) Set $x_{n+1} = x_n - \frac{f(x_n)}{f'(x_n)}$.
2) If $|x_{n+1} - x_n| < \varepsilon$, stop the iteration.

Output the solution x_{n+1}.

The main advantage of Newton's method is its typically very rapid convergence. However, the method does not always converge; there is no guarantee that an existing root will actually be found. Moreover, Newton's method requires that f is differentiable and that its derivative can be computed.

Example 4.30. Computing the root of $f(x) = x^2 - 2$ using Newton's method (with $\varepsilon = 2 \cdot 10^{-8}$) requires significantly fewer iterations than the previous two methods. The details of the computation are presented in the following table.

n	x_n	$f(x_n)$	$f'(x_n)$	$-\frac{f(x_n)}{f'(x_n)}$
0	4	14	8	−1.75
1	2.25	3.0625	4.5	−0.68055556
2	1.56944444	0.46315586	3.13888889	−0.14755408
3	1.42189036	0.02177221	2.84378073	−0.00765608
4	1.41423429	0.00005862	2.82846857	−0.00002072
5	1.41421356	0.00000000	2.82842713	−0.00000000
6	1.41421356	0.00000000	2.82842712	−0.00000000

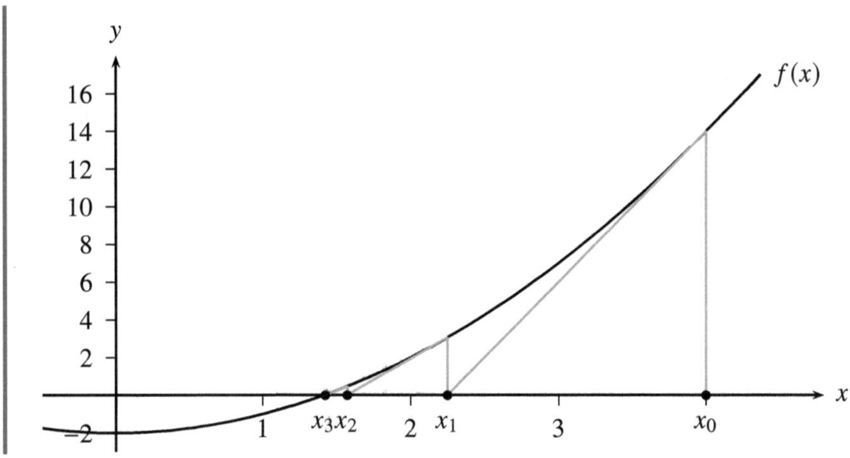

Newton's method can also be used for the numerical calculation of square roots. For this, one considers the function $f(x) = x^2 - a$, whose roots are $\pm\sqrt{a}$, and applies Newton's method. The iteration rule is given by

$$x_{n+1} = x_n - \frac{f(x_n)}{f'(x_n)} = x_n - \frac{x_n^2 - a}{2x_n} = \frac{1}{2}\left(x_n + \frac{a}{x_n}\right).$$

It can be shown that the sequence (x_n) converges to the square root of a for any initial value $x_0 > 0$. The convergence is very rapid, yielding highly accurate values after just a few iterations. This method is also known as **Heron's method**[14] or **Babylonian method**.

As already mentioned above, there is no guarantee that Newton's method will converge. For example, applying Newton's method to the function $f(x) = x^3 - 5x$, starting from the initial value $x_0 = 1$, produces the alternating sequence $x_n = (-1)^n$; see Figure 4.11.

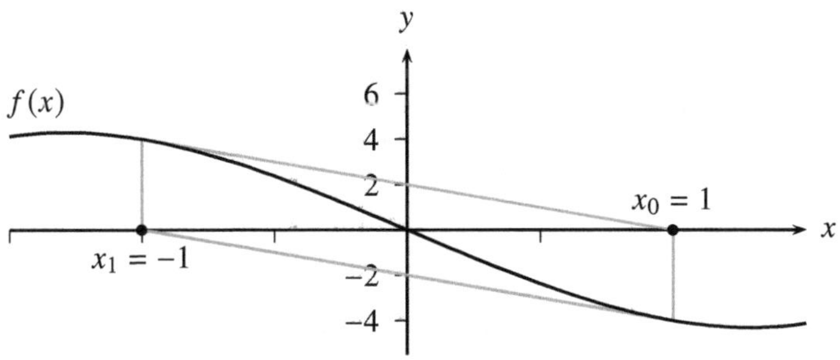

Figure 4.11: No convergence in Newton's method.

[14] Heron of Alexandria (1st century AD)

Secant method

The secant method can be seen as a modification of Newton's method. It has the advantage that the derivative of f does not need to be calculated. Instead of the tangent of f at x_n, the secant method uses the secant defined by two successive approximations x_{n-1} and x_n through the points $(x_{n-1}, f(x_{n-1}))$ and $(x_n, f(x_n))$. The new approximation value x_{n+1} is then obtained according to

$$x_{n+1} = x_{n-1} - f(x_{n-1})\frac{x_n - x_{n-1}}{f(x_n) - f(x_{n-1})}.$$

This results in the following algorithm:

Algorithm 4.4: Secant method

The function $f : D \to \mathbb{R}$ is continuous.

Start: Choose initial values $x_0, x_1 \in D$ arbitrarily.

Iteration: For $n = 1, 2, \ldots$ perform the following steps:

1) Set $x_{n+1} = x_{n-1} - f(x_{n-1})\frac{x_n - x_{n-1}}{f(x_n) - f(x_{n-1})}$.
2) If $|x_{n+1} - x_n| < \varepsilon$, end the iteration.

Output the solution x_{n+1}.

The convergence of the secant method is somewhat slower than that of Newton's method, but in most cases it is significantly faster than that of the regula falsi or bisection method. Unlike Newton's method, the secant method does not require the computation of derivatives. However, with an unfortunate choice of initial values, the secant method may also fail to converge.

Example 4.31. When applied to the function $f(x) = x^2 - 2$ with $\varepsilon = 2 \cdot 10^{-8}$, the secant method exhibits convergence behavior similar to that of Newton's method. The steps of the method are once again illustrated in the following table.

n	x_{n-1}	x_n	$f(x_{n-1})$	$f(x_n)$
1	0	4	−2	14
2	4	0.5	14	−1.75
3	0.5	0.88888889	−1.75	−1.20987654
4	0.88888889	1.76000000	−1.20987654	1.09760000
5	1.76000000	1.34563758	1.09760000	−0.18925949
6	1.34563758	1.40657821	−0.18925949	−0.02153774
7	1.40657821	1.41440381	−0.02153774	0.00053814
8	1.41440381	1.41421305	0.00053814	−0.00000146
9	1.41421305	1.41421356	−0.00000146	0.00000000
10	1.41421356	1.41421356	0.00000000	0.00000000

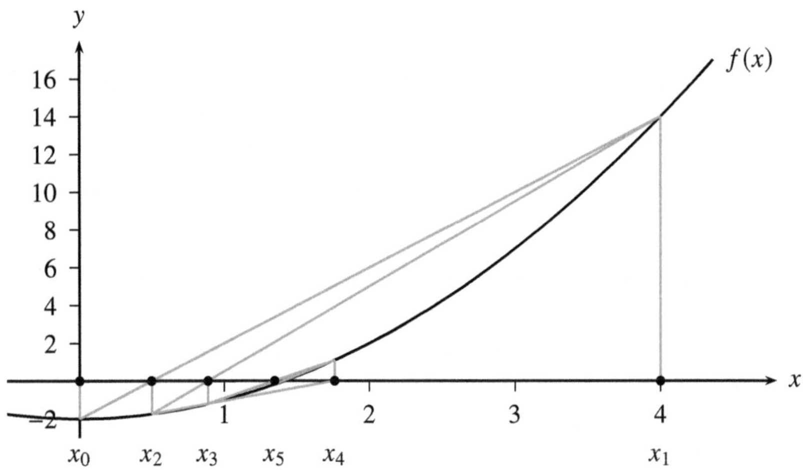

Important terms for review

After reading this chapter, the following terms should be familiar:

- Difference quotient and differential quotient
- Differentiability
- First and second derivative
- Tangent function
- Marginal change
- Differential
- Elasticity
- Differentiation rules: sum, difference, factor, product, quotient, and chain rule
- Concave upward, concave downward
- Derivative of the inverse function
- Zeros, local and global maximum and minimum, inflection and saddle point
- Curve sketching
- Necessary and sufficient condition for a local extremum
- Monotonicity, convexity and concavity of differentiable functions
- Higher order derivative
- Taylor polynomial
- L'Hôpital's rule
- Mean value theorem
- Bisection method, regula falsi, Newton's method, secant method

Self-test

You can check your understanding of this chapter using the following true-or-false questions. For each statement, indicate whether it is true (T) or false (F). Short solutions to these questions can be found in Appendix H.

I. The differential $df_a(dx) = f'(a)\,dx$ of the function $f(x)$ at $x = a$ represents the ...

	T	F
... approximate relative change in the function value when x increases by dx units starting from $x = a$.	☐	☐
... exact absolute change in the function value when x increases by dx units starting from $x = a$.	☐	☐
... approximate absolute change in the function value when x increases by dx percent starting from $x = a$.	☐	☐

II. The elasticity of a function $f : (0, \infty) \to \mathbb{R}$ ...

	T	F
... is constant for the function $f(x) = x^\beta$, $x > 0$, with $\beta \in \mathbb{R}$.	☐	☐
... represents the approximate percentage change in the function value when the argument is increased by one percent.	☐	☐
... is positive if the function is monotonically increasing.	☐	☐

III. Let $f : A \to \mathbb{R}$ be twice differentiable and $a \in A$ in the interior of the domain.

	T	F
If $f'(a) = 0$, then f has a minimum or a maximum at a.	☐	☐
If $f'(a) \neq 0$, then f has no extremum at a.	☐	☐
If $f'(a) = 0$ and $f''(a) = 0$, then f has no extremum at a.	☐	☐

IV. Are the following statements true or false?

	T	F
If a function f is convex and has a local minimum at $x = a$, then this is also a global minimum.	☐	☐
If a function f is concave and has a local minimum at $x = a$, then this is also a global minimum.	☐	☐
The second-degree Taylor polynomial of the function $f(x) = x^2 + 42$ at any point a in its domain exactly coincides with f.	☐	☐
Newton's method for finding roots can be applied to any function of a single variable.	☐	☐
If a differentiable function has exactly one root, Newton's method will converge to it regardless of the initial value.	☐	☐

Exercises

Exercise 1
Find the derivatives of the following functions:

a) $f_1(x) = 4x^2 - 7x + \frac{1}{x}$,

b) $f_2(x) = 5x^4 - 3x + 7$,

c) $f_3(x) = x(x-3)(x+3)(x-1)$,

d) $f_4(x) = x^{700} - x^{-5000} + 1$,

e) $f_5(x) = (x-4)^2(x+4)^2$.

Exercise 2
Find the derivatives of the following functions:

a) $g_1(x) = \ln(e^x)$,

b) $g_2(x) = \ln\big((x+7)^2\big)$,

c) $g_3(x) = \frac{1}{(x+5)^4}$,

d) $g_4(x) = \ln(e^x + 2)$,

e) $g_5(x) = \frac{1}{(e^{(x-\ln(x))}+2)^2}$.

Exercise 3
Find the derivatives of the following functions:

a) $h_1(x) = -4xe^{-\frac{x}{2}}$,

b) $h_2(x) = -e^x \ln(x-2)$,

c) $h_3(x) = \frac{e^x}{\ln(x)}$,

d) $h_4(x) = \frac{x^2 - \ln(e^x)}{-x^5}$.

Exercise 4
Let f be a twice differentiable function. Find the first and second derivatives (wherever they exist) of the following functions:

a) $g_1(x) = e^{f(x)}$,

b) $g_2(x) = \sqrt[3]{f(x)+7}$,

c) $g_3(x) = \frac{1+f(x)}{1-f(x)}$,

d) $g_4(x) = \big(\pi f(x) - \sqrt{f(x)} + \ln\big(f(x)\big) + e\big)^0$.

Exercise 5

The profit function of a small sparkling wine producer in a given period depends on the quantity sold x (in hectoliters). The exact functional relationship is unknown. However, the cellar master knows the following:

- If 1000 hectoliters or more are sold, then the company no longer incurs a loss, i.e., the break-even point is $x = 1000$.
- The fixed costs amount to 1 million €.
- For selling more than 4000 hectoliters, the costs of procuring sufficient raw materials become so high that no profit is made, i.e., profit is zero at $x = 4000$.

Model the profit function using a quadratic function and determine the quantity at which profit is maximized.

Exercise 6

Perform a curve sketching analysis (zeros, extrema, inflection points) for the following functions, and then draw the graphs:

a) $f_1(x) = x^3 - 3x^2 - x + 3$,

b) $f_2(x) = \frac{1}{e^{|x|}} - 0.5$. Hint: First consider $e^{|x|}$.

Exercise 7

Find all zeros of the following functions:

a) $f_1(x) = x^5 - 5x^3 + 4x$,

b) $f_2(x) = |x^3 - x|$,

c) $f_3(x) = e^{x^4+5x^2+7x} + \ln(2)$,

d) $f_4(x) = \ln(1)$,

e) $f_5(x) = \begin{cases} x^3 - x, & \text{if } x \leq 0, \\ -x, & \text{if } x > 0. \end{cases}$

Exercise 8

Find the derivatives of the following functions:

a) $f(x_1) = 4bcx_1^3 + b^2c^2x_1 + \frac{b\ln(c)}{d}$, $x_1 \in \mathbb{R}$,

b) $f(x_2) = 4a^3cx_2 + ac^2x_2^2 + \frac{x_2\ln(c)}{d}$, $x_2 \in \mathbb{R}$,

c) $f(x_3) = 4a^3bx_3 + ab^2x_3^2 + \frac{b\ln(x_3)}{d}$, $x_3 > 0$,

d) $f(x_4) = 4a^3bc + ab^2c^2 + \frac{b\ln(c)}{x_4}$, $x_4 \neq 0$,

where $a, b \in \mathbb{R}$, $c > 0$, $d \neq 0$.

Exercise 9

Consider differentiable functions f, g, h, where $g(x) > 0$ for all $x \in \mathbb{R}$.

a) Find the derivative of $k(x) = \frac{f(g(h(x)))}{g(f(x))}$.

b) Now let $f(x) = x^2 - 4$, $g(x) = e^x$, $h(x) = \sqrt{x}$. Find $k(x)$ and $k'(x)$.

Exercise 10

Find the derivatives of the following functions:

a) $f_1 : \mathbb{R} \to \mathbb{R}, f_1(x) = \dfrac{x^5}{5} + 4x^2 - 12,$

b) $f_2 : \mathbb{R} \to \mathbb{R}, f_2(x) = (x^2 - 1)^{10}(2 - x^2)^6,$

c) $f_3 : \mathbb{R} \to \mathbb{R}, f_3(x) = \dfrac{2x}{x^2 + 1},$

d) $f_4 : \mathbb{R} \to \mathbb{R}, f_4(x) = \cos^3(\sin(x)),$

e) $f_5 : \mathbb{R} \to \mathbb{R}, f_5(x) = \dfrac{e^{x+1} - 1}{e^{x+1} + 1},$

f) $f_6 : \mathbb{R}_+ \backslash \{0\} \to \mathbb{R}_+ \backslash \{0\}, f_6(x) = x^x.$

Exercise 11

Find the elasticities of the following functions:

a) $f_1(x) = 3x^{17},$

b) $f_2(x) = 3e^{5x},$

c) $f_3(x) = 3x^{17}e^{5x},$

d) $f_4(x) = \ln(x),$

e) $f_5(x) = \frac{1}{x^2},$

f) $f_6(x) = \frac{1}{x^2+5}.$

Exercise 9

Chapter 5

Differentiable functions of several variables

Most functions that appear in economic models, such as production functions and utility functions, are functions of several variables. In this chapter, we will discuss the differential calculus of real-valued functions of n variables, $n \geq 1$.

Example 5.1 (Production). A product is made from several raw materials or intermediate goods. The same questions arise as in Chapter 4: By how much does the output change – either in absolute or percentage terms – when one or more inputs are increased by a small unit or a small percentage?

In addition, there is the question of input substitution: By how much must one input be increased in order to reduce another input by one unit while keeping the output constant?

Finally, it is also of interest how the output changes if all inputs are varied by the same factor, for example, if they are all doubled.

These and other questions will be explored in this chapter. To answer them, we need some important mathematical tools: the concepts of partial and total derivatives of functions with several variables, the differential and elasticities, the degree of homogeneity of a function, and some useful rules for differentiating vector-valued and implicit functions.

K. Mosler et al., *Mathematical Methods for Business and Economics*, Mathematics Study Resources 20, https://doi.org/10.1007/978-3-662-73831-3_5

5.1 Derivative, differential, elasticities

In this chapter we consider real-valued functions $f(x_1, x_2, \ldots, x_n)$, where each variable $x_1, x_2, \ldots, x_n$ varies over some (possibly unbounded) interval. In other words: we study functions $f : A \to \mathbb{R}$, where A is an n-dimensional interval in $\mathbb{R}^n$. We begin with the bivariate case $n = 2$.

Example 5.2. As in Example 1.3, let

$$f: \begin{array}{ccc} \mathbb{R}_+^2 & \to & \mathbb{R}, \\ (x_1, x_2) & \mapsto & x_1\sqrt{x_2}. \end{array}$$

Now fix the first argument at a specific value, say $x_1 = 2$. This gives the "partial" function $x_2 \mapsto 2\sqrt{x_2}$, which has derivative $1/\sqrt{x_2}$. Figure 5.1 illustrates this example. The left panel shows the graph of the function $f(x_1, x_2) = x_1\sqrt{x_2}$, while the right panel shows the partial function $x_2 \mapsto f(2, x_2)$ together with its tangent at $x_2 = 1$. The slope of this tangent is the partial derivative of f with respect to x_2 at the point $(2, 1)$. The partial function is also sketched in the left panel.

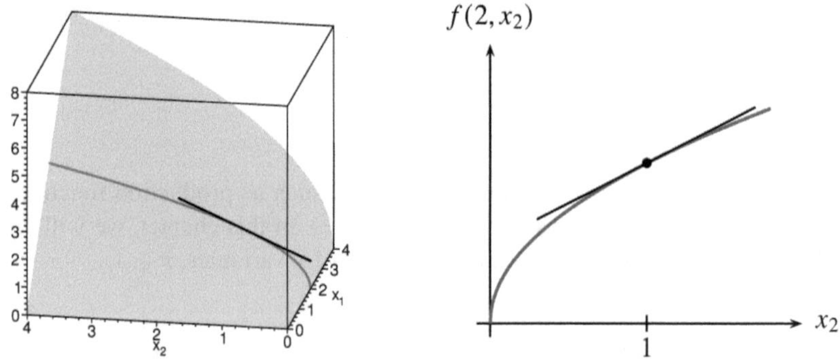

Figure 5.1: Graphs of $f(x_1, x_2) = x_1\sqrt{x_2}$ and partial function at $x_1 = 2$ with tangent.

In general, for a fixed value $x_1 = a_1$, we consider the univariate function $x_2 \mapsto f(a_1, x_2)$. If this function is differentiable at a point a_2, then its derivative is called the **partial derivative** of f with respect to x_2 at the point (a_1, a_2). We write this as

$$f_{x_2}(a_1, a_2) = \frac{\partial f}{\partial x_2}(a_1, a_2).$$

The symbol $\partial f/\partial x_2$ (as opposed to df/dx) indicates that f is differentiated only partially, namely with respect to the second component of x. By swapping the roles of x_1 and x_2, we get the partial derivative of f with respect to x_1 at the point (a_1, a_2),

$$f_{x_1}(a_1, a_2) = \frac{\partial f}{\partial x_1}(a_1, a_2).$$

In the above Example 5.2, for all $a_1, a_2 > 0$ we have

$$\frac{\partial f}{\partial x_1}(a_1, a_2) = \sqrt{a_2}, \quad \frac{\partial f}{\partial x_2}(a_1, a_2) = \frac{a_1}{2\sqrt{a_2}}.$$

The partial derivative of f with respect to x_1 at (a_1, a_2) indicates, to first order (see Section 4.1), how much the function value $y = f(a_1, a_2)$ changes when we increase the first variable x_1 by one unit while keeping the second variable $x_2 = a_2$ fixed. The same holds for the partial derivative of f with respect to x_2. The row vector of all partial derivatives is called the **gradient** or **derivative** of f at (a_1, a_2),

$$\operatorname{grad} f(a_1, a_2) = \left[\tfrac{\partial f}{\partial x_1}(a_1, a_2), \ \tfrac{\partial f}{\partial x_2}(a_1, a_2) \right].$$

For each of the variables x_i, $i = 1, 2$, the following holds. If we increase x_i by a small amount dx_i to $x_i + dx_i$, then y changes, to first order, by

$$\frac{\partial f}{\partial x_i}(a_1, a_2) \cdot dx_i .$$

This expression is called the **partial differential** with respect to the i-th variable. It approximately indicates the change in the function value when the i-th variable increases by dx_i while the other variables remain unchanged. Adding the partial differentials gives

$$\frac{\partial f}{\partial x_1}(a_1, a_2) \cdot dx_1 + \frac{\partial f}{\partial x_2}(a_1, a_2) \cdot dx_2 .$$

This is called the **total differential** of f at (a_1, a_2). It approximately quantifies the change in the function value when x_1 changes by dx_1 and, simultaneously, x_2 changes by dx_2.

If f has partial derivatives with respect to both x_1 and x_2 at (a_1, a_2) and if these partial derivatives depend continuously on (a_1, a_2), then we say that f is **totally differentiable** or simply **differentiable** at (a_1, a_2). Intuitively, this means that when (a_1, a_2) is shifted slightly in any direction, the gradient $\operatorname{grad}(a_1, a_2)$ – and thus the differential – changes smoothly without any "sudden jumps".

Derivative and differential for functions of n variables For a function in n variables, the concepts of partial derivative, total derivative, and gradient are defined in exactly the same way:

Definition 5.1: Partial derivatives

For given $\boldsymbol{a} = (a_1, \ldots, a_n) \in A$, consider the function $x_1 \mapsto f(x_1, a_2, \ldots, a_n)$. If this univariate function is differentiable at a_1, then its derivative is called the **partial derivative** of f at $\boldsymbol{a}$ with respect to x_1, written

$$f_{x_1}(\boldsymbol{a}) = \frac{\partial f}{\partial x_1}(\boldsymbol{a}) .$$

For $j = 2, 3, \ldots, n$, the partial derivative of f at $\boldsymbol{a}$ with respect to x_j is defined analogously. It is denoted by

$$f_{x_j}(a) = \frac{\partial f}{\partial x_j}(a).$$

If all partial derivatives exist at a, then f is **partially differentiable** at a. The function f is called **partially differentiable**, if it is partially differentiable at every $x \in A$.

The partial derivative of f with respect to x_1 at $(a_1, a_2, \ldots, a_n)$ thus indicates, to first order, how much the function value changes when a_1 is varied by a small amount while $a_2, \ldots, a_n$ remain unchanged.

Definition 5.2: Total differentiability, gradient

If f has partial derivatives with respect to all n variables at a and these depend continuously on a, then f is called **totally differentiable** at a. Often, one simply says: f is **differentiable** at a. In this case, the row vector

$$\operatorname{grad} f(a) = \left[\frac{\partial f(a)}{\partial x_1}, \ldots, \frac{\partial f(a)}{\partial x_n} \right]$$

is called the **gradient** or **derivative** of f at a. As with univariate functions, this derivative is notated in various ways:

$$f'(a) = \frac{df}{dx}(a) = \operatorname{grad} f(a).$$

f is called **totally differentiable** if it is differentiable at every $x \in A$.[1]

In what follows, by a **differentiable function** we always mean a totally differentiable function. The next theorem states that a differentiable function cannot have any "jumps": every differentiable function is continuous.

Theorem 5.1: Differentiability and continuity

Let $a \in A \subset \mathbb{R}^n$, $f : A \to \mathbb{R}$. Then,

$$f \text{ is differentiable at } a \quad \Longrightarrow \quad f \text{ is continuous at } a.$$

The statement of the theorem can also be formulated as follows: If f is differentiable at a and p is any unit vector, then the univariate function

$$t \mapsto f(a + tp), \quad \mathbb{R} \to \mathbb{R},$$

[1] In mathematical texts, the concept of total differentiability is somewhat broader. For $n = 1$, the above definition coincides with the notion of continuous differentiability for univariate functions from Section 4.3.

is continuous at $t = 0$. This means that a small step from $\boldsymbol{a}$ **in any direction** $\boldsymbol{p}$ changes the value of f only slightly.

Example 5.3. The partial derivatives of the function

$$f(\boldsymbol{x}) = f(x_1, x_2, x_3) = x_1 x_2^2 + \exp(5x_3), \quad \boldsymbol{x} \in \mathbb{R}^3,$$

at $\boldsymbol{x} = (x_1, x_2, x_3)$ are

$$\frac{\partial f}{\partial x_1}(\boldsymbol{x}) = x_2^2, \quad \frac{\partial f}{\partial x_2}(\boldsymbol{x}) = 2x_1 x_2, \quad \frac{\partial f}{\partial x_3}(\boldsymbol{x}) = 5\exp(5x_3).$$

They are continuous, so the function f is (totally) differentiable. The gradient is $\operatorname{grad} f(x_1, x_2, x_3) = [x_2^2,\ 2x_1x_2,\ 5\exp(5x_3)]$.

Definition 5.3: Total differential

If f is differentiable at $\boldsymbol{a}$, the linear function

$$df_{\boldsymbol{a}}(\boldsymbol{z}) = \operatorname{grad} f(\boldsymbol{a})\, \boldsymbol{z} = \sum_{i-1}^{n} \frac{\partial f}{\partial x_i}(\boldsymbol{a})\, z_i, \quad \boldsymbol{z} \in \mathbb{R}^n,$$

is called the **differential function** of f at $\boldsymbol{a}$. In particular, for $\boldsymbol{z} = d\boldsymbol{x} = [dx_1, \ldots, dx_n]^T$, the expression

$$df_{\boldsymbol{a}}(d\boldsymbol{x}) = \operatorname{grad} f(\boldsymbol{a})\, d\boldsymbol{x} = \sum_{j=1}^{n} \frac{\partial f(\boldsymbol{a})}{\partial x_j}\, dx_j$$

is called the **total differential** of f at $\boldsymbol{a}$.

The total differential indicates approximately (to first order) how much the function value of f changes when the variables $x_1, \ldots, x_n$ are changed by $dx_1, \ldots, dx_n$.

Usually, we omit the position $\boldsymbol{a}$ and simply write the total differential as

$$df = \sum_{i=1}^{n} \frac{\partial f}{\partial x_i}\, dx_i = \frac{\partial f}{\partial x_1}\, dx_1 + \cdots + \frac{\partial f}{\partial x_n}\, dx_n.$$

For $n = 2$, Figure 5.2 illustrates the approximate change in the function value of f. Starting from the point (x_1, x_2), the tangent function increases by $\frac{\partial f}{\partial x_1} dx_1$ in the x_1-direction when x_1 increases by dx_1, and by $\frac{\partial f}{\partial x_2} dx_2$ in the x_2-direction when x_2 increases by dx_2. In total, the tangent function[2] increases by $\frac{\partial f}{\partial x_1} dx_1 + \frac{\partial f}{\partial x_2} dx_2$ units.

[2]The multivariate tangent function is formally defined on page 165. For functions of two variables, its graph is called the tangent plane.

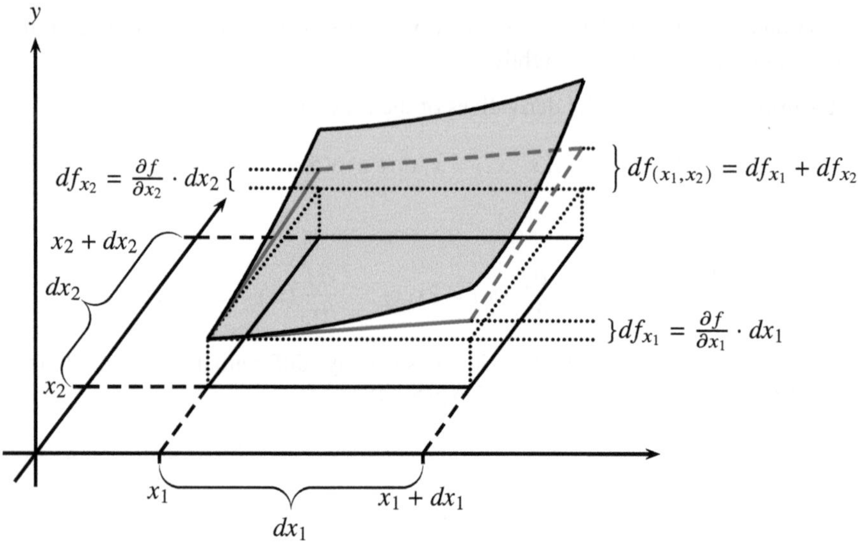

Figure 5.2: The approximate change in the function value of f: the total differential.

Example 5.4. Consider the function from Example 5.3. At the point $\boldsymbol{a} = (2, -1, 0)$, we have $f(\boldsymbol{a}) = 3$ and the differential

$$df_{(2,-1,0)}(dx_1, dx_2, dx_3) = dx_1 - 4\,dx_2 + 5\,dx_3\,.$$

If all three variables increase by 0.1 units, the function value increases approximately by

$$df_{(2,-1,0)}(0.1, 0.1, 0.1) = 0.1 - 0.4 + 0.5 = 0.2$$

units, from 3 to about 3.2.

If the first variable increases by 0.2 while the second and third decrease by 0.1, the approximate change is

$$df_{(2,-1,0)}(0.2, -0.1, -0.1) = 0.2 + 0.4 - 0.5 = 0.1$$

units, from 3 to 3.1.

For this simple function, the exact values can also be computed directly: $f(2.1, -0.9, 0.1) = 3.3497$ and $f(2.2, -1.1, -0.1) = 3.2685$. Thus, the approximations differ from the exact values by -0.1497 and -0.1685, respectively.

Elasticities So far, we have considered *absolute changes* in a function value y when one or more variables change by a certain amount. However, we are often more interested in *relative changes*. For example: By how many percent does y increase or decrease when the variables change by a certain percentage? Just as the absolute change in y is approximated by the partial derivatives, the relative change is determined by the **partial elasticities** of the function:[3]

Definition 5.4: Partial elasticities

Let the function $y = f(\boldsymbol{x})$ have partial derivatives $\frac{\partial f}{\partial x_1}(\boldsymbol{x}), \ldots, \frac{\partial f}{\partial x_n}(\boldsymbol{x})$ at $\boldsymbol{x} = (x_1, \ldots, x_n)$. Then

$$\varepsilon_{f,x_j}(\boldsymbol{x}) = \frac{x_j}{f(\boldsymbol{x})} \frac{\partial f(\boldsymbol{x})}{\partial x_j} = \frac{x_j}{y} \frac{\partial y}{\partial x_j}$$

is called the **partial elasticity** of f with respect to x_j at $\boldsymbol{x}$.

More intuitively, the partial elasticity is written as

$$\boxed{\varepsilon_{f,x_j}(\boldsymbol{x}) = \frac{\frac{\partial f(\boldsymbol{x})}{f(\boldsymbol{x})}}{\frac{\partial x_j}{x_j}}},$$

the relative change of the function value over that of the j-th argument. It thus represents the *approximate percentaged change* in the function value when the argument x_j increases by 1%. Simply put: a 1 % increase in x_j causes (approximately) an ε_{f,x_j} % change in the function value. See also Example 4.7 for the elasticity of univariate functions.

Partial elasticities play a major role in economic theory and empirics. Since they relate relative changes in the function value to relative changes in an argument, they have the key advantage of being *dimensionless*, unlike derivatives. They do not depend on the units chosen for the function value or the arguments. This is illustrated by two examples from micro- resp. macroeconomics.

Example 5.5 (Cobb-Douglas production function). The production of an output depending on two inputs is described by the bivariate Cobb-Douglas function (see Example 1.21, $n = 2$)

$$f(x_1, x_2) = c\, x_1^{\alpha} x_2^{\beta}, \quad x_1, x_2 \in \mathbb{R}_+, \ \alpha > 0, \ \beta > 0\,.$$

The partial derivatives of a production function are called **marginal productivities**. They indicate by how many units the output approximately increases when an input factor is increased by one unit. The Cobb-Douglas function has the marginal productivities

$$\frac{\partial f}{\partial x_1}(x_1, x_2) = \alpha\, c\, x_1^{\alpha-1} x_2^{\beta} \quad \text{and} \quad \frac{\partial f}{\partial x_2}(x_1, x_2) = \beta\, c\, x_1^{\alpha} x_2^{\beta-1}\,.$$

The partial differentials of a production function are referred to as **marginal**

[3] In what follows, we denote the point of differentiation by $\boldsymbol{x}$ instead of $\boldsymbol{a}$ (or by $\boldsymbol{y}$ instead of $\boldsymbol{b}$) when no confusion is possible.

products. Here, they read

$$\frac{\partial f}{\partial x_1}(x_1,x_2)\,dx_1 = \alpha\, c\, x_1^{\alpha-1} x_2^{\beta}\,dx_1 \quad\text{and}\quad \frac{\partial f}{\partial x_2}(x_1,x_2)\,dx_2 = \beta\, c\, x_1^{\alpha} x_2^{\beta-1}\,dx_2\,.$$

The elasticities ε_{f,x_1} and ε_{f,x_2} of a production function are called **output elasticities**. For the Cobb-Douglas production function, they are constant and positive:

$$\varepsilon_{f,x_1}(x_1,x_2) = \frac{x_1}{c\,x_1^{\alpha}x_2^{\beta}}\cdot \alpha\, c\, x_1^{\alpha-1}x_2^{\beta} = \alpha\,,$$
$$\varepsilon_{f,x_2}(x_1,x_2) = \frac{x_2}{c\,x_1^{\alpha}x_2^{\beta}}\cdot \beta\, c\, x_1^{\alpha}x_2^{\beta-1} = \beta\,.$$

Thus, the parameters α and β equal the output elasticities.

Conversely, the Cobb-Douglas function is the only production function having constant output elasticities.

Due to its simple functional form and the direct interpretation of its parameters as elasticities, the Cobb-Douglas production function is one of the most frequently used production functions.

The general Cobb-Douglas production function with n inputs (cf. Example 1.21) has the form

$$f(x_1,x_2,\ldots,x_n) = c\,x_1^{\alpha_1}x_2^{\alpha_2}\cdots x_n^{\alpha_n}\,.$$

Here, the output elasticity of each input i is constant and equal to α_i.

Example 5.6 (Macroeconomic production function). Of course, national income can also be viewed as the result of production where the total amount of labor, the capital stock, and technological progress serve as inputs. Here, Y denotes national income, L the amount of labor, K the capital stock, and t time (representing technological progress):

$$Y = f(L,K,t)\,.$$

The first two marginal productivities

$$\frac{\partial Y}{\partial L} = \frac{\partial f}{\partial L}(L,K,t) \quad\text{and}\quad \frac{\partial Y}{\partial K} = \frac{\partial f}{\partial K}(L,K,t)$$

approximately indicate the value added to national income by one additional unit of labor or capital.

Under perfect competition in factor markets, economic theory shows that a factor's marginal productivity equals its remuneration – the **wage rate** for labor resp. the **capital return** for capital.

The wage rate $\partial Y/\partial L$ times the labor amount L gives total labor remuneration. Divided by output Y, this yields the **wage share**:

$$\frac{\partial Y}{\partial L}\cdot\frac{L}{Y} = \varepsilon_L(L,K,t)\,.$$

Thus, the wage share equals the output elasticity with respect to labor.

Correspondingly, the output elasticity with respect to capital is the **profit share**:

$$\frac{\partial Y}{\partial K} \cdot \frac{K}{Y} = \varepsilon_K(L, K, t).$$

5.2 Differentiation rules

For the derivative of real-valued functions in several variables, similar rules apply as in the case of a single variable. In this section, we deal with multivariate versions of the sum, product, and quotient rules. As with differentiable functions of one variable, the terms of a sum can be differentiated individually, and a constant factor can be pulled out of the derivative.

Theorem 5.2: Sum rule, factor rule

Let $f, g : A \to \mathbb{R}$ be differentiable at $\boldsymbol{x} \in A$. Then,

(i) $f + g$ is differentiable at $\boldsymbol{x}$ with

$$(f + g)'(\boldsymbol{x}) = f'(\boldsymbol{x}) + g'(\boldsymbol{x}),$$

(ii) αf is differentiable at $\boldsymbol{x}$ with

$$(\alpha f)'(\boldsymbol{x}) = \alpha f'(\boldsymbol{x}).$$

To prove this, note that the gradient $(f + g)'(\boldsymbol{x})$ consists of n partial derivatives, for each of which the univariate sum rule applies. Similarly, the gradient $(\alpha f)'$ consists of n partial derivatives, each containing the constant factor α.

For univariate differentiable functions, the product rule has the form $(f \cdot g)' = gf' + fg'$, and the quotient rule is $(f/g)' = (gf' - fg')/g^2$. Both rules transfer directly to multivariate functions. However, note that f' and g' as well as $(fg)'$ and $(f/g)'$ are gradients, i.e., vectors.

Theorem 5.3: Product rule, quotient rule

Let $f, g : A \to \mathbb{R}$ be differentiable at $\boldsymbol{x} \in A$. Then,

(i) $f \cdot g$ is differentiable at $\boldsymbol{x}$ with

$$(f \cdot g)'(\boldsymbol{x}) = g(\boldsymbol{x})f'(\boldsymbol{x}) + f(\boldsymbol{x})g'(\boldsymbol{x}),$$

(ii) $\frac{f}{g}$ is differentiable at $\boldsymbol{x}$ with

$$\left(\frac{f}{g}\right)'(\boldsymbol{x}) = \frac{1}{(g(\boldsymbol{x}))^2}\Big(g(\boldsymbol{x})f'(\boldsymbol{x}) - f(\boldsymbol{x})g'(\boldsymbol{x})\Big),$$

provided that $g(\boldsymbol{x}) \neq 0$.

Again, note that the gradient $(f \cdot g)'$ consists of n partial derivatives, each of which can be calculated by the univariate product rule. Therefore,

$$\begin{aligned}(f \cdot g)'(\boldsymbol{x}) &= \left[\, g(\boldsymbol{x})\tfrac{\partial f(\boldsymbol{x})}{\partial x_1} + f(\boldsymbol{x})\tfrac{\partial g(\boldsymbol{x})}{\partial x_1}, \ldots, g(\boldsymbol{x})\tfrac{\partial f(\boldsymbol{x})}{\partial x_n} + f(\boldsymbol{x})\tfrac{\partial g(\boldsymbol{x})}{\partial x_n} \,\right] \\ &= g(\boldsymbol{x}) \operatorname{grad} f(\boldsymbol{x}) + f(\boldsymbol{x}) \operatorname{grad} g(\boldsymbol{x})\,.\end{aligned}$$

The gradient $(f/g)'(\boldsymbol{x})$ can be obtained analogously. We illustrate these rules with an example:

Example 5.7. Consider the function $f, g : \mathbb{R}^3 \to \mathbb{R}$, with

$$f(x_1,x_2,x_3) = x_1^2 + x_2^3 x_3 \quad \text{and} \quad g(x_1,x_2,x_3) = x_1x_2 - x_3\,.$$

Then,

$$\begin{aligned}(f+g)'(\boldsymbol{x}) &= f'(\boldsymbol{x}) + g'(\boldsymbol{x}) \\ &= [2x_1, 3x_2^2x_3, x_2^3] + [x_2, x_1, -1] \\ &= [2x_1 + x_2, 3x_2^2x_3 + x_1, x_2^3 - 1]\,, \\ (f \cdot g)'(\boldsymbol{x}) &= \underbrace{(x_1x_2 - x_3)}_{g(\boldsymbol{x})} \cdot \underbrace{[2x_1\,,\ 3x_2^2x_3\,,\ x_2^3]}_{f'(\boldsymbol{x})} + \underbrace{(x_1^2 + x_2^3x_3)}_{f(\boldsymbol{x})} \cdot \underbrace{[x_2\,,\ x_1\,,\ -1]}_{g'(\boldsymbol{x})}\,, \\ \left(\frac{f}{g}\right)'(\boldsymbol{x}) &= \frac{1}{(x_1x_2 - x_3)^2} \\ &\quad \cdot \Big(\underbrace{(x_1x_2 - x_3)}_{g(\boldsymbol{x})} \cdot \underbrace{[2x_1\,,\ 3x_2^2x_3\,,\ x_2^3]}_{f'(\boldsymbol{x})} - \underbrace{(x_1^2 + x_2^3x_3)}_{f(\boldsymbol{x})} \cdot \underbrace{[x_2\,,\ x_1\,,\ -1]}_{g'(\boldsymbol{x})}\Big)\,.\end{aligned}$$

In particular, for $\boldsymbol{x} = (1,0,1)$ we get $(f \cdot g)'(1,0,1) = [-2,1,-1]$ and $(f/g)'(1,0,1) = [-2,-1,1]$.

5.3 Derivative of vector-valued functions

Occasionally, it is necessary to consider several real-valued functions that depend on the same variables simultaneously.

Example 5.8 (Demand for several goods). Consider the demand for several goods, which depends on their prices and income. Let y denote income, q_i the demand for good i and p_i its price, for $i = 1, 2, \ldots, n$. The demand for the i-th good depends on its own price, the prices of the other goods, and income; thus, it is given by a real-valued function f_i,

$$q_i = f_i(p_1, p_2, \ldots, p_n, y) .$$

Overall, the demands for all n goods are described by n real-valued functions $f_1, \ldots, f_n$. To study these simultaneously rather than separately, we combine them into a column vector and define the vector-valued function $f : \mathbb{R}^{n+1} \to \mathbb{R}^n$,

$$\boldsymbol{q} = \begin{bmatrix} q_1 \\ q_2 \\ \vdots \\ q_n \end{bmatrix} = \begin{bmatrix} f_1(p_1, p_2, \ldots, p_n, y) \\ f_2(p_1, p_2, \ldots, p_n, y) \\ \vdots \\ f_n(p_1, p_2, \ldots, p_n, y) \end{bmatrix} = f(p_1, p_2, \ldots, p_n, y) .$$

Formally, we combine m real-valued functions $f_1, \ldots, f_m : A \to \mathbb{R}$ into a **vector-valued function** f which has domain A and codomain $\mathbb{R}^m$. The real-valued functions $f_1, \ldots, f_m : A \to \mathbb{R}$ are called the **component functions** of the vector-valued function f. The derivative of such a vector-valued function is a matrix whose entries are partial derivatives. In this section, we define and compute this matrix; it will also be needed in the following for a "chain rule" to derive composite multivariate functions.

Definition 5.5: Jacobian matrix, differentiability

Consider a function $f : A \to \mathbb{R}^m$,

$$f(\boldsymbol{x}) = \begin{bmatrix} f_1(\boldsymbol{x}) \\ \vdots \\ f_m(\boldsymbol{x}) \end{bmatrix} .$$

If each component function f_j is partially differentiable at $\boldsymbol{x} \in A$, then f is called partially differentiable at $\boldsymbol{x}$. The matrix of partial derivatives

$$J_f(\boldsymbol{x}) = \begin{bmatrix} \frac{\partial f_1}{\partial x_1}(\boldsymbol{x}) & \cdots & \frac{\partial f_1}{\partial x_n}(\boldsymbol{x}) \\ \vdots & & \vdots \\ \frac{\partial f_m}{\partial x_1}(\boldsymbol{x}) & \cdots & \frac{\partial f_m}{\partial x_n}(\boldsymbol{x}) \end{bmatrix}$$

is called the **Jacobian**[4] **matrix** or simply **Jacobian** of f at $\boldsymbol{x}$. f is called **(totally) differentiable** at $\boldsymbol{x}$, if each component f_j is totally differentiable at $\boldsymbol{x}$, i.e., if all partial derivatives are continuous.

[4] Carl Gustav Jacob Jacobi (1804–1851)

With $y_i = f_i(\boldsymbol{x})$, we write more concisely

$$J_f(\boldsymbol{x}) = \begin{bmatrix} \frac{\partial y_1}{\partial x_1} & \cdots & \frac{\partial y_1}{\partial x_n} \\ \vdots & & \vdots \\ \frac{\partial y_m}{\partial x_1} & \cdots & \frac{\partial y_m}{\partial x_n} \end{bmatrix}.$$

Other notations for the Jacobian are

$$f'(\boldsymbol{x}) = \frac{df}{d\boldsymbol{x}}(\boldsymbol{x}) = J_f(\boldsymbol{x}).$$

If $m = 1$, then the Jacobian equals the gradient. The structure of the Jacobian is easily remembered as follows:

Mnemonic 5.1: Jacobian matrix

The rows of the Jacobian are the gradients of the component functions.

Example 5.9. Consider the function $f : \mathbb{R}^3 \to \mathbb{R}^2$ with

$$f(\boldsymbol{x}) = \begin{bmatrix} x_1 x_2^2 + x_3 \\ x_1 x_2 e^{x_3} \end{bmatrix}.$$

Then, the Jacobian is

$$J_f(\boldsymbol{x}) = \begin{bmatrix} x_2^2 & 2x_1x_2 & 1 \\ x_2 e^{x_3} & x_1 e^{x_3} & x_1 x_2 e^{x_3} \end{bmatrix}.$$

Theorem 5.4: Sum rule, factor rule

Let $f, g : A \to \mathbb{R}^m$ be differentiable at $\boldsymbol{x} \in A$. Then,

(i) $f + g$ is differentiable at $\boldsymbol{x}$ with

$$(f + g)'(\boldsymbol{x}) = f'(\boldsymbol{x}) + g'(\boldsymbol{x}),$$

(ii) αf is differentiable at $\boldsymbol{x}$ with

$$(\alpha f)'(\boldsymbol{x}) = \alpha f'(\boldsymbol{x}).$$

This theorem almost literally agrees with Theorem 5.2. Note that the derivative $f'(\boldsymbol{x})$ is not the gradient, but rather the Jacobian $J_f(\boldsymbol{x})$.

The calculation of the Jacobian of an affine-linear function is particularly simple. Such a function is given by an $m \times n$ matrix $\boldsymbol{B}$ and a vector $\boldsymbol{c} \in \mathbb{R}^m$, and has the form

$$f(\boldsymbol{x}) = \boldsymbol{c} + \boldsymbol{B}\boldsymbol{x} = \begin{bmatrix} c_1 + b_{11}x_1 + \ldots + b_{1n}x_n \\ \vdots \\ c_m + b_{m1}x_1 + \ldots + b_{mn}x_n \end{bmatrix},$$

where the argument $\boldsymbol{x}$ is interpreted as a column vector $[x_1, \cdots, x_n]^T$. If $\boldsymbol{c}$ is the zero vector, the function is called **linear**, otherwise **affine-linear**. The following theorem states that the derivative of such a function equals the matrix $\boldsymbol{B}$.

Theorem 5.5: Derivative of an affine-linear function

Let $\boldsymbol{c} \in \mathbb{R}^m$, $\boldsymbol{B} \in \mathbb{R}^{m\times n}$, and $f(\boldsymbol{x}) = \boldsymbol{c}+\boldsymbol{B}\boldsymbol{x}$ for $\boldsymbol{x} \in \mathbb{R}^n$. Then f is differentiable and $f'(\boldsymbol{x}) = \boldsymbol{B}$ for all $\boldsymbol{x} \in \mathbb{R}^n$.

For example, consider the function $g : \mathbb{R}^3 \to \mathbb{R}^2$ with

$$g(\boldsymbol{x}) = g(x_1, x_2, x_3) = \begin{bmatrix} x_1 + x_2 \\ 3x_3 \end{bmatrix}.$$

It is linear, $g(\boldsymbol{x}) = \boldsymbol{B}\boldsymbol{x}$, with

$$\boldsymbol{B} = \begin{bmatrix} 1 & 1 & 0 \\ 0 & 0 & 3 \end{bmatrix}.$$

Hence, its derivative is $g'(\boldsymbol{x}) = J_g(\boldsymbol{x}) = \boldsymbol{B}$.

The **identity** on $\mathbb{R}^n$, $\mathrm{id}(\boldsymbol{x}) = \boldsymbol{x}$, which maps every point to itself, is a linear function, defined by $\boldsymbol{B} = \boldsymbol{I}_n$. Its derivative is therefore $\mathrm{id}'(\boldsymbol{x}) = J_{\mathrm{id}}(\boldsymbol{x}) = \boldsymbol{I}_n$.

Example 5.10 (Continuation: demand for several goods). We consider the demand for several goods as in Example 5.8. The elasticity

$$\varepsilon_{ii} = \frac{p_i}{q_i}\frac{\partial q_i}{\partial p_i}$$

is called the **price elasticity of demand** for good i. The quantity

$$\varepsilon_{ij} = \frac{p_j}{q_i}\frac{\partial q_i}{\partial p_j}$$

is called the **cross-price elasticity of demand** for good i with respect to the price of good j. If the cross-price elasticity is positive, this means that an increase in the price of good j leads to additional demand for good i; the two goods i and j are then called **substitutes**. In this case, a quantity of good j can be replaced by a quantity of good i, for example potatoes by pasta: when potato prices rise, the demand for pasta increases. On the other hand, a negative cross-price elasticity means that, as the price of good j rises, the demand for good i decreases; the two goods are then **complements**, and good j cannot be replaced by good i. For example, when gasoline prices rise in the summer, not only is less gasoline demanded, but also fewer vacation accommodations. The elasticity of demand for good i with respect to income,

$$\varepsilon_i = \frac{y}{q_i}\frac{\partial q_i}{\partial y},$$

is referred to as **income elasticity of demand** for good i.

5.4 Chain rule

For multivariate functions, analogous to Theorem 4.2, a chain rule applies that allows the derivative of composite functions to be computed. It has the same form as the univariate chain rule,

$$(g \circ f)'(x) = \big(g(f(x))\big)'(x) = g'\big(f(x)\big)f'(x),$$

except that the derivatives are matrices rather than numbers.

Theorem 5.6: Chain rule

Let $A \subset \mathbb{R}^n$, $B \subset \mathbb{R}^m$. Further let

$f : A \to B$ be differentiable at $\boldsymbol{x} \in A$,
$g : B \to \mathbb{R}^k$ be differentiable at $\boldsymbol{y} = f(\boldsymbol{x}) \in B$.

Then the composition $g \circ f : A \to \mathbb{R}^k$, $\boldsymbol{x} \mapsto g(f(\boldsymbol{x}))$, is differentiable at $\boldsymbol{x}$, and its derivative is

$$\underbrace{(g \circ f)'(\boldsymbol{x})}_{\in\mathbb{R}^{k\times n}} = \underbrace{g'(f(\boldsymbol{x}))}_{\in\mathbb{R}^{k\times m}}\,\underbrace{f'(\boldsymbol{x})}_{\in\mathbb{R}^{m\times n}}.$$

Note that this is an equation between matrices. On the left-hand side is the Jacobian matrix of format $k \times n$; on the right-hand side is the product of a $k \times m$ matrix with an $m \times n$ matrix, which again has format $k \times n$. With $\boldsymbol{y} = f(\boldsymbol{x})$ and $\boldsymbol{z} = g(\boldsymbol{y})$, the chain rule is briefly written as mnemonic.

Mnemonic 5.2: Chain rule

$$\frac{d\boldsymbol{z}}{d\boldsymbol{x}} = \frac{d\boldsymbol{z}}{d\boldsymbol{y}}\frac{d\boldsymbol{y}}{d\boldsymbol{x}}$$

Example 5.11. Consider the functions $f : \mathbb{R} \to \mathbb{R}^3$ and $g : \mathbb{R}^3 \to \mathbb{R}$ with

$$f : x \mapsto \begin{bmatrix} x^2 \\ x-1 \\ e^{-2x} \end{bmatrix}, \qquad g : \begin{bmatrix} y_1 \\ y_2 \\ y_3 \end{bmatrix} \mapsto y_1y_2 - y_3.$$

Here, $n = 1$, $m = 3$ and $k = 1$. The Jacobians of f and g are

$$f'(x) = \begin{bmatrix} 2x \\ 1 \\ -2e^{-2x} \end{bmatrix} \quad \text{and} \quad g'(\boldsymbol{y}) = [y_2\,,\ y_1\,,\ -1]\,.$$

With $\boldsymbol{y} = f(x)$, we obtain

$$g'(\boldsymbol{y}) = g'(f(x)) = g'\left(x^2, x-1, e^{-2x}\right) = [x-1, x^2, -1]\,.$$

From this it follows that

$$\begin{aligned}(g \circ f)'(x) = g'(f(x))f'(x) &= [x-1\,,\ x^2\,,\ -1]\begin{bmatrix} 2x \\ 1 \\ -2e^{-2x} \end{bmatrix} \\ &= 2x^2 - 2x + x^2 + 2e^{-2x} \\ &= 3x^2 - 2x + 2e^{-2x}\,.\end{aligned}$$

Alternatively, one can substitute $f(x)$ into g,

$$g(f(x)) = x^2(x-1) - e^{-2x} = x^3 - x^2 - e^{-2x}\,,$$

and differentiate with respect to x,

$$\frac{d}{dx}(x^3 - x^2 - e^{-2x}) = 3x^2 - 2x + 2e^{-2x}\,.$$

This is certainly easier than applying the chain rule. However, one often needs to differentiate compositions in general, without having the functions f and g numerically specified; then the application of the chain rule is unavoidable.

We have formulated the chain rule for a composite function in n variables with values in $\mathbb{R}^k$, where n and k are arbitrary natural numbers. An important special case, and one that frequently occurs, is $k = n = 1$. This is a function $h : \mathbb{R} \to \mathbb{R}, x \mapsto z$, which *indirectly* depends on m additional variables $y_1, \ldots, y_m$ through its actual variable x. The previous example contained such a function $h = g \circ f$ with $m = 3$. The derivative of h with respect to x accounts for the indirect dependence of the function value z on x mediated by the variables $y_1, \ldots, y_m$.

Theorem 5.7: Chain rule (special case $k = n = 1$)

Let $A \subset \mathbb{R}$ and $B \subset \mathbb{R}^m$. Further let $f : A \to B$ be differentiable at $x \in A$ and $g : B \to \mathbb{R}$ be differentiable at $\boldsymbol{y} = f(x) \in B$. Then the composition $g \circ f : A \to \mathbb{R}$, $x \mapsto g(f(x))$, is differentiable at x, and its derivative is

$$(g \circ f)'(x) = \frac{d}{dx} g\,(f(x)) = \operatorname{grad} g(f(x)) f'(x) = \sum_{i=1}^{m} \frac{\partial g}{\partial y_i}(f(x)) \cdot f_i'(x)\,.$$

With $\boldsymbol{y} = f(x)$ and $z = g(\boldsymbol{y})$ we get the following mnemonic:

Mnemonic 5.3

$$\frac{dz}{dx} = \sum_{i=1}^{m} \frac{\partial z}{\partial y_i}\frac{dy_i}{dx}.$$

Example 5.12. Consider the functions $f(x, y) = xe^y + x^2y$ as well as $x(t) = t^3$ and $y(t) = t^2$. Their first derivatives are $f_x(x, y) = e^y + 2xy$, $f_y(x, y) = xe^y + x^2$ as well as $x'(t) = 3t^2$ and $y'(t) = 2t$. Using the chain rule, we can compute the derivative of $z(t) := f(x(t), y(t))$:

$$\begin{aligned} z'(t) &= f_x(x(t), y(t)) \cdot x'(t) + f_y(x(t), y(t)) \cdot y'(t) \\ &= \left(e^{y(t)} + 2x(t)y(t)\right) \cdot 3t^2 + \left(x(t)e^{y(t)} + (x(t))^2\right) \cdot 2t \\ &= \left(e^{t^2} + 2t^3t^2\right) \cdot 3t^2 + \left(t^3e^{t^2} + (t^3)^2\right) \cdot 2t \\ &= 3t^2e^{t^2} + 2t^4e^{t^2} + 8t^7 . \end{aligned}$$

As already described in Example 5.11, since the functions are given explicitly, one could also write down the composite function $z(t)$ and then differentiate with respect to t.

5.5 Homogeneity

One of the initial questions was how the value of a function changes when all variables are doubled or, more generally, multiplied by a number $\lambda \in \mathbb{R}$. We introduce the concept of (positive) homogeneity for this purpose.

Definition 5.6: Homogeneous function

A function f of n variables is called **(positively) homogeneous of degree r**, where r is a real number, if for every number $\lambda > 0$ and all $\boldsymbol{x}$ it holds that

$$f(\lambda\boldsymbol{x}) = \lambda^r f(\boldsymbol{x}).$$

Remarks

1. In the economic literature, positively homogeneous functions are often simply mentioned as homogeneous functions, thus omitting the attribute "positively" – a convention we shall follow. In the sequel, we refer to homogeneous functions, but we actually mean positively homogeneous functions.

2. The domain A of a positively homogeneous function $f : A \to \mathbb{R}^n$ must have the property that with every point $\boldsymbol{x} \in A$, every positive multiple $\lambda\boldsymbol{x}$ with $\lambda > 0$

also lies in A. Such a set is called a **cone**. In most applications, the domain is either $\mathbb{R}^n$, $\mathbb{R}^n_+$ or $\mathbb{R}^n_{++} = \{(x_1, \ldots, x_n) \in \mathbb{R}^n \mid x_1, \ldots, x_n > 0\}$. It is easy to verify that each of these sets is indeed a cone.

With a homogeneous function of degree r, doubling all variables multiplies the function value by 2^r. A function that is homogeneous of degree $r = 1$ is also called **linearly homogeneous**. For example, the function $f(\boldsymbol{x}) = f(x_1, x_2) = 5x_1 - 2x_2$, is linearly homogeneous. For every number $\lambda > 0$, it holds that

$$f(\lambda \boldsymbol{x}) = f(\lambda x_1, \lambda x_2) = 5\lambda x_1 - 2\lambda x_2 = \lambda(5x_1 - 2x_2) = \lambda^1 f(\boldsymbol{x}).$$

If all variables are multiplied by the same factor λ, the function value is also multiplied by λ. More generally, every real-valued linear function $f(\boldsymbol{x}) = \alpha_1 x_1 + \ldots + \alpha_n x_n$ with constants $\alpha_i \in \mathbb{R}$, $i = 1, \ldots, n$, is linearly homogeneous.

Homogeneity of degree $r = 0$ means that the function value remains unchanged if all variables are multiplied by a factor λ. Such a function is called **scale-invariant**. For example, the function $f(\boldsymbol{x}) = f(x_1, x_2) = \ln(x_1/x_2)$, $x_1, x_2 > 0$, is homogeneous of degree 0, because for every $\lambda > 0$, it holds that

$$f(\lambda \boldsymbol{x}) = f(\lambda x_1, \lambda x_2) = \ln\left(\frac{\lambda x_1}{\lambda x_2}\right) = \ln\left(\frac{x_1}{x_2}\right) = 1 \cdot f(\boldsymbol{x}) = \lambda^0 f(\boldsymbol{x}).$$

Example 5.13 (Cobb-Douglas function). The Cobb-Douglas production function, $f : \mathbb{R}^2_+ \to \mathbb{R}$, $f(x_1, x_2) = c x_1^\alpha x_2^\beta$, is homogeneous of degree $r = \alpha + \beta$. For every $\lambda > 0$, we have

$$f(\lambda \boldsymbol{x}) = f(\lambda x_1, \lambda x_2) = c(\lambda x_1)^\alpha (\lambda x_2)^\beta = \lambda^\alpha \lambda^\beta c x_1^\alpha x_2^\beta = \lambda^{\alpha+\beta} f(\boldsymbol{x}).$$

Usually, it is also easy to judge if a function is not homogeneous:

Example 5.14. The function $f(x_1, x_2) = x_1 + x_1 x_2$ is not homogeneous. For all $r \in \mathbb{R}$, it holds

$$f(\lambda \boldsymbol{x}) = \lambda x_1 + \lambda x_1 \lambda x_2 = \lambda^1 x_1 + \lambda^2 x_1 x_2 \neq \lambda^r f(\boldsymbol{x}),$$

since the first summand requires factoring out λ^1, while the second requires λ^2.

The following theorem provides an important characterization of homogeneous functions.

Theorem 5.8: Euler's theorem

A differentiable function $f : \mathbb{R}^n \to \mathbb{R}$ is positively homogeneous of degree r if and only if

$$\sum_{i=1}^{n} x_i \frac{\partial f}{\partial x_i}(\boldsymbol{x}) = r f(\boldsymbol{x}) \tag{5.1}$$

holds for all $\boldsymbol{x} = (x_1, \ldots, x_n) \in \mathbb{R}^n$.

Proof: If f is homogeneous of degree r, then $f(\lambda \boldsymbol{x}) = \lambda^r f(\boldsymbol{x})$ for all $\boldsymbol{x}$ and all $\lambda > 0$. We differentiate both sides of this equation with respect to λ. For the left-hand side we get (according to the chain rule applied to the composite function $\lambda \mapsto \lambda \boldsymbol{x} \mapsto f(\lambda \boldsymbol{x})$)

$$\frac{d}{d\lambda} f(\lambda \boldsymbol{x}) = \sum_{i=1}^{n} \frac{\partial f}{\partial x_i}(\lambda \boldsymbol{x}) x_i \,,$$

and for the right-hand side $r\lambda^{r-1} f(\boldsymbol{x})$. In particular, for $\lambda = 1$ we get

$$\sum_{i=1}^{n} \frac{\partial f}{\partial x_i}(\boldsymbol{x}) x_i = r f(\boldsymbol{x}) \,.$$

Conversely, one can also show that if this equation holds, the function f is homogeneous of degree r. □

Dividing Equation (5.1) by $f(\boldsymbol{x})$ gives

$$\sum_{i=1}^{n} \varepsilon_{f,x_i}(\boldsymbol{x}) = r \,.$$

Therefore, Euler's[5] theorem can be formulated as follows:

Mnemonic 5.4: Euler's theorem

For a homogeneous function, the sum of the partial elasticities equals the degree of homogeneity.

The following two examples illustrate Euler's theorem.

Example 5.15. The function $f(x_1, x_2) = x_1^2 x_2$ is homogeneous of degree 3. The partial elasticities are

$$\varepsilon_{f,x_1}(x_1, x_2) = \frac{x_1}{f(x_1, x_2)} f_{x_1}(x_1, x_2) = \frac{x_1}{x_1^2 x_2} 2x_1 x_2 = 2$$

and

$$\varepsilon_{f,x_2}(x_1, x_2) = \frac{x_2}{f(x_1, x_2)} f_{x_2}(x_1, x_2) = \frac{x_2}{x_1^2 x_2} x_1^2 = 1 \,.$$

The sum of partial elasticities is $\varepsilon_{f,x_1}(x_1, x_2) + \varepsilon_{f,x_2}(x_1, x_2) = 2 + 1 = 3$, which equals the degree of homogeneity.

Example 5.16. The function $f(x_1, x_2) = x_1 + x_2^2$ has the elasticities

$$\varepsilon_{f,x_1}(x_1, x_2) = \frac{x_1}{x_1 + x_2^2} \quad \text{and} \quad \varepsilon_{f,x_2}(x_1, x_2) = \frac{2x_2^2}{x_1 + x_2^2} \,,$$

[5] Leonhard Euler (1707–1783)

and their sum is

$$\varepsilon_{f,x_1}(x_1,x_2)+\varepsilon_{f,x_2}(x_1,x_2)=\frac{x_1+2x_2^2}{x_1+x_2^2}=1+\frac{x_2^2}{x_1+x_2^2}\,.$$

The sum of partial elasticities depends on (x_1, x_2), so it is not constant. Therefore, according to Euler's theorem, the function f is not homogeneous.

For a production function that is homogeneous of degree r, a proportional increase in all inputs leads to an over-proportional increase in output if $r > 1$, and to an under-proportional increase if $r < 1$. However, even for nonhomogeneous production functions, it is of interest how a proportional increase in all inputs affects the output. This generally depends on the current level of inputs. To study this question, we consider the function

$$\tilde{f}:[0,\infty)\to\mathbb{R},\quad t\mapsto f(tx_1,tx_2,\ldots,tx_n)=f(t\boldsymbol{x})\,.$$

Here, $x_1, \ldots, x_n$ are the current levels of inputs. For $t = 1$, the function value $\tilde{f}(1)$ thus gives the current output. For a proportional increase of all inputs by, say 5%, let $t = 1.05$. The corresponding output is $\tilde{f}(1.05) = f(1.05\,x_1, \ldots, 1.05\,x_n)$. Thus, to find the (approximate) relative change of the output for a proportional increase of all inputs, one must find the elasticity of $\tilde{f}$ at $t = 1$, that is

$$\varepsilon_{\tilde{f}}(1)=\frac{1}{\tilde{f}(1)}\tilde{f}'(1)\,.$$

This quantity is called **elasticity of scale** and is denoted by $\varepsilon_{f,t}(\boldsymbol{x})$. To compute it, we consider the composite function

$$\tilde{f}(t)=f(tx_1,tx_2,\ldots,tx_n)=f(g(t))\qquad\text{with}\qquad g(t)=\begin{bmatrix}tx_1\\ \vdots\\ tx_n\end{bmatrix}.$$

Since $g_i'(t) = x_i$, using the chain rule, we get

$$\tilde{f}'(t)=\sum_{i=1}^{n}\frac{\partial f}{\partial x_i}(g(t))\,g_i'(t)=\sum_{i=1}^{n}\frac{\partial f}{\partial x_i}(t\boldsymbol{x})\,x_i\,.$$

For $t = 1$ we have

$$\tilde{f}'(1)=\sum_{i=1}^{n}\frac{\partial f}{\partial x_i}(\boldsymbol{x})\,x_i\,.$$

Hence, the elasticity of scale is

$$\varepsilon_{f,t}(\boldsymbol{x})=\frac{1}{\tilde{f}(1)}\tilde{f}'(1)=\frac{1}{f(\boldsymbol{x})}\sum_{i=1}^{n}\frac{\partial f}{\partial x_i}(\boldsymbol{x})\,x_i=\sum_{i=1}^{n}\frac{x_i}{f(\boldsymbol{x})}\frac{\partial f}{\partial x_i}(\boldsymbol{x})=\sum_{i=1}^{n}\varepsilon_{f,x_i}(\boldsymbol{x})\,.$$

Theorem 5.9: Elasticity of scale

The elasticity of scale $\varepsilon_{f,t}(\boldsymbol{x})$ equals the sum of partial elasticities:

$$\varepsilon_{f,t}(\boldsymbol{x}) = \sum_{i=1}^{n} \varepsilon_{f,x_i}(\boldsymbol{x}).$$

If the elasticity of scale is greater than one, a proportional increase of all inputs leads to an over-proportional increase in the output. This is referred to as **increasing returns to scale**. If the elasticity of scale is less than one, the output increases under-proportionally when all inputs increase proportionally. In this case, we speak of **decreasing returns to scale**. If the elasticity of scale is equal to one, this is mentioned as **constant returns to scale**.

For a homogeneous function of degree r, Euler's theorem implies that the sum of partial elasticities – and thus the elasticity of scale – equals r at every point $\boldsymbol{x}$. Consequently, homogeneous functions possess a constant elasticity of scale. Conversely, if a function has an elasticity of scale equal to r everywhere, then by Euler's theorem, the function is homogeneous of degree r.

Mnemonic 5.5

The functions with constant elasticity of scale r are precisely the homogeneous functions of degree r.

5.6 Implicit functions

In many situations, the relationship between two economic quantities x and y is given by an equation of the form

$$F(x, y) = 0.$$

In such cases, the main interest lies in how the variable y depends on the variable x.

Example 5.17. A company produces 100 units of a good from two inputs x and y according to the production function $P(x.y)$. So it holds

$$P(x, y) = 100, \quad \text{that is} \quad P(x, y) - 100 = 0.$$

If the price of the first input increases, the company wants to know the dependence of the quantity y of the second input on the quantity x of the first input used. Specifically, the company is interested in determining how many more units of the second input are needed to compensate for a certain reduction of the more expensive first input.

A functional relationship $y = f(x)$ may or may not be defined locally by a given equation $F(x, y) = 0$. This is illustrated by Figure 5.3.

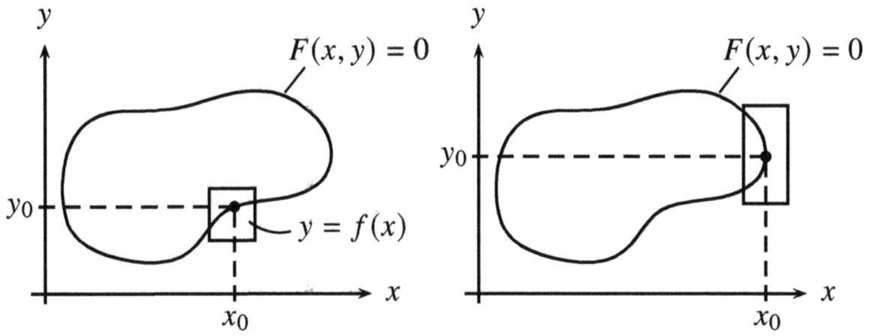

Figure 5.3: At (x_0, y_0) there is a local function $y = f(x)$ on the left, but not on the right.

Consider an equation $F(x, y) = 0$. Both panels show the set of all points satisfying this equation, as well as a specific point (x_0, y_0) on that set. In the left panel, intervals A around x_0 and B around y_0 exist such that the equation $F(x, y) = 0$ defines a function $f : A \to B$. Each $x \in A$ is mapped to the uniquely determined value $y = f(x) \in B$ for which $F(x, y) = 0$ holds. In the right panel, however, no intervals A around x_0 and B around y_0 exist such $F(x, y) = 0$ defines a function $f : A \to B$. To the left of x_0, the value of y satisfying $F(x, y) = 0$ is not unique, while to the right of x_0 no such y exists at all. This is apparently due to the fact that the curve defined by $F(x, y) = 0$ has a vertical tangent at the point (x_0, y_0).

However, if this situation does not occur, that is, if the tangent to the curve at (x_0, y_0) is not vertical, then the relation $F(x, y) = 0$ defines at least locally a function $f : A \to B$. In the following, we assume that this is always the case, i.e., that $\frac{\partial F}{\partial y}(x, y) \neq 0$.

A second issue is that this **implicit** function – defined by the equation $F(x, y) = 0$ – cannot always be written **explicitly** in the form $y = f(x)$. As examples, consider the two equations

$$x \ln(y) - 1 = 0 \qquad \text{and} \qquad y + \ln(y) - x^2 + 5 = 0\,.$$

The first equation can be solved explicitly for y as a function of x, namely $y = f(x) = \exp(1/x)$; the second one cannot. Nevertheless, calculus allows certain statements about this dependence, in particular about the derivative of y with respect to x. A function $x \mapsto y$ that is not given explicitly by a function rule $y = f(x)$, but only through an equation in x and y is called an **implicitly defined function** or simply an **implicit function**. In this section, we derive a formula for the derivative of an implicit function and apply it to the problem of substitution of production factors.

Definition 5.7: Implicit function

Consider a differentiable function $F : A \times B \to \mathbb{R}$, where A and B are open intervals in $\mathbb{R}$, and the equation

$$F(x, y) = 0\,.$$

If for each $x \in A$ there is exactly one $y \in B$ such that $F(x, y) = 0$, then $F(x, y) = 0$ is called an **implicit representation** of the function $x \mapsto y = f(x)$, or simply an **implicit function**.

Even if y cannot be expressed explicitly as a function of x, it is possible to obtain information about the behavior of the implicitly defined function $x \mapsto y = f(x)$. The following theorem shows how to determine the derivative of an implicit function.

Theorem 5.10: Derivative of an implicit function

Let F be differentiable at (x_0, y_0), with

$$F(x_0, y_0) = 0 \quad \text{and} \quad \frac{\partial F}{\partial y}(x_0, y_0) \neq 0\,.$$

Then there exist open intervals A around x_0 and B around y_0 such that the equation $F(x, y) = 0$ defines a differentiable function $f : A \to B$, satisfying

$$f'(x) = \frac{dy}{dx} = -\frac{\frac{\partial F}{\partial x}(x, y)}{\frac{\partial F}{\partial y}(x, y)}\,.$$

The theorem on the derivative of an implicit function is easy to derive: starting from the assumption $y = f(x)$, the equation becomes $F(x, f(x)) = 0$. Differentiating both sides with respect to x, the right-hand side yields 0. On the left we apply the chain rule to the composite function $x \mapsto F(x, f(x))$. It follows

$$\frac{\partial F}{\partial x}(x, y) \cdot 1 + \frac{\partial F}{\partial y}(x, y) \cdot f'(x) = 0\,,$$

and from this

$$f'(x) = -\frac{\frac{\partial F}{\partial x}(x, y)}{\frac{\partial F}{\partial y}(x, y)}\,.$$

Example 5.18. Consider the equation $y + \ln(y) = -x^3 + 9$ for $x \in \mathbb{R}$, $y > 0$, or equivalently, $F(x, y) = y + \ln(y) + x^3 - 9 = 0$. This equation cannot be solved explicitly for y. To find the derivative of the implicit function $y = f(x)$, compute the partial derivatives of F,

$$\frac{\partial F}{\partial x}(x, y) = 3x^2 \quad \text{and} \quad \frac{\partial F}{\partial y}(x, y) = 1 + \frac{1}{y}\,,$$

yielding

$$f'(x) = -\frac{\frac{\partial F}{\partial x}(x, y)}{\frac{\partial F}{\partial y}(x, y)} = -\frac{3x^2}{1 + \frac{1}{y}}\,.$$

Although the derivative can be found explicitly, it depends on both x and $y = f(x)$. Thus, computing $f'(2)$ requires knowing $y = f(2)$ first. Here, the point

$(x, y) = (2, 1)$ satisfies the equation, since

$$1 + \ln(1) = -2^3 + 9 \iff 1 = 1,$$

so $f(2) = 1$.

With this information, the derivative of f at $x = 2$ is

$$f'(2) = -\frac{3 \cdot 2^2}{1 + \frac{1}{1}} = -6.$$

We now interpret the above equation as an isoquant[6] of the production function $z = P(x, y) = y + \ln(y) + x^3$ for output level $z = 9$. As seen earlier, the input combination $(x, y) = (2, 1)$ lies on this isoquant, producing the output $z = 9$. The derivative $f'(2) = -6$ indicates that decreasing input y by approximately 6 units while increasing input x by 1 unit keeps output at 9 units. Conversely, reducing x by 1 unit requires increasing y by about 6 units to maintain production at 9 units. In general, a change dx in x must be compensated by a change $dy = f'(x)dx$ in y.

Factor substitution Consider a production function with two factors, such as the macroeconomic production function $Y = f(L, K)$ from Example 5.6, where L denotes labor and K capital. The contour line or isoquant at output level y_0 is given by

$$f(L, K) = y_0.$$

Using the implicit function theorem, we determine the substitution possibilities between factors by computing the slope of the isoquant dK/dL. Here, $F(L, K) = f(L, K) - y_0 = 0$. The derivative is

$$\frac{dK}{dL} = -\frac{\frac{\partial f}{\partial L}}{\frac{\partial f}{\partial K}}.$$

This derivative is always negative for the production functions typically used. Its absolute value, $|dK/dL|$, is referred to as the **marginal rate of technical substitution (MRTS)**.

Under perfect competition, the marginal rate of technical substitution therefore equals the quotient of the factor prices (wage rate divided by capital return). This quotient is also called the **factor price ratio.**

In particular, for the **Cobb-Douglas production function** $g(K, L) = c\,K^{\alpha} L^{\beta}$ the corresponding derivative is

$$\frac{dK}{dL} = -\frac{\beta}{\alpha}\frac{K}{L} < 0.$$

The marginal rate of technical substitution is thus

$$\left|\frac{dK}{dL}\right| = \frac{\beta}{\alpha}\frac{K}{L}.$$

[6] A contour line of a function is also called **isoquant**.

It increases with K and decreases with L. Hence, the higher the capital input (and lower the labor input), the more units of capital must be used to reduce labor by one unit while maintaining the same output. Contour lines of a Cobb-Douglas function are shown in Figure 5.4.

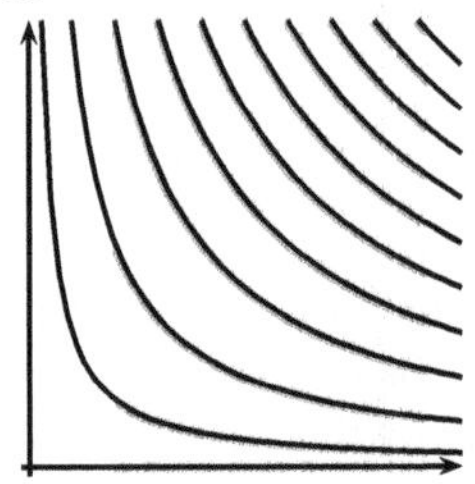

Figure 5.4: Contour lines (isoquants) of a Cobb-Douglas production function.

The **elasticity of substitution** σ of a production function $g(K, L)$ is the elasticity of the factor input ratio K/L with respect to the marginal rate of technical substitution,

$$\sigma = \frac{d\left(\frac{K}{L}\right)}{d\,MRTS} \frac{MRTS}{\frac{K}{L}}.$$

The elasticity of substitution describes how the factor input ratio changes when the marginal rate of substitution changes, which under perfect competition equals the factor price ratio.

To illustrate this, consider Figure 5.5. It shows an isoquant for two different production functions. At point A, both isoquants have a marginal rate of substitution of $|-1| = 1$ and a factor input ratio of 1. Now the marginal rate of substitution changes by the same amount on both isoquants, leading to point B on the dark gray isoquant and to point C on the light gray isoquant. At point C, the factor input ratio K/L is larger than at point B. Thus, the factor input ratio changes faster on the light gray isoquant than on the dark gray one. The elasticity of substitution σ is therefore larger on the light gray isoquant,as the factors are easier to substitute there.

The practical calculation of the elasticity of substitution will be illustrated with the Cobb-Douglas production function. We consider the marginal rate of technical substitution

$$MRTS = \frac{\beta}{\alpha} \frac{K}{L}$$

and express it as a function of the factor input ratio $z = K/L$,

$$MRTS(z) = \frac{\beta}{\alpha} z.$$

Hence, the derivative of the marginal rate of technical substitution with respect to the factor input ratio is

$$\frac{d\,MRTS}{dz}(z) = \frac{d}{dz}\left(\frac{\beta}{\alpha} z\right) = \frac{\beta}{\alpha}.$$

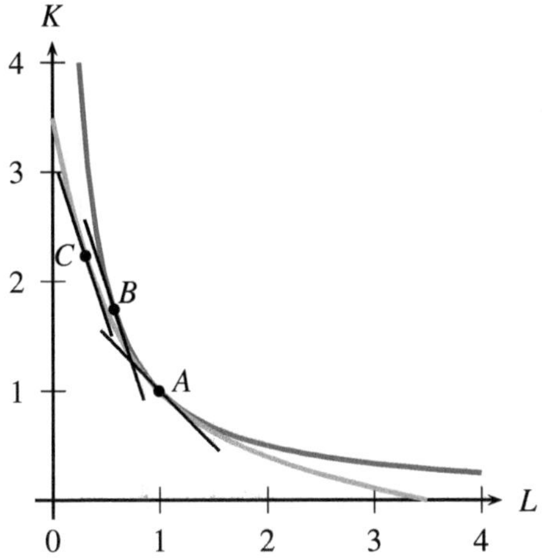

Figure 5.5: Illustration of elasticity of substitution.

Conversely, one can also differentiate the factor input ratio $z = K/L = (\alpha/\beta)MRTS$ with respect to $MRTS$,

$$\frac{dz}{d\,MRTS} = \frac{\alpha}{\beta}\,.$$

For the elasticity of substitution, one finally obtains

$$\sigma = \frac{d\left(\frac{K}{L}\right)}{d\,MRTS}\,\frac{MRTS}{\frac{K}{L}} = \frac{dz}{d\,MRTS}\,\frac{MRTS}{z} = \frac{\alpha}{\beta}\,\frac{\frac{\beta}{\alpha}z}{z} = \frac{\alpha}{\beta}\,\frac{\beta}{\alpha} = 1\,.$$

The elasticity of substitution of the Cobb-Douglas production function is therefore constant and equal to one.

A class of production functions whose elasticity of substitution is also constant, but not necessarily equal to one, is the class of CES production functions. A **CES function** (constant elasticity of substitution) has the form

$$Y = g(K, L) = c\,[\alpha K^{-\rho} + (1-\alpha)L^{-\rho}]^{\frac{-\gamma}{\rho}}$$

with $c > 0$, $\rho > -1$, $\rho \neq 0$, $0 < \alpha < 1$, $\gamma > 0$.

A CES production function has positive marginal productivities. The output elasticities are

$$\varepsilon_{Y,K} = \frac{\alpha}{c^{\rho}}\left(\frac{Y}{K}\right)^{\rho} \quad \text{and} \quad \varepsilon_{Y,L} = \frac{1-\alpha}{c^{\rho}}\left(\frac{Y}{L}\right)^{\rho},$$

and the substitution elasticity

$$\sigma = \frac{1}{1+\rho}\,.$$

In the limit case $\rho \to 0$, the CES function becomes the Cobb-Douglas production function. For $\rho \to \infty$, one obtains the so-called **Leontief production function**.

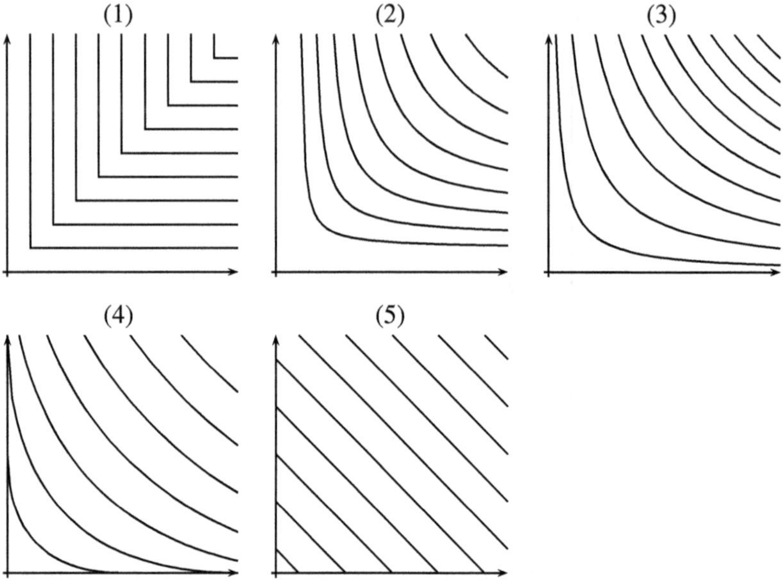

Figure 5.6: Isoquants of the production functions (1) to (5).

Figure 5.6 shows five CES functions having different elasticities of substitution. Specifically, these are the following five macroeconomic production functions:

(1) $\sigma = 0$: $Y = \min\{L, K\}$ (no substitutability)

(2) $\sigma = \frac{1}{2}$: $Y = \left[L^{-1} + K^{-1}\right]^{-1}$ (limited substitutability)

(3) $\sigma = 1$: $Y = LK$ (limited substitutability)

(4) $\sigma = 2$: $Y = \left[\sqrt{L} + \sqrt{K}\right]^2$ (good substitutability)

(5) $\sigma = \infty$: $Y = L + K$ (perfect substitutability)

Function (1) is a **Leontief production function**, where factors must be used in fixed proportions with no substitution possible. In function (2), factors exhibit **limited substitutability**, allowing partial replacement but requiring minimum quantities of each. Function (3) is a Cobb-Douglas production function. Here too, the factors exhibit limited substitutability in the sense that neither can be entirely eliminated. Unlike (2), however, no fixed minimum quantities of the factors are required to produce a given output. In function (4), factors show **good substitutability**, permitting complete replacement of one by the other. The linear production function (5) features **perfect substitutes**, with constant marginal rate of substitution.

The elasticity of substitution thus indicates how easily the factors can be substituted for each other. The larger σ is, the flatter the isoquants become and the slower the marginal rate of technical substitution decreases as K is replaced by L.

5.7 Directional derivative

The partial derivatives of a multivariate function indicate the slopes in the direction of the coordinate axes. Imagine the graph of a bivariate function f as a mountain rising above the x_1-x_2 plane: Then $f_{x_1}(\boldsymbol{x})$ gives the slope when walking from position $\boldsymbol{x}$ in the x_1 direction. Correspondingly, $f_{x_2}(\boldsymbol{x})$ gives the slope in the x_2 direction. But how steep is the slope when walking from $\boldsymbol{x}$ in any other direction, and in which direction is the slope steepest? To answer these and similar questions, one needs the concept of the **directional derivative**.

A direction in $\mathbb{R}^n$ is defined by a column vector $\boldsymbol{p} = [p_1, \cdots, p_n]^T$ of length $\|\boldsymbol{p}\| = 1$. When walking from a position $\boldsymbol{x}$ in a direction $\boldsymbol{p}$, one moves along the line $\{\boldsymbol{x} + t\boldsymbol{p} : t \in \mathbb{R}\}$. The heights (function values) that one reaches on this path are given by $f(\boldsymbol{x} + t\boldsymbol{p})$. Thus, to answer the initial question of the slope in a given direction, we need to calculate the derivative of the function

$$\tilde{f} : \mathbb{R} \to \mathbb{R}, \quad t \mapsto f(\boldsymbol{x} + t\boldsymbol{p}),$$

at $t = 0$. As in the calculation of the elasticity of scale, we use the chain rule. Consider the composite function

$$\tilde{f}(t) = f(x_1 + tp_1, \ldots, x_n + tp_n) = f(g(t)) \quad \text{with} \quad g(t) = \begin{bmatrix} x_1 + tp_1 \\ \vdots \\ x_n + tp_n \end{bmatrix}.$$

Since $g_i'(t) = p_i$, using the chain rule, gives

$$\tilde{f}'(t) = \sum_{i=1}^{n} \frac{\partial f}{\partial x_i}(g(t)) \cdot g_i'(t) = \sum_{i=1}^{n} \frac{\partial f}{\partial x_i}(\boldsymbol{x} + t\boldsymbol{p}) \cdot p_i .$$

For $t = 0$ we get

$$\tilde{f}'(0) = \sum_{i=1}^{n} \frac{\partial f}{\partial x_i}(\boldsymbol{x}) \cdot p_i = \operatorname{grad} f(\boldsymbol{x})\, \boldsymbol{p},$$

that is, the scalar product of the gradient with the direction vector $\boldsymbol{p}$.

Definition 5.8: Directional derivative

Let $f : A \to \mathbb{R}$ be differentiable, $\boldsymbol{x} \in A$, and $\boldsymbol{p} \in \mathbb{R}^n$ with $\|p\| = 1$. Then,

$$\frac{\partial f}{\partial \boldsymbol{p}}(\boldsymbol{x}) = \sum_{i=1}^{n} \frac{\partial f}{\partial x_i}(\boldsymbol{x}) \cdot p_i = \operatorname{grad} f(\boldsymbol{x})\, \boldsymbol{p}$$

is called the **directional derivative** of f at $\boldsymbol{x}$ in direction $\boldsymbol{p}$.

The concept of the directional derivative leads to an important interpretation of the gradient. The scalar product of two vectors $\boldsymbol{x}, \boldsymbol{y} \in \mathbb{R}^n$ is

$$\boldsymbol{x}^T \boldsymbol{y} = \|\boldsymbol{x}\| \cdot \|\boldsymbol{y}\| \cdot \cos\left(\alpha(\boldsymbol{x}, \boldsymbol{y})\right),$$

where $\alpha(\boldsymbol{x}, \boldsymbol{y})$ denotes the angle between $\boldsymbol{x}$ and $\boldsymbol{y}$ (see Section 2.6). Therefore, the directional derivative equals

$$\begin{aligned} \frac{\partial f}{\partial \boldsymbol{p}}(\boldsymbol{x}) &= \operatorname{grad} f(\boldsymbol{x})\, \boldsymbol{p} \\ &= \|\operatorname{grad} f(\boldsymbol{x})\| \cdot \|\boldsymbol{p}\| \cdot \cos\left(\alpha(\operatorname{grad} f(\boldsymbol{x}), \boldsymbol{p})\right) \\ &= \|\operatorname{grad} f(\boldsymbol{x})\| \cdot \cos\left(\alpha(\operatorname{grad} f(\boldsymbol{x}), \boldsymbol{p})\right). \end{aligned}$$

Since $|\cos(x)| \leq 1$, the directional derivative attains its maximum when the cosine is one, that is, when the angle between the gradient and the direction $\boldsymbol{p}$ is zero. This shows that the gradient points in the direction of steepest ascent. In that direction, the slope equals the norm of the gradient.

The following observation reveals another key property of the gradient. The directions in which the slope is zero are those orthogonal to the gradient. In the bivariate case, these define the tangent to the contour line passing through the point $\boldsymbol{x}$, since the contour line is characterized by constant function values along it. Thus, the gradient is always perpendicular to its contour line.[7]

Mnemonic 5.6: Interpretation of the gradient

The gradient points in the direction of steepest ascent, with its norm giving the slope there. It stands perpendicular to the contour line through that point.

5.8 Local linear approximation

Using the derivative, a differentiable function f can be locally approximated near a given point. The slope of the tangent to a univariate function f matches that of the function at the point of tangency. Near the point, f can be approximated by a linear function, the tangent function. Since the tangent function is a first-degree polynomial, we say that it provides a **first order approximation** of f. The same holds for a differentiable function f in several variables. It too can be locally approximated at any point by a multivariate tangent function. This tangent function is linear, with its gradient matching that of f. This can be seen from the following representation of a differentiable function $\mathbb{R}^n \to \mathbb{R}$:

[7] In the trivariate case, the level sets, i.e., the sets of all points having the same function value, are surfaces in $\mathbb{R}^3$. Here, similarly, the gradient is perpendicular to its level set. The same holds in higher dimensions.

Theorem 5.11: Linear approximation by the tangent function

Let $A \subset \mathbb{R}^n$ be open, $f : A \to \mathbb{R}$, and $\boldsymbol{a} \in A$. If f is (totally) differentiable at $\boldsymbol{a}$, then there exists a function $r : A \to \mathbb{R}$ with $\lim_{\boldsymbol{x}\to\boldsymbol{a}} r(\boldsymbol{x}) = 0$ such that for all $\boldsymbol{x} \in A$,

$$f(\boldsymbol{x}) = f(\boldsymbol{a}) + \operatorname{grad} f(\boldsymbol{a}) \cdot (\boldsymbol{x} - \boldsymbol{a}) + r(\boldsymbol{x})\|\boldsymbol{x} - \boldsymbol{a}\| \,. \tag{5.2}$$

The function

$$t_{f,\boldsymbol{a}}(\boldsymbol{x}) = f(\boldsymbol{a}) + \operatorname{grad} f(\boldsymbol{a})(\boldsymbol{x} - \boldsymbol{a}) = f(\boldsymbol{a}) + \sum_{i=1}^{n} \frac{\partial f}{\partial x_i}(\boldsymbol{a})(x_i - a_i)\,, \quad \boldsymbol{x} \in \mathbb{R}^n\,,$$

is called the **tangent function** of f at $\boldsymbol{a}$. Its graph is the **tangent hyperplane**. Through $t_{f,\boldsymbol{a}}(\boldsymbol{x})$, f is locally approximated at $\boldsymbol{a}$: the function values match at $\boldsymbol{a}$, but generally differ elsewhere, usually increasing as $\|\boldsymbol{x} - \boldsymbol{a}\|$ grows (i.e., the farther $\boldsymbol{x}$ is from $\boldsymbol{a}$). The approximation is exact at $\boldsymbol{a}$ and sufficiently accurate in a small neighborhood for applications.

Example 5.19. Consider the function $f(x_1,x_2) = 2 - x_1 + 3x_2$. The tangent function of f at $\boldsymbol{a} = (1,0)$ is

$$\begin{aligned} t_{f,(1,0)}(\boldsymbol{x}) &= f(\boldsymbol{a}) + \operatorname{grad} f(\boldsymbol{a})(\boldsymbol{x} - \boldsymbol{a}) = f(1,0) + \operatorname{grad} f(1,0) \begin{bmatrix} x_1 - 1 \\ x_2 - 0 \end{bmatrix} \\ &= 1 + \begin{bmatrix} -1 & 3 \end{bmatrix} \begin{bmatrix} x_1 - 1 \\ x_2 \end{bmatrix} = 1 - (x_1 - 1) + 3x_2 = 2 - x_1 + 3x_2\,. \end{aligned}$$

The graph of the function thus coincides with the tangent hyperplane. With the choice $r(x_1,x_2) = 0$ for all x_1,x_2, Equation (5.2) is satisfied.

Example 5.20. For the function $f(x_1,x_2) = -x_1^2 - x_2^2$, the partial derivatives are $f_{x_1}(x_1,x_2) = -2x_1$ and $f_{x_2}(x_1,x_2) = -2x_2$. At $\boldsymbol{a} = (1,1)$, the tangent function is

$$\begin{aligned} t_{f,(1,1)}(x_1,x_2) &= f(1,1) + f_{x_1}(1,1)(x_1 - 1) + f_{x_2}(1,1)(x_2 - 1) \\ &= -2 + (-2)(x_1 - 1) + (-2)(x_2 - 1) \\ &= 2 - 2x_1 - 2x_2\,. \end{aligned}$$

The graph of f together with the tangent plane at $\boldsymbol{a} = (1,1)$ is shown in Figure 5.7. With $r(x_1,x_2) = -\sqrt{(x_1 - 1)^2 + (x_2 - 1)^2}$, Equation (5.2) is satisfied:

$$\begin{aligned} &t_{f,\boldsymbol{a}}(\boldsymbol{x}) + r(\boldsymbol{x})\|\boldsymbol{x} - \boldsymbol{a}\| \\ &= (2 - 2x_1 - 2x_2) - \sqrt{(x_1 - 1)^2 + (x_2 - 1)^2}\,\|\boldsymbol{x} - \boldsymbol{a}\| \\ &= (2 - 2x_1 - 2x_2) - \sqrt{(x_1 - 1)^2 + (x_2 - 1)^2}\sqrt{(x_1 - 1)^2 + (x_2 - 1)^2} \\ &= 2 - 2x_1 - 2x_2 - \left((x_1 - 1)^2 + (x_2 - 1)^2\right) \\ &= 2 - 2x_1 - 2x_2 - x_1^2 + 2x_1 - 1 - x_2^2 + 2x_2 - 1 \\ &= -x_1^2 - x_2^2 = f(\boldsymbol{x}) \end{aligned}$$

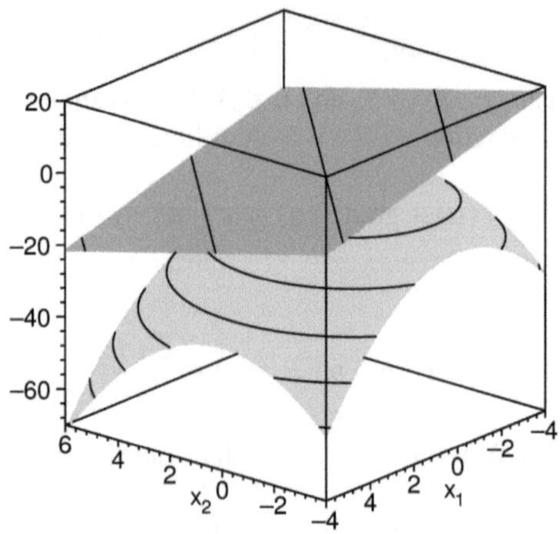

Figure 5.7: Graph of the function $f(x_1, x_2) = -x_1^2 - x_2^2$ and its tangent plane at $(1, 1)$.

Vector-valued functions, which are differentiable, can be approximated locally in the same way by an affine-linear function, the tangent function:

Theorem 5.12: Linear approximation (vector-valued functions)

Let $A \subset \mathbb{R}^n$ be open, $f : A \to \mathbb{R}^m$, and $\boldsymbol{a} \in A$. If f is (totally) differentiable at $\boldsymbol{a}$, then there exists a function $r : A \to \mathbb{R}^m$ with $\lim_{\boldsymbol{x}\to\boldsymbol{a}} r(\boldsymbol{x}) = \mathbf{0}$ such that for all $\boldsymbol{x} \in A$,

$$f(\boldsymbol{x}) = f(\boldsymbol{a}) + J_f(\boldsymbol{a})(\boldsymbol{x} - \boldsymbol{a}) + r(\boldsymbol{x})\|\boldsymbol{x} - \boldsymbol{a}\|,$$

where $J_f(\boldsymbol{a})$ denotes the Jacobian of f at $\boldsymbol{a}$.

The vector-valued function $t_{f,\boldsymbol{a}} : \mathbb{R}^n \to \mathbb{R}^m$, $\boldsymbol{x} \mapsto f(\boldsymbol{a}) + J_f(\boldsymbol{a})(\boldsymbol{x} - \boldsymbol{a})$ is the **tangent function** of f at $\boldsymbol{a}$. At any $\boldsymbol{x}$, its value deviates by $r(\boldsymbol{x})\|\boldsymbol{x} - \boldsymbol{a}\|$ from $f(\boldsymbol{x})$.

Important terms for review

After reading this chapter, the following terms should be familiar:

- Partial derivative, gradient
- Totally differentiable
- Total differential, partial differential
- Partial elasticity
- Elasticity of scale
- Jacobian matrix
- Chain rule
- Homogeneity of degree r, Euler's theorem
- Implicit function

- Marginal rate of technical substitution
- Elasticity of substitution, CES function
- Directional derivative
- Local linear approximation, tangent function

Self-test

You can check your understanding of this chapter using the following true-or-false questions. For each statement, indicate whether it is true (T) or false (F). Short solutions to these questions can be found in Appendix H.

I. Are the following statements true or false?

	T	F
The total differential is the sum of the partial differentials.	☐	☐
The elasticity of scale is the sum of the partial elasticities.	☐	☐
The gradient is the sum of the partial derivatives.	☐	☐

II. The Jacobian ...

	T	F
... of a real-valued differentiable function is the gradient.	☐	☐
... of the function $f(x, y, z) = \left[x + 3y + z^2, y^{42} + z, z\right]^T$ is a lower triangular matrix.	☐	☐
... of the function $f(x, y, z) = \left[x + 3y + z^2, y^{42} + z, z\right]^T$ is an upper triangular matrix.	☐	☐
... of the function $f(x, y, z)+g(x, y, z)$, where $f, g : \mathbb{R}^3 \to \mathbb{R}^3$, is the sum of the Jacobians of $f(x, y, z)$ and $g(x, y, z)$.	☐	☐

III. Consider the functions $f : \mathbb{R} \to \mathbb{R}^2$, $f(x) = \left[x^3, 2x\right]^T$, and $g : \mathbb{R}^2 \to \mathbb{R}$, $g(y_1, y_2) = y_1 - y_2$. Are the following statements true or false?

	T	F
The Jacobian of f is $f'(x) = \left[3x^2, 2\right]^T$.	☐	☐
The gradient of f is $f'(x) = \left[3x^2, 2\right]$.	☐	☐
The Jacobian of g is $g'(x) = [1, -1]^T$.	☐	☐
The gradient of g is $g'(x) = [1, -1]$.	☐	☐
According to the chain rule, the derivative of $g \circ f$ is $(g \circ f)'(x) = [1, -1] \left[3x^2, 2\right]^T$.	☐	☐

	T	F
According to the chain rule, the derivative of $g \circ f$ is $(g \circ f)'(x) = [1, -1]^T \left[3x^2, 2\right]$.	☐	☐

IV. Are the following statements about homogeneity of functions true or false?

	T	F
The sum of finitely many functions that are homogeneous of degree r is again homogeneous of degree r.	☐	☐
A function of the form $e^{f(x,y)}$ cannot be homogeneous.	☐	☐
A Cobb-Douglas production function with n inputs is homogeneous of degree equal to the sum of its exponents.	☐	☐

V. Are the following statements true or false?

	T	F
When computing the derivative of a function $y = f(x)$ defined implicitly by $F(x, y) = 0$, this derivative generally still depends on y, requiring a value $b = f(a)$ to evaluate $f'(a)$.	☐	☐
The greater the elasticity of substitution, the less we can substitute one input for the other.	☐	☐
The total differential of a differentiable function f gives the exact change in the function value of the tangent function.	☐	☐

Exercises

Exercise 1

Find the gradients of the following functions $f : A \to \mathbb{R}$:

a) $f(x, y, z) = x^3yz^2 + 2x^2y^2z + 5z^3$, $A = \mathbb{R}^3$,

b) $f(x, y, z) = e^{-x}\cos(yz)$, $A = \mathbb{R}^3$,

c) $f(x, y, z) = x\ln(y^2 + 2z^3 + 1)$, $A = \{(x, y, z)^T \in \mathbb{R}^3 : z \geq 0\}$.

Exercise 2

Find the partial elasticities of the following functions $f : \mathbb{R}^2 \to \mathbb{R}$:

a) $f(x, y) = xy$,

b) $f(x, y) = x^2y^5$,

c) $f(x, y) = x^n e^x y^n e^y$.

Exercise 3

Find all points, at which the function $f : \mathbb{R}^2 \to \mathbb{R}$ with

$$f(x, y) = y\sqrt{2x^2 + y^2}$$

is partially differentiable, and calculate the partial derivatives there.

Exercise 4

Consider the function $f : \mathbb{R}^2 \to \mathbb{R}$, $f(x, y) = x^3 - 4xy^2 - y^3 + x^2$.

a) Find the partial differential with respect to y.

b) Find the total differential.

c) Find the total differential at $(x, y) = (3, 1)$.

Exercise 5

Consider the production function

$$q = f(x, y, z) = \sqrt[3]{2x^2y + 3xz^2 + 4y^2z}\,.$$

Currently, production uses the inputs $(x, y, z) = (80, 100, 120)$.

a) Compute approximately the increase in output when all inputs increase by two units each.

b) By how many percent does output increase if each input increases by 30 percent?

Exercise 6

Consider the function $f : \mathbb{R}^2 \to \mathbb{R}$, $f(x, y) = x^2 - 4y^2$.

a) Sketch the contour lines for the values $-9, -4, -1, 0, 1, 4$, and 9.

b) Determine the tangent function $t_{f,\boldsymbol{a}}$ for $\boldsymbol{a} = (0, 1)^T$ and $\boldsymbol{a} = (2, 1)^T$.

Exercise 7

Which of the following functions are homogeneous? If so, determine the degree of homogeneity?

a) $f(x, y) = x^2 + xy - y^2$

b) $g(x, y, z) = xyz + 1$

c) $h(x, y) = \sqrt{x}y + \sqrt{y^3}$

d) $i(x, y, z) = z \cdot e^{\frac{x}{y}} - \sqrt{\frac{xy^2}{z}}$

c) $j(x, y) = \ln\left(\left(\frac{x}{y}\right)^x\right)$

Exercise 8

Consider the production function $f : \mathbb{R}^2 \to \mathbb{R}$, $f(x, y) = x^2y + 2xy^2$.

a) Is f homogeneous? If so, determine the degree of homogeneity.

b) By how many percent does the output decrease if each input is reduced by 10 percent?

Exercise 9

Consider the function $f : \mathbb{R}^2 \to \mathbb{R}$, $f(x, y) = 3x^2y^3 + 1$.

a) Determine the partial elasticities and the elasticity of scale.

b) Is f homogeneous? If so, determine the degree of homogeneity.

c) By how many percent does the function value approximately increase if both arguments – starting from $x = 2$ and $y = 1$ – are increased by 2 percent?

Exercise 10

Consider the functions

a) $F(x, y) = x^3 - 2xy + y^2$ and

b) $F(x, y) = e^{xy} + 2xy^2$.

The function $y = f(x)$ is defined implicitly by $F(x, y) = c$ for $c \in \mathbb{R}$. Determine the derivative $f'(x)$ in each case.

Exercise 11

Determine the derivative dz/dt in the following cases using the chain rule (i.e., first compute the partial derivatives of f with respect to x and y):

a) $z = f(x, y) = x + xy + y$, where $x = t$ and $y = t^3$,

b) $z = f(x, y) = xe^{2y}$, where $x = t^2$ and $y = \ln(t)$,

c) $z = f(x, y) = x^2y + y^2x$, where $x = e^t$ and $y = 2t$.

Exercise 12

The demand q of a household for a good is given by $q = 100p^{-0.3}y^{0.35}$ as a function of price p and income y. Assume the price grows by 2% and income by 3% per unit time, so $p = p_0 \cdot 1.02^t$ and $y = y_0 \cdot 1.03^t$, where p_0 and y_0 denote price and income at time $t = 0$. Using the chain rule, approximately calculate the relative change in demand per unit time.

Exercise 13

Compute the Jacobian of the function

$$f : \mathbb{R}^2 \to \mathbb{R}^3, \qquad f(x, y) = \begin{bmatrix} x^2 + y \\ x - y^2 \\ 2 - xy \end{bmatrix}.$$

What is the value of the derivative of f at $(2, 1)$?

Exercise 14

Consider the functions $f : \mathbb{R}^2 \to \mathbb{R}^2$, $g : \mathbb{R}^3 \to \mathbb{R}^2$ with

$$f(\boldsymbol{x}) = \begin{bmatrix} x_1^2 - x_2^2 \\ 2x_1x_2 \end{bmatrix} \quad \text{and} \quad g(\boldsymbol{x}) = \begin{bmatrix} x_1 + x_2 \\ 2x_3 \end{bmatrix}.$$

Which of the compositions $f \circ f$, $f \circ g$, $g \circ f$, and $g \circ g$ are defined? For the defined ones, calculate the derivative at $\boldsymbol{x} = [0, 1]^T$ or $\boldsymbol{x} = [1, 0, -1]^T$.

Exercise 15

a) Compute the directional derivative $\frac{\partial f}{\partial \boldsymbol{p}}(x, y)$ of the function

$$f(x, y) = \frac{\sqrt{3}}{2}x^2 + y$$

in direction $\boldsymbol{p} = \left[\frac{3}{5}, \frac{4}{5}\right]^T$ at any point $(x, y) \in \mathbb{R}^2$.

b) A small ball is placed at the point $(1, 1, 1 + \frac{1}{2}\sqrt{3})$ on the surface $z = \frac{\sqrt{3}}{2}x^2 + y$. In which direction does the ball start to roll?

Chapter 6

Optimization of functions of several variables

In this chapter, we determine maxima and minima of functions of several variables using calculus.

Example 6.1 (Maximum profit). The profit of a company depends on a multitude of decision variables, such as quantities produced, inputs used, and advertising expenditures. The goal is to find a global maximum of the profit function. The decision variables cannot take arbitrary values, but are subject to constraints on the domain. Therefore, a profit maximum must be sought in the permissible domain.

Example 6.2 (Production at minimum cost). The output of a production is given. It should be created at minimal cost. The cost function is to be minimized with respect to the inputs used, but only for those input combinations that deliver the given output. This is a problem of cost minimization subject to an output constraint.

The following sections deal with the determination of extrema in the interior of the domain of a multivariate function, on the boundary of its domain, and subject to constraints. In the interior of the domain, local maxima and minima are characterized by the first two derivatives of the function, similarly to functions of one variable. Searching for extrema on the boundary requires special considerations. Maxima and minima subject to some given constraints are determined using either Lagrange's method or the elimination method.

K. Mosler et al., *Mathematical Methods for Business and Economics*, Mathematics Study Resources 20, https://doi.org/10.1007/978-3-662-73831-3_6

6.1 Extrema in the interior of the domain

Maxima and minima of a multivariate function are defined almost literally as in the univariate case. Consider a domain $A \subset \mathbb{R}^n$, a function $f : A \to \mathbb{R}$, and a point $\boldsymbol{a} \in A$.

Definition 6.1: Maximum, minimum

- f has a **global maximum** at $\boldsymbol{a}$, if $f(\boldsymbol{a}) \geq f(\boldsymbol{x})$ for all $\boldsymbol{x} \in A$.
- f has a **global minimum** at $\boldsymbol{a}$, if $f(\boldsymbol{a}) \leq f(\boldsymbol{x})$ for all $\boldsymbol{x} \in A$.
- f has a **local maximum** at $\boldsymbol{a}$, if there exists an $\varepsilon > 0$ such that $f(\boldsymbol{a}) \geq f(\boldsymbol{x})$ for all $\boldsymbol{x} \in A$, which are in the ε-neighborhood of $\boldsymbol{a}$, i.e., for which $||\boldsymbol{x} - \boldsymbol{a}|| < \varepsilon$ holds.
- f has a **local minimum** at $\boldsymbol{a}$, if there exists an $\varepsilon > 0$ such that $f(\boldsymbol{a}) \leq f(\boldsymbol{x})$ for all $\boldsymbol{x} \in A$, which are in the ε-neighborhood of $\boldsymbol{a}$.

A local extremum is also called **relative**, and a global extremum called **absolute**.

Every global maximum or minimum is also a local maximum or minimum, but the reverse is not true.

We first examine a given function $f : A \to \mathbb{R}$ for possible extrema in the interior of its domain, i.e., at the points of A except the boundary points.[1]

In the following, we will restrict ourselves to domains $A \subset \mathbb{R}^n$ that are n-dimensional intervals. In the case $n = 2$, A is a rectangle, in the case $n = 3$ a **cuboid**. An n-dimensional interval can have finite or infinite boundaries; for example, $A = \mathbb{R}^n_+ = [0, \infty) \times \ldots \times [0, \infty)$ and $A = \mathbb{R}^n$ are allowed. The interval can be open, closed, or half-open. For example, an upper half-open interval has the form

$$A = [\boldsymbol{b}, \boldsymbol{c}) = [b_1, c_1) \times \ldots \times [b_n, c_n) \quad \text{with} \quad \boldsymbol{b}, \boldsymbol{c} \in \mathbb{R}^n .$$

The interior of A is the corresponding open interval,

$$(\boldsymbol{b}, \boldsymbol{c}) = (b_1, c_1) \times \ldots \times (b_n, c_n) .$$

The function $f : A \to \mathbb{R}$ is assumed to be (totally) differentiable. If f has a local maximum at a point in the interior, $\boldsymbol{a} = (a_1, \cdots, a_n)$, then f must be maximal there with respect to each individual variable $x_1, \ldots, x_n$ and have a horizontal tangent at $\boldsymbol{a}$. It follows that the partial derivatives at $\boldsymbol{a}$ are all zero. The same applies if f has a local minimum at $\boldsymbol{a}$. Therefore, the following theorem holds:

[1] A general definition of the terms "interior" and "boundary points" can be found in Section 2.7.

Theorem 6.1: Necessary condition for a local extremum

If f has a local maximum or minimum at point $\boldsymbol{a}$ in the interior of its domain, then

$$\frac{\partial f}{\partial x_i}(\boldsymbol{a}) = 0 \text{ for all } i = 1, \ldots, n, \quad \text{i.e., } \operatorname{grad} f(\boldsymbol{a}) = \mathbf{0}\,.$$

The condition of the theorem is necessary, but generally not sufficient for an extremum of f at $\boldsymbol{a}$. If the partial derivatives at $\boldsymbol{a}$ are all zero, $\boldsymbol{a}$ is called a **stationary point**. There may be a maximum or minimum at this point, but there may also be a **saddle point**. $\boldsymbol{a}$ is a saddle point of f if $\operatorname{grad} f(\boldsymbol{a}) = \mathbf{0}$ and for every $\varepsilon > 0$ exist points $\boldsymbol{x}_-$ and $\boldsymbol{x}_+$ in the ε-neighborhood of $\boldsymbol{a}$ such that $f(\boldsymbol{x}_-) < f(\boldsymbol{a}) < f(\boldsymbol{x}_+)$.

Example 6.3 (Profit function). The function $g(x_1, x_2) = 4x_1x_2 - x_1^4 - x_2^4$ (Figure 6.1) describes the profit of a company. The variables x_1 and x_2 can take any values in $\mathbb{R}$ and are to be chosen so that the profit is maximized. We first examine the necessary conditions for a local extremum. The gradient of g is

$$\operatorname{grad} g(\boldsymbol{x}) = \operatorname{grad} g(x_1, x_2) = [4x_2 - 4x_1^3\,,\; 4x_1 - 4x_2^3]\,.$$

Setting the gradient to zero yields the two equations $4x_2 - 4x_1^3 = 0$ and $4x_1 - 4x_2^3 = 0$, which have three solutions: $(0, 0), (1, 1)$ and $(-1, -1)$. The corresponding function values are $g(0, 0) = 0$ and $g(1, 1) = g(-1, -1) = 2$.

It is easy to see that $g(0, 0)$ can be neither maximum nor minimum since g has both positive and negative values in every neighborhood of the origin: For any number $\alpha \in (0, \sqrt{2})$, we have $g(\alpha, \alpha) = 2\alpha^2(2 - \alpha^2) > 0$ and $g(-\alpha, \alpha) = -4\alpha^2 - 2\alpha^4 < 0$. Whether or not local extrema exist at the solutions $(1, 1)$ and $(-1, -1)$, we will check later, after introducing a sufficient criterion.

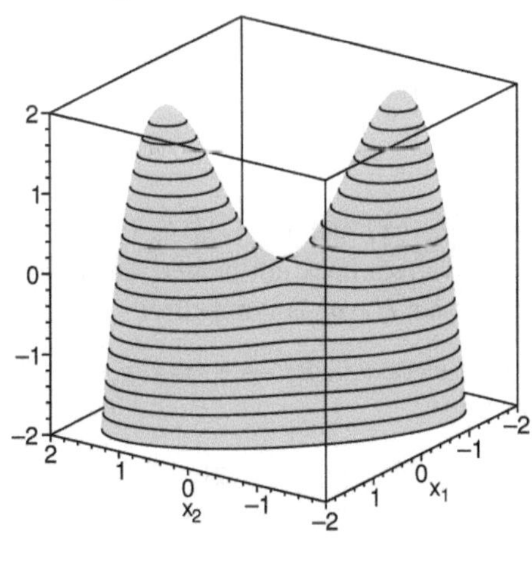

Figure 6.1: Graph of the bivariate function $g(x_1, x_2) = 4x_1x_2 - x_1^4 - x_2^4$.

Like a univariate function, a multivariate one has a second derivative that indicates its curvature at the point in question. By this, at a stationary point one can conclude a minimum or maximum. The second derivative of a function in n variables consists of its n^2 second order partial derivatives, which form the socalled Hessian matrix:

Definition 6.2: Higher partial derivatives, Hessian matrix

If $f : A \longrightarrow \mathbb{R}$ is partially differentiable and the partial derivatives of f are themselves partially differentiable, then one can form the second order partial derivatives. For $i, j = 1, 2, \ldots, n$ they are denoted by

$$\frac{\partial^2 f}{\partial x_i \partial x_j}(\boldsymbol{a}) = \frac{\partial}{\partial x_i}\frac{\partial f}{\partial x_j}(\boldsymbol{a}), \quad \frac{\partial^2 f}{\partial x_i^2}(\boldsymbol{a}) = \frac{\partial}{\partial x_i}\frac{\partial f}{\partial x_i}(\boldsymbol{a}).$$

The second order partial derivatives are summarized in a matrix, the **Hessian**[2] **matrix** (short **Hessian**) at point $\boldsymbol{a}$,

$$H_f(\boldsymbol{a}) = \left[\, \frac{\partial^2 f}{\partial x_i \partial x_j}(\boldsymbol{a}) \,\right]_{i,j=1,\ldots,n}.$$

If all second order partial derivatives exist and are continuous, then we call f **twice continuously differentiable** or in short: **twice differentiable.**

For simplification we omit the argument $\boldsymbol{a}$ when no ambiguity arises, and write for the partial derivatives

$$f_{x_i} = \frac{\partial f}{\partial x_i}, \qquad f_{x_i x_j} = \frac{\partial^2 f}{\partial x_i \partial x_j}$$

and for the Hessian

$$H_f = \begin{bmatrix} f_{x_1x_1} & \cdots & f_{x_1x_n} \\ \vdots & & \vdots \\ f_{x_nx_1} & \cdots & f_{x_nx_n} \end{bmatrix}.$$

If the functions $\boldsymbol{x} \mapsto \frac{\partial^2 f}{\partial x_i \partial x_j}(\boldsymbol{x})$ and $\boldsymbol{x} \mapsto \frac{\partial^2 f}{\partial x_j \partial x_i}(\boldsymbol{x})$ are both continuous on A, the order of the two derivatives does not matter. Then at every point $\boldsymbol{a}$ it holds $f_{x_ix_j} = f_{x_jx_i}$, and the Hessian $H_f(\boldsymbol{a})$ is symmetric.

Example 6.4. For the function $f : \mathbb{R}^3 \to \mathbb{R}$, $f(x, y, z) = x^2y + y^3z + 3xz^2 - 4yz$, the Hessian is given by

$$H_f(x, y, z) = \begin{bmatrix} 2y & 2x & 6z \\ 2x & 6yz - 4z & 3y^2 - 4x \\ 6z & 3y^2 - 4x & 6x - 4y \end{bmatrix}.$$

[2]Ludwig Otto Hesse (1811–1874)

Theorem 6.2: Sufficient condition for a local extremum

Let $f : A \to \mathbb{R}$ and $\boldsymbol{a}$ be a point in the interior of A, where f is twice continuously differentiable. Let $f_{x_i}(\boldsymbol{a}) = 0$ for $i = 1, 2, \dots, n$.

(i) In the case $n = 2$ we consider the expressions

$$\Delta_1 := f_{x_1x_1} \quad \text{and} \quad \Delta_2 := f_{x_1x_1} f_{x_2x_2} - (f_{x_1x_2})^2 .$$

If at the point $\boldsymbol{a}$

$\Delta_1 > 0$ and $\Delta_2 > 0$, then f has a local minimum at $\boldsymbol{a}$,
$\Delta_1 < 0$ and $\Delta_2 > 0$, then f has a local maximum at $\boldsymbol{a}$.

(ii) In the case $n = 3$, we additionally consider the expression

$$\begin{aligned}\Delta_3 \quad := \quad & f_{x_1x_1} f_{x_2x_2} f_{x_3x_3} + 2 f_{x_1x_2} f_{x_2x_3} f_{x_3x_1} \\ & - f_{x_1x_1}(f_{x_2x_3})^2 - f_{x_2x_2}(f_{x_3x_1})^2 - f_{x_3x_3}(f_{x_1x_2})^2 .\end{aligned}$$

If at the point $\boldsymbol{a}$

$\Delta_1 > 0$, $\Delta_2 > 0$ and $\Delta_3 > 0$, then f has a local minimum at $\boldsymbol{a}$,
$\Delta_1 < 0$, $\Delta_2 > 0$ and $\Delta_3 < 0$, then f has a local maximum at $\boldsymbol{a}$.

(iii) For $n \geq 2$: If $\Delta_2 < 0$ at the point $\boldsymbol{a}$, then f has a saddle point at $\boldsymbol{a}$.

Example 6.5 (Profit function). In Example 6.3 we calculate the Hessian $H_g(\boldsymbol{x})$ of the profit function first for general $\boldsymbol{x}$:

$$H_g(\boldsymbol{x}) = \begin{bmatrix} -12x_1^2 & 4 \\ 4 & -12x_2^2 \end{bmatrix} .$$

Substituting the points $\boldsymbol{x} = (1, 1)$ and $\boldsymbol{x} = (-1, -1)$ yields

$$H_g(1, 1) = H_g(-1, -1) = \begin{bmatrix} -12 & 4 \\ 4 & -12 \end{bmatrix} .$$

In both points, $\Delta_1 = -12 < 0$ and $\Delta_2 = (-12)^2 - 4^2 = 128 > 0$. Thus, the sufficient condition of the theorem is satisfied. We have found two maxima of the profit functon, at $(1, 1)$ and at $(-1, -1)$.

For $\boldsymbol{x} = (0, 0)$ we get

$$H_g(0, 0) = \begin{bmatrix} 0 & 4 \\ 4 & 0 \end{bmatrix} .$$

At the point $\boldsymbol{x} = (0, 0)$, $\Delta_2 = 0^2 - 4^2 = -16 < 0$. So g does not have an extremum there, but a saddle point.

Example 6.6. Let $f : \mathbb{R}^2 \to \mathbb{R}$, $f(x_1, x_2) = x_1x_2$. At the point $(x_1, x_2) = (0, 0)$ the necessary condition is satisfied, $f_{x_1}(0, 0) = f_{x_2}(0, 0) = 0$. However, since

$\Delta_2(0,0) = -1 < 0$, no extremum exists at $(0,0)$, but a saddle point; see Figure 6.2.

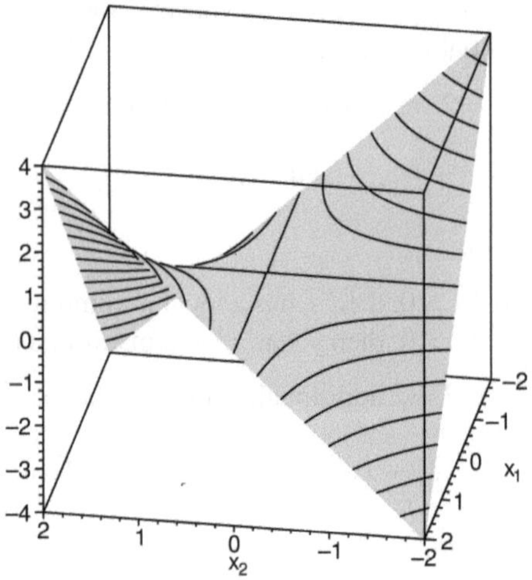

Figure 6.2: Graph and contour lines of the function $f(x_1, x_2) = x_1 x_2$.

In the cases $n = 2$ and $n = 3$, Δ_2 resp. Δ_3 is the determinant (see Chapter 10) of the Hessian of f. In general, Δ_i is the determinant of the top-left $i \times i$ submatrix of the Hessian. For $n \geq 4$, sufficient conditions analogous to Theorem 6.2 (i) and (ii) are formulated with corresponding determinants Δ_i, $i \leq n$; see Section 6.6 .

6.2 Extrema on the boundary of the domain

In most applications, economic variables cannot take arbitrary real numbers as values. If the domain A of a function f does not include all of $\mathbb{R}^n$, then local and global extrema can also occur on the boundary of A.However, the above necessary and sufficient conditions on the derivatives of f only capture the local extrema **in the interior** of the domain A. Local extrema that occur **on the boundary** are usually not found in this way.

Their determination requires a separate examination of the values that the function takes at the boundary points and in their vicinity. Depending on the shape of the boundary of A, very different approaches are considered.

If A is an n-dimensional interval, its boundary can be described relatively simply. For example, the boundary of a bounded closed rectangle in $\mathbb{R}^2$, $[b_1, c_1] \times [b_2, c_2]$, consists of four line segments.

Example 6.7. Consider the following infinite **open cuboid** $A \subset \mathbb{R}^3$, $A = (b_1, c_1) \times (b_2, \infty) \times (b_3, c_3)$. Its boundary consists of five faces, four of which are unbounded rectangles.

We illustrate the analysis of boundary extrema with the following example.

Example 6.8. Let $f(x_1,x_2) = x_1^3 x_2 - x_1^2 - x_2$, $(x_1,x_2) \in \mathbb{R}_+^2$ (Figure 6.3).
The first partial derivatives are

$$f_{x_1}(x_1,x_2) = 3x_1^2 x_2 - 2x_1\,, \quad f_{x_2}(x_1,x_2) = x_1^3 - 1\,.$$

Setting them to zero yields the unique solution $x_1 = 1, x_2 = 2/3$. If an extremum exists in the interior of $\mathbb{R}_+^2$, it must lie at the point $(1, 2/3)$. The Hessian gives more information:

$$H_f(x_1,x_2) = \begin{bmatrix} 6x_1x_2 - 2 & 3x_1^2 \\ 3x_1^2 & 0 \end{bmatrix}, \quad H_f\left(1, \tfrac{2}{3}\right) = \begin{bmatrix} 2 & 3 \\ 3 & 0 \end{bmatrix},$$

$$\Delta_1 = 2 > 0\,, \quad \Delta_2 = -9 < 0\,.$$

Thus, there is no extremum in the interior of $\mathbb{R}^2$. Next, we examine the boundary of the domain $A = \mathbb{R}_+^2$: The boundary $\mathrm{bd}(A)$ consists of the two non-negative semi-axes,

$$\mathrm{bd}(A) = \{(x_1,x_2) \in \mathbb{R}_+^2 \,:\, x_1 = 0 \text{ or } x_2 = 0\}\,.$$

On the part of the boundary with $x_2 = 0$, we look for possible extremes of the function f by setting the second variable $x_2 = 0$ and determining the local extremes of the univariate function $[0,\infty) \to \mathbb{R}$, $x \mapsto f(x,0)$ as in Section 4.4. However, this only yields *candidates* for extremes. To decide whether such a candidate really represents a local extremum, one must also examine the behavior of the function in the adjacent interior of the domain.

On the horizontal semi-axis ($x_2 = 0$), $f(x_1,0) = -x_1^2 \le 0$ reaches its maximum at the point $x_1 = 0$. On the vertical semi-axis ($x_1 = 0$), $f(0,x_2) = -x_2 \le 0$ is maximal at $x_2 = 0$.

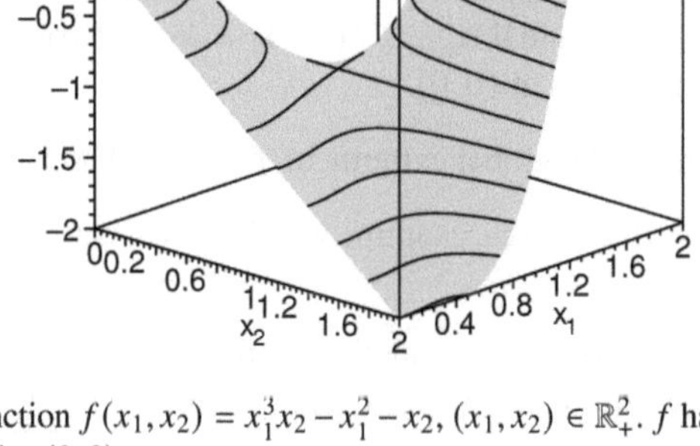

Figure 6.3: The function $f(x_1,x_2) = x_1^3 x_2 - x_1^2 - x_2$, $(x_1,x_2) \in \mathbb{R}_+^2$. f has a local maximum at the boundary point $(0,0)$.

The only candidate for an extremum is therefore the point $(0,0)$. We check whether there exists a maximum at this point by looking at the behavior of the function in a small neighborhood of the point $(0,0)$. It holds

$$f(x_1,x_2) = \underbrace{x_2}_{\geq 0} \cdot \underbrace{(x_1^3 - 1)}_{\substack{<0 \\ \text{for } x_1 < 1}} - \underbrace{x_1^2}_{\geq 0} \leq 0 \quad \text{for } 0 \leq x_1 < 1,\ 0 \leq x_2\,.$$

Hence, there is a local maximum at the point $(0,0)$; the corresponding function value is $f(0,0) = 0$.

6.3 Global extrema

With the methods described so far, **local** maxima and minima of a multivariate function are calculated. However, in most applications, one is interested in **global** extrema. As before, let f be a differentiable function $A \to \mathbb{R}$ and A an n-dimensional interval in $\mathbb{R}^n$.

We first determine the global extrema of the functions from Examples 6.3 and 6.8.

Example 6.9 (Profit function). We again consider the profit function from Example 6.3, $g(x_1,x_2) = 4x_1x_2 - x_1^4 - x_2^4$ having unrestricted domain $\mathbb{R}^2$. As we have already calculated, the profit function attains two local maxima, at points $(1,1)$ and $(-1,-1)$. The function value is $g(1,1) = g(-1,-1) = 2$. Since the gradient vanishes nowhere except at the points already examined, there can be no further local maxima of g in $\mathbb{R}^2$. Boundary extremes do not need to be considered here, as $\mathbb{R}^2$ has no boundary points. However, the behavior of $g(\boldsymbol{x})$ needs to be examined when $\boldsymbol{x}$ grows beyond all limits in any direction. For $||\boldsymbol{x}|| \to \infty$, i.e., when the distance to the origin tends to infinity, we get $g(\boldsymbol{x}) \to -\infty$. The two local maxima are therefore also global maxima. There are no local minima, and consequently no global minima.

Example 6.10. In Example 6.8 there is a local maximum at $(0,0)$; it has the function value $f(0,0) = 0$. This is not a global maximum, as for example $f(2,1) = 3 > 0$. Overall, the function on $\mathbb{R}_+^2$ has neither a global maximum nor a global minimum. The function takes arbitrarily large and small values, for example $\lim_{x_1\to\infty} f(x_1,0) = -\infty$ and $\lim_{x_1\to\infty} f(x_1,1) = \infty$. It also has no local minimum, but only a local maximum at $(0,0)$.

In general, to determine the global extrema of f, proceed as follows:

- Using the first two derivatives, examine whether local extrema exist in the interior of the domain A.
- Consider the restriction of the function to the boundary of A (or parts of the boundary) and determine the local extrema there.
- Check whether these extrema are also extreme with respect to the adjacent points

of the interior of A.

- If the domain (as in Example 6.3) is unbounded, examine the behavior of the function for $||\boldsymbol{x}|| \to \infty$.
- If a global maximum exists, it must be one of the local maxima (the one with the largest function value).
- An analogous statement holds for a global minimum.

In Example 6.8 there is neither a global maximum nor a global minimum, while in Example 6.3 there are two global maxima, but no such minimum. In general, the question arises under what circumstances a function has global extrema. The following theorem applies (cf. Theorem 3.32):

Theorem 6.3: Sufficient condition for the existence of global extrema

If A is bounded and closed[3], then a continuous function on A has at least one global maximum and one global minimum.

Every differentiable function is continuous according to Theorem 5.1. Therefore, a differentiable function that is defined on a bounded closed set has both a global maximum and a global minimum.

In particular, if A is a bounded closed interval, i.e., of the form

$$A = [b_1, c_1] \times [b_2, c_2] \times \ldots \times [b_n, c_n],$$

then f has a global minimum and maximum on A, either at a point $\boldsymbol{a}$ in the interior with grad $f(\boldsymbol{a}) = 0$ or at a point on the boundary of A.

The determination of global maxima and minima is considerably simplified if f is a concave or a convex function. In this case, as with univariate functions (Theorem 4.8), it is sufficient to search the local maxima or minima. The concavity and convexity of a function of several variables is defined literally as for a function of one variable; cf. Definition 1.4 and Section 1.5. The following theorem characterizes the location of global extrema in concave and convex functions.

Theorem 6.4: Global extrema of concave and convex functions

Let A be an n-dimensional interval, $\boldsymbol{a} \in A$, and let $f : A \to \mathbb{R}$ be either a concave or a convex function. Then the following statements hold:

(i) If f is concave, then every local maximum is also a global maximum.

(ii) If f is convex, then every local minimum is also a global minimum.

(iii) If f is concave and has a global minimum, then it lies one on the

[3] I.e., all boundary points are contained in A; cf. Section 2.7.

boundary of A.

(iv) If f is convex and has a global maximum, then it lies on the boundary of A.

Figure 1.15 shows the graph of a bivariate convex function, which has a single local, and thus global minimum. The function shown in Figure 6.4 is also convex, but has more than one local (and global) minimum: It has global minima in all points where $x_1^2 + x_2^2 \leq 1$. In addition, it has four global maxima at the corners of the domain, for example at the point $(2, 2)$.

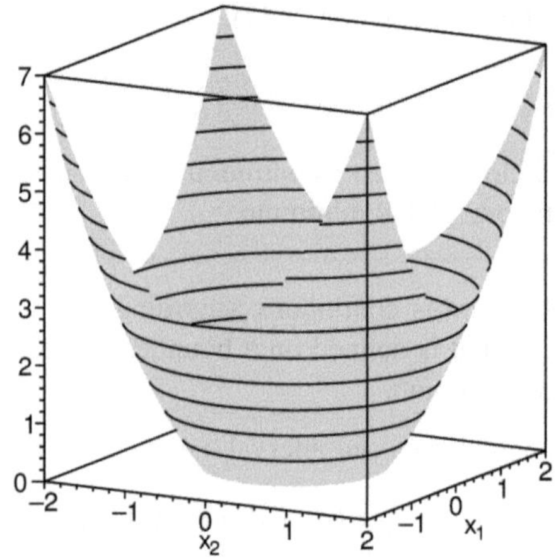

Figure 6.4: Graph of $f(x_1, x_2) = \max\{(x_1^2 + x_2^2 - 1), 0\}, -2 \leq x_1, x_2 \leq 2$.

If a given function is concave or convex, one only needs to maximize and minimize it locally to obtain global extrema. Whether a function is concave or convex or neither can be checked with the Hessian:

Theorem 6.5: Convexity and concavity of a differentiable function

Let $f : A \to \mathbb{R}$ be twice continuously differentiable, where A is convex.

(i) If $n = 2$, then:

- If $\Delta_1 > 0, \Delta_2 > 0$ holds at every $\boldsymbol{x} \in A$, then f is convex.
- If $\Delta_1 < 0, \Delta_2 > 0$ holds at every $\boldsymbol{x} \in A$, then f is concave.

(ii) If $n = 3$, then:

- If $\Delta_1 > 0, \Delta_2 > 0, \Delta_3 > 0$ holds at every $\boldsymbol{x} \in A$, then f is convex.
- If $\Delta_1 < 0, \Delta_2 > 0, \Delta_3 < 0$ holds at every $\boldsymbol{x} \in A$, then f is concave.

Example 6.11. We consider the function $f : \mathbb{R}^2 \to \mathbb{R}$,

$$f(x_1, x_2) = \frac{(x_1 + 2)^2}{4} + \frac{(x_2 + 1)^2}{9} - 1$$

from Example 1.19 (Figure 1.15). It has the gradient

$$\operatorname{grad} f(\boldsymbol{x}) = \left[\tfrac{x_1+2}{2}, \tfrac{2(x_2+1)}{9} \right]$$

and the Hessian

$$H_f(\boldsymbol{x}) = \begin{bmatrix} \frac{1}{2} & 0 \\ 0 & \frac{2}{9} \end{bmatrix} \quad \text{for all } \boldsymbol{x} \in \mathbb{R}^2 .$$

At every point $\boldsymbol{x} \in \mathbb{R}^2$ it holds

$$\Delta_1 = f_{x_1x_1} = \frac{1}{2} > 0, \qquad \Delta_2 = f_{x_1x_1} f_{x_2x_2} - (f_{x_1x_2})^2 = \frac{1}{2} \cdot \frac{2}{9} - 0^2 = \frac{1}{9} > 0,$$

so f is convex. Setting the gradient to zero yields exactly one solution, $\boldsymbol{x} = (-2, -1)$. Due to the convexity of f, this is the only local and thus global minimum. It holds $f(-2, 1) = -1 \leq f(x_1, x_2)$ for all $(x_1, x_2) \in \mathbb{R}^2$.

If one finds a local minimum (or local maximum) by Theorem 6.2 and is able to prove with the help of Theorem 6.5 that the function is convex (resp. concave), then Theorem 6.4 yields that the local minimum (resp. local maximum) is a global one. This is summarized as the following sufficient conditions for global extrema in the interior of the domain:

Theorem 6.6: Sufficient condition for a global extremum

Let $f : A \longrightarrow \mathbb{R}$ be twice continuously differentiable, where A is convex. Let further $\boldsymbol{a}$ be in the interior of A and assume that $f_{x_i}(\boldsymbol{a}) = 0$ for $i = 1, 2, \ldots, n$.

(i) In the case $n = 2$, if for all $\boldsymbol{x} \in A$ holds

$\Delta_1 > 0$ and $\Delta_2 > 0$, then f has a global minimum at $\boldsymbol{a}$,
$\Delta_1 < 0$ and $\Delta_2 > 0$, then f has a global maximum at $\boldsymbol{a}$.

(ii) In the case $n = 3$, if for all $\boldsymbol{x} \in A$ holds

$\Delta_1 > 0, \Delta_2 > 0$ and $\Delta_3 > 0$, then f has a global minimum at $\boldsymbol{a}$,
$\Delta_1 < 0, \Delta_2 > 0$ and $\Delta_3 < 0$, then f has a global maximum at $\boldsymbol{a}$.

Example 6.12. We consider the function $f : \mathbb{R}^2 \to \mathbb{R}^2$,

$$f(x_1, x_2) = x_1^4 + x_1^2 + x_2^2 .$$

For the function f holds

$$\operatorname{grad} f(\boldsymbol{x}) = \left[4x_1^3 + 2x_1 , 2x_2 \right] \quad \text{and} \quad H_f(\boldsymbol{x}) = \begin{bmatrix} 12x_1^2 + 2 & 0 \\ 0 & 2 \end{bmatrix} .$$

Setting the partial derivatives equal to zero, we get

$$4x_1^3 + 2x_1 \stackrel{!}{=} 0 \iff 2x_1(2x_1^2 + 1) = 0 \iff x_1 = 0\,,$$
$$2x_2 \stackrel{!}{=} 0 \iff x_2 = 0\,,$$

and thus the only stationary point $\boldsymbol{a} = (0, 0)$. Since

$$\Delta_1 = 12x_1^2 + 2 > 0 \quad \text{for all } \boldsymbol{x} \in \mathbb{R}^2\,,$$
$$\Delta_2 = 24x_1^2 + 4 > 0 \quad \text{for all } \boldsymbol{x} \in \mathbb{R}^2$$

it follows from Theorem 6.6 that there is a global minimum at $\boldsymbol{a} = (0, 0)$.

6.4 Constrained optimization

Often one is looking for a solution $\boldsymbol{x}$ of the maximization or minimization problem which satisfies certain side conditions. Such a condition on $\boldsymbol{x}$ is referred to as a **constraint**. The points of the domain that satisfy the constraint are called **feasible solutions** of the optimization problem. When constraints are present, the previous approach generally does not provide a feasible solution to the problem.

In this section, we introduce a method, the so-called **Lagrange's[4] method**, with which one can determine extrema in constrained optimization problems: Instead of the gradient of the objective function f, the gradient of an extended objective function, the **Lagrangian function** or short **Lagrangian**, which takes into account the constraint, is set to zero. This ensures that only feasible solutions are included in the optimization. We first formulate the method, then apply it to the cost minimization problem (Example 6.2), and to further examples.

We are again searching the extrema of a function $f : A \to \mathbb{R}$ having domain $A \subset \mathbb{R}^n$. However, now m constraints $g_j(\boldsymbol{x}) = 0$, $j = 1, \ldots, m$, are given, where $m \leq n$. Now, we are not looking for the extrema of f regarding the entire domain A, but for the extrema of f with respect to those points of the domain that satisfy the constraints $g_j(\boldsymbol{x}) = 0$, $j = 1, \ldots, m$. If $n = 2$ and $m = 1$, this search can be visualized on the graph of f: one does not want to determine the summit of the mountain defined by the graph of f, but the highest point on a "hiking trail" determined by the constraint.

To solve the optimization problem, one forms an auxiliary function from the objective function f and the constraints, the so-called **Lagrangian function** or short **Lagrangian** L. It has the form

$$L(x_1, \ldots, x_n, \lambda_1, \ldots, \lambda_m) = f(\boldsymbol{x}) + \lambda_1 g_1(\boldsymbol{x}) + \ldots + \lambda_m g_m(\boldsymbol{x})$$

and contains besides the original variables $x_1, \ldots, x_n$ the auxiliary variables $\lambda_1, \ldots, \lambda_m$; they are called **Lagrange multipliers**. The stationary points of the Lagrangian are determined w.r.t. all variables (including the Lagrange multipliers). These points

[4] Joseph Louis de Lagrange (1736–1813)

are the possible candidates for extrema of the original function f taking into account the constraints. As in optimization without constraints, there is a necessary condition based on the first derivatives and a sufficient condition that includes the second derivatives. The necessary condition for a local extremum subject to constraints is:

Theorem 6.7: Lagrange's method: necessary condition

Let $f(\boldsymbol{a})$ be a local minimum or maximum of the function $f(\boldsymbol{x})$ subject to the constraints $g_j(\boldsymbol{x}) = 0$ for $j = 1, \dots, m$. Assume that f as well as the constraint functions g_j are differentiable, and that the gradients of the constraints, $\operatorname{grad} g_1(\boldsymbol{a}), \dots, \operatorname{grad} g_m(\boldsymbol{a})$, are linearly independent[5] at $\boldsymbol{a}$. Then there are numbers $\lambda_1^*, \dots, \lambda_m^* \in \mathbb{R}$, such that

$$L_{x_i}(\boldsymbol{a}, \lambda_1^*, \dots, \lambda_m^*) = \frac{\partial L}{\partial x_i}(\boldsymbol{a}, \lambda_1^*, \dots, \lambda_m^*) = 0 \quad \text{for } i = 1, \dots, n$$

and

$$L_{\lambda_j}(\boldsymbol{a}, \lambda_1^*, \dots, \lambda_m^*) = \frac{\partial L}{\partial \lambda_j}(\boldsymbol{a}, \lambda_1^*, \dots, \lambda_m^*) = 0 \quad \text{for } j = 1, \dots, m\,.$$

A point at which this necessary condition is met is called a **stationary** or **critical point**. The restriction on the derivatives L_{λ_j} is mentioned as **constraint qualification**.

Since $L_{\lambda_j}(x_1, \dots, x_n, \lambda_1, \dots, \lambda_m) = g_j(\boldsymbol{x})$, setting L_{λ_j} to zero means that the j-th constraint $g_j(\boldsymbol{x}) = 0$ must be met at the point $\boldsymbol{a}$. Note that the constraints should always be brought into the form $g_j(\boldsymbol{x}) = 0$ before setting up the Lagrangian. If, for example, $x_1 = 2x_3$ is given as the j-th constraint, this is transformed to $x_1 - 2x_3 = 0$. Thus, $g_j(\boldsymbol{x}) = x_1 - 2x_3$. Just as well, this constraint can be brought into the form $\widetilde{g}_j(\boldsymbol{x}) = 2x_3 - x_1$. This corresponds to a change of sign of λ_j, but leads to the same optimal values of $x_1, \dots, x_n$. As long as one is only interested in the values of the original variables, the signs of the Lagrange multipliers do not matter.

In summary, the procedure is as follows:

- The constraints are brought into the form $g_j(\boldsymbol{x}) = 0$.
- The Lagrangian is formulated.
- The partial derivatives of the Lagrangian are calculated and set to zero.
- The resulting system of equations in the variables $x_1, \dots, x_n, \lambda_1, \dots, \lambda_m$ is solved. The solutions $(x_1, \dots, x_n)$ are the stationary points.

Example 6.13 (Production at minimum cost). In a production process, a given output is to be produced from two inputs. We use the following notation:

[5] For the notion of linear independence see Definition 9.1.

x_1, x_2	the inputs,
p_1, p_2	the input prices,
F	the production function,
q_0	the given output.

Both inputs and the given output are non-negative, $x_1, x_2, q_0 \geq 0$, and the input prices are positive, $p_1, p_2 > 0$.

The inputs x_1 and x_2 are to be determined in such a way that on the one hand the given output q_0 is produced, i.e., the constraint

$$F(x_1, x_2) = q_0$$

is met, and on the other hand the costs of the two inputs,

$$c(x_1, x_2) = p_1 x_1 + p_2 x_2 \,,$$

are minimized. Here, the number of variables is $n = 2$ and the number of constraints is $m = 1$. Using Lagrange's method, we first transform the constraint in the form

$$g(x_1, x_2) = q_0 - F(x_1, x_2) = 0$$

and formulate the Lagrangian,

$$L(x_1, x_2, \lambda) = p_1 x_1 + p_2 x_2 + \lambda(q_0 - F(x_1, x_2)) \,.$$

Then we calculate the partial derivatives of the Lagrangian and set them equal to zero,

$$\begin{aligned} L_{x_1}(x_1, x_2, \lambda) &= p_1 - \lambda F_{x_1}(x_1, x_2) \stackrel{!}{=} 0 \,, \\ L_{x_2}(x_1, x_2, \lambda) &= p_2 - \lambda F_{x_2}(x_1, x_2) \stackrel{!}{=} 0 \,, \\ L_\lambda(x_1, x_2, \lambda) &= q_0 - F(x_1, x_2) \stackrel{!}{=} 0 \,. \end{aligned}$$

The last equation (partial derivative with respect to λ) corresponds to the constraint. We solve the first two equations for λ and obtain

$$\lambda = \frac{p_1}{F_{x_1}(x_1, x_2)} \quad \text{and} \quad \lambda = \frac{p_2}{F_{x_2}(x_1, x_2)} \,.$$

Equating these two equations yields

$$\frac{p_1}{F_{x_1}(x_1, x_2)} = \frac{p_2}{F_{x_2}(x_1, x_2)} \,,$$

and we obtain as a necessary condition

$$\frac{p_1}{p_2} = \frac{F_{x_1}(x_1, x_2)}{F_{x_2}(x_1, x_2)} = \frac{\text{marginal productivity of the first input}}{\text{marginal productivity of the second input}} \,,$$

$$\lambda = \frac{p_1}{F_{x_1}(x_1, x_2)}.$$

This is an important statement of microeconomic theory: In the cost minimum, the marginal productivities of the two inputs behave like their prices. In other words: **The ratio of the marginal productivities is equal to the price ratio.**

We now apply the general statement to a specific production function and numerically specified prices. Let the production function be $F(x_1, x_2) = x_1 x_2$, $x_1, x_2 \geq 0$, the output $q_0 = 1$, the input prices $p_1 = 1$ and $p_2 = 2$. Thus the costs $c(x_1, x_2) = x_1 + 2x_2$ are minimized subject to the constraint $1 = x_1 x_2$.

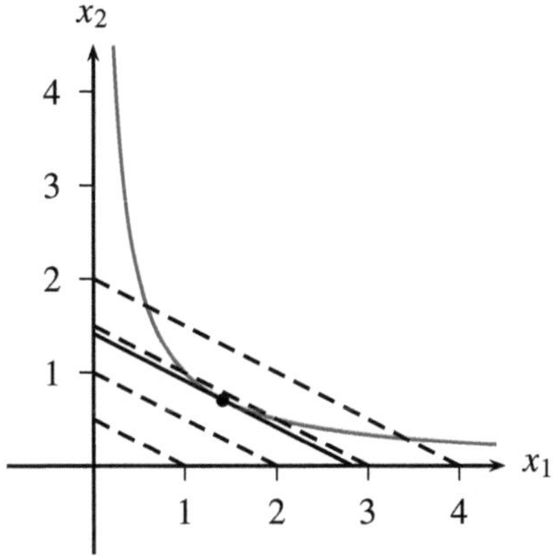

The hyperbola in the above figure corresponds to the constraint. Each point on the hyperbola denotes a factor combination with which an output of 1 can be produced. The straight lines are isocost lines, $x_1 + 2x_2 = c$, for different c. The further to the top right the isocost line is, the higher the costs. We are looking for the isocost line that meets the isoquant of the production function at the smallest value c. This yields the input combination for which the specified output can be produced at the lowest cost.

It is $F_{x_1} = x_2$ and $F_{x_2} = x_1$. So in the cost optimum the following equation holds:

$$\frac{p_1}{p_2} = \frac{1}{2} = \frac{x_2}{x_1} = \frac{F_{x_1}}{F_{x_2}}.$$

From $x_2 = \frac{x_1}{2}$ and $x_1 x_2 = 1$ we obtain the optimal, i.e., cost-minimal inputs,

$$x_1 = \sqrt{2}, \qquad x_2 = \frac{\sqrt{2}}{2} = \frac{1}{\sqrt{2}}.$$

It is assumed in Theorem 6.7 that the gradients of the constraints at the point $\boldsymbol{a}$ be linearly independent. In an optimization problem having a *single constraint* g, this means that $\operatorname{grad} g(\boldsymbol{a}) \neq \mathbf{0}$. Otherwise, extrema may exist at which the gradient of the Lagrangian is not zero, which is illustrated by the following example:

Example 6.14. The extrema of the objective function $f(x_1, x_2) = x_1 + 2$ is are searched subject to the constraint $(x_1^2 + x_2^2 - 1)^2 = 0$. The following figure shows that there is a maximum at the point $(1, 0)$ and a minimum at the point $(-1, 0)$.

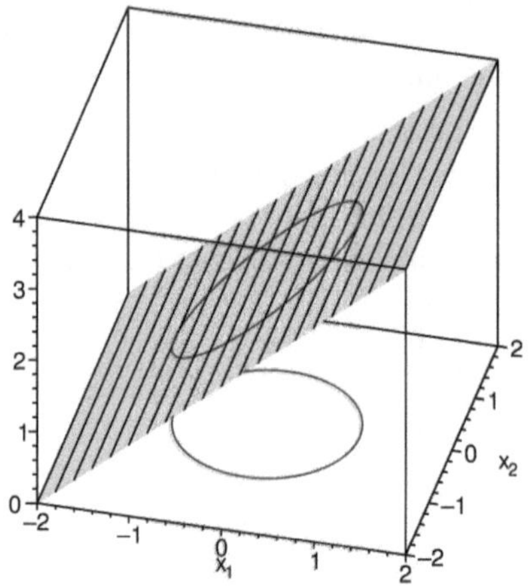

However, by using Lagrange's method we will not find the two extrema. The Lagrangian is

$$L(x_1, x_2, \lambda) = x_1 + 2 + \lambda(x_1^2 + x_2^2 - 1)^2,$$

and the necessary conditions are

$$\begin{aligned} 1 + 4x_1\lambda(x_1^2 + x_2^2 - 1) &= 0, \\ 4x_2\lambda(x_1^2 + x_2^2 - 1) &= 0, \\ (x_1^2 + x_2^2 - 1)^2 &= 0. \end{aligned}$$

The last equation is satisfied when $x_1^2 + x_2^2 - 1 = 0$. If we insert this into the first equation, we get $1 = 0$. Therefore, the system of equations cannot be solved.

The method does not work here because the gradient of $(x_1^2 + x_2^2 - 1)^2$ is equal to $[4x_1(x_1^2 + x_2^2 - 1), 4x_2(x_1^2 + x_2^2 - 1)]$. For all points that satisfy the constraint the gradient is equal to $\mathbf{0}$. Thus the constraint qualification is not satisfied.

It should be noted that the problem described in the previous example is caused solely by the formulation of the constraint. If the constraint is formulated in an equivalent way as $x_1^2 + x_2^2 - 1 = 0$, then Lagrange's method provides both extrema.

Solution by elimination If an optimization problem under constraints is numerically specified and each constraint can be solved for a different variable, instead of employing Lagrange's method, the constraints may be used to eliminate these variables in the objective function. Then, the constrained optimization problem is solved by optimizing the objective function with respect to the remaining variables.

We return to Example 6.13 and minimize the function $c(x_1, x_2) = x_1 + 2x_2$ subject to the constraint $1 = x_1 \cdot x_2$.

Example 6.15 (Example 6.13 continued). Solved for x_1, the constraint reads $x_1 = 1/x_2$. By substituting into $c(x_1, x_2)$ we obtain a cost function that depends only on x_2,

$$\widetilde{c}(x_2) = c\left(\frac{1}{x_2}, x_2\right) = \frac{1}{x_2} + 2x_2 .$$

Then, the derivative is $\widetilde{c}'(x_2) = -\frac{1}{x_2^2} + 2$. Setting this equal to zero yields $x_2 = \frac{1}{\sqrt{2}}$ and $x_1 = \frac{1}{x_2} = \sqrt{2}$ as before.

However, note that such an elimination is not possible when the functions are not numerically specified.

To evaluate the necessary condition of Lagrange's method, a generally nonlinear system of equations in $\lambda_1, \ldots, \lambda_m$ and $x_1, \ldots, x_n$ must be solved. Usually, it is advisable to first eliminate the Lagrange multipliers λ_j, as these only appear in linear form in the equations.

If constraints of the type "≤" instead of "=" are present, the Karush-Kuhn-Tucker conditions provide a necessary characterization of the solution; they are described in Section 6.7.

As with unconstrained optimization, there are also sufficient conditions, which are based on second derivatives. We first provide them in the case of two variables and one constraint. The general case of n variables and m constraints is treated in Section 6.6.

Theorem 6.8: Sufficient conditions if $n = 2$, $m = 1$

Let $n = 2$ and $m = 1$, let f and g be twice continuously differentiable and let the necessary condition from Theorem 6.7 be satisfied at (a_1, a_2). Sufficient for a local minimum or maximum of f in (a_1, a_2) subject to the constraint $g(x_1, x_2) = 0$ is:

$$\Delta := 2L_{x_1x_2}g_{x_1}g_{x_2} - L_{x_1x_1}(g_{x_2})^2 - L_{x_2x_2}(g_{x_1})^2 \begin{cases} < 0 & \text{(minimum)} \\ > 0 & \text{(maximum)} \end{cases}$$

at the point (λ^*, a_1, a_2).

The expression Δ is a determinant (see Chapter 10), namely the determinant of the Hessian matrix of the Lagrangian $L(\lambda, x_1, x_2)$,[6]

$$\Delta = \det \begin{bmatrix} 0 & g_{x_1} & g_{x_2} \\ g_{x_1} & L_{x_1x_1} & L_{x_1x_2} \\ g_{x_2} & L_{x_1x_2} & L_{x_2x_2} \end{bmatrix} .$$

at the point (λ^*, a_1, a_2).

[6] The Hessian matrix of a Lagrangian function is usually written with the Lagrange multipliers in the first places of variables, as we do it here and in the following two sections. It facilitates the formulation of sufficient conditions for constrained extrema by means of determinants.

Example 6.16. Given are an objective function in three variables and two constraints ($n = 3, m = 2$),

$$\begin{aligned} f(x_1,x_2,x_3) &= x_1^2 + x_2 - 2x_3\,, \\ g_1(x_1,x_2,x_3) &= x_1 + x_2 = 0\,, \\ g_2(x_1,x_2,x_3) &= 2x_2 + x_3 = 0\,. \end{aligned}$$

We are looking for the minima and maxima of f subject to the two constraints. The associated Lagrangian is

$$L(x_1,x_2,x_3,\lambda_1,\lambda_2) = x_1^2 + x_2 - 2x_3 + \lambda_1(x_1 + x_2) + \lambda_2(2x_2 + x_3)\,.$$

Its partial derivatives are set to zero,

$$\begin{aligned} L_{x_1} &= 2x_1 + \lambda_1 \overset{!}{=} 0\,, \\ L_{x_2} &= 1 + \lambda_1 + 2\lambda_2 \overset{!}{=} 0\,, \\ L_{x_3} &= -2 + \lambda_2 \overset{!}{=} 0\,, \\ L_{\lambda_1} &= x_1 + x_2 \overset{!}{=} 0\,, \\ L_{\lambda_2} &= 2x_2 + x_3 \overset{!}{=} 0\,. \end{aligned}$$

From the third equation, $\lambda_2 = 2$. Substituting this into the second equation yields $\lambda_1 = -5$. Then, from the first equation follows $x_1 = \frac{5}{2}$, and substituting this into the fourth equation yields $x_2 = -\frac{5}{2}$. Finally from the fifth equation we get $x_3 = 5$.

Alternative solution with elimination and subsequent Lagrange's method. We first use the second constraint, $2x_2 + x_3 = 0$, to eliminate x_3. By substituting $x_3 = -2x_2$ we obtain

$$\widetilde{f}(x_1,x_2) := f(x_1,x_2,-2x_2) = x_1^2 + x_2 + 4x_2 = x_1^2 + 5x_2\,,$$

and the task is reduced to an optimization in two variables subject to one constraint:

$$\begin{aligned} \widetilde{f}(x_1,x_2) &= x_1^2 + 5x_2\,, \\ \widetilde{g}_1(x_1,x_2) &= x_1 + x_2\,. \end{aligned}$$

We solve this reduced task with Lagrange's method. The Lagrangian is

$$\widetilde{L}(x_1,x_2,\lambda_1) = x_1^2 + 5x_2 + \lambda_1(x_1 + x_2)\,,$$

and the partial derivatives of the Lagrangian are

$$\begin{aligned} \widetilde{L}_{x_1} &= 2x_1 + \lambda_1 \overset{!}{=} 0\,, \\ \widetilde{L}_{x_2} &= 5 + \lambda_1 \overset{!}{=} 0\,, \\ \widetilde{L}_{\lambda_1} &= x_1 + x_2 \overset{!}{=} 0\,. \end{aligned}$$

As above, the solution is $\lambda_1 = -5$, $x_1 = \frac{5}{2}$, $x_2 = -\frac{5}{2}$, $x_3 = 5$. To evaluate the sufficient condition, we calculate the second derivatives of the Lagrangian and the first derivatives of the constraint function,

$$L_{x_1x_1} = 2\,, \quad L_{x_1x_2} = 0\,, \quad L_{x_2x_2} = 0\,, \quad g_{x_1} = 1\,, \quad g_{x_2} = 1\,.$$

Then, $\Delta = 2L_{x_1x_2}g_{x_1}g_{x_2} - L_{x_1x_1}(g_{x_2})^2 - L_{x_2x_2}(g_{x_1})^2 = -2 \cdot 1^2 = -2 < 0$. Thus, there is a local minimum at the point $(\frac{5}{2}, -\frac{5}{2}, 5)$.

We now return to the general problem of producing a given output at minimal cost (cf. Example 6.13) and ask what conditions a function F must fulfill to be considered an economically sensible production function.

We restrict ourselves to two inputs, x_1 and $x_2 \in \mathbb{R}_+$. First, we assume that both inputs are scarce and the output increases with each input. In other words, the marginal productivities of F are positive everywhere,

$$F_{x_1} > 0\,, \quad F_{x_2} > 0\,. \tag{6.1}$$

We seek a zero of the gradient of the Lagrangian

$$L(x_1, x_2, \lambda) = p_1x_1 + p_2x_2 + \lambda(q_0 - F(x_1, x_2))$$

where the costs are minimum (and neither maximum nor saddle point is located). The sufficient conditions for a local minimum are satisfied if and only if $\Delta < 0$. It holds (cf. Example 6.13)

$$g_{x_1} = -F_{x_1} < 0\,, \quad g_{x_2} = -F_{x_2} < 0 \quad \text{and} \quad L_{x_ix_j} = -\lambda F_{x_ix_j}\,.$$

Therefore,

$$\begin{aligned}\Delta &= 2L_{x_1x_2}g_{x_1}g_{x_2} - L_{x_1x_1}(g_{x_2})^2 - L_{x_2x_2}(g_{x_1})^2 \\ &= -2\lambda F_{x_1x_2}F_{x_1}F_{x_2} + \lambda F_{x_1x_1}(F_{x_2})^2 + \lambda F_{x_2x_2}(F_{x_1})^2 \\ &= \lambda\big(-2F_{x_1x_2}F_{x_1}F_{x_2} + F_{x_1x_1}(F_{x_2})^2 + F_{x_2x_2}(F_{x_1})^2\big)\,.\end{aligned}$$

Since in the cost minimum $\lambda = \frac{p_1}{F_{x_1}} > 0$, the condition $\Delta < 0$ is satisfied if and only if

$$\boxed{F_{x_1x_1}(F_{x_2})^2 + F_{x_2x_2}(F_{x_1})^2 - 2F_{x_1x_2}F_{x_1}F_{x_2} < 0\,.} \tag{6.2}$$

If the inequality (6.2) is satisfied in the entire domain of F, F is called **strictly quasiconcave**. This is equivalent to the contour lines of F having a certain shape (see Figure 1.17 left): For any two points on the same contour line, the connecting line is "northeast" of the contour line. A production function that satisfies the conditions (6.1) and (6.2) is referred to as a **neoclassical production function**.

Specifically, we consider the Cobb-Douglas production function,

$$F(x_1, x_2) = cx_1^{\alpha}x_2^{\beta}\,, \qquad x_1 \geq 0,\ x_2 \geq 0\,,$$

and first discuss the signs of the three parameters α, β and c. Since positive inputs should yield a positive output, $c > 0$ has to be assumed. The marginal productivities are $F_{x_1} = c\alpha x_1^{\alpha-1} x_2^{\beta}$ and $F_{x_2} = c\beta x_1^{\alpha} x_2^{\beta-1}$. For them to be positive, $\alpha > 0$ and $\beta > 0$ must hold. It follows for all $x_1, x_2 \in \mathbb{R}_+$

$$F_{x_1x_1}(F_{x_2})^2 + F_{x_2x_2}(F_{x_1})^2 - 2F_{x_1x_2}F_{x_1}F_{x_2}$$
$$= \underbrace{c^3\alpha\beta x_1^{3\alpha-2}x_2^{3\beta-2}}_{>0} \underbrace{(\beta(\alpha-1) + \alpha(\beta-1) - 2\alpha\beta)}_{=-\alpha-\beta<0} < 0\,,$$

so (6.2) is satisfied. Consequently, the Cobb-Douglas function is a neoclassical production function.

6.5 Envelope theorem

In many applications, the objective function of an optimization problem and/or its constraints depend on a parameter[7]. It is then interesting to see how the optimum value and the point at which it is achieved change as the parameter varies. In the Production Example 6.13, one might ask, for example, what additional costs are caused by an increase in the required production quantity q_0, or how a change in input prices affects the optimal factor combination and thus the production costs.

In general, the following constrained optimization problem is considered:

$$\boxed{\min_{\boldsymbol{x}} f(\boldsymbol{x}, \alpha) \quad \text{subject to} \quad g_j(\boldsymbol{x}, \alpha) = 0 \quad \text{for } j = 1, \ldots, m\,.}$$

Here, α is a parameter of the objective function f. The point at which f takes an extremum now also depends on the value of the parameter α. This point is denoted by $\boldsymbol{x}^*(\alpha)$. The optimal value $y^*(\alpha)$ of the objective function depends on α in two ways, directly and indirectly through $\boldsymbol{x}^*(\alpha)$,

$$y^*(\alpha) = f(\boldsymbol{x}^*(\alpha), \alpha)\,.$$

Firstly, if the objective function changes, its value may change, which is the *direct* effect of α. Secondly, an *indirect* effect has to be considered: Due to the change in the objective function, the optimum is attained at a different location, which also may change the objective function value.

The function $y^* : \alpha \mapsto y^*(\alpha)$ is referred to as the **optimal value function.**[8] The Lagrangian of the optimization problem is

$$L(\boldsymbol{x}, \boldsymbol{\lambda}, \alpha) \quad = \quad L(x_1, \ldots, x_n, \lambda_1, \ldots, \lambda_m, \alpha)$$

[7]We start with one parameter, being a real number. For two and more parameters, see Example 6.18 below.

[8]We assume in the following that the functions $\alpha \mapsto \boldsymbol{x}^*(\alpha)$ and $\alpha \mapsto y^*(\alpha)$ are defined and continuously differentiable for all α in an interval.

$$= \quad f(x_1,\ldots,x_n,\alpha) + \sum_{j=1}^{m} \lambda_j g_j(x_1,\ldots,x_n,\alpha)\,.$$

Since $\boldsymbol{x}^*(\alpha)$ is extremal, there are Lagrange multipliers $\lambda_1^*(\alpha),\cdots,\lambda_m^*(\alpha)$, such that the partial derivatives of the Lagrangian at $\boldsymbol{x}^*(\alpha)$ and $\boldsymbol{\lambda}^*(\alpha) = (\lambda_1^*(\alpha),\cdots,\lambda_m^*(\alpha))$ vanish. I.e., for all $i = 1,\ldots,n$ it holds[9]

$$\frac{\partial L}{\partial x_i} = \frac{\partial f}{\partial x_i} + \sum_{j=1}^{m} \lambda_j^* \frac{\partial g_j}{\partial x_i} = 0$$

or

$$\frac{\partial f}{\partial x_i} = -\sum_{j=1}^{m} \lambda_j^* \frac{\partial g_j}{\partial x_i}\,. \tag{6.3}$$

Since at $\boldsymbol{x}^*(\alpha)$ all constraints are satisfied, it is $g_j\big(\boldsymbol{x}^*(\alpha),\alpha\big) = 0$ for all j. Therefore, the derivative of the function $\alpha \mapsto g_j\big(\boldsymbol{x}^*(\alpha),\alpha\big)$ with respect to α is also zero. From the chain rule it follows

$$\frac{d}{d\alpha} g_j\big(\boldsymbol{x}^*(\alpha),\alpha\big) = \sum_{i=1}^{n} \frac{\partial g_j}{\partial x_i}\frac{dx_i^*}{d\alpha} + \frac{\partial g_j}{\partial \alpha} = 0\,,$$

or equivalently

$$\sum_{i=1}^{n} \frac{\partial g_j}{\partial x_i}\frac{dx_i^*}{d\alpha} = -\frac{\partial g_j}{\partial \alpha}\,. \tag{6.4}$$

To determine the approximate change of the optimal value function with a change of the parameter α, we calculate the derivative of $y^*(\alpha) = f(\boldsymbol{x}^*(\alpha),\alpha)$ with respect to the parameter α. Using the chain rule we get

$$\frac{dy^*(\alpha)}{d\alpha} = \frac{d}{d\alpha} f(\boldsymbol{x}^*(\alpha),\alpha) = \sum_{i=1}^{n} \frac{\partial f}{\partial x_i}\frac{dx_i^*}{d\alpha} + \frac{\partial f}{\partial \alpha}\,.$$

With Equation (6.3) we get

$$\frac{dy^*(\alpha)}{d\alpha} = \sum_{i=1}^{n}\left(-\sum_{j=1}^{m}\lambda_j^*\frac{\partial g_j}{\partial x_i}\right)\frac{dx_i^*}{d\alpha} + \frac{\partial f}{\partial \alpha} = -\sum_{j=1}^{m}\lambda_j^*\left(\sum_{i=1}^{n}\frac{\partial g_j}{\partial x_i}\frac{dx_i^*}{d\alpha}\right) + \frac{\partial f}{\partial \alpha}\,.$$

Using Equation (6.4) finally results in

$$\frac{dy^*(\alpha)}{d\alpha} = -\sum_{j=1}^{m}\lambda_j^*\left(-\frac{\partial g_j}{\partial \alpha}\right) + \frac{\partial f}{\partial \alpha} = \sum_{j=1}^{m}\lambda_j^*\frac{\partial g_j}{\partial \alpha} + \frac{\partial f}{\partial \alpha}\,.$$

But this is nothing else than the partial derivative of the Lagrangian with respect to the parameter α,

$$\frac{dy^*(\alpha)}{d\alpha} = \frac{\partial f}{\partial \alpha} + \sum_{j=1}^{m}\lambda_j^*\frac{\partial g_j}{\partial \alpha} = \frac{\partial L}{\partial \alpha}\,.$$

[9] We omit the arguments and write simply λ_j^* and $\frac{\partial L}{\partial x_i}$ instead of $\lambda_j^*(\alpha)$ and $\frac{\partial L}{\partial x_i}(\boldsymbol{x}^*(\alpha),\boldsymbol{\lambda}^*(\alpha),\alpha)$.

We formulate this result as a theorem:

Theorem 6.9: Envelope theorem

If the functions f and g_j, $j = 1, \dots, m$, as well as the optimal value function y^* are differentiable, it holds

$$\frac{dy^*(\alpha)}{d\alpha} = \frac{\partial f(\boldsymbol{x}^*(\alpha), \alpha)}{\partial \alpha} + \sum_{j=1}^{m} \lambda_j^*(\alpha) \frac{\partial g_j(\boldsymbol{x}^*(\alpha), \alpha)}{\partial \alpha} = \frac{\partial L}{\partial \alpha}(\boldsymbol{x}^*(\alpha), \boldsymbol{\lambda}^*(\alpha), \alpha).$$

In short (and more memorable) we write

$$\boxed{\frac{dy^*}{d\alpha} = \frac{\partial L}{\partial \alpha}.}$$

The formula and its derivation also apply when *no constraint* is to be considered ($m = 0$). In this case, the formula simplifies to

$$\boxed{\frac{dy^*}{d\alpha} = \frac{\partial f}{\partial \alpha}.}$$

Note that in the case of an **unconstrained** optimization problem the envelope theorem states that in deriving the optimal value with respect to the parameter the indirect effect of the parameter can be ignored.

Example 6.17 (Production of a given output at minimum cost). We return to Example 6.13:

$$\text{Minimize } c(x_1, x_2) = p_1 x_1 + p_2 x_2 \quad \text{subject to} \quad q - F(x_1, x_2) = 0.$$

First, we want to examine how the costs change when the given output q is changed. Since $\frac{\partial c}{\partial q} = 0$, the objective function does not directly depend on the produced quantity. The derivative of the constraint with respect to q is $\frac{\partial g}{\partial q} = 1$, so it follows

$$\frac{dy^*(q)}{dq} ha = \frac{\partial c}{\partial q} + \lambda^* \frac{\partial g}{\partial q} = \lambda^*.$$

The envelope theorem provides an interesting interpretation of the Lagrange multiplier. It equals the approximate change in minimum cost for producing an additional unit of output. By this, the Lagrange multiplier is referred to as the **shadow price** of the good. Thus, for the production of an additional unit to be worthwhile for the company at cost-optimal production, the market must achieve at least the shadow price of the good.

Next, we consider how the minimum costs depend on the prices of the two inputs. The derivative of the objective function with respect to the price p_i is $\frac{\partial c}{\partial p_1} = x_1$;

furthermore, $\frac{\partial g}{\partial p_i} = 0$. With the envelope theorem, we thus obtain

$$\frac{dy^*(p_i)}{dp_i} = \frac{\partial c}{\partial p_i} + \lambda^* \frac{\partial g}{\partial p_i} = x_i \,.$$

Since the constraint is independent of the prices, we only need to consider the direct effect of a price change.

Example 6.18 (Production of a given output at minimum cost). Consider again Example 6.13 with prices $p_1 = 1$, $p_2 = 2$, a given output quantity $q = 1$, and a production function $F(x_1, x_2) = x_1 x_2$.

We have already seen that in this case the minimum cost combination is $x_1^* = \sqrt{2}$, $x_2^* = \frac{\sqrt{2}}{2}$. The Lagrange multiplier is then $\lambda^* = \sqrt{2}$ and the minimum costs amount to $2\sqrt{2} = 2.8284$.

If you want to know how the minimum costs change with a change in the given output, consider the quantity to be produced as parameter q. The Lagrangian is then

$$L(x_1, x_2, \lambda, q) = x_1 + 2x_2 + \lambda(q - x_1 x_2)$$

and its partial derivative with respect to q is

$$\frac{\partial L}{\partial q}(x_1, x_2, \lambda, q) = \lambda \,.$$

If the given output quantity is to be increased from 1 unit to 1.1 units, the costs thus increase approximately by $\lambda^* dq$ units. With $\lambda^* = \sqrt{2}$ and $dq = 0.1$ we get $\lambda^* dq = \sqrt{2} \cdot 0.1 = 0.1414$, i.e., the costs increase by approximately 0.1414 monetary units.

If one is interested in how the minimum costs react to a change in the price p_2, consider p_2 as a parameter. This gives you the Lagrangian

$$L(x_1, x_2, \lambda, p_2) = x_1 + p_2 x_2 + \lambda(1 - x_1 x_2)$$

and its partial derivative with respect to p_2 is

$$\frac{\partial L}{\partial p_2}(x_1, x_2, \lambda, p_2) = x_2 \,.$$

If the price of the second good increases from 2 monetary units to 2.1 units, i.e., by $dp_2 = 0.1$ units, the costs increase approximately by $x_2^* dp_2$. With $x_2^* = \frac{\sqrt{2}}{2}$ and $dp_2 = 0.1$ we get $x_2^* dp_2 = \frac{\sqrt{2}}{2} \cdot 0.1 \approx 0.07071$, i.e., the costs increase by approximately 0.07071 units.

Similarly, it may be of interest how a change in the production function affects the minimum costs. If an optimization in the production process changes the production function to $\tilde{F}(x_1, x_2) = x_1^{1.1} x_2$, one can regard the exponent of x_1 in the production function as a parameter α. The Lagrangian is then

$$L(x_1, x_2, \lambda, \alpha) = x_1 + 2x_2 + \lambda(1 - x_1^{\alpha} x_2)$$

and its partial derivative with respect to α is

$$\frac{\partial L}{\partial p_2}(x_1, x_2, \lambda, \alpha) = -\lambda x_1^{\alpha} \ln(x_1) x_2 .$$

If, as stated above, the production function changes from $F(x_1, x_2) = x_1 x_2$ to $\tilde{F}(x_1, x_2) = x_1^{1.1} x_2$, the exponent increases by $d\alpha = 0.1$. The minimum costs change approximately by $-\lambda^*(x_1^*)^{\alpha} \ln(x_1^*) x_2^* d\alpha$. With $x_1^* = \sqrt{2}$, $x_2^* = \frac{\sqrt{2}}{2}$, $\lambda^* = \sqrt{2}$ and $\alpha = 1$ we get

$$-\lambda^*(x_1^*)^{\alpha} \ln(x_1^*) x_2^* \, d\alpha = -\sqrt{2}(\sqrt{2})^1 \ln(\sqrt{2}) \frac{\sqrt{2}}{2} \cdot 0.1 = -0.04901 ,$$

i.e., the costs decrease approximately by 0.04901 monetary units.

If all the described changes occur simultaneously, the optimal value function depends not only on one, but on several (here three) parameters. The above calculated approximate changes in the minimum costs represent in this case the corresponding partial differentials. When all three parameters change simultaneously, the total differential of the optimal value function must be calculated, i.e., the sum of the corresponding three partial differentials. In this case, we have

$$\begin{aligned} dy^* &= \frac{\partial L}{\partial q} dq + \frac{\partial L}{\partial x_2} dx_2 + \frac{\partial L}{\partial \alpha} d\alpha \\ &= \lambda^* dq + x_2^* dp_2 - \lambda^*(x_1^*)^{\alpha} \ln(x_1^*) x_2^* \, d\alpha \\ &= 0.1414 + 0.07071 - 0.04901 \\ &= 0.1631 , \end{aligned}$$

i.e., the minimum costs increase approximately by 0.1631 monetary units.

6.6 Sufficient conditions in constrained optimization

The function to be optimized is again $f : \mathbb{R}^n \to \mathbb{R}$ subject to the m constraints $g_j(\boldsymbol{x}) = 0$, $j = 1, \ldots, m$. Theorem 6.7 provides a necessary condition for an optimum: It must be attained at a stationary point of the Lagrangian[10]

$$L(\boldsymbol{\lambda}, \boldsymbol{x}) = f(\boldsymbol{x}) + \sum_{j=1}^{m} \lambda_j g_j(\boldsymbol{x}) .$$

[10] We write here $L(\boldsymbol{\lambda}, \boldsymbol{x})$ instead of $L(\boldsymbol{x}, \boldsymbol{\lambda})$, as this makes the following theorem easier to formulate.

So far, a sufficient condition was only given for one constraint and two variables, $m = 1$ and $n = 2$. In the case of several constraints, we consider the Hessian of $L(\boldsymbol{\lambda}, \boldsymbol{x})$,

$$H_L(\boldsymbol{\lambda}, \boldsymbol{x}) = \begin{bmatrix} 0 & \cdots & 0 & \frac{\partial g_1}{\partial x_1} & \cdots & \frac{\partial g_1}{\partial x_n} \\ \vdots & & \vdots & \vdots & & \vdots \\ 0 & \cdots & 0 & \frac{\partial g_m}{\partial x_1} & \cdots & \frac{\partial g_m}{\partial x_n} \\ \frac{\partial g_1}{\partial x_1} & \cdots & \frac{\partial g_m}{\partial x_1} & \frac{\partial^2 L}{\partial x_1^2} & \cdots & \frac{\partial^2 L}{\partial x_1 \partial x_n} \\ \vdots & & \vdots & \vdots & & \vdots \\ \frac{\partial g_1}{\partial x_n} & \cdots & \frac{\partial g_m}{\partial x_n} & \frac{\partial^2 L}{\partial x_n \partial x_1} & \cdots & \frac{\partial^2 L}{\partial^2 x_n} \end{bmatrix},$$

as well as its upper left $i \times i$ submatrices. The determinants[11] of these submatrices are called **principal minors** of the matrix and are denoted by Δ_i.

Theorem 6.10: Sufficient conditions in constrained optimization

Let $(\boldsymbol{\lambda}^*, \boldsymbol{a})$ be a stationary point of the Lagrangian

$$L(\boldsymbol{\lambda}, \boldsymbol{x}) = f(\boldsymbol{x}) + \sum_{j=1}^{m} \lambda_j g_j(\boldsymbol{x}).$$

Then, subject to the constraints $g_1(\boldsymbol{x}) = 0, \ldots, g_m(\boldsymbol{x}) = 0$, at the stationary point $\boldsymbol{a}$, the objective function f has

(i) a local minimum, if $(-1)^m \Delta_i(\boldsymbol{\lambda}^*, \boldsymbol{a}) > 0$ for $i = 2m + 1, \ldots, m + n$,

(ii) a local maximum, if $(-1)^{i-m} \Delta_i(\boldsymbol{\lambda}^*, \boldsymbol{a}) > 0$ for $i = 2m + 1, \ldots, m + n$.

In other words: The sufficient conditions depend on the signs of the last $n - m$ determinants Δ_i. Sufficient for a local minimum with an even (or odd) number of constraints is that they all have the same positive (or negative) sign. Sufficient for a local maximum with an even (or odd) number of variables is that the sign alternates so that the last determinant has a positive (or negative) sign.

In particular, for $n = 2$ variables and $m = 1$ constraint, a minimum (maximum) is attained at $\boldsymbol{a}$ if $-\Delta_3(\boldsymbol{a}) > 0$ $(-\Delta_3(\boldsymbol{a}) < 0)$. This special case has been covered by Theorem 6.8.

In the case of **unconstrained optimization**, $m = 0$, the known sufficient conditions are obtained. The Δ_i are then the principal minors of the Hessian:

[11] For the notion of a determinant and the concept of a principal minor see Chapter 10.

Theorem 6.11: Sufficient conditions in unconstrained optimization

Let $\boldsymbol{a}$ be a stationary point of f. Then, at the point $\boldsymbol{a}$, f has

(i) a local minimum, if $\Delta_i(\boldsymbol{a}) > 0$ for $i = 1, \dots, n$,

(ii) a local maximum, if $(-1)^i \Delta_i(\boldsymbol{a}) > 0$ for $i = 1, \dots, n$.

If all principal minors are positive, then f has a local minimum. If the odd principal minors are negative and the even principal minors are positive, then f has a local maximum.

Unfortunately, as the sign conditions for the principal minors Δ_i depend on the number of constraints and the number of variables, the sufficient conditions of Theorem 6.10 are not easy to remember. To formulate the sufficient conditions in a more memorable form, we consider the so-called **modified Hessian**. The modified Hessian is derived from the Hessian of $L(\boldsymbol{\lambda}, \boldsymbol{x})$ by reversing the signs in all columns corresponding to Lagrange multipliers, i.e., in the first m columns. We denote the modified Hessian with $\widetilde{H}_L$,

$$\widetilde{H}_L(\boldsymbol{\lambda}, \boldsymbol{x}) = \begin{bmatrix} 0 & \cdots & 0 & \frac{\partial g_1}{\partial x_1} & \cdots & \frac{\partial g_1}{\partial x_n} \\ \vdots & & \vdots & \vdots & & \vdots \\ 0 & \cdots & 0 & \frac{\partial g_m}{\partial x_1} & \cdots & \frac{\partial g_m}{\partial x_n} \\ -\frac{\partial g_1}{\partial x_1} & \cdots & -\frac{\partial g_m}{\partial x_1} & \frac{\partial^2 L}{\partial x_1^2} & \cdots & \frac{\partial^2 L}{\partial x_1 \partial x_n} \\ \vdots & & \vdots & \vdots & & \vdots \\ -\frac{\partial g_1}{\partial x_n} & \cdots & -\frac{\partial g_m}{\partial x_n} & \frac{\partial^2 L}{\partial x_n \partial x_1} & \cdots & \frac{\partial^2 L}{\partial^2 x_n} \end{bmatrix}.$$

If we further denote the i-th principal minor (i.e., the determinant of the top left $i \times i$ submatrix) of the modified Hessian with $\widetilde{\Delta}_i$, we can reformulate Theorem 6.10 as follows.

Theorem 6.12: Sufficient conditions (alternative formulation)

Let $(\boldsymbol{\lambda}^*, \boldsymbol{a})$ be a stationary point of the Lagrangian

$$L(\boldsymbol{\lambda}, \boldsymbol{x}) = f(\boldsymbol{x}) + \sum_{j=1}^{m} \lambda_j g_j(\boldsymbol{x}).$$

Then, subject to the constraints $g_1(\boldsymbol{x}) = 0, \dots, g_m(\boldsymbol{x}) = 0$, at the stationary point $\boldsymbol{a}$, the objective function f has

(i) a local minimum, if $\widetilde{\Delta}_i(\boldsymbol{\lambda}^*, \boldsymbol{a}) > 0$ for $i = 2m+1, \dots, m+n$,

(ii) a local maximum, if $(-1)^i \widetilde{\Delta}_i(\boldsymbol{\lambda}^*, \boldsymbol{a}) > 0$ for $i = 2m+1, \dots, m+n$.

This formulation has the advantage that the sign conditions on the principal minors $\widetilde{\Delta}_i(\boldsymbol{\lambda}^*, \boldsymbol{a})$ are independent of the number of constraints and also of the number of variables. In particular, for unconstrained and constrained optimization problems the sufficient conditions can be formulated in the same, easy to memorize, way:

Mnemonic 6.1: Checking the sufficient conditions

1. Calculate the Hessian $H_L(\boldsymbol{\lambda}^*, \boldsymbol{a})$.
2. Reverse the sign in the columns of the Hessian belonging to Lagrange multipliers. This yields the modified Hessian $\widetilde{H}_L(\boldsymbol{\lambda}^*, \boldsymbol{a})$.
3. Calculate the last $n - m$ principal minors $\widetilde{\Delta}_i(\boldsymbol{\lambda}^*, \boldsymbol{a})$.
4. If these $n-m$ principal minors are all positive, then there is a local minimum.
5. If the odd principal minors are negative and the even principal minors are positive, then there is a local maximum.

As an illustration, we will revisit Example 6.16.

Example 6.19. In Example 6.16, we considered the following optimization problem in three variables and two constraints ($n = 3, m = 2$):

$$\begin{aligned} f(x_1,x_2,x_3) &= x_1^2 + x_2 - 2x_3\,, \\ g_1(x_1,x_2,x_3) &= x_1 + x_2 = 0\,, \\ g_2(x_1,x_2,x_3) &= 2x_2 + x_3 = 0\,. \end{aligned}$$

The only stationary point identified was $(x_1,x_2,x_3) = (\frac{5}{2}, -\frac{5}{2}, 5)$ with the associated Lagrange multipliers $\lambda_1 = -5$ and $\lambda_2 = 2$. The Hessian of the Lagrangian is

$$H_L(\boldsymbol{\lambda}, \boldsymbol{x}) = \begin{bmatrix} 0 & 0 & 1 & 1 & 0 \\ 0 & 0 & 0 & 2 & 1 \\ 1 & 0 & 2 & 0 & 0 \\ 1 & 2 & 0 & 0 & 0 \\ 0 & 1 & 0 & 0 & 0 \end{bmatrix}.$$

Since the Hessian is constant here, we also get the above Hessian at the point $(\boldsymbol{\lambda}, \boldsymbol{x}) = (-5, 2, \frac{5}{2}, -\frac{5}{2}, 5)$. As there are two constraints, the signs in the first two columns must be reversed to get the modified Hessian. Thus, the modified Hessian is

$$\widetilde{H}_L(\boldsymbol{\lambda}, \boldsymbol{x}) = \begin{bmatrix} 0 & 0 & 1 & 1 & 0 \\ 0 & 0 & 0 & 2 & 1 \\ -1 & 0 & 2 & 0 & 0 \\ -1 & -2 & 0 & 0 & 0 \\ 0 & -1 & 0 & 0 & 0 \end{bmatrix}.$$

Since we have $n = 3$ variables and $m = 2$ constraints, $n - m = 3 - 2 = 1$ minors need to be checked, i.e., only the largest minor $\widetilde{\Delta}_5$. Expanding it along "favorable" columns, we get

$$\det\begin{bmatrix} 0 & 0 & 1 & 1 & 0 \\ 0 & 0 & 0 & 2 & 1 \\ -1 & 0 & 2 & 0 & 0 \\ -1 & -2 & 0 & 0 & 0 \\ 0 & -1 & 0 & 0 & 0 \end{bmatrix} = (-1)\cdot\det\begin{bmatrix} 0 & 0 & 1 & 1 \\ -1 & 0 & 2 & 0 \\ -1 & -2 & 0 & 0 \\ 0 & -1 & 0 & 0 \end{bmatrix}$$

$$= (-1)\cdot(-1)\cdot\det\begin{bmatrix} -1 & 0 & 2 \\ -1 & -2 & 0 \\ 0 & -1 & 0 \end{bmatrix} = (-1)\cdot(-1)\cdot 2\cdot\det\begin{bmatrix} -1 & -2 \\ 0 & -1 \end{bmatrix}$$

$$= (-1)\cdot(-1)\cdot 2\cdot(-1)^2 = 2\,.$$

Thus, $\widetilde{\Delta}_5 = 2 > 0$. Since the only principal minor to be checked is positive, a local minimum is attained at $(\frac{5}{2}, -\frac{5}{2}, 5)$.

6.7 Karush-Kuhn-Tucker conditions

In many optimization problems, not only **equality constraints** arise but also **inequality constraints**. In this section we consider the **general minimization problem** with equality constraints of the form $g_i(\boldsymbol{x}) = 0$ and inequality constraints of the form $h_j(\boldsymbol{x}) \leq 0$:

$$\begin{array}{rl} \min & f(\boldsymbol{x}) \\ \text{subject to} & g_i(\boldsymbol{x}) = 0\,,\ i = 1,\dots,m\,, \\ & h_j(\boldsymbol{x}) \leq 0\,,\ j = 1,\dots,l\,. \end{array} \tag{6.5}$$

We assume that all occurring functions $f, g_1, \dots, g_m, h_1, \dots, h_l : \mathbb{R}^n \to \mathbb{R}$ are differentiable.

Accordingly, the **general maximization problem** is

$$\begin{array}{rl} \max & f(\boldsymbol{x}) \\ \text{subject to} & g_i(\boldsymbol{x}) = 0\,,\ i = 1,\dots,m\,, \\ & h_j(\boldsymbol{x}) \leq 0\,,\ j = 1,\dots,l\,. \end{array} \tag{6.6}$$

Any optimization problem whose constraints are given in equality or inequality form can be represented as a general minimization problem (6.5). If there are constraints of the form $h_j(\boldsymbol{x}) \geq 0$, they can be multiplied by -1, yielding $-h_j(\boldsymbol{x}) \leq 0$. Similarly, the problem of maximizing a function can be reduced to a minimization problem by multiplying the objective function by -1, since $\max_{\boldsymbol{x}} f(\boldsymbol{x}) = -\min_{\boldsymbol{x}}(-f(\boldsymbol{x}))$. It is therefore enough to develop solution criteria for the minimization problem (6.5).

The Lagrangian of the minimization problem (6.5) (and likewise of the maximization problem (6.6)) is defined as

$$L(\boldsymbol{x}, \boldsymbol{\lambda}, \boldsymbol{\mu}) = f(\boldsymbol{x}) + \sum_{i=1}^{m} \lambda_i g_i(\boldsymbol{x}) + \sum_{j=1}^{l} \mu_j h_j(\boldsymbol{x})\,.$$

A point $\boldsymbol{x}$ that satisfies all constraints is called a **feasible solution**. If $\boldsymbol{x}$ is a feasible solution, then $g_i(\boldsymbol{x}) = 0$ for all $i = 1, \ldots, l$. Regarding the inequality constraints, we distinguish between those for which $h_j(\boldsymbol{x}) = 0$ and those for which $h_j(\boldsymbol{x}) < 0$ holds. A constraint for which $h_j(\boldsymbol{x}) = 0$ is called **active** at $\boldsymbol{x}$, the remaining constraints are **inactive** at $\boldsymbol{x}$. The indices of the inequality constraints that are active at a feasible point $\boldsymbol{x}$ are denoted by $I(\boldsymbol{x})$,

$$I(\boldsymbol{x}) = \{j \, : \, h_j(\boldsymbol{x}) = 0\}\,.$$

The following theorem, the so-called **Karush-Kuhn-Tucker theorem**[12], or short KKT theorem, gives a necessary condition for the existence of a local minimum.

Theorem 6.13: KKT theorem

If $\boldsymbol{a}$ is a local minimum of the problem (6.5) and the gradients of the active constraints are linearly independent, then there are constants $\boldsymbol{\lambda}^* = (\lambda_1^*, \ldots, \lambda_m^*)$, $\boldsymbol{\mu}^* = (\mu_1^*, \ldots, \mu_l^*)$ such that the following conditions hold:

(1) $f'(\boldsymbol{a}) + \sum_{i=1}^{m} \lambda_i^* g_i'(\boldsymbol{a}) + \sum_{j=1}^{l} \mu_j^* h_j'(\boldsymbol{a}) = \boldsymbol{0}\,,$ **stationarity**

(2) $g_i(\boldsymbol{a}) = 0, i = 1, \ldots, m, \quad h_j(\boldsymbol{a}) \leq 0, j = 1, \ldots, l\,,$ **primal feasibility**

(3) $\mu_j^* \geq 0, j = 1, \ldots, l\,,$ **dual feasibility**

(4) $\mu_j^* h_j(\boldsymbol{a}) = 0\,, j = 1, \ldots, l\,.$ **complementary slackness**

For the maximization problem (6.6), a corresponding statement applies, in which condition (3) is replaced by the condition

(3′) $\mu_j^* \leq 0, j = 1, \ldots, l.$

Conditions (1) to (4) are referred to as Karush-Kuhn-Tucker conditions or briefly KKT conditions. The condition of **stationarity** (1) states that at a minimum $\boldsymbol{a}$ the partial derivatives of the Lagrangian with respect to the variables $x_1, \ldots, x_n$ are zero, i.e., $\boldsymbol{a}$ is a stationary point of the Lagrangian. The second condition, also known as **primal feasibility**, means that all constraints are satisfied. If all constraints are in equality form, i.e., if $l = 0$, then conditions (3) and (4) do not apply. In this case, conditions (1) and (2) coincide with the necessary condition from Theorem 6.7. When inequality constraints are present, conditions (3) and (4) must additionally hold. Condition (3) states that the Lagrange multipliers of the inequality constraints are non-negative. This condition is also referred to as **dual feasibility**. Condition (4) is referred to as **complementary slackness**. It states that each of the inequality constraints is either active or has a Lagrange multiplier equal to zero.

With the KKT theorem, the problem of boundary extrema can be elegantly solved. Let us reconsider Example 6.8.

[12] William Karush (1917–1997), Harold W. Kuhn (1925–2014) and Albert William Tucker (1905–1995)

Example 6.20. We seek the extrema of the function $f : \mathbb{R}^2_+ \to \mathbb{R}$ with $f(x_1, x_2) = x_1^3 x_2 - x_1^2 - x_2$. This problem can be formulated as a constrained optimization problem: Find all extrema of the objective function f subject to the constraints $x_1, x_2 \geq 0$. To bring this problem into the form of the general minimization problem, the constraints are transformed to $-x_1 \leq 0$ and $-x_2 \leq 0$. The Lagrangian is then

$$L(x_1, x_2, \mu_1, \mu_2) = x_1^3 x_2 - x_1^2 - x_2 - \mu_1 x_1 - \mu_2 x_2 ,$$

and the KKT conditions for a local *minimum* are:

$$\begin{aligned}
\text{Stationarity:} \quad & 3x_1^2 x_2 - 2x_1 - \mu_1 = 0 , \\
& x_1^3 - 1 - \mu_2 = 0 , \\
\text{primal feasibility:} \quad & x_1 \geq 0 , \quad x_2 \geq 0 , \\
\text{dual feasibility:} \quad & \mu_1 \geq 0 , \quad \mu_2 \geq 0 , \\
\text{complementary slackness:} \quad & \mu_1 x_1 = 0 , \quad \mu_2 x_2 = 0 .
\end{aligned}$$

If $\mu_2 > 0$, then from complementary slackness, it follows that $x_2 = 0$. The first stationarity condition then implies $-2x_1 - \mu_1 = 0$. Since both x_1 and μ_1 must be greater than or equal to zero, it follows that $x_1 = \mu_1 = 0$. From the second stationarity condition, it then follows $-1 - \mu_2 = 0$, which contradicts $\mu_2 > 0$.

Therefore, $\mu_2 = 0$ must hold. Then, the second stationarity condition is $x_1^3 - 1 = 0$ and thus $x_1 = 1$. Because of complementary slackness, $\mu_1 = 0$ has to hold. The first stationarity condition yields $3x_2 - 2 = 0$ and thus $x_2 = \frac{2}{3}$. For the point $(x_1, x_2) = (1, \frac{2}{3})$ and $\mu_1 = \mu_2 = 0$ the KKT conditions are satisfied. The only possible location of a minimum is thus $(1, \frac{2}{3})$.

If one wants to find candidates for a local *maximum*, the condition of dual feasibility is: $\mu_1 \leq 0$, $\mu_2 \leq 0$. If $\mu_2 = 0$, it follows as before that $(x_1, x_2) = (1, \frac{2}{3})$ is also a candidate for a maximum.

If $\mu_2 < 0$, it again follows from complementary slackness, that $x_2 = 0$. The first stationarity condition then yields $-2x_1 - \mu_1 = 0$ or $\mu_1 = -2x_1$. Because of complementary slackness, at least one of the quantities μ_1 and x_1 must be equal to zero. Hence, we get $x_1 = \mu_1 = 0$. With the second stationarity condition, it follows that $-1 - \mu_2 = 0$ or $\mu_2 = -1$. For $(x_1, x_2) = (0, 0)$ and $\mu_1 = 0$, $\mu_2 = -1$, the KKT conditions are therefore satisfied. Candidates for maxima are the two points $(1, \frac{2}{3})$ and $(0, 0)$.

Sufficient conditions For the two general optimization problems, one can also provide sufficient conditions for the existence of an extremum at a given point $\boldsymbol{a}$. These resemble the conditions given in Theorem 6.10 and Theorem 6.12. Essentially, one proceeds as follows: One ignores the inactive inequality constraints in $\boldsymbol{a}$ and sets up a Lagrangian that only takes into account the active constraints in $\boldsymbol{a}$. Then one applies the conditions from Theorem 6.10 or 6.12 to the thus modified problem. However, this "recipe" fails when there are active constraints for which the Lagrange multiplier is zero. Active inequality constraints, whose associated Lagrange multiplier is not zero, are referred to as **strictly active**; those with Lagrange multiplier zero are referred to as **weakly active**. In a given point $\boldsymbol{a}$, at most $n-m$ constraints can be strictly active. The set of all strictly active constraints in $\boldsymbol{a}$ is denoted by $I^*(\boldsymbol{a})$: $I^*(\boldsymbol{a}) = \{j \in I(\boldsymbol{a}) \, : \, \mu_j \neq 0\}$.

The **restricted Lagrangian** $\widetilde{L}$ is a Lagrangian that, in addition to the equality constraints, takes into account the strictly active inequality constraints (but neither the weakly active nor inactive ones):

$$\widetilde{L}(\boldsymbol{\lambda}, \boldsymbol{\mu}, \boldsymbol{x}) = f(\boldsymbol{x}) + \sum_{i=1}^{m} \lambda_i g_i(\boldsymbol{x}) + \sum_{j \in I^*(\boldsymbol{x})} \mu_j h_j(\boldsymbol{x}) \,.$$

If exactly the first k inequality constraints are strictly active at $\boldsymbol{x}$, i.e., $I^*(\boldsymbol{x}) = \{1, \ldots, k\}$, then the Hessian of the restricted Lagrangian is given as follows:

$$H_{\widetilde{L}}(\boldsymbol{\lambda}, \boldsymbol{\mu}, \boldsymbol{x}) = \begin{bmatrix} 0 & \cdots & 0 & 0 & \cdots & 0 & \frac{\partial g_1}{\partial x_1} & \cdots & \frac{\partial g_1}{\partial x_n} \\ \vdots & & \vdots & \vdots & & \vdots & \vdots & & \vdots \\ 0 & \cdots & 0 & 0 & \cdots & 0 & \frac{\partial g_m}{\partial x_1} & \cdots & \frac{\partial g_m}{\partial x_n} \\ 0 & \cdots & 0 & 0 & \cdots & 0 & \frac{\partial h_1}{\partial x_1} & \cdots & \frac{\partial h_1}{\partial x_n} \\ \vdots & & \vdots & \vdots & & \vdots & \vdots & & \vdots \\ 0 & \cdots & 0 & 0 & \cdots & 0 & \frac{\partial h_k}{\partial x_1} & \cdots & \frac{\partial h_k}{\partial x_n} \\ \frac{\partial g_1}{\partial x_1} & \cdots & \frac{\partial g_m}{\partial x_1} & \frac{\partial h_1}{\partial x_1} & \cdots & \frac{\partial h_k}{\partial x_1} & \frac{\partial^2 L}{\partial x_1^2} & \cdots & \frac{\partial^2 L}{\partial x_1 \partial x_n} \\ \vdots & & \vdots & \vdots & & \vdots & \vdots & & \vdots \\ \frac{\partial g_1}{\partial x_n} & \cdots & \frac{\partial g_m}{\partial x_n} & \frac{\partial h_1}{\partial x_n} & \cdots & \frac{\partial h_k}{\partial x_n} & \frac{\partial^2 L}{\partial x_n \partial x_1} & \cdots & \frac{\partial^2 L}{\partial^2 x_n} \end{bmatrix} .$$

If other than the first k inequality constraints are strictly active, one forms $H_{\widetilde{L}}$ with $h_j, j \in I^*(\boldsymbol{x})$, instead of $h_j, j \in \{1, \ldots, k\}$. As before, we denote by Δ_i the determinant of the upper left $i \times i$ submatrix of $H_{\widetilde{L}}$.

Theorem 6.14: Sufficient conditions

(i) Assume that for the point $\boldsymbol{a}$ and the Lagrange multipliers $\boldsymbol{\lambda}^*, \boldsymbol{\mu}^*$, the necessary KKT conditions for the **minimization problem** (6.5) are satisfied, and that the number of strictly active constraints at $\boldsymbol{a}$ is k. Then f has a local **minimum** subject to the constraints $g_i(\boldsymbol{x}) = 0$, $i = 1, \ldots, m$, $h_j(\boldsymbol{x}) \leq 0$, $j = 1, \ldots, l$, if

$$(-1)^{m+k} \Delta_i(\boldsymbol{\lambda}^*, \boldsymbol{\mu}^*, \boldsymbol{a}) > 0 \quad \text{for} \quad i = 2(m+k)+1, \ldots, m+k+n \,.$$

(ii) Assume that for the point $\boldsymbol{a}$ and the Lagrange multipliers $\boldsymbol{\lambda}^*, \boldsymbol{\mu}^*$, the necessary KKT conditions for the **maximization problem** (6.6) are satisfied, and that the number of strictly active constraints in $\boldsymbol{a}$ is k. Then f has a local **maximum** at the point $\boldsymbol{a}$ subject to the constraints $g_i(\boldsymbol{x}) = 0, i = 1, \ldots, m, h_j(\boldsymbol{x}) \leq 0, j = 1, \ldots, l$, if

$$(-1)^{i-(m+k)} \Delta_i(\boldsymbol{\lambda}^*, \boldsymbol{\mu}^*, \boldsymbol{a}) > 0 \quad \text{for} \quad i = 2(m+k)+1, \ldots, m+k+n\,.$$

As in Section 6.6, the sufficient conditions can also be formulated using the modified Hessian $\widetilde{H}_{\widetilde{L}}$. As there, the modified Hessian is obtained by reversing the sign in the columns corresponding to Lagrange multipliers. We denote the principal minors of the modified Hessian with $\widetilde{\Delta}_i$. The alternative formulation of the sufficient conditions is then as follows:

Theorem 6.15: Sufficient conditions (alternative formulation)

(i) Assume that for the point $\boldsymbol{a}$ and the Lagrange multipliers $\boldsymbol{\lambda}^*, \boldsymbol{\mu}^*$, the necessary KKT conditions for the **minimization problem** (6.5) are satisfied, and that the number of strictly active constraints in $\boldsymbol{a}$ is k. Then f has a local **minimum** subject to the constraints $g_i(\boldsymbol{x}) = 0$, $i = 1, \ldots, m, h_j(\boldsymbol{x}) \leq 0, j = 1, \ldots, l$, if

$$\widetilde{\Delta}_i(\boldsymbol{\lambda}^*, \boldsymbol{\mu}^*, \boldsymbol{a}) > 0 \quad \text{for} \quad i = 2(m+k)+1, \ldots, m+k+n\,.$$

(ii) Assume that for the point $\boldsymbol{a}$ and the Lagrange multipliers $\boldsymbol{\lambda}^*, \boldsymbol{\mu}^*$, the necessary KKT conditions for the **maximization problem** (6.6) are satisfied, and that the number of strictly active constraints in $\boldsymbol{a}$ is k. Then f has a local **maximum** subject to the constraints $g_i(\boldsymbol{x}) = 0$, $i = 1, \ldots, m, h_j(\boldsymbol{x}) \leq 0, j = 1, \ldots, l$, if

$$(-1)^i \widetilde{\Delta}_i(\boldsymbol{\lambda}^*, \boldsymbol{\mu}^*, \boldsymbol{a}) > 0 \quad \text{for} \quad i = 2(m+k)+1, \ldots, m+k+n\,.$$

The Mnemonic 6.1 in Section 6.6 for checking the sufficient conditions is *mutatis mutandis* still valid.

We illustrate the sufficient conditions using Example 6.20.

Example 6.21. In Example 6.20, we identified two points as candidates for extrema of the function $f(x_1, x_2) = x_1^3 x_2 - x_1^2 - x_2$ subject to $x_1, x_2 \geq 0$. At the point $(1, \frac{2}{3})$, a local minimum or maximum may exist, while at the point $(0, 0)$ only a local maximum can exist.

We first check the point $(1, \frac{2}{3})$. At this point, both constraints are inactive. The restricted Lagrangian $\widetilde{L}$ is therefore just the objective function itself. Therefore, the Hessian of $\widetilde{L}$ is the Hessian H_f of the objective function and the conditions from Theorem 6.14 coincide with the sufficient conditions from Theorem 6.2. However, we have already checked these in Example 6.8 with the result that neither a minimum nor a maximum exists at the point $(1, \frac{2}{3})$.

The only point left to check is $(0,0)$, where a local maximum may exist. The Lagrange multipliers are $\mu_1 = 0$ and $\mu_2 = -1$. Since only the second inequality is strictly active, only the constraint $-x_2 \le 0$ needs to be considered in the restricted Lagrangian:

$$\widetilde{L}(\mu_2, x_1, x_2) = x_1^3 x_2 - x_1^2 - x_2 - \mu_2 x_2 \,.$$

The Hessian of the restricted Lagrangian is then

$$H_{\widetilde{L}}(\mu_2, x_1, x_2) = \begin{bmatrix} 0 & 0 & -1 \\ 0 & 6x_1x_2 - 2 & 3x_1^2 \\ -1 & 3x_1^2 & 0 \end{bmatrix} .$$

Since only one constraint is strictly active, the modified Hessian is obtained by reversing the signs in the first column:

$$\widetilde{H}_{\widetilde{L}}(\mu_2, x_1, x_2) = \begin{bmatrix} 0 & 0 & -1 \\ 0 & 6x_1x_2 - 2 & 3x_1^2 \\ 1 & 3x_1^2 & 0 \end{bmatrix} .$$

At the point $(\mu_2, x_1, x_2) = (-1, 0, 0)$ we have

$$\widetilde{H}_{\widetilde{L}}(-1, 0, 0) = \begin{bmatrix} 0 & 0 & -1 \\ 0 & -2 & 0 \\ 1 & 0 & 0 \end{bmatrix} .$$

According to Theorem 6.15, the number of principal minors to be checked is the difference between the number n of variables and the number $m + k$ of constraints considered in the restricted Lagrangian, here $n - (m + k) = 2 - (0 + 1) = 1$. Thus, only the largest principal minor, i.e., $\widetilde{\Delta}_3$, needs to be checked. Due to $\widetilde{\Delta}_3(-1, 0, 0) = -2 < 0$, all odd minors to be considered are negative and all even minors to be considered are positive. According to Mnemonic 6.1 on page 199, there is therefore a local maximum at the point $(0, 0)$.

Example 6.22. Consider the objctive function $f(x_1, x_2) = x_1^2 + x_2^2$ subject to $2x_1 + x_2 + 1 \ge 0$ and $x_1 + 4x_2 + 2 \ge 0$.

First, the constraints are brought into the form of $\le$ constraints by multiplying by -1, resulting in

$$\begin{aligned} \text{minimize/maximize} \quad & f(x_1, x_2) = x_1^2 + x_2^2 \,, \\ \text{subject to} \quad & -2x_1 - x_2 - 1 \le 0 \,, \\ & -x_1 - 4x_2 - 2 \le 0 \,. \end{aligned}$$

The Lagrangian is

$$L(x_1, x_2, \mu_1, \mu_2) = x_1^2 + x_2^2 - \mu_1(2x_1 + x_2 + 1) - \mu_2(x_1 + 4x_2 + 2) \,,$$

and the KKT conditions for a local *minimum* are:

$$\begin{aligned} \text{stationarity:} \quad & 2x_1 - 2\mu_1 - \mu_2 = 0\,, \\ & 2x_2 - \mu_1 - 4\mu_2 = 0\,, \\ \text{primal feasibility:} \quad & 2x_1 + x_2 + 1 \geq 0\,, \\ & x_1 + 4x_2 + 2 \geq 0\,, \\ \text{dual feasibility:} \quad & \mu_1 \geq 0\,, \quad \mu_2 \geq 0\,, \\ \text{complementary slackness:} \quad & \mu_1(-2x_1 - x_2 - 1) = 0\,, \\ & \mu_2(-x_1 - 4x_2 - 2) = 0\,. \end{aligned}$$

The KKT conditions for a maximum are obtained by reversing the inequality signs in the dual feasibility condition.

Depending on the active constraints, we distinguish four cases:

Case 1: No constraint is active, i.e., $\mu_1 = \mu_2 = 0$.

Then dual feasibility and complementary slackness are satisfied. From stationarity, it follows that $(x_1, x_2) = (0, 0)$. This point is also primal feasible. The point $(0, 0)$ is therefore a candidate for a maximum or minimum.

Case 2: Only the first constraint is active, i.e., $\mu_2 = 0$.

The stationarity conditions then simplify to

$$2x_1 - 2\mu_1 = 0 \qquad \text{and} \qquad 2x_2 - \mu_1 = 0\,,$$

from which $\mu_1 = x_1 = 2x_2$ follows. Since the first constraint is active, it follows

$$2x_1 + x_2 + 1 = 2 \cdot (2x_2) + x_2 + 1 = 5x_2 + 1 = 0\,,$$

and thus $x_2 = -\frac{1}{5}$ and $x_1 = -\frac{2}{5}$. The point $(-\frac{2}{5}, -\frac{1}{5})$ also satisfies the second constraint, so it is primal feasible. Furthermore, $\mu_1 = x_1 = -\frac{2}{5} \leq 0$, so the point is a candidate for a maximum.

Case 3: Only the second constraint is active, i.e., $\mu_1 = 0$.

In this case, the stationarity conditions yield

$$2x_1 - \mu_2 = 0 \qquad \text{and} \qquad 2x_2 - 4\mu_2 = 0\,,$$

from which $\mu_2 = 2x_1 = \frac{1}{2}x_2$ follows, i.e., $x_2 = 4x_1$. Since the second constraint is active, we get

$$x_1 + 4x_2 + 2 = x_1 + 4 \cdot (4x_1) + 2 = 17x_1 + 2 = 0$$

and from this $x_1 = -\frac{2}{17}$ and $x_2 = -\frac{8}{17}$. The point $(-\frac{2}{17}, -\frac{8}{17})$ is primal feasible, as it also satisfies the first constraint. Furthermore, $\mu_2 = 2x_1 = -\frac{4}{17} \leq 0$, so the point is a candidate for a maximum.

Case 4: Both constraints are active.

Then $2x_1+x_2+1=0$ and $x_1+4x_2+2=0$. Solving the system of equations yields $x_1=-\frac{2}{7}$ and $x_2=-\frac{3}{7}$. Due to stationarity, we have

$$2\cdot\left(-\frac{2}{7}\right)-2\mu_1-\mu_2=0 \quad\text{and}\quad 2\cdot\left(-\frac{3}{7}\right)-\mu_1-4\mu_2=0\,,$$

i.e.,

$$2\mu_1+\mu_2+\frac{4}{7}=0 \quad\text{and}\quad \mu_1+4\mu_2+\frac{6}{7}=0\,.$$

Solving the system of equations yields the Lagrange multipliers $\mu_1=-\frac{10}{49}$ and $\mu_2=-\frac{8}{49}$. Since both Lagrange multipliers are negative, the point $(-\frac{2}{7},-\frac{3}{7})$ is a candidate for a maximum.

So there are four candidates for extreme points, where the sufficient conditions have to be checked.

Case 1: Since no constraint is active, the restricted Lagrangian coincides with the objective function. The Hessian is therefore

$$H_{\widetilde{L}}(x_1,x_2)=\begin{bmatrix}2&0\\0&2\end{bmatrix}=H_f(0,0)\,.$$

Because $\Delta_1(0,0)=2>0$ and $\Delta_2(0,0)=4>0$, there is a local minimum at the point $(0,0)$.

Case 2: Since only the first constraint is strictly active at $(-\frac{2}{5},-\frac{1}{5})$, the restricted Lagrangian there is

$$\widetilde{L}(\mu_1,x_1,x_2)=x_1^2+x_2^2-\mu_1(2x_1+x_2+1)\,.$$

Its modified Hessian is

$$\widetilde{H}_{\widetilde{L}}(\mu_1,x_1,x_2)=\begin{bmatrix}0&-2&-1\\2&2&0\\1&0&2\end{bmatrix}=\widetilde{H}_{\widetilde{L}}\left(-\frac{2}{5},-\frac{2}{5},-\frac{1}{5}\right).$$

This point is a candidate for a maximum only. Since there are $n=2$ variables and only one constraint is considered in the restricted Lagrangian, only $2-1=1$ minors need to be checked. Due to $\widetilde{\Delta}_3=10>0$, the sufficient condition for a maximum is not satisfied.

Case 3: The point $(-\frac{2}{17},-\frac{8}{17})$ is again only a candidate for a maximum. Since only the second constraint is strictly active, the restricted Lagrangian is

$$\widetilde{L}(\mu_2,x_1,x_2)=x_1^2+x_2^2-\mu_2(x_1+4x_2+2)$$

and its modified Hessian

$$\widetilde{H}_{\widetilde{L}}(\mu_2,x_1,x_2)=\begin{bmatrix}0&-1&-4\\1&2&0\\4&0&2\end{bmatrix}=\widetilde{H}_{\widetilde{L}}\left(-\frac{4}{17},-\frac{2}{17},-\frac{8}{17}\right).$$

Sufficient for a maximum at this point is $\widetilde{\Delta}_3 < 0$. However, $\widetilde{\Delta}_3 = 34 > 0$ holds, and the condition is not satisfied.

Case 4: Since both constraints are strictly active, according to Theorem 6.15 there is a maximum at the point $(-\frac{2}{7}, -\frac{3}{7})$ if

$$(-1)^i \widetilde{\Delta}_i(\boldsymbol{\lambda}^*, \boldsymbol{\mu}^*, \boldsymbol{a}) > 0 \quad \text{for} \quad i = 2 \cdot 2 + 1, \ldots, 2 + 2$$

where $k = 2, m = 0$, and $n = 2$ have been inserted. Since two constraints are considered in the restricted Lagrangian, $2 - 2 = 0$ minors need to be checked. The sufficient condition is therefore trivially satisfied. Hence, there is a local maximum at the point $(-\frac{2}{7}, -\frac{3}{7})$.

The optimization problem therefore has two local extrema: A local minimum in $(0, 0)$ – which is obviously also a global minimum – and a local maximum at the point $(-\frac{2}{7}, -\frac{3}{7})$.

For certain optimization problems, the KKT conditions are not only necessary, but also sufficient. This is the case when, in the minimization problem, the objective function f and the functions defining the inequality constraints $h_1, \ldots, h_l$ are convex, and the functions defining the equality constraints $g_1, \ldots, g_m$ are affine-linear[13].

Theorem 6.16: Sufficient KKT conditions

(i) Assume that in the minimization problem (6.5)
- the objective function f is convex,
- the constraints g_i are affine-linear,
- the constraints h_j are convex,

and let $\boldsymbol{a}$ be a point where the KKT conditions from Theorem 6.13 are satisfied. Then f has a **global minimum** subject to the given constraints at the point $\boldsymbol{a}$.

(ii) Assume that in the maximization problem (6.6)
- the objective function f is concave,
- the constraints g_i are affine-linear,
- the constraints h_j are convex,

and let $\boldsymbol{a}$ be a point where the KKT conditions from Theorem 6.13 are satisfied. Then f has a **global maximum** subject to the given constraints at the point $\boldsymbol{a}$.

Since an affine-linear function is both convex and concave, the conditions in part (i) and (ii) of the above theorem are satisfied in particular if both the objective function and all constraints are affine-linear. Such an optimization problem is referred to as a **linear optimization problem** or **linear program**. Together with Theorem 6.13,

[13] A function $g : \mathbb{R}^n \to \mathbb{R}$ is affine-linear if it has the form $g(x_1, \ldots, x_n) = a + b_1 x_1 + \ldots + b_n x_n$; see also Section 2.5.

Theorem 6.16 thus implies that for linear programs, the KKT conditions are both necessary and sufficient for the existence of global extrema. We will revisit this when investigating linear programs in Chapter 11.

6.8 Taylor polynomial

A multivariate differentiable function f can be locally approximated by its tangent function. However, this approximation is sometimes too inaccurate. The tangent function is linear and based on the gradient, i.e., the first partial derivatives. Therefore, the approximation by the tangent function is also called **first order approximation**.

To improve the quality of the approximation, one can – as with univariate functions (Section 4.7) – include higher derivatives and approximate f by a multivariate Taylor polynomial of second or higher order. This is a polynomial in the variables $x_1, \dots, x_n$, whose coefficients depend on the partial derivatives of the function f.

A third order Taylor polynomial, for example, contains the variables $x_1, \dots, x_n$ up to the third power, which also includes mixed terms like $x_i x_j x_k$ and $x_i^2 x_j$. We give the formula for the second degree Taylor polynomial:

Definition 6.3: Second degree Taylor polynomial

Let A be an open set in $\mathbb{R}^n$, $f : A \to \mathbb{R}$, $\boldsymbol{a} \in A$, and let f be twice continuously differentiable. The polynomial

$$T_{f,\boldsymbol{a},2}(\boldsymbol{x}) = f(\boldsymbol{a}) + \sum_{i=1}^{n} \frac{\partial f(\boldsymbol{a})}{\partial x_i}(x_i - a_i) + \frac{1}{2} \sum_{i,j=1}^{n} \frac{\partial^2 f(\boldsymbol{a})}{\partial x_i \partial x_j}(x_i - a_i)(x_j - a_j)$$

is called **second degree Taylor polynomial** of f at the point $\boldsymbol{a}$.

The sufficient criteria for the existence of extrema are based on the fact that the function can be approximated locally by the second degree Taylor polynomial.

Omitting the second sum results in the tangent function, which is the first degree Taylor polynomial.

With the help of matrix and vector notation, Taylor polynomials can be written more compactly. In particular, the second degree Taylor polynomial is

$$f(\boldsymbol{x}) = f(\boldsymbol{a}) + \operatorname{grad} f(\boldsymbol{a})(\boldsymbol{x} - \boldsymbol{a}) + \frac{1}{2}(\boldsymbol{x} - \boldsymbol{a})^T H_f(\boldsymbol{a})(\boldsymbol{x} - \boldsymbol{a}) .$$

Example 6.23. We calculate the second degree Taylor polynomial of the function from Example 5.3,

$$f(\boldsymbol{x}) = f(x_1, x_2, x_3) = x_1 x_2^2 + \exp(5x_3) , \quad \boldsymbol{x} \in \mathbb{R}^3 ,$$

at the point $\boldsymbol{a} = (1,1,0)$. It is $f(1,1,0) = 2$. The gradient of f is

$$\left[\frac{\partial f(\boldsymbol{a})}{\partial x_1}, \frac{\partial f(\boldsymbol{a})}{\partial x_2}, \frac{\partial f(\boldsymbol{a})}{\partial x_3}\right] = \left[a_2^2,\ 2a_1 \cdot a_2,\ 5\exp(5a_3)\right] = \left[1,2,5\right].$$

The Hessian is equal to

$$\left[\frac{\partial^2 f(\boldsymbol{a})}{\partial x_i \partial x_j}\right]_{ij} = \begin{bmatrix} 0 & 2a_2 & 0 \\ 2a_2 & 2a_1 & 0 \\ 0 & 0 & 25e^{5a_3} \end{bmatrix} = \begin{bmatrix} 0 & 2 & 0 \\ 2 & 2 & 0 \\ 0 & 0 & 25 \end{bmatrix}.$$

The second degree Taylor polynomial at the point $\boldsymbol{a} = (1,1,0)$ is therefore

$$\begin{aligned} T_{f,\boldsymbol{a},2}(\boldsymbol{x}) &= 2 + 1 \cdot (x_1 - 1) + 2 \cdot (x_2 - 1) + 5 \cdot (x_3 - 0) \\ &\quad + \frac{1}{2}[2 \cdot 2 \cdot (x_1 - 1)(x_2 - 1) + 2 \cdot (x_2 - 1)^2 + 25 \cdot x_3^2] \\ &= 2 - 2x_1 - x_2 + 5x_3 + 2x_1x_2 + x_2^2 + \frac{25}{2}x_3^2. \end{aligned}$$

Important terms for review

After reading this chapter, the following terms should be familiar:

- Extremum in the interior of the domain
- Extremum on the boundary
- Higher partial derivatives, Hessian matrix
- Necessary and sufficient condition for a local extremum in the interior
- Constrained optimization
- Lagrange's method, Lagrange multipliers, Lagrangian function
- Stationary point, saddle point
- Envelope Theorem
- Karush-Kuhn-Tucker conditions

Self-test

You can check your understanding of this chapter using the following true-or-false questions. For each statement, indicate whether it is true (T) or false (F). Short solutions to these questions can be found in Appendix H.

I. Are the following statements true or false?

	T	F
For twice differentiable functions, the Hessian coincides with the Jacobian matrix.	☐	☐

	T	F
If the Hessian of a function $f : \mathbb{R}^3 \to \mathbb{R}$ is a diagonal matrix with only positive entries on the main diagonal, then the function is convex.	☐	☐
If the Hessian of a function $f : \mathbb{R}^3 \to \mathbb{R}$ is a diagonal matrix with only positive entries on the main diagonal, then the function can only have global maxima as global extreme points.	☐	☐
If one considers $f(x, y)$ subject to $g(x, y) = 0$, then with the Lagrangian $L(x, y) = f(x, y) + \lambda g(x, y)$ one obtains the same local extreme points (x, y) as with $\widetilde{L}(x, y) = f(x, y) - \lambda g(x, y)$.	☐	☐

II. Consider the envelope theorem and Example 6.15 (production of a given output at minimal cost). Assume that one additional unit of the output has to be produced. Are the following statements true or false?

	T	F
The shadow price approximately indicates the additional costs.	☐	☐
If the price p_1 of the input x_1 increases, the shadow price also increases.	☐	☐
If the price p_1 of the input x_1 increases, there are production functions where the minimal costs remain unchanged.	☐	☐

III. Are the following statements about the extrema of a multivariate function true or false?

	T	F
Local extrema on the boundary of the domain cannot always be found with Theorem 6.1 and 6.2.	☐	☐
Every convex function $f : A \to \mathbb{R}$ with $A \subset \mathbb{R}$ has a global minimum.	☐	☐
Since the function $f(x, y) = x^2 + y^2$, $-1 \leq x, y \leq 1$, is convex, it has no local maxima.	☐	☐
If one considers the function $f(x, y) = x^2 + y^2$ on the domain $[-1, 1] \times [-1, 1]$, there are four global maxima, which lie on the vertices of the domain.	☐	☐
The function $f : \mathbb{R}^3 \to \mathbb{R}$ with $f(x, y, z) = x^2 y^4 z^6$ has no global extreme points.	☐	☐
The function $f : \mathbb{R}^3 \to \mathbb{R}$ with $f(x, y, z) = x^2 y^4 z^6$ has infinitely many global extreme points.	☐	☐
The function $f : \mathbb{R}^3 \to \mathbb{R}$ with $f(x, y, z) = x^2 + y^4 + z^6$ has no global extreme points.	☐	☐

The function $f : \mathbb{R}^3 \to \mathbb{R}$ with $f(x, y, z) = x^2 + y^4 + z^6$ has infinitely many global extreme points. ☐ ☐

IV. Consider the KKT theorem (Theorem 6.13, the necessary condition for the existence of a local minimum). Are the following statements true or false?

	T	F
The KKT conditions of stationarity and primal feasibility are exactly the necessary conditions for extrema in optimization problems with only equality constraints.	☐	☐
Dual feasibility states that the Lagrange multipliers of the inequality constraints must be negative.	☐	☐

Exercises

Exercise 1

Find all local extrema of the function $f : \mathbb{R}^2 \to \mathbb{R}$,

$$f(x, y) = \frac{1}{4}y^4 + \frac{1}{3}x^3 - \frac{1}{2}y^2 - x\,.$$

Are these local extrema also global extrema?

Exercise 2

Find all local extrema of the function $f : \mathbb{R}^2 \to \mathbb{R}$,

$$f(x, y) = x^6 - \frac{3}{2}x^4 + y^2\,.$$

Exercise 3

Find all local and global extrema of the function $f : \mathbb{R}^2 \to \mathbb{R}$,

$$f(x_1, x_2) = \frac{1}{3}x_1^3 - 4x_1 + x_1x_2^2 + x_1^2x_2\,.$$

Exercise 4

Find all local and global extrema of the function $f : \mathbb{R}_+^2 \to \mathbb{R}$,

$$f(x, y) = x^3 - 27x - 4y + y^2\,.$$

In particular, also examine the boundary of the domain.

Exercise 5

At which points does the function $f(x_1, x_2, x_3) = -x_1^2x_2^2x_3^4$ have local extrema? Are these minima or maxima? Also check whether the function has global extrema.

Exercise 6
A monopolist produces two goods. The demand quantities x_1 and x_2 for the two goods are given as a function of the prices p_1 and p_2 by

$$x_1 = 135 - 3p_1 + 2p_2 \quad \text{and} \quad x_2 = 110 + 2p_1 - 4p_2 .$$

The production costs of the monopolist are given by $c(x_1, x_2) = 3x_1 + 5x_2$. At which price combination will the profit (= revenue – production costs) be maximum? How many units of the two goods must then be produced, and what is the corresponding profit?

Exercise 7
Check whether the function $f : \mathbb{R}^2_+ \to \mathbb{R}$,

$$f(x, y) = x^3 + 2x^2y + \frac{3}{2}xy^2 + 2y^3 + y^2 - 2x - 3y ,$$

is convex or concave. Does the result change if the function f is defined on the entire $\mathbb{R}^2$?

Exercise 8
Check the following functions $f : \mathbb{R}^2 \to \mathbb{R}$ for convexity or concavity:

a) $f(x, y) = x^2 + 4xy + 8y^2 - 4x - 6y + 10$,
b) $f(x, y) = -2x^2 + 4xy - 8y^2 + 4x - 6y - 12$,
c) $f(x, y) = x^2 + 4xy + y^2 + 2x + 2y + 20$.

Exercise 9
Find all stationary points of the function $f : \mathbb{R}^2 \to \mathbb{R}$, $f(x, y) = xy$, subject to $\frac{1}{8}x^2 + \frac{1}{2}y^2 = 1$.

Exercise 10
Find all stationary points of the function $f(x, y, z) = x + y + z$ subject to $x^2 + y^2 = 2$ and $x + z = 1$.

Exercise 11
Find all extrema of the function $f(x, y, z) = x^2 + 3y^2 + 2z^2$ subject to $x + 3y = 30$ and $y + 2z = 20$.

Exercise 12
A production process is considered in which a good is produced from two production factors. The production function is $F(x_1, x_2) = 5x_1^2x_2$. The costs of production are $c(x_1, x_2) = 6x_1 + 12x_2$.

a) Find the minimum cost combination for producing 80 units of this good.
b) By what amount do the costs of the minimum cost combination approximately change if the quantity to be produced is increased by one unit?

Exercise 13

Find all extrema of the function $f(x, y) = x^2 + 2y$ subject to $x^2 + y^2 \le 5$ and $y \ge 0$.

Exercise 14

Find all extrema of the function $f(x, y) = x^3 - 27x - 4y + y^2$ subject to $x \ge 0$ and $y \ge 0$. (See also Exercise 4.)

Chapter 7

Integral calculus

Integral calculus has two main purposes. First, integrating a function serves as the inverse operation of differentiation. Thus, the primary goal of integral calculus is to determine a function from its rate of change, as described by its first derivative. Second, integral calculus is used to calculate areas and volumes.

For example, one may know the growth behavior of a function and be interested in the function itself. The function may depict the evolution of a capital y over time, when the growth behaviour is given by a known interest yield. Similarly, y may denote a population size whose growth rate is given.

The goal is to determine the value of y as a function of time. Alternatively, one may be interested only in the increase of y from a certain time α to another time β.

The first question leads to the problem of finding an antiderivative or an indefinite integral, while the second concerns the calculation of a definite integral with limits α and β.

The definite integral is also used to calculate areas. Specifically, one seeks the area of a region in the x-y plane whose boundary is defined by lines or curves. For example, the region may be bounded by the graph of the function $f(x) = 1/x$, the x-axis, and the vertical lines at $x = 1$ and $x = 3$. Its area is calculated using the definite integral of the function $1/x$ from 1 to 3. Areas like this need to be determined in the analysis of production and consumption.

K. Mosler et al., *Mathematical Methods for Business and Economics*, Mathematics Study Resources 20, https://doi.org/10.1007/978-3-662-73831-3_7

7.1 Antiderivatives

Let A be an interval in $\mathbb{R}$ and $f : A \to \mathbb{R}$ a function. We are looking for functions $A \to \mathbb{R}$, whose derivative is f.

Definition 7.1: Antiderivative

A function $F : A \to \mathbb{R}$ is called an **antiderivative** of f, if

(i) F is continuous at all $x \in A$ and

(ii) $F'(x) = f(x)$ for all x except for at most countably[1] many $x \in A$.

If F is an antiderivative of f, then the function

$$\widetilde{F}(x) = F(x) + c\,, \quad x \in A\,,$$

is also an antiderivative of f. Here, c represents an arbitrarily chosen real constant. $\widetilde{F}$ is indeed continuous and $\widetilde{F}'(x) = F'(x) = f(x)$ for all $x \in A$ except at the same countably many exceptional points. Obviously, one can add an arbitrary constant to an antiderivative F and get another antiderivative.

Many antiderivatives are differentiable at all points of their domain and thus satisfy the condition $F'(x) = f(x)$ for all $x \in A$. In general, however, there may be a countable number of points, finite or infinite, at which this condition does not hold. We first give examples of antiderivatives that are differentiable everywhere.

Example 7.1.

a) For $n \in \mathbb{Z}$, $n \neq -1$, we consider the power function

$$f(x) = x^n\,, \quad x \in [a, b]\,,$$

where $0 \notin [a, b]$ when $n < 0$. An antiderivative is

$$F(x) = \frac{1}{n+1} x^{n+1}\,, \quad x \in [a, b]\,.$$

It is differentiable everywhere. Again, $\widetilde{F}(x) = \frac{1}{n+1} x^{n+1} + c$ with any $c \in \mathbb{R}$ is an antiderivative as well. In general, c is mentioned as the **constant of integration**.

b) For the function

$$f(x) = \frac{1}{x}\,, \quad x \in [a, b]\,,$$

where $0 \notin [a, b]$, we obtain the antiderivatives

$$F(x) = \ln(|x|) + c\,.$$

[1] A set of numbers is called **countable**, if it is either finite or its elements can be written as an infinite sequence; see Appendix B.

Here, c is an arbitrary real number. To see this, consider the two cases $x > 0$ and $x < 0$ and calculate the derivative of F.

c) The function

$$f(x) = \begin{cases} 2x, & \text{if } x \le 1, \\ 2x^2, & \text{if } x > 1, \end{cases}$$

has two branches. An antiderivative is determined separately for each branch,

$$F(x) = \begin{cases} x^2 + c_1, & \text{if } x \le 1, \\ \frac{2}{3}x^3 + c_2, & \text{if } x > 1, \end{cases}$$

with constants of integration c_1 and c_2. Since F as an antiderivative is continuous, both branches must meet at the point $x = 1$. It follows that

$$F(1) = 1 + c_1 \stackrel{!}{=} \frac{2}{3} + c_2 = \lim_{x \searrow 1} F(x), \quad \text{thus} \quad c_2 = \frac{1}{3} + c_1.$$

The constant c_1 can be chosen freely. Therefore, an antiderivative is

$$F(x) = \begin{cases} x^2 + c_1, & \text{if } x \le 1, \\ \frac{2}{3}x^3 + \frac{1}{3} + c_1, & \text{if } x > 1. \end{cases}$$

The graphs of f and its antiderivative F are shown in Figure 7.1.

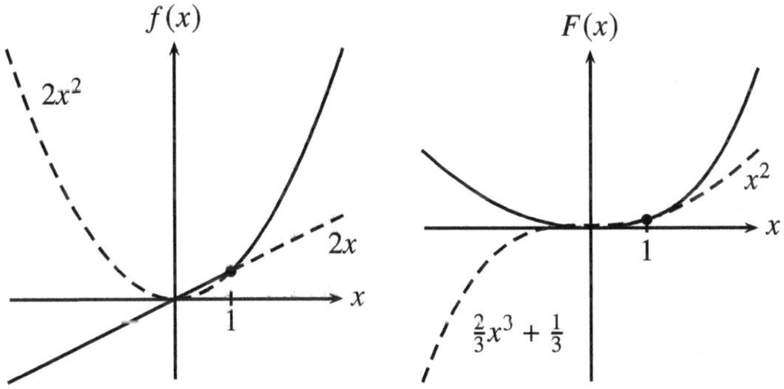

Figure 7.1: Function f and antiderivative F (with $c_1 = 0$).

We now present two examples of antiderivatives that are not differentiable at some exceptional points.

Example 7.2.

a) The function

$$f(x) = 1_{(a,b]}(x) = \begin{cases} 0 & \text{for } x \le a, \\ 1 & \text{for } a < x \le b, \\ 0 & \text{for } x > b, \end{cases}$$

takes the value 1 on the interval $(a, b]$; otherwise it equals 0. The function indicates whether its argument x is in $(a, b]$ or not. It is therefore called the **indicator function**[2] of the interval $(a, b]$. An antiderivative of $1_{(a,b]}$ is

$$F(x) = \begin{cases} 0 & \text{for } x \le a\,, \\ x - a & \text{for } a < x \le b\,, \\ b - a & \text{for } x > b\,. \end{cases}$$

F is continuous and piecewise linear, but not differentiable at the two exceptional points $x = a$ and $x = b$.

b) Let f be the **floor function** from Example 1.8. It assigns to each $x \in \mathbb{R}$ the largest integer $f(x) = \lfloor x \rfloor$ that is less than or equal to x. For each integer $n \in \mathbb{Z}$ we therefore have

$$\begin{aligned} f(x) &= n\,, && \text{if } n \le x < n+1\,, \quad \text{and thus} \\ F(x) &= nx + c_n\,, && \text{if } n \le x < n+1\,. \end{aligned}$$

The constants c_n have to be determined such that F is continuous at the points $x = n$, i.e., that $F(n)$ matches the left-hand limit $F(n-)$. Thus,

$$F(n) = F(n-) \iff n \cdot n + c_n = (n-1)n + c_{n-1} \iff c_n = c_{n-1} - n\,.$$

Obviously, one of these constants, say c_0, can be chosen freely as the constant of integration, while the other c_n are determined by the choice of c_0 as follows:[3]

$$c_n = c_{n-1} - n = \ldots = c_0 - 1 - 2 - \ldots - n = c_0 - \frac{n(n+1)}{2}\,,$$

and thus

$$F(x) = nx - \frac{n(n+1)}{2} + c_0\,, \quad \text{if } n \le x < n+1\,, \quad n \in \mathbb{Z}\,.$$

The antiderivative F is nondifferentiable at each $x \in \mathbb{Z}$, i.e., it has an infinite number of exceptional points.

In general, two questions arise: How do the various antiderivatives of a given function f differ, and which functions f have an antiderivative at all? It holds:

- An antiderivative of a function $f : A \to \mathbb{R}$ is *unique up to an additive constant*, i.e., if F and $\widetilde{F}$ are antiderivatives of f, then there is a constant $c \in \mathbb{R}$ such that $\widetilde{F}(x) = F(x) + c$ for all $x \in A$.
- A function f defined on a finite interval has an antiderivative F, if f is a **regulated**

[2] The indicator function of an interval is a special step function, see Definition 7.4.

[3] The equation $\sum_{i=1}^{n} i = n(n+1)/2$ can be proved by mathematical induction; see for example the proof of Theorem 3.11 in Chapter 3.

function i.e., if the one-sided limits exist at all points x of the interval.[4] An example of a regulated function f is shown in Figure 7.2.

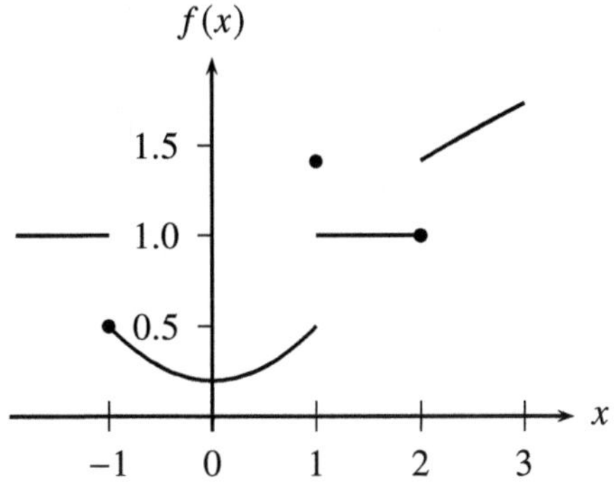

Figure 7.2: Regulated function with jumps at the points $-1, 1$ and 2.

In particular, every continuous function has an antiderivative. Nearly all functions that are of interest in business or economics are continuous or at least regulated functions. Hence, they have antiderivatives.

The function values of a regulated function at its jump discontinuities do not matter at all. In fact, the antiderivative F does not change if f is altered at at most countably infinitely many points, as long as f remains regulated.

Antiderivative and derivative are inverse operations:

- F is an antiderivative of f if and only if f is a derivative of F.

Table 7.1 presents the antiderivatives of important and often used functions. Note that a table of functions and their derivatives (like the overview in Section 4.3) can also be *read backwards* as a table of functions and their antiderivatives.

7.2 Indefinite integrals

Definition 7.2: Indefinite integral

Let F be an antiderivative of f. The function

$$x \mapsto \int f(x)\,dx = F(x)\,, \quad x \in A\,,$$

is called **indefinite integral** of f. The function f is referred to as the **integrand**, x as the **variable of integration**.

[4]It can be shown that such a function has at most countably infinitely many jumps in the interval.

$f(x)$	defined for	$F(x)$
b	$x \in \mathbb{R}$	$bx + c$
x^n	$x \in \mathbb{R},\ n \in \mathbb{N}$	$\frac{1}{n+1}x^{n+1} + c$
x^n	$x \neq 0,\ n \in \mathbb{Z} \setminus \{-1\}$	$\frac{1}{n+1}x^{n+1} + c$
x^α	$x > 0,\ \alpha \neq -1$	$\frac{1}{\alpha+1}x^{\alpha+1} + c$
x^{-1}	$x \neq 0$	$\ln(\lvert x\rvert) + c$
$\exp(x)$	$x \in \mathbb{R}$	$\exp(x) + c$
a^x	$x \in \mathbb{R},\ a > 0$	$\frac{a^x}{\ln(a)} + c$
$\sin(x)$	$x \in \mathbb{R}$	$-\cos(x) + c$
$\cos(x)$	$x \in \mathbb{R}$	$\sin(x) + c$
$\tan(x)$	$x \neq (2n-1)\frac{\pi}{2}, n \in \mathbb{Z}$	$-\ln(\lvert\cos(x)\rvert) + c$
$\cot(x)$	$x \neq 2n\frac{\pi}{2}, n \in \mathbb{Z}$	$\ln(\lvert\sin(x)\rvert) + c$
$\frac{1}{\cos^2(x)}$	$x \neq (2n-1)\frac{\pi}{2}, n \in \mathbb{Z}$	$\tan(x) + c$
$\frac{1}{\sin^2(x)}$	$x \neq 2n\frac{\pi}{2}, n \in \mathbb{Z}$	$-\cot(x) + c$
$\frac{1}{\sqrt{1-x^2}}$	$x \in [0, 1]$	$\arcsin(x) + c$
$\frac{1}{1+x^2}$	$x \in \mathbb{R}$	$\arctan(x) + c$

Table 7.1: Antiderivative of commonly used functions

The indefinite integral of a function f is nothing else than an antiderivative of f. It is uniquely determined up to the **constant of integration** c. The symbol $\int f(x)dx$ always denotes *one* version of the indefinite integral.

To calculate indefinite integrals, rules are applied, which correspond to the rules of differentiation. For real functions f and g defined on A, the following holds

Theorem 7.1: Rules for integrals

$$\int (f(x) + g(x))\, dx = \int f(x)\, dx + \int g(x)\, dx,$$

$$\int \lambda f(x)\, dx = \lambda \int f(x)\, dx, \quad \lambda \in \mathbb{R}.$$

The integral of a sum of functions is equal to the sum of the integrals, and the integral of a multiple of a function is equal to the multiple of the integral.[5]

Example 7.3. It holds

$$\int \left(x^2 + \frac{3}{x}\right) dx = \int x^2 dx + 3 \int \frac{1}{x}\, dx = \frac{x^3}{3} + 3\ln(|x|) + c\,, \quad c \in \mathbb{R}\,.$$

Example 7.4 (Integration of a quotient of polynomials). We are looking for the indefinite integral

$$\int \underbrace{\frac{x^4 - 2x^2 + 2}{x^2 + 1}}_{f(x)} dx\,.$$

The integrand is simplified by polynomial long division:

$$\begin{array}{l} (x^4{-}2x^2{+}2) : (x^2 + 1) = x^2 - 3 + \dfrac{5}{x^2 + 1} = f(x) \\ \underline{-x^4 - x^2} \\ \quad -3x^2{+}2 \\ \quad\ \ \underline{3x^2{+}3} \\ \qquad\quad 5 \end{array}$$

With $\int \frac{1}{x^2+1} dx = \arctan(x)$ (see Table 7.1) it follows

$$\int f(x) dx = \frac{x^3}{3} - 3x + 5\arctan(x) + c\,, \quad c \in \mathbb{R}\,.$$

If g is a differentiable function and nowhere zero, the expression

$$\boxed{\frac{d}{dx} \ln\left(|g(x)|\right) = \frac{g'(x)}{g(x)}}$$

is called the **logarithmic derivative** of g. This derivative provides a useful integration formula,

$$\boxed{\int \frac{g'(x)}{g(x)}\, dx = \ln(|g(x)|) + c\,.}$$

Example 7.5. It holds

$$\int \frac{2x}{x^2 + 5}\, dx = \ln(|x^2 + 5|) = \ln(x^2 + 5) + c\,, \quad c \in \mathbb{R}\,,$$

with $g(x) = x^2 + 5$ and $g'(x) = 2x$.

[5] Strictly speaking the right and left sides of the equations can differ by a constant; but this can be avoided by a suitable choice of the constant of integration.

Example 7.6 (Continuous compounding of capital). A capital K earns interest which is compounded continuously. The interest is credited to the account at every instant, not just at the end of a period. How does K develop as a function of time? Let $K_x = g(x)$ be the capital at time x and let $p > 0$ be the interest rate per unit of time, for example, $p = 0.05$ per annum. The marginal change of the capital is then

$$\frac{d}{dx}g(x) = p \cdot g(x) \quad \text{or} \quad \frac{g'(x)}{g(x)} = p\,.$$

Since $g(x) > 0$, the above integration formula gives

$$\int \frac{g'(x)}{g(x)}\,dx = \ln(|g(x)|) + c = \ln(g(x)) + c\,, \qquad c \in \mathbb{R}\,.$$

On the other hand,

$$\int \frac{g'(x)}{g(x)}\,dx = \int p\,dx = px\,.$$

It follows $K_x = g(x) = \exp(px - c)$. If the initial capital is denoted by K_0, then $K_0 = g(0) = e^{-c}$ and the capital at time x is

$$K_x = K_0\,e^{px}\,.$$

Solving for K_0 yields the **present value** of the capital K_x available at time x under continuous compounding,

$$K_0 = e^{-px} K_x\,.$$

For example, let the initial capital be $K_0 = 12\,000$ €. It earns interest at a continuous rate of five percent per annum (i.e., $p = 0.05$). After seven years the capital amounts to

$$K_7 = 12\,000 \cdot e^{0.05\cdot 7} = 17\,028.81\text{ €}\,.$$

Next, we compare continuous compounding with **periodic compounding**, where the interest is credited to the account at the end of each period. What periodic interest rate is needed to achieve the same future capital as with continuous compounding? Let $\overline{p}$ be the periodic interest rate and p be the continuous interest rate. After n periods, the final capital should be the same for both methods:

$$K_0 e^{pn} = K_0(1 + \overline{p})^n\,.$$

This leads to $e^p = 1 + \overline{p}$ and thus $\overline{p} = e^p - 1$. It follows that $p < \overline{p}$: The equivalent continuous interest rate p is always lower than the periodic interest rate $\overline{p}$ required to achieve the same final amount.

Further methods for calculating indefinite integrals are integration by parts and integration by substitution. They are discussed in Sections 7.6 and 7.7.

7.3 Definite integrals

Suppose we are not interested in the entire development of a capital as a function of time, but only in the increase of the capital from a certain time a to another time b. This question leads us to the concept of the definite integral.

Definition 7.3: Definite iIntegral

Let F be an antiderivative of f. For $\alpha, \beta \in A$ the number

$$\int_{\alpha}^{\beta} f(x)\,dx = F(\beta) - F(\alpha)$$

is called the **definite integral** of f from α to β.

While an indefinite integral is a function, a definite integral represents a number. To calculate a definite integral, one first determines an antiderivative, i.e., the indefinite integral, and then inserts the limits.

Example 7.7. We have

$$\begin{aligned}\int_0^2 \underbrace{(6x^2 - 1 + e^x)}_{f(x)}\,dx &= \underbrace{\frac{6}{3}x^3 - x + e^x}_{F(x)}\Bigg|_0^2 \\ &= F(2) - F(0) \\ &= (16 - 2 + e^2) - (0 - 0 + e^0) \\ &= 13 + e^2 .\end{aligned}$$

A definite integral does not depend on the choice of the antiderivative F, since any other antiderivative $\widetilde{F}$ differs from F only by a constant c, $\widetilde{F}(x) = F(x) + c$ for all x. Therefore, $\widetilde{F}(\beta) - \widetilde{F}(\alpha) = F(\beta) - F(\alpha)$. Since in the calculation of definite integrals, the constant of integration is irrelevant, it will therefore be mostly omitted.

In addition to the above rules for the indefinite integral, the following rules for the definite integral are useful. Note that, in general, the lower limit of a definite integral can be greater than the upper limit.

Theorem 7.2: Rules for definite integrals

For $\alpha, \beta, \delta \in A$ we have

$$\int_{\alpha}^{\beta} f(x)\,dx = -\int_{\beta}^{\alpha} f(x)\,dx ,$$

$$\int_{\alpha}^{\alpha} f(x)\,dx = 0,$$

$$\int_{\alpha}^{\beta} f(x)\,dx = \int_{\alpha}^{\delta} f(x)\,dx + \int_{\delta}^{\beta} f(x)\,dx\,.$$

If we consider the upper limit of the definite integral as a variable, we obtain a function h_1 defined on A,[6]

$$h_1 : x \mapsto \int_{\alpha}^{z} f(t)\,dt = F(x) - F(\alpha)\,, \quad x \in A\,.$$

h_1 is a continuous function and obviously an antiderivative of f. Similarly, we obtain a continuous function h_2 depending on the lower limit,

$$h_2 : x \mapsto \int_{x}^{\beta} f(t)\,dt = F(\beta) - F(x)\,, \quad x \in A\,.$$

It is the negative of an antiderivative of f. Conversely, the derivatives of the two functions are given by

$$\frac{d}{dx}\int_{\alpha}^{x} f(t)\,dt = f(x)\,,$$

$$\frac{d}{dx}\int_{x}^{\beta} f(t)\,dt = -f(x)\,.$$

7.4 Further rules for definite integrals

Let F and G be antiderivatives of f and g, respectively. In addition to the rules for definite integrals presented so far, the following rules are useful. The first rule states that the definite integral of a non-negative function is always greater than or equal to zero.

Theorem 7.3: Non-negative integrand

If $f(x) \geq 0$ for $\alpha < x < \beta$, then

$$\int_{\alpha}^{\beta} f(x)\,dx \geq 0\,.$$

Proof: If $f(x)$ is non-negative for $\alpha < x < \beta$, then its antiderivative $F(x)$ increases monotonically on the interval (α, β). Therefore, $F(\beta) \geq F(\alpha)$ and thus $\int_{\alpha}^{\beta} f(x)\,dx = F(\beta) - F(\alpha) \geq 0\,.$ □

From this it follows that the definite integral is monotonic. If two functions are pointwise ordered, their integrals are also ordered:

[6] The name of the integration variable does not matter!

Theorem 7.4: Monotonicity of the integral

If $f(x) \leq g(x)$ for all $x \in (\alpha, \beta)$, then

$$\int_\alpha^\beta f(x)\,dx \leq \int_\alpha^\beta g(x)\,dx\,.$$

Proof: For $x \in (\alpha, \beta)$ the difference $h(x) = g(x) - f(x) \geq 0$ according to the assumption. From Theorem 7.3 it follows that

$$\int_\alpha^\beta g(x)\,dx - \int_\alpha^\beta f(x)\,dx = \int_\alpha^\beta \big(g(x) - f(x)\big)\,dx = \int_\alpha^\beta h(x)\,dx \geq 0\,.$$

□

From Theorem 7.4 follows that the absolute value of an integral is always less than or equal to the integral of the absolute value of the integrand.

Theorem 7.5: Absolute value of an integral

It holds

$$\left|\int_\alpha^\beta f(x)\,dx\right| \leq \int_\alpha^\beta \big|f(x)\big|\,dx\,.$$

Proof: For all x, obviously $f(x) \leq |f(x)|$ and $-f(x) \leq |f(x)|$. From Theorem 7.4 follows

$$\int_\alpha^\beta f(x)\,dx \leq \int_\alpha^\beta \big|f(x)\big|\,dx \quad \text{and} \quad -\int_\alpha^\beta f(x)\,dx \leq \int_\alpha^\beta \big|f(x)\big|\,dx$$

Now the absolute value of the integral is either $\int_\alpha^\beta f(x)\,dx$ or $-\int_\alpha^\beta f(x)\,dx$. In both cases, it is less than or equal to the integral of the absolute value $\int_\alpha^\beta |f(x)|\,dx$. □

7.5 Calculation of areas

A central task of integral calculus is the calculation of areas. The rectangle shown in gray in Figure 7.3 has the area $v = (\beta - \alpha) \cdot \gamma$ (= width times height). The top edge of the rectangle is given by the constant function $f(x) = \gamma$. The area is bounded below by the x-axis and laterally by the lines $x = \alpha$ and $x = \beta$. An antiderivative of $f(x)$ is $F(x) = \gamma x + c$. So the definite integral of f from α to β equals

$$\int_\alpha^\beta f(x)dx = F(\beta) - F(\alpha) = \gamma \cdot (\beta - \alpha)\,,$$

hence the area of the rectangle.

The same holds for areas that are bounded above by a regulated function $f(x) \geq 0$:

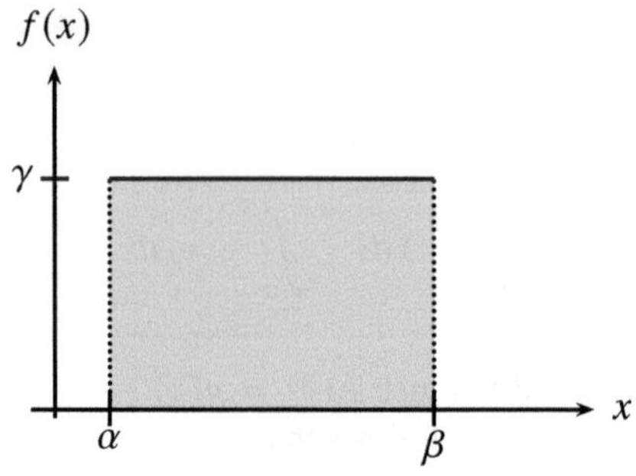

Figure 7.3: Area of a rectangle.

Mnemonic 7.1: Area calculation by integration

If $f(x) \geq 0$ for $\alpha \leq x \leq \beta$, then

$$\int_{\alpha}^{\beta} f(x)\,dx = \text{Area between the graph of } f \text{ and the } x\text{-axis between the limits } \alpha \text{ and } \beta\,.$$

Next we will illustrate this fact using step functions.

Definition 7.4: Step function

Let $\alpha_0, \ldots, \alpha_k \in \mathbb{R}$, $\alpha_0 < \alpha_1 < \alpha_2 < \ldots < \alpha_k$ and $\gamma_1, \ldots, \gamma_k \in \mathbb{R}_+$. The function

$$f(x) = \begin{cases} 0, & \text{if } x \leq \alpha_0\,, \\ \gamma_i\,, & \text{if } \alpha_{i-1} < x \leq \alpha_i\,, \quad i = 1, \ldots, k\,, \\ 0, & \text{if } x > \alpha_k\,. \end{cases}$$

is called a **step function**.

From Example 7.2 we already know the indicator function of an interval, which is a special step function. Conversely, a step function can be written as a sum of indicator functions,

$$f(x) = \sum_{i=1}^{k} \gamma_i\, 1_{(\alpha_{i-1}, \alpha_i]}(x)\,.$$

An antiderivative of the indicator function $1_{(a_{i-1}, a_i]}$ is

$$F_i(x) = \begin{cases} 0 & \text{for } x \leq a_{i-1}\,, \\ x - a_{i-1} & \text{for } a_{i-1} < x \leq a_i\,, \\ a_i - a_{i-1} & \text{for } x > a_i\,. \end{cases}$$

The definite integral of the step function f from α_0 to α_k is therefore

$$\int_{\alpha_0}^{\alpha_k} f(x)\,dx = \int_{\alpha_0}^{\alpha_1} \gamma_1\,dx + \int_{\alpha_1}^{\alpha_2} \gamma_2\,dx + \ldots + \int_{\alpha_{k-1}}^{\alpha_k} \gamma_n\,dx$$
$$= \sum_{i=1}^{k} \gamma_i(\alpha_i - \alpha_{i-1})\,.$$

The gray area in Figure 7.4 consists of k rectangles having areas $\gamma_i(\alpha_i - \alpha_{i-1})$, $i = 1, 2, \ldots k$. Hence the definite integral of the step function is equal to the area "under" its graph.

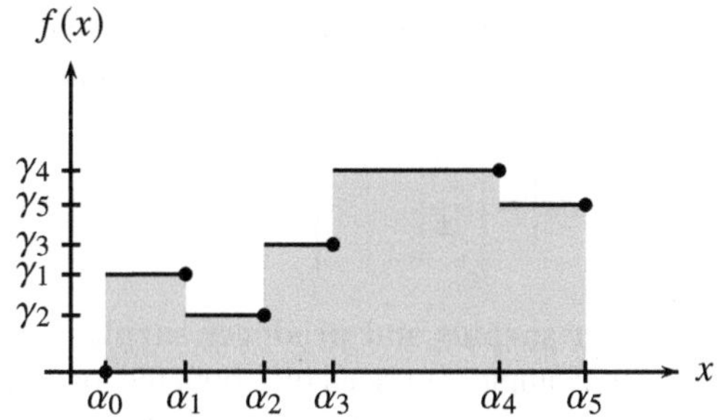

Figure 7.4: Step function with $k = 5$.

Any non-negative regulated function f defined on a bounded interval can be approximated arbitrarily well by a sequence of step functions with an increasing number of steps. The area under the graph of f is then equal to the limit of the areas under these step functions. Consequently, the area "under" a regulated function can also be computed as a definite integral.

An area in the x-y plane, which is not bounded below by the x-axis, but by a second function $g(x) \geq 0$, is calculated as the difference of the two areas with respect to f and g. This is equal to the definite integral of $f - g$ from a to b. In this way, one obtains the content of areas that extend completely above the x-axis; see Figure 7.5.

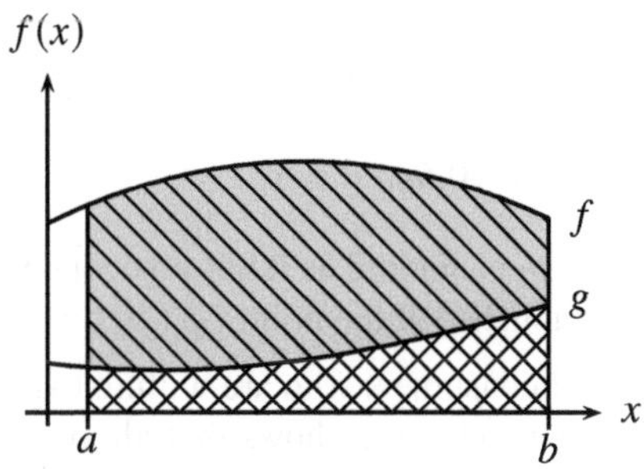

Figure 7.5: Area between the graphs of f and g over the interval $[a, b]$.

To calculate areas between the x-axis and the graph of a function $f(x) \leq 0$, one uses the definite integral of its negative $-f(x)$. If a function f takes both positive and negative values, the areas above and below the x-axis are integrated separately and the areas are added to give the total area.

Example 7.8.

$$\int_{-1}^{1} x^3\,dx = \frac{x^4}{4}\bigg|_{-1}^{1} = \frac{1}{4} - \frac{1}{4} = 0$$

As the graph of f is point symmetric to $(0,0)$, the area between f and the x-axis in the range $[0,1]$ is as large as in the range $[-1,0]$. To actually obtain the area between the x-axis and the function in the range $[-1,1]$, one must integrate from $x = -1$ to the zero $x = 0$ and then again from $x = 0$ to $x = 1$ and calculate the sum of the absolute values of the integrals. Here one obtains

$$\left|\int_{-1}^{0} x^3\,dx\right| + \left|\int_{0}^{1} x^3\,dx\right| = \left|\frac{x^4}{4}\bigg|_{-1}^{0}\right| + \left|\frac{x^4}{4}\bigg|_{0}^{1}\right| = \left|0 - \frac{1}{4}\right| + \left|\frac{1}{4} - 0\right| = \frac{1}{2}.$$

Example 7.9 (Consumer surplus and producer surplus). On a market for a homogeneous good (for example, rapeseed oil), both the total quantity demanded and the total quantity supplied depend on the price p. Let $p_D(q)$ denote the price at which the total demand is q, and $p_S(q)$ the price at which the total supply is q. Typically, the **demand function** p_D is monotonically decreasing, while the **supply function** p_S is monotonically increasing. The equilibrium price is the price p_0 at which the market clears, i.e., where the supply and demand curves intersect (see Figure 7.6).

For example, consider the following supply and demand functions:

$$p_D(q) = 10 - 0.5q\,, \quad q \geq 0\,,$$
$$p_S(q) = 4 + 0.02q^2, \quad q \geq 0\,.$$

Equilibrium price and quantity are found by equating the two functions:

$$10 - 0.5q = 4 + 0.02q^2\,,$$
$$q_0 = -12.5 \pm \sqrt{456.25} = 8.86 \quad \text{or} \quad -33.86\,,$$
$$p_0 = 10 - 0.5q_0 = 5.57\,.$$

The negative solution for q_0 is ignored as it has no economic meaning. Thus, the market equilibrium is established at the point $(q_0, p_0) = (8.86, 5.57)$.

At the market price p_0, all consumers who are willing to pay at least p_0 will buy. The strictly decreasing demand curve shows that almost all consumers pay less than they would be willing to pay. This difference, aggregated over consumers, is called the **consumer surplus** CS. The consumer surplus is represented as the light gray area in Figure 7.6. Similarly, all producers sell at the market price and hence at a higher price than almost all of them would be willing to sell at. The resulting difference in payment is called the **producer surplus** PS. It corresponds to the dark gray area in Figure 7.6.

The consumer surplus is given by

$$CS = \int_0^{q_0} p_D(q)\, dq - p_0 q_0 .$$

The producer surplus is calculated as

$$PS = p_0 q_0 - \int_0^{q_0} p_S(q)\, dq .$$

For the given demand and supply functions, we obtain

$$CS = \int_0^{8.86} (10 - 0.5q)\, dq - 5.57 \cdot 8.86 = 19.63 ,$$

$$PS = 5.57 \cdot 8.86 - \int_0^{8.86} (4 + 0.02q^2)\, dq = 9.27 .$$

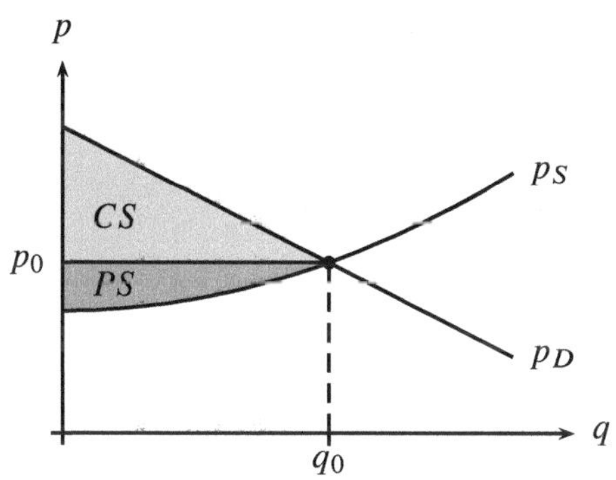

Figure 7.6: Market equilibrium with consumer and producer surplus.

7.6 Integration by parts

The following two sections discuss additional techniques for calculating both indefinite and definite integrals. They are based on two rules of differentiation: the product rule and the chain rule.

Let F and G be antiderivatives of f and $g : A \to \mathbb{R}$, respectively. Then the following formula[7], known as **integration by parts**, holds,

[7]Here, too, the two sides of the equation may differ by a constant, which can be eliminated by an appropriate choice of the constant of integration.

$$\int f(x)G(x)\,dx = F(x)G(x) - \int F(x)g(x)\,dx\,.$$

The formula follows directly from the product rule. It holds (except at possible exceptional points) that $(F \cdot G)'(x) = f(x)G(x) + F(x)g(x)$.

Integration by parts is used to integrate a product of the form $f(x)G(x)$ when the antiderivative F of f and $\int F(x)g(x)dx$ are known or can be more easily calculated.

By inserting limits α, β, the corresponding formula for definite integrals is obtained,

$$\int_\alpha^\beta f(x)G(x)\,dx = F(x)G(x)\Big|_\alpha^\beta - \int_\alpha^\beta F(x)g(x)\,dx\,.$$

Example 7.10.

a) The definite integral $\int_\alpha^\beta x\,e^x\,dx$ is to be determined.

One factor of the integrand, x, has the constant derivative 1, the other factor, e^x, is equal to its antiderivative e^x. We therefore use integration by parts with $G(x) = x$ and $f(x) = e^x$ to first obtain the indefinite integral

$$\begin{aligned}\int e^x \cdot x\,dx &= e^x \cdot x - \int e^x \cdot 1\,dx \qquad \left|\begin{array}{l|l} f(x) = e^x & F(x) = e^x \\ G(x) = x & g(x) = 1 \end{array}\right| \\ &= e^x x - e^x \\ &= e^x(x-1)\,,\end{aligned}$$

Now, the definite integral is

$$\int_\alpha^\beta xe^x\,dx = e^\beta(\beta - 1) - e^\alpha(\alpha - 1)\,.$$

For example, with limits $\alpha = 0$ and $\beta = 1$ we obtain

$$\int_0^1 xe^x\,dx = e^1(1-1) - e^0(0-1) = 1\,.$$

One might be tempted to set $f(x) = x$ and $G(x) = e^x$. However, this would result in the integral $\int \frac{x^2}{2} e^x\,dx$ on the right-hand side, which is even more complicated than the original integral.

b) Calculate $\int \ln(x)\,dx$ and $\int_1^5 \ln(x)\,dx$.

Although this integrand is not a product, it can be written as one, namely in the form $\ln(x) = 1 \cdot \ln(x)$. Integration by parts gives

$$\int \ln(x)\,dx = \int 1 \cdot \ln(x)\,dx \qquad \left|\begin{array}{l|l} f(x) = 1 & F(x) = x \\ G(x) = \ln(x) & g(x) = \frac{1}{x} \end{array}\right|$$

$$= x\ln(x) - \int x \cdot \frac{1}{x}\,dx$$
$$= x\ln(x) - x$$
$$= x(\ln(x) - 1)\,.$$

The definite integral from 1 to 5 is

$$\int_1^5 \ln(x)\,dx = 5(\ln(5) - 1) - 1(0 - 1) = 5\ln(5) - 4\,.$$

c) Calculate $\int_0^{2\pi} \sin^2(x)\,dx$ using integration by parts.

It holds

$$\int \sin^2(x)\,dx = \int \sin(x)\cdot\sin(x)\,dx \quad \left|\begin{array}{c|c} f(x) = \sin(x) & F(x) = -\cos(x) \\ G(x) = \sin(x) & g(x) = \cos(x) \end{array}\right|$$
$$= -\sin(x)\cos(x) + \int \underbrace{\cos^2(x)}_{1-\sin^2(x)}\,dx$$
$$= -\sin(x)\cos(x) + \int 1dx - \int \sin^2(x)\,dx\,.$$
$$= -\sin(x)\cos(x) + x - \int \sin^2(x)\,dx\,.$$

By rearranging, we obtain

$$2\int \sin^2(x)\,dx = -\sin(x)\cos(x) + x\,,$$

and thus

$$\int \sin^2(x)\,dx = \frac{1}{2}\Big(x - \sin(x)\cos(x)\Big)\,.$$

For the given limits we get

$$\int_0^{2\pi} \sin^2(x)\,dx = \frac{1}{2}(2\pi - 0) - \frac{1}{2}(0 - 0) = \pi\,.$$

7.7 Integration by substitution

Another important technique is integration by substitution. It is based on the chain rule of differentiation and used for integrating composite functions. Consider a composite function defined on an interval $A \in \mathbb{R}$,

$$x \mapsto f(x) \mapsto g(f(x))\,, \quad x \in A\,.$$

Let g have an antiderivative G, and let f is differentiable. Then the first formula of **integration by substitution (I)** holds,

$$\int g(f(x))f'(x)\,dx = \int g(y)\,dy\Big|_{y=f(x)} = G(f(x))\,.$$

The formula says that replacing $y = f(x)$ in the original integral transforms it into an equivalent integral with respect to y. The transformed integral is evaluated as a function of y. Then $y = f(x)$ is substituted to obtain $G(f(x))$.

The formula is a direct consequence of the chain rule of differentiation. According to the usual chain rule, $(G \circ f)'(x) = g(f(x)) \cdot f'(x)$ holds for all $x \in A$. This implies

$$\int g(f(x))f'(x)\,dx = (G \circ f)(x) = G(f(x)) = G(y)|_{y=f(x)} = \int g(y)\,dy\Big|_{y=f(x)}\,.$$

Example 7.11. The task is to find the definite integral $\int_{-1}^{2} 2x\sqrt{x^2+1}\,dx$.

We first calculate the indefinite integral by substituting $y = f(x) = x^2 + 1$,

$$\int \underbrace{2x}_{f'(x)}\ \underbrace{\sqrt{x^2+1}}_{g(f(x))}\,dx = \int g(y)\,dy = \int \sqrt{y}\,dy = \int y^{\frac{1}{2}}\,dy = \frac{2}{3}y^{\frac{3}{2}} = \frac{2}{3}\sqrt{(x^2+1)^3}\,.$$

The definite integral is then

$$\int_{-1}^{2} 2x\sqrt{x^2+1}\,dx = \frac{2}{3}\left(\sqrt{125} - \sqrt{8}\right) = 5.5679\,.$$

If the *inner function* f is invertible, a second formula of **integration by substitution (II)** holds,

$$\int g(f(x))\,dx = \int g(y)\frac{1}{f'(f^{-1}(y))}\,dy\,.$$

To see this, note that the derivative of the inverse function f^{-1} is given by

$$\frac{d}{dy}f^{-1}(y) = \frac{1}{f'(f^{-1}(y))}\,, \tag{7.1}$$

for which we simply write $\frac{dx}{dy}$. The substitution formula (II) simplifies to

$$\int g(f(x))\,dx = \int g(y)\frac{d}{dy}f^{-1}(y)\,dy = \int g(y)\frac{dx}{dy}dy\,.$$

This formula follows from the first formula of substitution (I), by substituting $h(y) = g(y)\frac{d}{dy}f^{-1}(y)$ for $g(y)$ on its right side:

$$\int \underbrace{g(y)\frac{d}{dy}f^{-1}(y)}_{h(y)}\, dy\Big|_{y=f(x)} = \int h(y)\,dy\Big|_{y=f(x)} \overset{(I)}{=} \int h(f(x))f'(x)\,dx$$

$$\overset{(7.1)}{=} \int g(f(x)) \cdot \underbrace{\frac{d}{dy}\left(f^{-1}(f(x))\right) \cdot f'(x)}_{=1 \text{ for all } x}\,dx$$

$$= \int g(f(x))dx$$

The substitution rules for indefinite integrals lead to corresponding rules for definite integrals:

$$\boxed{\int_\alpha^\beta g(f(x))f'(x)\,dx = \int_{f(\alpha)}^{f(\beta)} g(y)\,dy\,,}$$

$$\boxed{\int_\alpha^\beta g(f(x))\,dx = \int_{f(\alpha)}^{f(\beta)} g(y)\frac{1}{f'(f^{-1}(y))}\,dy\,, \quad \text{if } f^{-1} \text{ exists.}}$$

Example 7.12.

a) The task is to find the indefinite integral $\int e^{-px}\,dx$.

We use the second formula and substitute $y = f(x) = -px$. Then,

$$f^{-1}(y) = -\frac{y}{p} \quad \text{and} \quad \frac{d}{dy}f^{-1}(y) = -\frac{1}{p}\,.$$

Hence,

$$\int \underbrace{e^{-px}}_{g(f(x))}\,dx = \int e^y\left(-\frac{1}{p}\right)dy = -\frac{1}{p}\int e^y\,dy = -\frac{1}{p}e^y = -\frac{1}{p}e^{-px}\,.$$

In practice, integration by substitution is often carried out as follows: We know that an indefinite integral of e^x is e^x. Since the given integrand is not e^x, but e^{-px}, we substitute $-px$ with the new variable y. The derivative of y with respect to x is $\frac{dy}{dx} = -p$. From this we get the “equation” $dx = -\frac{1}{p}dy$. This is used to substitute dx in the original integral. In this way, we obtain the equation $\int e^{-px}dx = \int e^y(-\frac{1}{p})\,dy$, from which the integral can be easily calculated.

b) The task is to find the definite integral $\int_0^1 e^{2x+1}dx$.

We substitute $y = 2x + 1$. The derivative of y with respect to x is given by $\frac{dy}{dx} = 2$. Hence, $dx = \frac{1}{2}dy$. Now, we also have to substitute the limits. For

$x = 0$ we get $y = 2 \cdot 0 + 1 = 1$ and for $x = 1$ we get $y = 2 \cdot 1 + 1 = 3$. Substituting $2x + 1 = y$, $dx = \frac{1}{2}y$ and inserting the limits $y = 1$ and $y = 3$ yields

$$\int_0^1 e^{2x+1} dx = \int_1^3 e^y \frac{1}{2}\, dy = \frac{1}{2} \int_1^3 e^y\, dy = \frac{1}{2}(e^3 - e)\,.$$

7.8 Improper integrals

In numerous applications, such as financial mathematics and probability theory, intervals of integration occur that are not bounded, i.e., one or both limits of integration are at (minus or plus) infinity. Such integrals are called – by a rather peculiar name – **improper integrals**. We start with an application:

Example 7.13 (Present value of a cash flow stream). Consider an account that receives a continuous cash flow of a monetary units per unit time, compounded continuously at an interest rate p from time 0 to T. The present value at time 0 is then given by (see Figure 7.7)

$$PV(T) = \int_0^T e^{-pt} a\, dt = -\frac{a}{p} e^{-pt} \Big|_0^T = \frac{a}{p}\left[1 - e^{-pT}\right].$$

The monetary inflow a received at time t is discounted by the factor e^{-pt}. If we let the time horizon T go to infinity, we get

$$\lim_{T\to\infty} PV(T) = \lim_{T\to\infty} \frac{a}{p}\left[1 - e^{-pT}\right] = \frac{a}{p}\,.$$

This is the present value of the perpetuity a under continuous compounding at rate p.

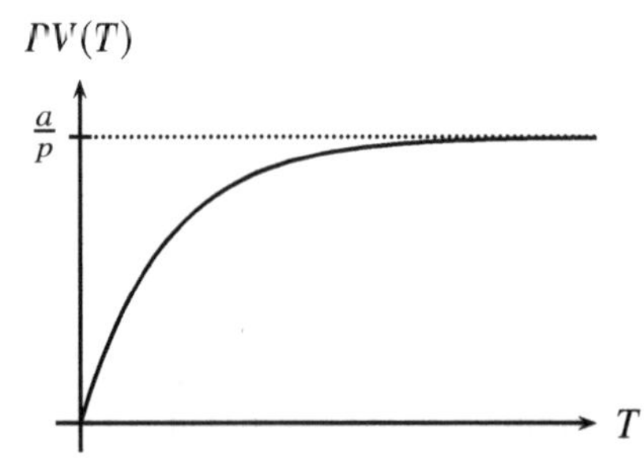

Figure 7.7: Present value of a monetary stream with continuous compounding.

In general, a definite integral whose upper and/or lower limit is infinite is defined as the limit of corresponding integrals with finite limits.

Definition 7.5: Improper integrals

Let $f : A \longrightarrow \mathbb{R}$, and F be an antiderivative of f. We define

$$\int_{\alpha}^{\infty} f(x)\,dx = \lim_{\beta\to\infty} F(\beta) - F(\alpha)\,,$$

provided $[\alpha, \infty) \subset A$ and the limit exists in $\mathbb{R} \cup \{-\infty, \infty\}$, that is, it either converges to a real number or diverges to $+\infty$ or $-\infty$.[8] Analogously, we define

$$\int_{-\infty}^{\beta} f(x)\,dx = F(\beta) - \lim_{\alpha\to-\infty} F(\alpha)\,,$$

provided $(-\infty, \beta] \subset A$ and the limit exists in $\mathbb{R} \cup \{-\infty, \infty\}$.

Further, let

$$\int_{-\infty}^{\infty} f(x)\,dx = \lim_{\beta\to\infty} F(\beta) - \lim_{\alpha\to-\infty} F(\alpha)\,,$$

provided $A = \mathbb{R}$ and both limits exist in $\mathbb{R} \cup \{-\infty, \infty\}$, but are not both $+\infty$ or both $-\infty$.[9] Then, for every $\gamma \in \mathbb{R}$ (in particular for $\gamma = 0$) holds

$$\int_{-\infty}^{\infty} f(x)\,dx = \int_{-\infty}^{\gamma} f(x)\,dx + \int_{\gamma}^{\infty} f(x)\,dx\,.$$

Example 7.14.

a) Find the value of the improper integral $\int_{-1}^{\infty} x\,dx$.

For $f(x) = x$ an antiderivative is $F(x) = \frac{1}{2}x^2$. Thus,

$$\int_{-1}^{\infty} x\,dx = \lim_{\beta\to\infty} \frac{\beta^2}{2} - \frac{(-1)^2}{2} = \infty\,.$$

This improper integral is divergent to ∞.

b) Find the value of the improper integral $\int_{-\infty}^{0} xe^x\,dx = \lim_{\alpha\to-\infty} \int_{\alpha}^{0} xe^x\,dx$.

$F(x) = e^x(x-1)$ is an antiderivative of $f(x) = xe^x$, see Example 7.10 a). Thus,

$$\int_{\alpha}^{0} xe^x\,dx = -1 - e^{\alpha}(\alpha - 1)\,.$$

For $\alpha \to -\infty$ this yields an "indeterminate form" $-\infty/\infty$. Hence, the limit can be found using l'Hôpital's rule[10]:

$$\lim_{\alpha\to-\infty} e^{\alpha}(\alpha - 1) = \lim_{\alpha\to-\infty} \frac{\alpha - 1}{e^{-\alpha}} = \lim_{\alpha\to-\infty} \frac{1}{-e^{-\alpha}} = 0\,.$$

[8] For divergence to $\pm\infty$ see Definition 3.3.

[9] When one or both limits are infinite, their difference is determined as follows: $\infty - c = \infty$, $c - \infty = -\infty$, $\infty - (-\infty) = \infty$, etc. Not defined are $\infty - \infty$ and $-\infty - (-\infty)$.

[10] See Section 4.8.

Thus, $\int_{-\infty}^{0} xe^x dx = -1$.

c) Find the value of the improper integral $\int_1^\infty \frac{1}{x^2} dx$.

An antiderivative of $f(x) = \frac{1}{x^2}$ is $F(x) = -\frac{1}{x}$. Thus (left panel in Figure 7.8),

$$\int_1^\infty \frac{1}{x^2} dx = \lim_{\beta\to\infty} -\frac{1}{x}\bigg|_1^\beta = \lim_{\beta\to\infty} -\frac{1}{\beta} + 1 = 1 .$$

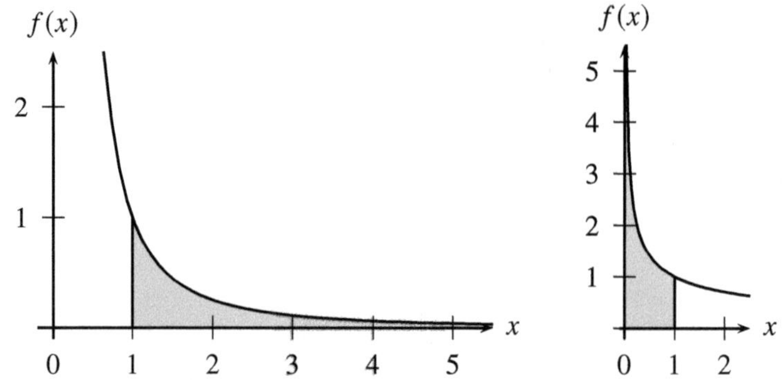

Figure 7.8: Illustration of Examples 7.14 c) and 7.15 a).

Improper integrals also include definite integrals having finite limits, where the integrand approaches positive or negative infinity at the interval's boundary, as illustrated in the following example.

Example 7.15.

a) Calculate the integral $\int_0^1 \frac{1}{\sqrt{x}} dx$.

Here, the integrand $f(x) = \frac{1}{\sqrt{x}}$ approaches infinity as x approaches the lower limit 0. However, there is an antiderivative, $F(x) = 2\sqrt{x}$, which has finite values at both limits. We get (right panel in Figure 7.8)

$$\int_0^1 \frac{1}{\sqrt{x}} dx = 2\sqrt{x}\bigg|_0^1 = 2 - 0 = 2 .$$

b) Calculate the integral $\int_1^\infty x^q dx$ for arbitrary $q \in \mathbb{R}$.

We first distinguish the two cases $q = -1$ and $q \neq -1$. For $q = -1$ we get

$$\int_1^\beta x^{-1} dx = \ln(\beta) - \ln(1) = \ln(\beta) \xrightarrow{\beta\to\infty} \infty .$$

For $q \neq -1$, $F(x) = \frac{1}{q+1}x^{q+1}$ is an antiderivative of $f(x) = x^q$, and we get

$$\int_1^\beta x^q dx = \frac{1}{q+1}\beta^{q+1} - \frac{1}{q+1}1^{q+1} = \frac{1}{q+1}(\beta^{q+1} - 1) .$$

For $q > -1$ holds $q + 1 > 0$ and $\lim_{\beta\to\infty} \beta^{q+1} = \infty$.

For $q < -1$ holds $q + 1 < 0$ and $\lim_{\beta\to\infty} \beta^{q+1} = 0$.

We summarize the three cases $q = -1$, $q < -1$ and $q > -1$ as follows:

$$\int_1^\infty x^q\,dx = \begin{cases} -\frac{1}{q+1}, & \text{if } q < -1\,, \\ \infty, & \text{if } q \geq -1\,. \end{cases}$$

In the case $q \geq -1$, the integral is divergent to ∞.

The following general definition of the definite integral includes both types of improper integrals as well as the ordinary ("proper") definite integrals.

Definition 7.6: Definite integral, existence, integrability

Let $a, b \in \mathbb{R} \cup \{-\infty, \infty\}$, $f : (a, b) \to \mathbb{R}$, and f has an antiderivative F. We define

$$\int_a^b f(x)\,dx := \lim_{\beta\to b} F(\beta) - \lim_{\alpha\to a} F(\alpha)\,,$$

provided both limits exist in $\mathbb{R} \cup \{-\infty, \infty\}$, but are not both $+\infty$ or both $-\infty$. In this case, we say that the **integral exists**. A function f is called **integrable** over (a, b), if the integral $\int_a^b f(x)\,dx$ exists and is finite.

Example 7.16.

a) Calculate the integral $\int_0^\infty \frac{1}{x^2} dx$.

$F(x) = -\frac{1}{x}$ is an antiderivative of $f(x) = \frac{1}{x^2}$. However, $f(x)$ and $F(x)$ are not defined for $x = 0$. For the improper integral we get

$$\int_0^\infty \frac{1}{x^2}\,dx = \lim_{\beta\to\infty} -\frac{1}{\beta} - \lim_{\alpha\to 0} -\frac{1}{\alpha} = 0 + \infty = \infty\,.$$

Hence, the integral exists, but is divergent to ∞. The function $f(x) = \frac{1}{x^2}$ is not integrable over $(0, \infty)$.

b) Calculate the integral $\int_{-1}^0 \frac{1}{\sqrt{x+1}}\,dx$.

The integrand $f(x) = \frac{1}{\sqrt{x+1}}$ is not defined for $x = -1$. An antiderivative of $f(x)$ is given by $F(x) = 2\sqrt{x+1}$. Thus,

$$\int_{-1}^0 \frac{1}{\sqrt{x+1}}\,dx = 2\sqrt{0+1} - \lim_{\alpha\to -1} 2\sqrt{\alpha+1} = 2 - 0 = 2\,.$$

The integral exists and is finite. The function f is integrable over $(-1, 0)$.

c) The integral $\int_0^1 x^q dx$ exists and is finite, if $q > -1$. It can be shown that it is divergent to ∞ if $q \leq -1$.

7.9 Multiple integrals

In many applications, functions of several variables arise that need to be integrated within certain bounds. Let $A \subset \mathbb{R}^2$, $f : A \to \mathbb{R}$, $(x_1, x_2) \mapsto f(x_1, x_2)$ be continuous. We consider a rectangle $[\alpha_1, \beta_1] \times [\alpha_2, \beta_2] \subset A$ and the univariate functions

$$g_1 : [\alpha_1, \beta_1] \to \mathbb{R}, x_1 \mapsto \int_{\alpha_2}^{\beta_2} f(x_1, x_2)\, dx_2\,,$$

$$g_2 : [\alpha_2, \beta_2] \to \mathbb{R}, x_2 \mapsto \int_{\alpha_1}^{\beta_1} f(x_1, x_2)\, dx_1\,.$$

Then g_1 and g_2 are continuous. Just as the definite integral of a univariate function over an interval is calculated, the definite integral of a multivariate function over a rectangle can be calculated. However, the function has to be integrated with respect to each of the variables. In the case of a function of two variables, this leads to the concept of a double integral, which is defined as follows.

Definition 7.7: Double integral

$$\int_{\alpha_2}^{\beta_2} \int_{\alpha_1}^{\beta_1} f(x_1, x_2)\, dx_1\, dx_2 = \int_{\alpha_2}^{\beta_2} g_2(x_2)\, dx_2$$

is called the **(double) integral** of f over the rectangle $[\alpha_1, \beta_1] \times [\alpha_2, \beta_2]$.

Remarks

- A double integral is evaluated from the inside out. One holds x_2 constant and computes the inner integral $\int_{\alpha_1}^{\beta_1} f(x_1, x_2) dx_1$. Its value is a function of x_2, namely the function $g_2(x_2)$. Then one evaluates the outer integral $\int_{\alpha_2}^{\beta_2} g_2(x_2) dx_2$.
- The order of integration can be interchanged,[11]

$$\int_{\alpha_2}^{\beta_2} \int_{\alpha_1}^{\beta_1} f(x_1, x_2)\, dx_1\, dx_2 = \int_{\alpha_1}^{\beta_1} \int_{\alpha_2}^{\beta_2} f(x_1, x_2)\, dx_2\, dx_1\,.$$

- The double integral of a non-negative function f over on the rectangle $[\alpha_1, \beta_1] \times [\alpha_2, \beta_2]$ represents a **volume**, specifically the volume of the "rectangular column" whose base is the rectangle $[\alpha_1, \beta_1] \times [\alpha_2, \beta_2]$ and which is bounded above by the surface $z = f(x_1, x_2)$; see Figure 7.9.
- Even if f is discontinuous, double integrals can be formed according to definition 7.7. Jump discontinuities on finitely or countably many isolated lines parallel to the coordinate axes are allowed.

[11] However, this straightforward approach only works when integrating over a rectangle, that is, when the limits of integration are constant. For more complicated regions of integration, changing the order of integration generally also changes the limits.

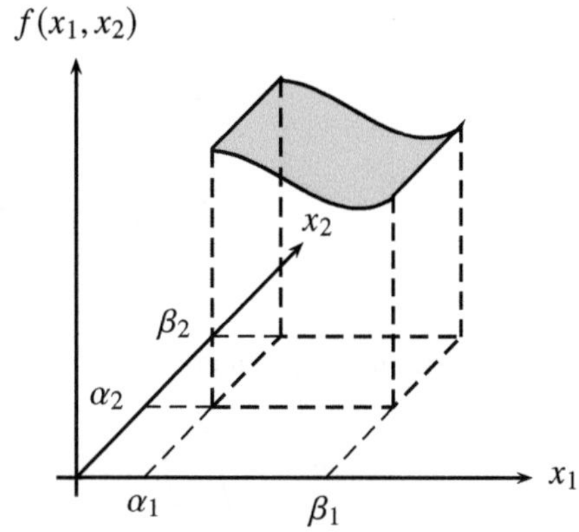

Figure 7.9: The volume of the rectangular column under the graph of f is the double integral of f over the rectangle $[\alpha_1, \beta_1] \times [\alpha_2, \beta_2]$.

- For a function f in n variables, $n > 2$, the n-fold integral is defined in the same way:

$$\underbrace{\int_{\alpha_n}^{\beta_n} \cdots \int_{\alpha_2}^{\beta_2} \int_{\alpha_1}^{\beta_1}}_{n} f(x_1, x_2, \ldots, x_n)\, dx_1\, dx_2 \ldots dx_n\,.$$

It is calculated in n steps from the inside out.

Example 7.17.

a) Calculate $\int_{-1}^{0}\int_{2}^{3}(x_1 - x_2)^2 dx_1 dx_2$.

We transform the integrand using the binomial formula,

$$\int_{-1}^{0}\int_{2}^{3}(x_1 - x_2)^2 dx_1 dx_2 = \int_{-1}^{0}\int_{2}^{3}(x_1^2 - 2x_1x_2 + x_2^2)\, dx_1 dx_2\,.$$

First, calculate the inner integral

$$\begin{aligned}\int_{2}^{3}(x_1^2 - 2x_1x_2 + x_2^2)\, dx_1 &= \frac{x_1^3}{3} - 2\frac{x_1^2x_2}{2} + x_1x_2^2\Bigg|_{x_1=2}^{x_1=3} \\ &= 9 - 9x_2 + 3x_2^2 - \frac{8}{3} + 4x_2 - 2x_2^2 \\ &= \frac{19}{3} - 5x_2 + x_2^2 = g_2(x_2)\,.\end{aligned}$$

Then, insert it into the outer integral:

$$\int_{-1}^{0}\int_{2}^{3}(x_1 - x_2)^2 dx_1 dx_2 = \int_{-1}^{0}\left(\frac{19}{3} - 5x_2 + x_2^2\right) dx_2$$

$$= \frac{19}{3}x_2 - \frac{5}{2}x_2^2 + \frac{x_2^3}{3}\Bigg|_{x_2=-1}^{x_2=0}$$

$$= \frac{19}{3} + \frac{5}{2} + \frac{1}{3} = \frac{55}{6}.$$

b) Calculate $\int_0^3\int_0^2 e^{3x_1+2x_2}\,dx_1dx_2$.
First, the integrand is written as a product,

$$\int_0^3\int_0^2 e^{3x_1+2x_2}\,dx_1dx_2 = \int_0^3\int_0^2 e^{3x_1}e^{2x_2}\,dx_1dx_2\,.$$

Then, the inner integral is calculated,

$$\int_0^2 e^{3x_1}e^{2x_2}\,dx_1 = e^{2x_2}\cdot\frac{1}{3}e^{3x_1}\Bigg|_{x_1=0}^{x_1=2} = \frac{1}{3}e^{2x_2}\,[e^6 - e^0] = g_2(x_2)\,.$$

Inserting this into the outer integral gives the solution:

$$\begin{aligned}\int_0^3\int_0^2 e^{3x_1}e^{2x_2}\,dx_1dx_2 &= \int_0^3 \frac{1}{3}(e^6-1)e^{2x_2}\,dx_2\\ &= \frac{1}{3}(e^6-1)\frac{1}{2}e^{2x_2}\Bigg|_{x_2=0}^{x_2=3}\\ &= \frac{1}{6}(e^6-1)(e^6-1)\\ &= \frac{1}{6}(e^6-1)^2 = 26991.49\,.\end{aligned}$$

If the integrand, as in the previous example, can be written as a product of functions that each depend on only one of the variables, the double integral simplifies by separation of variables.

Separation of variables: If there are real functions h_1, h_2 with $f(x_1,x_2) = h_1(x_1)\cdot h_2(x_2)$ then

$$\begin{aligned}\int_{\alpha_2}^{\beta_2}\int_{\alpha_1}^{\beta_1} f(x_1,x_2)\,dx_1dx_2 &= \int_{\alpha_2}^{\beta_2}\int_{\alpha_1}^{\beta_1} h_1(x_1)h_2(x_2)\,dx_1dx_2\\ &= \int_{\alpha_2}^{\beta_2} h_2(x_2)dx_2\cdot\int_{\alpha_1}^{\beta_1} h_1(x_1)dx_1\,.\end{aligned}$$

Example 7.18. The integrand in Example 7.17 b) allows a separation of variables with $h_1(x_1) = e^{3x_1}$, $h_2(x_2) = e^{2x_2}$, so that

$$\int_0^2 e^{3x_1}\,dx_1 = \frac{1}{3}(e^6 - e^0),$$

$$\int_0^3 e^{2x_2}dx_2 = \frac{1}{2}(e^6 - e^0),$$

$$\int_0^3\int_0^2 e^{3x_1}e^{2x_2}dx_1dx_2 = \frac{1}{3}(e^6 - e^0)\cdot\frac{1}{2}(e^6 - e^0).$$

Multiple improper integrals are particularly used in probability theory. They are calculated as limits of ordinary multiple integrals.

Example 7.19.

a) See Example 7.17 b). We calculate the improper integral

$$\begin{aligned}\int_{-\infty}^3\int_{-\infty}^2 e^{3x_1+2x_2}dx_1dx_2 &= \lim_{\alpha_2\to-\infty}\int_{\alpha_2}^3\left(\lim_{\alpha_1\to-\infty}\int_{\alpha_1}^2 e^{3x_1+2x_2}\,dx_1\right)dx_2\\ &= \lim_{\alpha_2\to-\infty}\int_{\alpha_2}^3 e^{2x_2}\left(\lim_{\alpha_1\to-\infty}\int_{\alpha_1}^2 e^{3x_1}\,dx_1\right)dx_2\\ &= \lim_{\alpha_2\to-\infty}\int_{\alpha_2}^3 e^{2x_2}\left(\lim_{\alpha_1\to-\infty}\frac{1}{3}\left[e^{3\cdot 2}-e^{3\cdot\alpha_1}\right]\right)dx_2\\ &= \lim_{\alpha_2\to-\infty}\int_{\alpha_2}^3 e^{2x_2}\frac{1}{3}\left[e^6-0\right]\,dx_2\\ &= \frac{1}{3}e^6\lim_{\alpha_2\to-\infty}\frac{1}{2}\left[e^{2\cdot 3}-e^{2\cdot\alpha_2}\right]\,dx_2\\ &= \frac{1}{6}e^6\left[e^6-0\right] = \frac{1}{6}e^{12}.\end{aligned}$$

b) Calculate

$$\int_{-\infty}^{\infty}\int_{-\infty}^{\infty} f(x_1,x_2)\,dx_1dx_2$$

with

$$f(x_1,x_2) = \begin{cases}\frac{1}{72}\,x_1x_2^2, & \text{if } 0\le x_1\le 4 \text{ and } 0\le x_2\le 3,\\ 0, & \text{otherwise}.\end{cases}$$

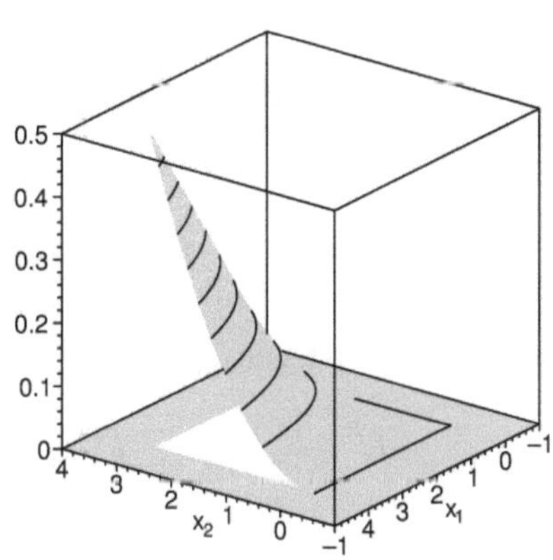

Since outside the rectangle $[0,4] \times [0,3]$ the integrand is equal to zero, it follows that

$$\begin{aligned}\int_{-\infty}^{\infty}\int_{-\infty}^{\infty} f(x_1,x_2)\,dx_1dx_2 &= \int_0^3\int_0^4 f(x_1,x_2)\,dx_1dx_2\\ &= \frac{1}{72}\int_0^3\int_0^4 x_1x_2^2\,dx_1dx_2\\ &= \frac{1}{72}\int_0^3\left(\frac{4^2}{2}-0\right)x_2^2\,dx_2\\ &= \frac{1}{72}\frac{4^2}{2}\left(\frac{3^3}{3}-0\right) = 1\,.\end{aligned}$$

Instead of integrating over a rectangle, one can also integrate over a more general region, such as

$$B = \left\{\begin{pmatrix}x_1\\x_2\end{pmatrix} \in \mathbb{R}^2 \,:\, u(x_1) \le x_2 \le v(x_1), \alpha_1 \le x_1 \le \beta_1\right\}$$

Here, u and v are continuous functions; see Figure 7.10.

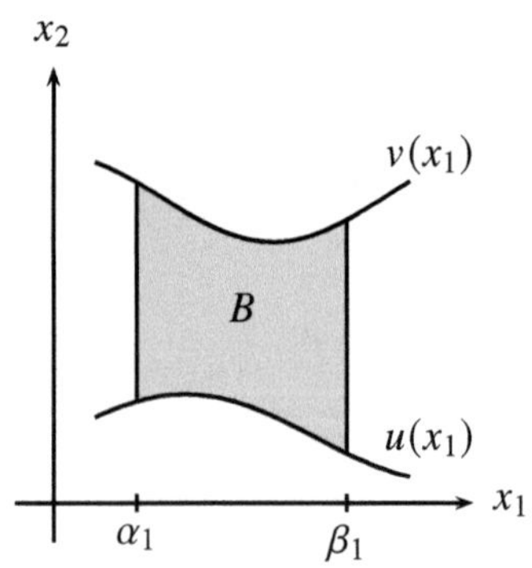

Figure 7.10: An integration region bounded above and below by two functions u and v.

The limits of integration for x_2 are not constant, but depend on x_1. Therefore, choose the inner integral as the expression

$$g_1(x_1) = \int_{u(x_1)}^{v(x_1)} f(x_1,x_2)dx_2$$

with limits depending on x_1. The integral over the region B is then equal to

$$\iint_B f(x_1,x_2)\,dx_1dx_2 = \int_{\alpha_1}^{\beta_1} g_1(x_1)\,dx_1 = \int_{\alpha_1}^{\beta_1}\int_{u(x_1)}^{v(x_1)} f(x_1,x_2)\,dx_2dx_1\,.$$

Example 7.20. Let B be a triangle as shown below. Then the limits for x_1 are $u(x_1) = 0$ and $v(x_1) = \beta_2 - (\beta_2/\beta_1)x_1$. The limits for x_1 are 0 and β_1.

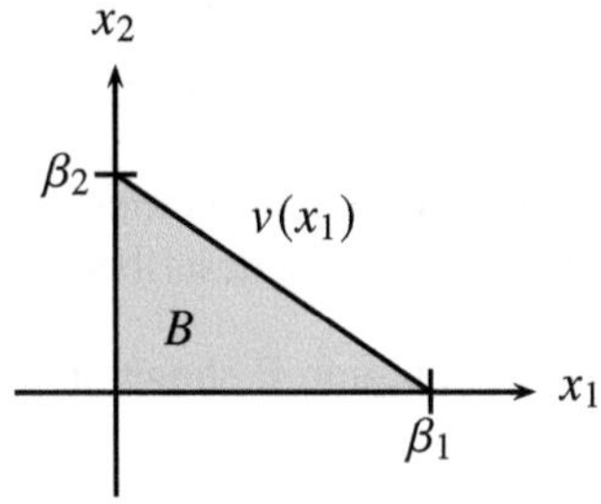

So, we calculate the integral

$$\iint_B f(x_1,x_2)\,dx_1dx_2 = \int_0^{\beta_1}\int_0^{\beta_2-\frac{\beta_2}{\beta_1}x_1} f(x_1,x_2)\,dx_2dx_1\,.$$

For $\beta_1, \beta_2 = 1$ and $f(x_1,x_2) = 2x_1x_2$ we get

$$\begin{aligned}\int_0^1\int_0^{1-x_1} 2x_1x_2\,dx_2\,dx_1 &= \int_0^1 2x_1\left(\frac{x_2^2}{2}\bigg|_{x_2=0}^{x_2=1-x_1}\right)dx_1 = \int_0^1 x_1(1-x_1)^2dx_1\\ &= \int_0^1 (x_1 - 2x_1^2 + x_1^3)\,dx_1 = \frac{x_1^2}{2} - \frac{2x_1^3}{3} + \frac{x_1^4}{4}\bigg|_{x_1=0}^{x_1=1}\\ &= \frac{1}{2} - \frac{2}{3} + \frac{1}{4} = \frac{1}{12}\,.\end{aligned}$$

Example 7.21. Let B be the disk with radius r centered at the origin. It is

$$B = \left\{(x_1,x_2) \in \mathbb{R}^2 : x_1^2 + x_2^2 \le r^2\right\}\,.$$

B can be described as the area between the two functions $v(x_1) = \sqrt{r^2 - x_1^2}$ and $u(x_1) = -\sqrt{r^2 - x_1^2}$, where x_1 runs between $-r$ and r.

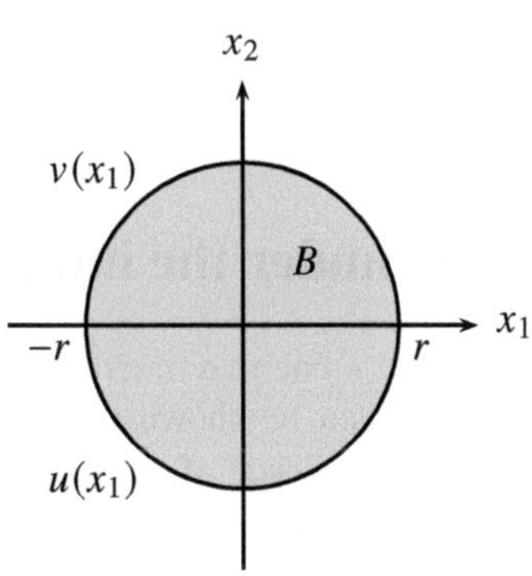

Hence, one would calculate

$$\iint_B f(x_1,x_2)\,dx_1dx_2 = \int_{-r}^{r}\int_{-\sqrt{r^2-x_1^2}}^{\sqrt{r^2-x_1^2}} f(x_1,x_2)\,dx_2dx_1\,.$$

Remark: Sometimes, the region of integration must be divided into several subregions $B_1, B_2, \dots B_m$, each of which can be described in the same manner as B.

Example 7.22. The integration region B shown in the following figure must be divided into $m = 2$ parts:

$$B_1 \text{ with } u(x_1) = x + 1\,,\ v(x_1) = \sqrt{1-x_1^2}\,,$$

$$B_2 \text{ with } \tilde{u}(x_1) = -\sqrt{1-x_1^2}\,,\ \tilde{v}(x_1) = -x - 1\,,$$

where the limits of integration for x_1 are -1 and 0.

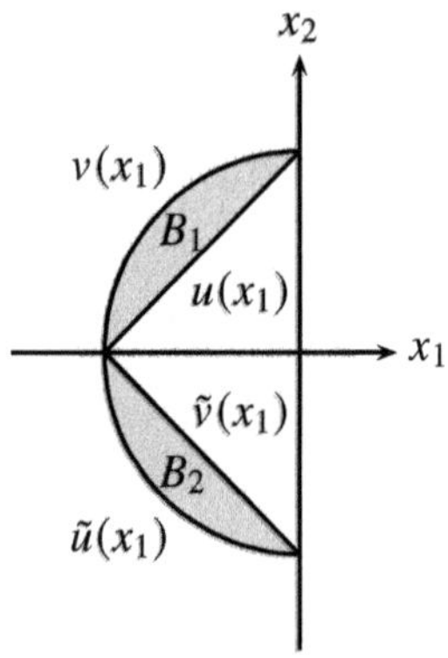

Therefore, we calculate

$$\iint_B f(x_1,x_2)\,dx_1dx_2 = \iint_{B_1} f(x_1,x_2)\,dx_1dx_2 + \iint_{B_2} f(x_1,x_2)\,dx_1dx_2$$
$$= \int_{-1}^{0}\int_{x+1}^{\sqrt{1-x_1^2}} f(x_1,x_2)\,dx_2dx_1 + \int_{-1}^{0}\int_{-\sqrt{1-x_1^2}}^{-x-1} f(x_1,x_2)\,dx_2dx_1\,.$$

7.10 Differentiation under the integral

In certain cases – but not always! – one may interchange an integral with a limit, specifically with a derivative. It can be shown that, if a sequence of functions converges pointwise to zero, the sequence of definite integrals also converges to zero. Moreover, a similar statement holds for a sequence of functions that converges pointwise in a monotonic way:

Theorem 7.6: Monotone limit

Let $(f_k)_{k\in\mathbb{N}}$ be a monotone (decreasing or increasing) sequence of functions with antiderivatives $(F_k)_{k\in\mathbb{N}}$, and let f_0 be a function with antiderivative F_0. If for some α and $\beta \in \mathbb{R}$

$$\lim_{k\to\infty} f_k(x) = f_0(x) \quad \text{for all } x \in (\alpha, \beta),$$

then

$$\lim_{k\to\infty} \int_\alpha^\beta f_k(x)\,dx = \int_\alpha^\beta \lim_{k\to\infty} f_k(x)\,dx = \int_\alpha^\beta f_0(x)\,dx = F_0(\beta) - F_0(\alpha).$$

The next statement refers to a possible interchange of integration and differentiation. The interchange is always allowed when the integrand is continuously partially differentiable.

Theorem 7.7: Differentiation under the integral

If the integrand f is continuously partially differentiable with respect to x_1, then the integral $g_1 = \int_{\alpha_2}^{\beta_2} f(x_1, x_2)\,dx_2$ is also differentiable with respect to x_1 and it holds

$$\frac{d}{dx_1} \int_{\alpha_2}^{\beta_2} f(x_1, x_2)\,dx_2 = \int_{\alpha_2}^{\beta_2} \frac{\partial}{\partial x_1} f(x_1, x_2)\,dx_2.$$

The differentiation under the integral is done with respect to the "parameter" x_1. The same applies for x_2.

Important terms for review

After reading this chapter, the following terms should be familiar:

- Antiderivative
- Indefinite integral, constant of integration
- Definite integral
- Calculation of areas
- Integration by parts
- Integration by substitution
- Improper integrals
- Integrability, existence of the integral
- Double integral
- Calculation of volumes
- Separation of variables

Self-test

You can check your understanding of this chapter using the following true-or-false questions. For each statement, indicate whether it is true (T) or false (F). Short solutions to these questions can be found in Appendix H.

I. Are the following statements true or false?

	T	F
There are infinitely many antiderivatives of a function, each differing by a constant of integration.	☐	☐
To calculate a definite integral with integrand f, any antiderivative of f can be used.	☐	☐
The absolute value of the integral $\int_a^b f(x)dx$ is greater than or equal to the area between the graph of the function f and the x-axis on the interval $[a, b]$.	☐	☐
An improper integral is calculated by calculating a definite integral and then forming a limit that makes the integration region smaller and smaller.	☐	☐

II. Are the following statements true or false?

	T	F
$\int_0^1 (e^x + \sin(x) + x^5)dx = \int_0^1 e^x dx - \int_1^0 (\sin(x) - x^5)dx + 2\int_0^1 x^5 dx$	☐	☐
If the integrand is the product of two functions (of the same variable), integration by parts is used when this leads to an integral that is easier to evaluate.	☐	☐
$\int_3^4 (x^2 - 1)dx \geq \int_3^4 x^2 dx$	☐	☐
The limits of a definite integral remain unchanged after substituting the integration variable with a monotonically increasing function.	☐	☐

III. Are the following statements true or false?

	T	F
To integrate $\int_0^1 \frac{6x^2+4x}{\sqrt{x^3+x^2}} dx$ using substitution, the appropriate substitution is $u := 6x^2 + 4x$.	☐	☐
To integrate $\int_0^1 \frac{6x^2+4x}{\sqrt{x^3+x^2}} dx$ using substitution, the appropriate substitution is $u := x^3 + x^2$.	☐	☐
$\int_0^n \lfloor x \rfloor \, dx = \sum_{i=0}^{n-1} i$	☐	☐
Every continuous function is regulated.	☐	☐

IV. Are the following statements true or false?

	T	F
In a double integral, the two integrals are calculated from the outside in.	☐	☐
$\int_0^1 \int_{-1}^1 f(x,y)\,dy\,dx = \int_{-1}^1 \int_0^1 f(x,y)\,dx\,dy$	☐	☐
The integral $\int_0^4 \int_0^5 6\,dy\,dx$ can be interpreted as the volume of the rectangular box with a base area of four by five and a height of six.	☐	☐
The integral $\int_{-1}^1 \int_{-1}^1 x^3 y^3\,dy\,dx$ can be interpreted as the volume between the rectangle $[-1,1] \times [-1,1]$ in the x-y plane and the surface defined by the graph of the function $x^3 y^3$.	☐	☐

Exercises

Exercise 1
Determine all antiderivatives of the functions $f_i : A_i \to \mathbb{R}$, $i = 1, \ldots, 5$, where

$$f_1(x) = x^3 - 2x + 5, \qquad A_1 = \mathbb{R},$$
$$f_2(x) = \frac{1}{x^2} + \frac{3}{2x} - (x+1)^2, \qquad A_2 = \mathbb{R} \setminus \{0\},$$
$$f_3(x) = \sqrt{x+1} + \frac{1}{\sqrt{x}}, \qquad A_3 = \mathbb{R}_+ \setminus \{0\},$$
$$f_4(x) = \frac{x^2}{1+x^2}, \qquad A_4 = \mathbb{R},$$
$$f_5(x) = 2xe^{x^2}, \qquad A_5 = \mathbb{R}.$$

Exercise 2
Calculate the following definite and indefinite integrals using (possibly multiple) integration by parts:

a) $\displaystyle\int_0^2 xe^{-x}\,dx$,

b) $\displaystyle\int_1^e \frac{\ln(x)}{x}\,dx$,

c) $\displaystyle\int \sin(x)e^x\,dx$.

Exercise 3
Calculate the following definite and indefinite integrals using integration by substitution:

a) $\int_{-1}^{8} x\sqrt{1+x}\,dx$,

b) $\int_{0}^{\frac{\pi}{2}} \sin^2(x)\cos(x)\,dx$,

c) $\int \frac{x+3}{(x^2+6x)^4}\,dx$.

Exercise 4
Calculate, if possible, the following improper integrals:

a) $\int_{0}^{\infty} \frac{x}{(1+x^2)^2}\,dx$,

b) $\int_{0}^{1} \ln(x)\,dx$,

c) $\int_{-\infty}^{1} (e^x+1)^2 e^x\,dx$,

d) $\int_{0}^{\infty} \frac{e^x}{x}\,dx$.

Exercise 5
Calculate the following double integrals:

a) $\int_{1}^{2}\int_{-1}^{1} (x^2y - xy^2)\,dxdy$,

b) $\int_{0}^{\infty}\int_{0}^{\infty} e^{-x-2y}\,dydx$,

c) $\int_{-1}^{1}\int_{0}^{x} (xy^2 + x)\,dydx$,

d) $\int_{-\pi}^{\pi}\int_{-y}^{y} \sin(x)\,dxdy$.

Exercise 6
Calculate the following double integrals:

a) $\int_{0}^{\infty}\int_{0}^{x} e^{-x-y}\,dy\,dx$,

b) $\int_{-2}^{2}\int_{0}^{\sqrt{4-x^2}} x^2y\,dy\,dx$.

Chapter 8

Linear equations

In this chapter, we will discuss the solution of linear equations. A system of m linear equations in n unknowns can be considered as an equation between vectors of $\mathbb{R}^m$, $\boldsymbol{A}\boldsymbol{x} = \boldsymbol{b}$, where the unknowns are combined into the vector $\boldsymbol{x} \in \mathbb{R}^n$, $\boldsymbol{A}$ is an $m \times n$ matrix, and $\boldsymbol{b} \in \mathbb{R}^m$.

8.1 Solution to a linear equation

Given $\boldsymbol{A} = (a_{ij}) \in \mathbb{R}^{m\times n}$ and $\boldsymbol{b} \in \mathbb{R}^m$. We are looking for the set

$$L(\boldsymbol{A}, \boldsymbol{b}) = \{\boldsymbol{x} \in \mathbb{R}^n \, : \, \boldsymbol{A}\boldsymbol{x} = \boldsymbol{b}\}\,.$$

$L(\boldsymbol{A}, \boldsymbol{b})$ is called the **solution** (or **general solution**) of the **linear equation** $\boldsymbol{A}\boldsymbol{x} = \boldsymbol{b}$. Each element $\boldsymbol{x} \in L(\boldsymbol{A}, \boldsymbol{b})$ is called a **particular solution** of $\boldsymbol{A}\boldsymbol{x} = \boldsymbol{b}$. The $m \times n$ matrix $\boldsymbol{A}$ and the m-component column vector $\boldsymbol{b}$ are combined into an $m \times (n + 1)$ matrix; it is called the **augmented matrix** of the linear equation.

Example 8.1. Given is the system of linear equations

$$\begin{array}{llllllll}
\text{I:} & 3x_1 & + & 5x_2 & + & 1x_3 & = & 1 \\
\text{II:} & 2x_1 & + & 4x_2 & + & 5x_3 & = & 2 \\
\text{III:} & 1x_1 & + & 2x_2 & + & 2x_3 & = & 0 \\
\text{IV:} & 1x_1 & + & 2x_2 & + & 3x_3 & = & 2
\end{array}$$

In matrix-vector notation, the system of equations can be written as a linear

K. Mosler et al., *Mathematical Methods for Business and Economics*, Mathematics Study Resources 20, https://doi.org/10.1007/978-3-662-73831-3_8

equation,

$$\begin{bmatrix} 3 & 5 & 1 \\ 2 & 4 & 5 \\ 1 & 2 & 2 \\ 1 & 2 & 3 \end{bmatrix} \begin{bmatrix} x_1 \\ x_2 \\ x_3 \end{bmatrix} = \begin{bmatrix} 1 \\ 2 \\ 0 \\ 2 \end{bmatrix}.$$

The augmented matrix of this linear equation is

$$\left[\begin{array}{ccc|c} 3 & 5 & 1 & 1 \\ 2 & 4 & 5 & 2 \\ 1 & 2 & 2 & 0 \\ 1 & 2 & 3 & 2 \end{array}\right].$$

The last column of the augmented matrix contains the vector of the right-hand side of the equation. For clarity, it is separated by a vertical line .

8.2 Elementary row operations

A linear equation is solved by transforming the equation through "elementary" operations that do not change the solution to the equation, so that the solution to the transformed equation is more easily found. Given a system of linear equations, the following transformations do obviously not change its solution:

- swapping two rows of the system,
- replacing an equation by its sum or difference with a second equation,
- multiplying an equation with a non-zero number.

The latter two transformations may be combined to

- replacing an equation by its sum or difference with a multiple of a second equation,

If these transformations are performed on the augmented matrix of the system of equations, we speak of **elementary row operations**[1]. Elementary row operations of a matrix are therefore the following transformations:

ERO 1: swapping two rows,

ERO 2: replacing a row by its sum or difference with a multiple of another row,

ERO 3: multiplying a row with a non-zero number.

If an $m \times n$ matrix $\widetilde{\boldsymbol{A}}$ can be formed from a matrix $\boldsymbol{A}$ by a finite number of elementary row operations, we write $\boldsymbol{A} \xrightarrow{ERO} \widetilde{\boldsymbol{A}}$.

It is easy to see that the reversal of each elementary row operation is also an elementary row operations. Thus, the swapping of two rows is undone by swapping the same rows again. Accordingly, the addition of one row to a second row is reversed

[1] Analogously, one can define **elementary column operations** of a matrix $\boldsymbol{A}$. However, apart from swapping two columns, these are not relevant for the following.

by subtracting it, and the multiplication of a row with $\alpha \neq 0$ by the multiplication with the reciprocal $1/\alpha$.

If the augmented matrix $[\boldsymbol{A} \,|\, \boldsymbol{b}]$ can be transformed into $[\widetilde{\boldsymbol{A}} \,|\, \widetilde{\boldsymbol{b}}]$ via elementary row operations, then $[\boldsymbol{A} \,|\, \boldsymbol{b}]$ can also be obtained from $[\widetilde{\boldsymbol{A}} \,|\, \widetilde{\boldsymbol{b}}]$ using elementary row operations, and vice versa, i.e.,

$$[\boldsymbol{A} \,|\, \boldsymbol{b}] \overset{ERO}{\longrightarrow} [\widetilde{\boldsymbol{A}} \,|\, \widetilde{\boldsymbol{b}}] \quad \Longleftrightarrow \quad [\widetilde{\boldsymbol{A}} \,|\, \widetilde{\boldsymbol{b}}] \overset{ERO}{\longrightarrow} [\boldsymbol{A} \,|\, \boldsymbol{b}] .$$

Since a single elementary row operation of the augmented matrix does not change the general solution to the corresponding equation, the same holds for a sequence of consecutive elementary row operations. We formulate this important result as a theorem.

Theorem 8.1: General solution and elementary row operations

If the augmented matrix $[\boldsymbol{A} \,|\, \boldsymbol{b}]$ of a linear equation $\boldsymbol{A}\boldsymbol{x} = \boldsymbol{b}$ is transformed by elementary row operations, the general solution of the associated linear equation does not change:

$$[\boldsymbol{A} \,|\, \boldsymbol{b}] \overset{ERO}{\longrightarrow} [\widetilde{\boldsymbol{A}} \,|\, \widetilde{\boldsymbol{b}}] \quad \Longrightarrow \quad L(\boldsymbol{A}, \boldsymbol{b}) = L(\widetilde{\boldsymbol{A}}, \widetilde{\boldsymbol{b}}) .$$

8.3 Gauss-Jordan elimination

We describe a systematic method for solving a linear equation, the so-called **Gauss[2]-Jordan[3] elimination**. This method can not only be used to solve linear equations, but also to calculate the inverse of a matrix and to computationally solve many other problems that we will encounter in later chapters. In the Gauss-Jordan elimination, the augmented matrix is transformed by elementary row operations until the solution to the linear equation can be read off directly.

Consider the linear equation $\boldsymbol{A}\boldsymbol{x} = \boldsymbol{b}$, where $\boldsymbol{A}$ is an $m \times n$ matrix and $\boldsymbol{x}$ and $\boldsymbol{b}$ are column vectors with n and m components, respectively.

Gauss-Jordan Elimination

1. In the first step, elementary row operations are chosen such that the first column of $\boldsymbol{A}$ is transformed from

$$\begin{bmatrix} a_{11} \\ a_{21} \\ \vdots \\ a_{m1} \end{bmatrix} \quad \text{to} \quad \begin{bmatrix} 1 \\ 0 \\ \vdots \\ 0 \end{bmatrix} = \boldsymbol{e}_1 .$$

[2] Carl Friedrich Gauß (1777–1855)
[3] Wilhelm Jordan (1842–1899)

If $a_{11} \neq 0$, this is achieved by first dividing the first row of the extended matrix by a_{11}. The element a_{11} is then called **pivot element**. After that, for $i = 2, \ldots, m$, the a_{i1}-multiple of the first row is subtracted from the i-th row. Of course, if already $a_{i1} = 0$, this operation does not need to be performed. This step is called **elimination step** for the first column.

2. The elimination step is repeated for the other columns of $\boldsymbol{A}$. So in the second step, a_{22} (if $\neq 0$) is chosen as pivot element, and the second column is transformed into the second canonical unit vector. In the third step, a_{33} (if $\neq 0$), is used as pivot element, etc.

2a. If a diagonal element a_{jj} equals zero, the elimination step for the j-th column cannot be performed directly. However, if one of the elements $a_{j+1,j}, \ldots, a_{mj}$ *below the diagonal* is non-zero, the corresponding *row* is first swapped with the j-th row and then the elimination step for the new j-th row is performed as described.

2b. If a_{jj} and all elements below the diagonal in the j-th column are zero, subsequent columns are checked for a non-zero element in rows $j, \ldots, m$. In a non-zero element is found, the corresponding **column** is swapped with the j-th column. Then Step 2a. is performed. In this case, however, one must keep track of the column interchanges, since swapping columns corresponds to permuting the variables x_i, see Example 8.4.

2c. The procedure ends when no further pivot element can be chosen. The current augmented matrix $[\widetilde{\boldsymbol{A}}, \widetilde{\boldsymbol{b}}]$, together with the variables $x_1, \ldots, x_n$ is presented as a **final tableau**.

If the algorithm ends after k elimination steps, the final tableau has the following form.

$$
\begin{array}{ccc|ccc|c}
x_1 & \cdots & x_k & x_{k+1} & \cdots & x_n & \widetilde{\boldsymbol{b}} \\
\hline
1 & \cdots & 0 & * & \cdots & * & * \\
\vdots & \ddots & \vdots & \vdots & & \vdots & \vdots \\
0 & \cdots & 1 & * & \cdots & * & * \\
\hline
0 & \cdots & 0 & 0 & \cdots & 0 & * \\
\vdots & & \vdots & \vdots & & \vdots & \vdots \\
0 & \cdots & 0 & 0 & \cdots & 0 & *
\end{array}
$$

The matrix $\widetilde{\boldsymbol{A}}$ of the final tableau (without $\widetilde{\boldsymbol{b}}$) thus consists of three submatrices. On the top left is a $k \times k$ identity matrix, to the right a matrix of format $k \times (n-k)$, and below the zero matrix $\boldsymbol{0}_{m-k,n}$.

The solution to the linear equation can be directly read from the final tableau. Several cases must be distinguished, which we explain by examples.

Case 1 ($k = m = n$)**:** In this case, $\widetilde{\boldsymbol{A}}$ is an identity matrix. For example, a final tableau may look like this:

x_1	x_2	x_3	$\widetilde{\boldsymbol{b}}$
1	0	0	18
0	1	0	−11
0	0	1	2

Here, the first equation reads $1 \cdot x_1 + 0 \cdot x_2 + 0 \cdot x_3 = 18$, so $x_1 = 18$. Similarly, one obtains $x_2 = -11$ and $x_3 = 2$. Thus, the linear equation $\boldsymbol{Ax} = \boldsymbol{b}$ has a **unique solution** $L(\widetilde{\boldsymbol{A}}, \widetilde{\boldsymbol{b}}) = L(\boldsymbol{A}, \boldsymbol{b}) = \{ [18, -11, 2]^T \}$.

Case 2 ($k = m < n$)**:** In this case, the first m columns of $\widetilde{\boldsymbol{A}}$ form an identity matrix, followed by further columns, e.g.,

x_1	x_2	x_3	x_4	$\widetilde{\boldsymbol{b}}$
1	0	0	4	18
0	1	0	0	−11
0	0	1	−2	2

Here, the solution to the linear equation is **not unique**, as one variable can be chosen freely. In particular, for every choice of x_4, the variables x_1, x_2 and x_3 are uniquely determined. If we set $x_4 = -\lambda \in \mathbb{R}$, the solution is given by $x_1 - 4\lambda = 18 \iff x_1 = 18 + 4\lambda$, $x_2 = -11$, $x_3 + 2\lambda = 2 \iff x_3 = 2 - 2\lambda$. Thus, the general solution is

$$L(\boldsymbol{A}, \boldsymbol{b}) = \left\{ \begin{bmatrix} 18 + 4\lambda \\ -11 \\ 2 - 2\lambda \\ -\lambda \end{bmatrix} : \lambda \in \mathbb{R} \right\}$$

or, more clearly written[4],

$$L(\boldsymbol{A}, \boldsymbol{b}) = \left\{ \begin{bmatrix} 18 \\ -11 \\ 2 \\ 0 \end{bmatrix} + \lambda \begin{bmatrix} 4 \\ 0 \\ -2 \\ -1 \end{bmatrix} : \lambda \in \mathbb{R} \right\}.$$

For each $\lambda \in \mathbb{R}$, one thus obtains a particular solution to the linear equation. For example, for $\lambda = 0$ the solution $[18, -11, 2, 0]^T$ is obtained, for $\lambda = 1$ the particular solution $[14, -11, 4, 1]^T$ and for $\lambda = 5$ the particular solution $[-2, -11, 12, 5]^T$.

In general, in Case 2 ($k = m < n$), the last $n - k$ columns cannot be transformed into unit vectors. Then the last $n - k$ variables can be chosen freely, and the solution is written analogously; see Example 8.3.

[4] The rationale for assigning $x_4 = -\lambda$ (instead of λ) now becomes evident: This choice allows the components of the vectors in the general solution to be directly extracted from the final tableau without sign adjustments. Thus, in subsequent steps, free variables will be similarly designated as $-\lambda$, $-\mu$, and so on.

Case 3 ($k < m$)**:** Here the last $m - k$ rows of $\widetilde{\boldsymbol{A}}$ are zero rows. Whether a solution exists or not, depends on the last $m - k$ components of $\widetilde{\boldsymbol{b}}$. If all of them are zero, the last $m - k$ rows of $[\widetilde{\boldsymbol{A}}, \widetilde{\boldsymbol{b}}]$ can be discarded and the solution is determined according to Case 1 or 2. If one of the last $m - k$ components of $\widetilde{\boldsymbol{b}}$ is non-zero, then the linear equation has **no solution**.

For example, in the final tableau

x_1	x_2	x_3	x_4	$\widetilde{\boldsymbol{b}}$
1	0	0	4	18
0	1	0	0	−11
0	0	1	−2	2
0	0	0	0	0
0	0	0	0	0

the last two rows of $[\widetilde{\boldsymbol{A}}, \widetilde{\boldsymbol{b}}]$ are zero, and therefore deleted. We obtain the tableau

x_1	x_2	x_3	x_4	$\widetilde{\boldsymbol{b}}$
1	0	0	4	18
0	1	0	0	−11
0	0	1	−2	2

,

whose solution has been already discussed in Case 2.

In the final tableau

x_1	x_2	x_3	x_4	$\widetilde{\boldsymbol{b}}$
1	0	0	4	18
0	1	0	0	−11
0	0	1	−2	2
0	0	0	0	0
0	0	0	0	1

the last row corresponds to the equation $0 \cdot x_1 + 0 \cdot x_2 + 0 \cdot x_3 + 0 \cdot x_4 = 1$, i.e., $0 = 1$, which has no solution. The general solution to the linear equation is thus empty.

Example 8.2. Find the general solution to the linear equation

$$\begin{bmatrix} 1 & 2 & 4 \\ 3 & 6 & 13 \\ 1 & 5 & 7 \end{bmatrix} \begin{bmatrix} x_1 \\ x_2 \\ x_3 \end{bmatrix} = \begin{bmatrix} 20 \\ 63 \\ 35 \end{bmatrix}.$$

We solve the linear equation using Gauss-Jordan elimination. First, we subtract three times the first row from the second row and the first row from the third row:

x_1	x_2	x_3	$\widetilde{\boldsymbol{b}}$		x_1	x_2	x_3	$\widetilde{\boldsymbol{b}}$
1	2	4	20	$\longrightarrow$	1	2	4	20
3	6	13	63	II−3·I III−I	0	0	1	3
1	5	7	35		0	3	3	15

Since a_{22} is zero, we swap the second and third rows and continue with the new second column.

x_1	x_2	x_3	$\widetilde{b}$
1	2	4	20
0	0	1	3
0	3	3	15

$$\xrightarrow[\text{II}\leftrightarrow\text{III}]{}$$

x_1	x_2	x_3	$\widetilde{b}$
1	2	4	20
0	3	3	15
0	0	1	3

$$\xrightarrow[\substack{\text{I}-2/3\cdot\text{II}\\ \text{II}/3}]{}$$

x_1	x_2	x_3	$\widetilde{b}$
1	0	2	10
0	1	1	5
0	0	1	3

The following elimination step for the third column yields our final tableau:

x_1	x_2	x_3	$\widetilde{b}$
1	0	2	10
0	1	1	5
0	0	1	3

$$\xrightarrow[\substack{\text{I}-2\cdot\text{III}\\ \text{II}-\text{III}}]{}$$

x_1	x_2	x_3	$\widetilde{b}$
1	0	0	4
0	1	0	2
0	0	1	3

The linear equation thus has the unique solution

$$L(A,b) = \left\{ \begin{bmatrix} 4 \\ 2 \\ 3 \end{bmatrix} \right\}.$$

Example 8.3. We consider the linear equation

$$\begin{bmatrix} 1 & 2 & 4 & 8 & 9 \\ 3 & 6 & 13 & 25 & 29 \\ 1 & 5 & 7 & 14 & 15 \end{bmatrix} \begin{bmatrix} x_1 \\ x_2 \\ x_3 \\ x_4 \\ x_5 \end{bmatrix} = \begin{bmatrix} 20 \\ 63 \\ 35 \end{bmatrix}.$$

As in the previous example, the linear equation is transformed using Gauss-Jordan elimination:

x_1	x_2	x_3	x_4	x_5	$\widetilde{b}$
1	2	4	8	9	20
3	6	13	25	29	63
1	5	7	14	15	35

$$\xrightarrow[\substack{\text{II}-3\cdot\text{I}\\ \text{III}-\text{I}}]{}$$

x_1	x_2	x_3	x_4	x_5	$\widetilde{b}$
1	2	4	8	9	20
0	0	1	1	2	3
0	3	3	6	6	15

$$\xrightarrow[\text{II}\leftrightarrow\text{III}]{}$$

x_1	x_2	x_3	x_4	x_5	$\widetilde{b}$
1	2	4	8	9	20
0	3	3	6	6	15
0	0	1	1	2	3

$$\xrightarrow[\substack{\text{I}-2/3\cdot\text{II}\\ \text{II}/3}]{}$$

x_1	x_2	x_3	x_4	x_5	$\widetilde{b}$
1	0	2	4	5	10
0	1	1	2	2	5
0	0	1	1	2	3

$$\xrightarrow[\text{II}-\text{III}]{\text{I}-2\cdot\text{III}}\quad \begin{array}{ccccc|c} x_1 & x_2 & x_3 & x_4 & x_5 & \widetilde{b} \\ \hline 1 & 0 & 0 & 2 & 1 & 4 \\ 0 & 1 & 0 & 1 & 0 & 2 \\ 0 & 0 & 1 & 1 & 2 & 3 \end{array}$$

Here, the variables x_4 and x_5 can be chosen freely. We set $x_4 = -\lambda$ and $x_5 = -\mu$ and obtain the general solution

$$L(\boldsymbol{A}, \boldsymbol{b}) = \left\{ \begin{bmatrix} 4 \\ 2 \\ 3 \\ 0 \\ 0 \end{bmatrix} + \lambda \begin{bmatrix} 2 \\ 1 \\ 1 \\ -1 \\ 0 \end{bmatrix} + \mu \begin{bmatrix} 1 \\ 0 \\ 2 \\ 0 \\ -1 \end{bmatrix} : \lambda, \mu \in \mathbb{R} \right\}.$$

Example 8.4. Now we consider the linear equation

$$\begin{bmatrix} 1 & 2 & 4 & 8 \\ 3 & 6 & 13 & 25 \\ 1 & 2 & 7 & 11 \end{bmatrix} \begin{bmatrix} x_1 \\ x_2 \\ x_3 \\ x_4 \end{bmatrix} = \begin{bmatrix} 20 \\ 63 \\ 29 \end{bmatrix}.$$

The elimination step for the first column gives:

$$\begin{array}{cccc|c} x_1 & x_2 & x_3 & x_4 & \widetilde{b} \\ \hline 1 & 2 & 4 & 8 & 20 \\ 3 & 6 & 13 & 25 & 63 \\ 1 & 2 & 7 & 11 & 29 \end{array} \quad \xrightarrow[\text{III}-\text{I}]{\text{II}-3\cdot\text{I}} \quad \begin{array}{cccc|c} x_1 & x_2 & x_3 & x_4 & \widetilde{b} \\ \hline 1 & 2 & 4 & 8 & 20 \\ 0 & 0 & 1 & 1 & 3 \\ 0 & 0 & 3 & 3 & 9 \end{array}$$

In the second column, all entries on and below the diagonal are zero, so the elimination step for this column cannot be performed and a column swap is needed. We exchange the second and third columns:

$$\begin{array}{cccc|c} x_1 & x_2 & x_3 & x_4 & \widetilde{b} \\ \hline 1 & 2 & 4 & 8 & 20 \\ 0 & 0 & 1 & 1 & 3 \\ 0 & 0 & 3 & 3 & 9 \end{array} \quad \xrightarrow[x_2 \leftrightarrow x_3]{} \quad \begin{array}{cccc|c} x_1 & x_3 & x_2 & x_4 & \widetilde{b} \\ \hline 1 & 4 & 2 & 8 & 20 \\ 0 & 1 & 0 & 1 & 3 \\ 0 & 3 & 0 & 3 & 9 \end{array}$$

Since now a_{22} is non-zero, we proceed with the elimination step for the second column.

$$\begin{array}{cccc|c} x_1 & x_3 & x_2 & x_4 & \widetilde{b} \\ \hline 1 & 4 & 2 & 8 & 20 \\ 0 & 1 & 0 & 1 & 3 \\ 0 & 3 & 0 & 3 & 9 \end{array} \quad \xrightarrow[\text{III}-3\cdot\text{II}]{\text{I}-4\cdot\text{II}} \quad \begin{array}{cccc|c} x_1 & x_3 & x_2 & x_4 & \widetilde{b} \\ \hline 1 & 0 & 2 & 4 & 8 \\ 0 & 1 & 0 & 1 & 3 \\ 0 & 0 & 0 & 0 & 0 \end{array}$$

Deleting the last zero row gives the final tableau:

$$\begin{array}{cccc|c} x_1 & x_3 & x_2 & x_4 & \widetilde{b} \\ \hline 1 & 0 & 2 & 4 & 8 \\ 0 & 1 & 0 & 1 & 3 \end{array}.$$

Since in the final tableau the third and fourth columns belong to the variables x_2 and x_4, these can be freely chosen. We set $x_2 = -\lambda$ and $x_4 = -\mu$ and obtain the general solution

$$L(\boldsymbol{A}, \boldsymbol{b}) = \left\{ \begin{bmatrix} 8 \\ 0 \\ 3 \\ 0 \end{bmatrix} + \lambda \begin{bmatrix} 2 \\ -1 \\ 0 \\ 0 \end{bmatrix} + \mu \begin{bmatrix} 4 \\ 0 \\ 1 \\ -1 \end{bmatrix} : \lambda, \mu \in \mathbb{R} \right\} .$$

Example 8.5. As a final example, we modify Example 8.4 by setting the third component of $\boldsymbol{b}$ to 30 instead of 29,

$$\begin{bmatrix} 1 & 2 & 4 & 8 \\ 3 & 6 & 13 & 25 \\ 1 & 2 & 7 & 11 \end{bmatrix} \begin{bmatrix} x_1 \\ x_2 \\ x_3 \\ x_4 \end{bmatrix} = \begin{bmatrix} 20 \\ 63 \\ 30 \end{bmatrix} .$$

As in the previous example, the following elementary row operations are performed:

x_1	x_2	x_3	x_4	$\widetilde{b}$
1	2	4	8	20
3	6	13	25	63
1	2	7	11	30

$\xrightarrow[\text{III}-\text{I}]{\text{II}-3\cdot\text{I}}$

x_1	x_2	x_3	x_4	$\widetilde{b}$
1	2	4	8	20
0	0	1	1	3
0	0	3	3	10

$\xrightarrow{x_2 \leftrightarrow x_3}$

x_1	x_3	x_2	x_4	$\widetilde{b}$
1	4	2	8	20
0	1	0	1	3
0	3	0	3	10

$\xrightarrow[\text{III}-3\cdot\text{II}]{\text{I}-4\cdot\text{II}}$

x_1	x_3	x_2	x_4	$\widetilde{b}$
1	0	2	4	8
0	1	0	1	3
0	0	0	0	1

In the last row, we have zeros on the left-hand side, while the corresponding component on the right-hand side is non-zero. Since this corresponds to the equation $0 = 1$, the general solution is empty,

$$L(\boldsymbol{A}, \boldsymbol{b}) = \emptyset .$$

A key feature of Gauss-Jordan elimination is that the row operations required depend solely on the coefficient matrix $\boldsymbol{A}$. This means that if you need to solve several linear equations with the *same coefficient matrix*, but with different right-hand sides, such as $\boldsymbol{Ax} = \boldsymbol{b}$, $\boldsymbol{Ax} = \boldsymbol{c}$, and $\boldsymbol{Ax} = \boldsymbol{d}$, you can efficiently solve all of them simultaneously by applying the same sequence of row operations to each right-hand side.

Example 8.6. Solve the linear equations $\boldsymbol{Ax} = \boldsymbol{b}$ and $\boldsymbol{Ax} = \boldsymbol{c}$, where

$$\boldsymbol{A} = \begin{bmatrix} 1 & 4 & 2 \\ 2 & 2 & 10 \\ 1 & 3 & 0 \end{bmatrix}, \quad \boldsymbol{b} = \begin{bmatrix} 1 \\ 8 \\ 5 \end{bmatrix}, \quad \boldsymbol{c} = \begin{bmatrix} 2 \\ -2 \\ 7 \end{bmatrix}.$$

Both systems are solved simultaneously using Gauss-Jordan elimination.

x_1	x_2	x_3	$\widetilde{b}$	$\widetilde{c}$
1	4	2	1	2
2	2	10	8	−2
1	3	0	5	7

$\xrightarrow[\text{III}-\text{I}]{\text{II}-2\cdot\text{I}}$

x_1	x_2	x_3	$\widetilde{b}$	$\widetilde{c}$
1	4	2	1	2
0	−6	6	6	−6
0	−1	−2	4	5

$\xrightarrow[\text{II}/(-6)]{}$

x_1	x_2	x_3	$\widetilde{b}$	$\widetilde{c}$
1	4	2	1	2
0	1	−1	−1	1
0	−1	−2	4	5

$\xrightarrow[\text{III}+\text{II}]{\text{I}-4\cdot\text{II}}$

x_1	x_2	x_3	$\widetilde{b}$	$\widetilde{c}$
1	0	6	5	−2
0	1	−1	−1	1
0	0	−3	3	6

$\xrightarrow[\text{III}/(-3)]{}$

x_1	x_2	x_3	$\widetilde{b}$	$\widetilde{c}$
1	0	6	5	−2
0	1	−1	−1	1
0	0	1	−1	−2

$\xrightarrow[\text{II}+\text{III}]{\text{I}-6\cdot\text{III}}$

x_1	x_2	x_3	$\widetilde{b}$	$\widetilde{c}$
1	0	0	11	10
0	1	0	−2	−1
0	0	1	−1	−2

Hence, the solutions of the two linear equations are

$$L(\boldsymbol{A}, \boldsymbol{b}) = \left\{ \begin{bmatrix} 11 \\ -2 \\ -1 \end{bmatrix} \right\} \quad \text{and} \quad L(\boldsymbol{A}, \boldsymbol{c}) = \left\{ \begin{bmatrix} 10 \\ -1 \\ -2 \end{bmatrix} \right\}.$$

8.4 Inversion of a matrix

Gauss-Jordan elimination can also be used to calculate the inverse of a matrix, provided it exists. The key to this lies in Example 8.6, where we have solved simultaneously two linear equations $\boldsymbol{Ax} = \boldsymbol{b}$ and $\boldsymbol{Ax} = \boldsymbol{c}$ having the same coefficient matrix $\boldsymbol{A}$. If we combine the two right-hand sides $\boldsymbol{b}$ and $\boldsymbol{c}$ into a matrix $\boldsymbol{B}$, $\boldsymbol{B} = [\boldsymbol{b}, \boldsymbol{c}]$, we have also found the solution to the **matrix equation** $\boldsymbol{AX} = \boldsymbol{B}$. The solution to the matrix equation

$$\begin{bmatrix} 1 & 4 & 2 \\ 2 & 2 & 10 \\ 1 & 3 & 0 \end{bmatrix} \begin{bmatrix} x_{11} & x_{12} \\ x_{21} & x_{22} \\ x_{31} & x_{32} \end{bmatrix} = \begin{bmatrix} 1 & 2 \\ 8 & -2 \\ 5 & 7 \end{bmatrix}$$

is therefore given by

$$\boldsymbol{X} = [x_{ij}] = \begin{bmatrix} 11 & 10 \\ -2 & -1 \\ -1 & -2 \end{bmatrix} .$$

To find the inverse of a square $n \times n$ matrix $\boldsymbol{A}$, one has to solve the matrix equation $\boldsymbol{A}\boldsymbol{X} = \boldsymbol{I}_n$. As discussed above this can be done by setting up the augmented matrix with the matrix $\boldsymbol{A}$ to be inverted on the left-hand side and the identity matrix $\boldsymbol{I}_n$ on the right-hand side. Gauss-Jordan elimination is carried out until the left-hand side is transformed to the identity matrix. Then, the matrix on the right-hand side is the inverse of the original matrix $\boldsymbol{A}$. We illustrate this procedure with an example.

Example 8.7. Find the inverse of the matrix

$$\boldsymbol{A} = \begin{bmatrix} 1 & 4 & 2 \\ 2 & 2 & 10 \\ 1 & 3 & 0 \end{bmatrix} .$$

Gauss-Jordan elimination to find the inverse is carried out as follows:

$$\begin{array}{rrr|rrrl}
1 & 4 & 2 & 1 & 0 & 0 & \\
2 & 2 & 10 & 0 & 1 & 0 & \\
1 & 3 & 0 & 0 & 0 & 1 & \\
\hline
1 & 4 & 2 & 1 & 0 & 0 & \\
0 & -6 & 6 & -2 & 1 & 0 & \mathrm{II} - 2 \cdot \mathrm{I} \\
0 & -1 & -2 & -1 & 0 & 1 & \mathrm{III} - \mathrm{I} \\
\hline
1 & 4 & 2 & 1 & 0 & 0 & \\
0 & 1 & -1 & \frac{1}{3} & -\frac{1}{6} & 0 & \mathrm{II}/(-6) \\
0 & -1 & -2 & -1 & 0 & 1 & \\
\hline
1 & 0 & 6 & -\frac{1}{3} & \frac{2}{3} & 0 & \mathrm{I} - 4 \cdot \mathrm{II} \\
0 & 1 & -1 & \frac{1}{3} & -\frac{1}{6} & 0 & \\
0 & 0 & -3 & -\frac{2}{3} & -\frac{1}{6} & 1 & \mathrm{III} + \mathrm{II} \\
\hline
1 & 0 & 6 & -\frac{1}{3} & \frac{2}{3} & 0 & \\
0 & 1 & -1 & \frac{1}{3} & -\frac{1}{6} & 0 & \\
0 & 0 & 1 & \frac{2}{9} & \frac{1}{18} & -\frac{1}{3} & \mathrm{III}/(-3) \\
\hline
\end{array}$$

$$\begin{array}{ccc|ccc l} 1 & 0 & 0 & -\frac{5}{3} & \frac{1}{3} & 2 & \mathrm{I}-6\cdot\mathrm{III} \\ 0 & 1 & 0 & \frac{5}{9} & -\frac{1}{9} & -\frac{1}{3} & \mathrm{II}+\mathrm{III} \\ 0 & 0 & 1 & \frac{2}{9} & \frac{1}{18} & -\frac{1}{3} & \\ \hline \end{array}$$

Thus, the inverse of $\boldsymbol{A}$ is

$$\boldsymbol{A}^{-1} = \begin{bmatrix} -\frac{5}{3} & \frac{1}{3} & 2 \\ \frac{5}{9} & -\frac{1}{9} & -\frac{1}{3} \\ \frac{2}{9} & \frac{1}{18} & -\frac{1}{3} \end{bmatrix}.$$

The method just described computes the inverse of a matrix $\boldsymbol{A}$, provided $\boldsymbol{A}$ is regular. But what happens if the matrix $\boldsymbol{A}$ is singular, i.e., if it has no inverse? Do we first need to check this before using Gauss-Jordan elimination? Fortunately, this is not the case. If one tries to invert a singular matrix using Gauss-Jordan elimination, a zero row (or column) will appear at some step of the algorithm on the left-hand side, making it impossible to transform the left-hand side into an identity matrix. Gauss-Jordan elimination thus serves two purposes. Firstly, it checks whether a matrix $\boldsymbol{A}$ is regular or not, and secondly it calculates the inverse if it exists.

We summarize:

- ***A* singular:** At some step of the algorithm, a zero row appears on the left-hand side and the algorithm stops.
- ***A* regular:** The left-hand side can be transformed into an identity matrix. Then, the inverse of $\boldsymbol{A}$ is the matrix on the right-hand side of the final tableau.

If the inverse $\boldsymbol{A}^{-1}$ is known, the solution to a linear equation $\boldsymbol{A}\boldsymbol{x} = \boldsymbol{b}$ is simply given by $\boldsymbol{x} = \boldsymbol{A}^{-1}\boldsymbol{b}$, because

$$\boldsymbol{A}\boldsymbol{x} = \boldsymbol{b} \iff \boldsymbol{A}^{-1}\boldsymbol{A}\boldsymbol{x} = \boldsymbol{A}^{-1}\boldsymbol{b} \iff \boldsymbol{I}\boldsymbol{x} = \boldsymbol{A}^{-1}\boldsymbol{b} \iff \boldsymbol{x} = \boldsymbol{A}^{-1}\boldsymbol{b}\,.$$

However, it does not make sense to first compute the inverse in order to solve a linear equation, since this is computationally more expensive than just solving the linear equation by Gauss-Jordan elimination.

Important terms for review

After reading this chapter, the following terms should be familiar:

- Linear equation
- General solution, particular solution
- Existence of a solution
- Uniqueness of a solution
- Gauss-Jordan elimination for solving a linear equation

- Gauss-Jordan elimination for inverting a matrix
- Regular matrix, singular matrix

Self-test

You can check your understanding of this chapter using the following true-or-false questions. For each statement, indicate whether it is true (T) or false (F). Short solutions to these questions are found in Appendix H.

I. Consider a linear equation $\boldsymbol{A}\boldsymbol{x} = \boldsymbol{b}$, where $\boldsymbol{A} \in \mathbb{R}^{n\times n}$, $\boldsymbol{b} \in \mathbb{R}^n$. Are the following statements true or false?

	T	F
If $\boldsymbol{A}$ is invertible, the solution is unique.	☐	☐
If $\boldsymbol{A}$ is not invertible, has no solution.	☐	☐
If $\boldsymbol{c}$ and $\boldsymbol{d}$ are two distinct solutions to the equation, then $\frac{1}{2}\boldsymbol{c}+\frac{1}{2}\boldsymbol{d}$ is also a solution.	☐	☐
If $\boldsymbol{c}$ is a solution to $\boldsymbol{A}\boldsymbol{x} = \boldsymbol{b}$ and $\boldsymbol{d}$ is a solution to $\boldsymbol{A}\boldsymbol{x} = \boldsymbol{0}$, then $\boldsymbol{c} + \boldsymbol{d}$ is a solution to $\boldsymbol{A}\boldsymbol{x} = \boldsymbol{b}$.	☐	☐

II. Evaluate the following statements about Gauss-Jordan elimination for solving a linear equation $\boldsymbol{A}\boldsymbol{x} = \boldsymbol{b}$, $\boldsymbol{A} \in \mathbb{R}^{m\times n}$, $\boldsymbol{b} \in \mathbb{R}^m$.

	T	F
If after some steps of Gauss-Jordan elimination on $\boldsymbol{A}\vert\boldsymbol{b}$ you get a zero row, then the equation has no solution.	☐	☐
If elementary row operations are performed in an unsystematic way, it is possible to end up back at the initial tableau.	☐	☐
If $\boldsymbol{A}$ is not invertible, it is not possible to transform it into an identity matrix by elementary row operations.	☐	☐
If the vector $\boldsymbol{x}$ has more entries than $\boldsymbol{A}$ has rows, then the equation has no solution.	☐	☐

Exercises

Exercise 1
Find the general solution of the linear equation $\boldsymbol{A}\boldsymbol{x} = \boldsymbol{0}$, where

$$\boldsymbol{A} = \begin{bmatrix} 1 & 0 & 3 & -1 & 2 \\ 3 & -2 & 1 & 0 & 1 \\ -1 & 2 & 5 & -2 & 5 \end{bmatrix}.$$

Exercise 2
Solve the following linear equations:

$$\text{a)}\ \begin{bmatrix} 2 & 12 & 6 \\ 2 & 6 & 4 \\ 2 & 0 & 2 \end{bmatrix} \begin{bmatrix} x_1 \\ x_2 \\ x_3 \end{bmatrix} = \begin{bmatrix} 12 \\ 9 \\ 6 \end{bmatrix} \quad \text{and} \quad \text{b)}\ \begin{bmatrix} 2 & 12 & 6 \\ 2 & 6 & 4 \\ 2 & 0 & 2 \end{bmatrix} \begin{bmatrix} x_1 \\ x_2 \\ x_3 \end{bmatrix} = \begin{bmatrix} 6 \\ 3 \\ 3 \end{bmatrix}.$$

Exercise 3
Below are three final tableaus from Gauss-Jordan elimination. For each, provide the general solution.

a)

x_1	x_2	x_3	x_4	
1	0	0	−2	1
0	1	0	1	4
0	0	1	4	0

b)

x_1	x_2	x_3	x_4	x_5	
1	0	0	−1	4	0
0	1	0	2	1	3
0	0	1	0	1	1

c)

x_3	x_5	x_1	x_4	x_2	
1	0	0	3	2	2
0	1	0	0	−6	4
0	0	1	−1	3	8

Exercise 4
Compute the inverse to the given matrix $\boldsymbol{A}$ and solve the linear equation $\boldsymbol{A}\boldsymbol{x} = \boldsymbol{b}$, where

$$\boldsymbol{A} = \begin{bmatrix} 2 & 1 & 1 \\ 0 & 1 & 1 \\ 1 & 0 & 1 \end{bmatrix} \quad \text{and} \quad \boldsymbol{b} = \begin{bmatrix} 1 \\ 3 \\ 3 \end{bmatrix}.$$

Exercise 5
Are the following matrices invertible? If yes, calculate their inverse.

$$\boldsymbol{A} = \begin{bmatrix} 1 & 1 & 2 \\ 2 & 0 & 2 \\ 4 & -6 & 0 \end{bmatrix} \qquad \boldsymbol{B} = \begin{bmatrix} 1 & -6 & -7 \\ 2 & -3 & -5 \\ 1 & 2 & 1 \end{bmatrix}$$

Chapter 9

Fundamentals of linear algebra

Linear algebra is a subfield of mathematics that deals with the structure of so-called vector spaces. A vector space is a structure in which one can "essentially" perform calculations as we know them from $\mathbb{R}^n$. In this chapter, we will introduce the basic concepts of linear algebra. For simplicity and because it is sufficient for most economic applications, we will largely limit ourselves to the case where the vector space is the n-dimensional Euclidean space $\mathbb{R}^n$. The concepts carry over straightforwardly from $\mathbb{R}^n$ to more general vector spaces. Those spaces will be briefly treated in Section 9.7.

Numerous properties of the Euclidean space $\mathbb{R}^n$ have already been introduced in Chapter 2. There, the concepts of a matrix and a vector were introduced and it was shown that the points of $\mathbb{R}^n$ can be viewed as vectors having n components.

9.1 Linear combinations and span

The basic arithmetic operations that can be performed with vectors are addition of vectors and multiplication of a vector with a scalar (cf. Chapter 2).

With these two operations, any vector of $\mathbb{R}^2$ is generated from the two canonical unit vectors $\boldsymbol{e}_1$ and $\boldsymbol{e}_2$. In fact,

$$\boldsymbol{z} = \begin{bmatrix} z_1 \\ z_2 \end{bmatrix} = z_1 \begin{bmatrix} 1 \\ 0 \end{bmatrix} + z_2 \begin{bmatrix} 0 \\ 1 \end{bmatrix} = z_1 \boldsymbol{e}_1 + z_2 \boldsymbol{e}_2 .$$

Given n (not necessarily distinct) vectors $\boldsymbol{x}_1, \boldsymbol{x}_2, \ldots, \boldsymbol{x}_n$, a vector of the form

$$\lambda_1 \boldsymbol{x}_1 + \lambda_2 \boldsymbol{x}_2 + \ldots + \lambda_n \boldsymbol{x}_n ,$$

K. Mosler et al., *Mathematical Methods for Business and Economics*, Mathematics Study Resources 20, https://doi.org/10.1007/978-3-662-73831-3_9

where $\lambda_1, \ldots, \lambda_n \in \mathbb{R}$, is referred to as a **linear combination** of the vectors $\boldsymbol{x}_1, \ldots, \boldsymbol{x}_n$.

In this concept, every vector in $\mathbb{R}^2$ is a linear combination of the vectors $\boldsymbol{e}_1$ and $\boldsymbol{e}_2$. Conversely, given vectors $\boldsymbol{x}_1, \ldots, \boldsymbol{x}_n$, one may ask which vectors can be generated by linear combinations of these. The set of all linear combinations of a given **family**[1] of vectors is referred to as the **linear hull**, **linear span** or simply **span** of these vectors. The span of a (finite or infinite) family M of vectors is denoted by $\mathrm{sp}(M)$. We will restrict ourselves to the case where M has finitely many vectors $\boldsymbol{x}_1, \boldsymbol{x}_2, \ldots, \boldsymbol{x}_n$ and write

$$\begin{aligned}\mathrm{sp}(\boldsymbol{x}_1, \boldsymbol{x}_2, \ldots, \boldsymbol{x}_n) &= \{\, \lambda_1\boldsymbol{x}_1 + \ldots + \lambda_n\boldsymbol{x}_n \,:\, \lambda_1, \ldots, \lambda_n \in \mathbb{R}\,\} \\ &= \left\{ \sum_{i=1}^{n} \lambda_i \boldsymbol{x}_i \,:\, \lambda_1, \ldots, \lambda_n \in \mathbb{R} \right\}.\end{aligned}$$

The span of an empty family is defined to be the set containing only the zero vector, i.e., $\mathrm{sp}(\emptyset) = \{\mathbf{0}\}$.

Example 9.1. For the unit vectors $\boldsymbol{e}_1$ and $\boldsymbol{e}_2$ in $\mathbb{R}^2$, we have

$$\mathrm{sp}(\boldsymbol{e}_1, \boldsymbol{e}_2) = \mathbb{R}^2 .$$

More generally, for any n, we have

$$\mathrm{sp}(\boldsymbol{e}_1, \ldots, \boldsymbol{e}_n) = \mathbb{R}^n .$$

Example 9.2. For the vector $[2, 1]^T$, we have

$$\mathrm{sp}\left(\begin{bmatrix} 2 \\ 1 \end{bmatrix}\right) = \left\{ \lambda \begin{bmatrix} 2 \\ 1 \end{bmatrix} \,:\, \lambda \in \mathbb{R} \right\}.$$

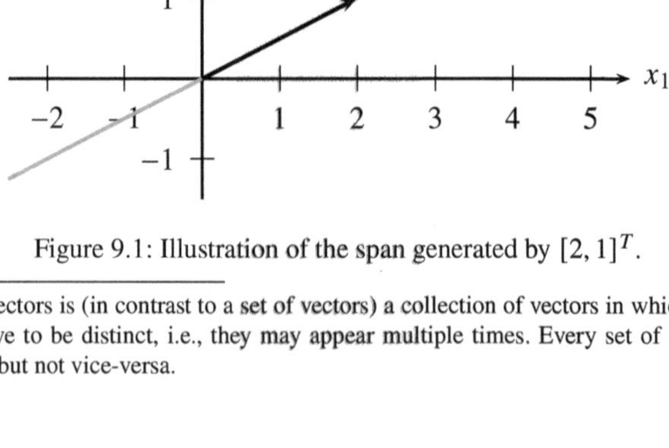

Figure 9.1: Illustration of the span generated by $[2, 1]^T$.

[1] A family of vectors is (in contrast to a set of vectors) a collection of vectors in which the individual vectors do not have to be distinct, i.e., they may appear multiple times. Every set of vectors is thus a family of vectors, but not vice-versa.

Generally, for a vector $\boldsymbol{x} \in \mathbb{R}^n$, we have

$$\operatorname{sp}(\boldsymbol{x}) = \{ \lambda \boldsymbol{x} \, : \, \lambda \in \mathbb{R} \} .$$

The span of a single vector is the set of all multiples of the vector. Geometrically, this corresponds to a line through the origin; see Figure 9.1.

Example 9.3. The span of the two vectors $[1, 0, 1]^T$ and $[0, 1, 1]^T$ is

$$\operatorname{sp}\left(\begin{bmatrix} 1 \\ 0 \\ 1 \end{bmatrix}, \begin{bmatrix} 0 \\ 1 \\ 1 \end{bmatrix} \right) = \left\{ \lambda \begin{bmatrix} 1 \\ 0 \\ 1 \end{bmatrix} + \mu \begin{bmatrix} 0 \\ 1 \\ 1 \end{bmatrix} : \lambda, \mu \in \mathbb{R} \right\} = \left\{ \begin{bmatrix} \lambda \\ \mu \\ \lambda + \mu \end{bmatrix} : \lambda, \mu \in \mathbb{R} \right\} .$$

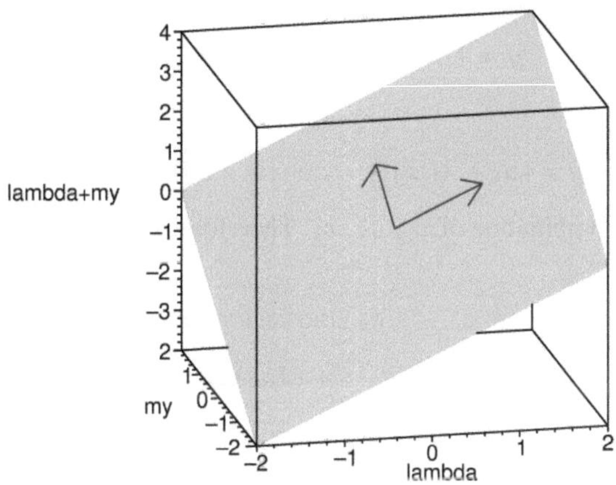

Figure 9.2: The plane generated by $[1, 0, 1]^T$ and $[0, 1, 1]^T$.

Geometrically, this corresponds to a plane through the origin and the two points $[1, 0, 1]^T$ and $[0, 1, 1]^T$, see Figure 9.2.

9.2 Linear subspaces

A **linear subspace** or simply **subspace** of $\mathbb{R}^n$ is a non-empty subset U of $\mathbb{R}^n$, in which the usual operations for vectors can be performed without restriction, more precisely, which has the following properties:

U0: $U \neq \emptyset$.

U1: If $\boldsymbol{x}$ and $\boldsymbol{y}$ are in U, then their sum $\boldsymbol{x} + \boldsymbol{y}$ is also in U.

U2: If $\boldsymbol{x}$ is in U, then for every real number λ, $\lambda \boldsymbol{x}$ is also in U.

In other words: Addition and scalar multiplication do not lead out of the subspace. This is also referred to by saying that a subspace is closed under vector addition and scalar multiplication. Every subspace contains the origin $\mathbf{0}$. This follows immediately from property **U2**, when choosing $\lambda = 0$. The space $\mathbb{R}^n$ contains two trivial subspaces:

the entire space $\mathbb{R}^n$ itself and the zero subspace $\{\mathbf{0}\}$. An important result is the following theorem.

Theorem 9.1: Span is a subspace

For $\boldsymbol{z}_1, \ldots, \boldsymbol{z}_k \in \mathbb{R}^n$, the span $\operatorname{sp}(\boldsymbol{z}_1, \ldots, \boldsymbol{z}_k)$ is a subspace of $\mathbb{R}^n$.

Proof: By definition, $x_1, \ldots, x_k \in \operatorname{sp}(x_1, \ldots, x_k)$. Therefore, $\operatorname{sp}(\boldsymbol{z}_1, \ldots, \boldsymbol{z}_k) \neq \emptyset$ and **U0** holds.

Assume $\boldsymbol{x}, \boldsymbol{y}$ are two vectors in $\operatorname{sp}(\boldsymbol{z}_1, \ldots, \boldsymbol{z}_k)$. Then, $\boldsymbol{x}$ and $\boldsymbol{y}$ are linear combinations of the vectors $\boldsymbol{z}_1, \ldots, \boldsymbol{z}_k$, i.e., there are numbers $\mu_1, \ldots, \mu_k$ and $\lambda_1, \ldots, \lambda_k$ such that

$$\begin{aligned} \boldsymbol{x} &= \mu_1 \boldsymbol{z}_1 + \ldots + \mu_k \boldsymbol{z}_k \,, \\ \boldsymbol{y} &= \lambda_1 \boldsymbol{z}_1 + \ldots + \lambda_k \boldsymbol{z}_k \,. \end{aligned}$$

Then, it holds that

$$\boldsymbol{x} + \boldsymbol{y} = (\mu_1 + \lambda_1)\boldsymbol{z}_1 + \ldots + (\mu_k + \lambda_k)\boldsymbol{z}_k \,.$$

Thus, $\boldsymbol{x}+\boldsymbol{y}$ is a linear combination of $\boldsymbol{z}_1, \ldots, \boldsymbol{z}_k$. Therefore, $\boldsymbol{x}+\boldsymbol{y} \in \operatorname{sp}(\boldsymbol{z}_1, \ldots, \boldsymbol{z}_k)$, and **U1** holds.

For $\boldsymbol{x} = \mu_1 \boldsymbol{z}_1 + \ldots + \mu_k \boldsymbol{z}_k \in \operatorname{sp}(\boldsymbol{z}_1, \ldots, \boldsymbol{z}_k)$ the vector $\lambda \boldsymbol{x}$ is given by

$$\lambda \boldsymbol{x} = (\lambda \mu_1)\boldsymbol{z}_1 + \ldots + (\lambda \mu_k)\boldsymbol{z}_k \,.$$

Therefore, $\lambda \boldsymbol{x}$ is a linear combination of $\boldsymbol{z}_1, \ldots, \boldsymbol{z}_k$. Hence, $\lambda \boldsymbol{x} \in \operatorname{sp}(\boldsymbol{z}_1, \ldots, \boldsymbol{z}_k)$, and **U2** holds. □

9.3 Linear independence

The span of the first two canonical unit vectors in $\mathbb{R}^3$ is given by

$$\operatorname{sp}(\boldsymbol{e}_1, \boldsymbol{e}_2) = \operatorname{sp}\left(\begin{bmatrix} 1 \\ 0 \\ 0 \end{bmatrix}, \begin{bmatrix} 0 \\ 1 \\ 0 \end{bmatrix} \right) = \left\{ \begin{bmatrix} \lambda \\ \mu \\ 0 \end{bmatrix} \;:\; \lambda, \mu \in \mathbb{R} \right\} .$$

If we add, for example, the vector $[1, 1, 0]^T$, the span of the three vectors $\boldsymbol{e}_1$, $\boldsymbol{e}_2$ and $[1, 1, 0]^T$ cannot be smaller than before. On the other hand, it is not larger than before, since a linear combination of the three vectors has the form

$$\lambda \begin{bmatrix} 1 \\ 0 \\ 0 \end{bmatrix} + \mu \begin{bmatrix} 0 \\ 1 \\ 0 \end{bmatrix} + \nu \begin{bmatrix} 1 \\ 1 \\ 0 \end{bmatrix} = \begin{bmatrix} \lambda + \nu \\ \mu + \nu \\ 0 \end{bmatrix} .$$

Thus, every linear combination of the three vectors has its third component zero, and every such vector can already be generated by $\boldsymbol{e}_1$ and $\boldsymbol{e}_2$. Hence,

$$\operatorname{sp}(\boldsymbol{e}_1, \boldsymbol{e}_2) = \operatorname{sp}(\boldsymbol{e}_1, \boldsymbol{e}_2, [1, 1, 0]^T) \,.$$

What is the reason that adding the vector $[1, 1, 0]^T$ does not increase the span? Obviously, this is due to the fact that $[1, 1, 0]^T$ itself is a linear combination of e_1 and e_2. If for a family of vectors, one of the vectors is a linear combination of the other vectors, we say that the vectors are **linearly dependent**. Conversely, if none of the vectors is a linear combination of the others, the vectors are called **linearly independent**. Usually, the concept of linear independence is defined as follows:

Definition 9.1: Linear independence

The vectors $z_1, \dots, z_k$, $k \geq 1$, are called **linearly independent** if the equation

$$\lambda_1 z_1 + \dots + \lambda_k z_k = \mathbf{0} \tag{9.1}$$

can only be satisfied by $\lambda_1 = \dots = \lambda_k = 0$.

The vectors are called **linearly dependent** if they are not linearly independent.

Remark: The above definition does not state that the vectors are linearly independent if for $\lambda_1 = \dots = \lambda_k = 0$ Equation (9.1) is satisfied. This is always the case. Rather, it is required that $\lambda_1 = \dots = \lambda_k = 0$ is the *only* solution to (9.1).

From the definition follows, that a single vector z is linearly independent if $z \neq \mathbf{0}$. The vector $\mathbf{0}$ is linearly dependent.

If the vectors $z_1, \dots, z_k$ are linearly dependent and $k \geq 2$, then there are numbers $\lambda_1, \dots, \lambda_k$, not all of which are zero, such that

$$\lambda_1 z_1 + \dots + \lambda_k z_k = \mathbf{0}\,.$$

If, for example, $\lambda_1 \neq 0$, this equation can be solved for z_1 and we get

$$z_1 = -\frac{\lambda_2}{\lambda_1} z_2 - \dots - \frac{\lambda_k}{\lambda_1} z_k\,.$$

Thus, z_1 is a linear combination of $z_2, \dots, z_k$. Similarly, each z_j can be expressed as a linear combination of the other vectors, provided the associated coefficient $\lambda_j \neq 0$. This shows, that for linearly dependent vectors, at least *one* of the vectors is a linear combination of the others. However, this does *not mean that every vector* is a linear combination of the others.

Conversely, if one of the vectors is a linear combination of the others, then, the vectors are linearly dependent.

Example 9.4. The unit vectors $e_1, e_2, \dots, e_n$ in $\mathbb{R}^n$ are linearly independent.

Example 9.5. Consider the vectors

$$x = \begin{bmatrix} 1 \\ 1 \\ 0 \end{bmatrix}, \quad y = \begin{bmatrix} 1 \\ 2 \\ 0 \end{bmatrix}, \quad z = \begin{bmatrix} 2 \\ 1 \\ 0 \end{bmatrix} \quad \text{and} \quad u = \begin{bmatrix} 1 \\ 1 \\ 1 \end{bmatrix}.$$

$\boldsymbol{x}, \boldsymbol{y}, \boldsymbol{z}$ and $\boldsymbol{u}$ are linearly dependent, because

$$3\boldsymbol{x} + (-1)\boldsymbol{y} + (-1)\boldsymbol{z} + 0\boldsymbol{u} = \boldsymbol{0}\,.$$

Here, $\boldsymbol{u}$ is no linear combination of the first three vectors. However, each of the first three vectors is a linear combination of the others.

If we only consider $\boldsymbol{x}$, $\boldsymbol{y}$ and $\boldsymbol{z}$, they are still linearly dependent. This can be seen if we omit the summand "$0\boldsymbol{u}$" in the above equation.

$\boldsymbol{x}$, $\boldsymbol{y}$, $\boldsymbol{u}$ are linearly independent. This is because $\lambda\boldsymbol{x} + \mu\boldsymbol{y} + \nu\boldsymbol{u} = \boldsymbol{0}$ if and only if

$$\begin{aligned} \lambda + \mu + \nu &= 0\,, \\ \lambda + 2\mu + \nu &= 0\,, \\ \nu &= 0\,. \end{aligned}$$

The third equation implies $\nu = 0$. Subtracting the first equation from the second yields $\mu = 0$ and thus also $\lambda = 0$. $\lambda = \mu = \nu = 0$ is therefore the only solution to the equation.

Similarly, $\boldsymbol{x}$, $\boldsymbol{z}$ and $\boldsymbol{u}$ as well as $\boldsymbol{y}$, $\boldsymbol{z}$ and $\boldsymbol{u}$ are each linearly independent. The same applies if we consider only one or only two of the above four vectors.

Let the vectors $\boldsymbol{v}_1, \ldots, \boldsymbol{v}_k$ all have norm one and let any two of these vectors be orthogonal. Then, the vectors are called an **orthonormal system** or short **orthonormal**. An orthonormal system of vectors is always linearly independent. If $\lambda_1\boldsymbol{v}_1 + \ldots + \lambda_k\boldsymbol{v}_k = \boldsymbol{0}$, then by multiplying with $\boldsymbol{v}_i^T$ we get

$$\begin{aligned} &\boldsymbol{v}_i^T(\lambda_1\boldsymbol{v}_1 + \ldots + \lambda_i\boldsymbol{v}_i + \ldots + \lambda_k\boldsymbol{v}_k) \\ &\qquad = \lambda_1 \underbrace{\boldsymbol{v}_i^T\boldsymbol{v}_1}_{=0} + \ldots + \lambda_i \underbrace{\boldsymbol{v}_i^T\boldsymbol{v}_i}_{=1} + \ldots + \lambda_k \underbrace{\boldsymbol{v}_i^T\boldsymbol{v}_k}_{=0} = \lambda_i = 0\,, \end{aligned}$$

i.e., all λ_i must be zero.

9.4 Basis and dimension

We have already seen that the span of a finite number of vectors is a linear subspace. One might now ask whether the converse is also true, i.e., if every subspace of $\mathbb{R}^n$ is the span of finitely many vectors. To see that this is indeed the case we introduce the concept of a basis.

Definition 9.2: Basis

Let U be a subspace of $\mathbb{R}^n$. A set $\{\boldsymbol{z}_1, \ldots, \boldsymbol{z}_k\}$ of vectors is a **basis** of U, if

- $\boldsymbol{z}_1, \ldots, \boldsymbol{z}_k$ are linearly independent, and
- $\mathrm{sp}(\boldsymbol{z}_1, \ldots, \boldsymbol{z}_k) = U$.

Example 9.6. The canonical unit vectors $e_1, \ldots, e_n$ are linearly independent and span $\mathbb{R}^n$, hence form a basis. This basis is called the **canonical basis** of $\mathbb{R}^n$.

Example 9.7. We again consider the vectors $\boldsymbol{x}$, $\boldsymbol{y}$, $\boldsymbol{z}$ and $\boldsymbol{u}$ from Example 9.5. They do not form a basis of $\mathbb{R}^3$, as they are linearly dependent.

We already know that $\boldsymbol{x}, \boldsymbol{y}, \boldsymbol{u}$ are linearly independent. To check whether $\boldsymbol{x}, \boldsymbol{y}, \boldsymbol{u}$ form a basis of $\mathbb{R}^3$, we still need to investigate whether $\mathrm{sp}(\boldsymbol{x}, \boldsymbol{y}, \boldsymbol{u}) = \mathbb{R}^3$, i.e., whether every vector of $\mathbb{R}^3$ can be represented as a linear combination of these three vectors. Let $\boldsymbol{a} = [a_1, a_2, a_3]^T$ be an arbitrary vector of $\mathbb{R}^3$. We need to show that $\boldsymbol{a}$ is a linear combination of $\boldsymbol{x}, \boldsymbol{y}, \boldsymbol{u}$. It holds that $\lambda\boldsymbol{x} + \mu\boldsymbol{y} + \nu\boldsymbol{u} = \boldsymbol{a}$ if and only if

$$\begin{aligned} \lambda + \mu + \nu &= a_1 \,, \\ \lambda + 2\mu + \nu &= a_2 \,, \\ \nu &= a_3 \,. \end{aligned}$$

From the third equation it immediately follows that $\nu = a_3$. Subtracting the first equation from the second gives $\mu = a_2 - a_1$, and from this it immediately follows $\lambda = 2a_1 - a_2 - a_3$. Thus,

$$(2a_1 - a_2 - a_3)\boldsymbol{x} + (a_2 - a_1)\boldsymbol{y} + a_3\boldsymbol{u} = \boldsymbol{a} \,,$$

i.e., $\mathrm{sp}(\boldsymbol{x}, \boldsymbol{y}, \boldsymbol{u}) = \mathbb{R}^3$. The set of vectors $\{\boldsymbol{x}, \boldsymbol{y}, \boldsymbol{u}\}$ satisfies both conditions of Definition 9.2 and is therefore a basis.

The vectors $\boldsymbol{x}, \boldsymbol{y}$ are linearly independent. However, the span of $\boldsymbol{x}, \boldsymbol{y}$ is not $\mathbb{R}^3$, as for example the vector $[0, 0, 1]^T$ is no linear combination of $\boldsymbol{x}$ and $\boldsymbol{y}$. Here, the second condition of Definition 9.2 is violated. Thus, $\{\boldsymbol{x}, \boldsymbol{y}\}$ is no basis of $\mathbb{R}^3$.

Several questions now arise. Does every subspace of $\mathbb{R}^n$ have a basis or are there subspaces that have no basis? Is the basis of a subspace unique or may there exist several bases? The second question has been already answered by the two preceding examples. $\mathbb{R}^3$ has the canonical basis, but also the basis $\{\boldsymbol{x}, \boldsymbol{y}, \boldsymbol{u}\}$ from Example 9.7. Thus, a subspace can have different bases. However, both these bases of $\mathbb{R}^3$ consist of three elements. That this is not coincidental states the following theorem, which also provides an answer to our first question.

Theorem 9.2: Existence of a basis

Every subspace U of $\mathbb{R}^n$ has a basis. Furthermore, all bases of U have the same number of vectors.

One can even show that every subspace has an **orthonormal basis**, i.e., a basis whose vectors form an orthonormal system.

The number of vectors contained in a basis is an important characteristic of a subspace. It is called **dimension**.

Definition 9.3: Dimension

Let U be a subspace of $\mathbb{R}^n$. The number k of elements of a basis is called the **dimension** of the subspace. We write $\dim U = k$.

Example 9.8. $\dim \mathbb{R}^n = n$, as the canonical basis $\{e_1, \ldots, e_n\}$ consists of n elements.

Example 9.9. As previously mentioned, the zero vector $\mathbf{0}$ is linearly dependent. Consequently, the subspace $\{\mathbf{0}\}$ cannot contain any non-empty basis. By definition, its unique basis is the empty set, and therefore $\dim\{\mathbf{0}\} = 0$.

Why do we actually require a basis to be linearly independent? If we were to omit the first condition from Definition 9.2, then every vector from U would still be representable as a linear combination of the basis vectors. However, this representation would no longer be unique. If, on the other hand, we require a basis to be linearly independent, we obtain the following theorem.

Theorem 9.3: Uniqueness of basis representation

Let $\{\boldsymbol{z}_1, \ldots, \boldsymbol{z}_k\}$ be a basis of U. Then every vector $\boldsymbol{x}$ from U has a (up to the order of the summands) unique representation $\boldsymbol{x} = \lambda_1 \boldsymbol{z}_1 + \ldots + \lambda_k \boldsymbol{z}_k$ as a linear combination of the basis vectors. λ_i is called the **coordinate** of $\boldsymbol{x}$ with respect to the basis vector $\boldsymbol{z}_i$.

Proof: If $\lambda_1 \boldsymbol{z}_1 + \ldots + \lambda_k \boldsymbol{z}_k$ and $\mu_1 \boldsymbol{z}_1 + \ldots + \mu_k \boldsymbol{z}_k$ are two representations of a vector $x \in U$ as a linear combination of the basis vectors, then it holds

$$\begin{aligned}\mathbf{0} = \boldsymbol{x} - \boldsymbol{x} &= (\lambda_1 \boldsymbol{z}_1 + \ldots + \lambda_k \boldsymbol{z}_k) - (\mu_1 \boldsymbol{z}_1 + \ldots + \mu_k \boldsymbol{z}_k) \\ &= (\lambda_1 - \mu_1)\boldsymbol{z}_1 + \ldots + (\lambda_k - \mu_k)\boldsymbol{z}_k\,.\end{aligned}$$

Due to the linear independence of the basis vectors, it follows that $\lambda_i - \mu_i = 0$, i.e., $\lambda_i = \mu_i$ for all i. The two representations are therefore equal. □

The following theorem contains useful statements about the dimension of subspaces.

Theorem 9.4: Dimensions of subspaces

The following holds:

(i) If U_1, U_2 are subspaces with $U_1 \subset U_2$, then $\dim U_1 \leq \dim U_2$.

(ii) If U_1, U_2 are subspaces with $U_1 \subset U_2$ and $\dim U_1 = \dim U_2$, then $U_1 = U_2$.

(iii) If U_1, U_2 are subspaces with $U_1 \subset U_2$ and $\dim U_1 < \dim U_2$, then $U_1 \neq U_2$.

(iv) If U is a subspace of $\mathbb{R}^n$, then $\dim U \leq n$.

(v) For arbitrary $z_1, \ldots, z_k \in \mathbb{R}^n$, $\dim \operatorname{sp}(z_1, \ldots, z_k) \leq k$.

(vi) $\dim \operatorname{sp}(z_1, \ldots, z_k) = k$ if and only if the vectors $z_1, \ldots, z_k$ are linearly independent.

(vii) If $k > \dim U$, then any k vectors in U are linearly dependent.

(viii) If $\dim U = k$, then any k linearly independent vectors in U are a basis of U.

So, to find a basis of a k-dimensional subspace, it is sufficient to find k linearly independent vectors in this subspace.

Theorem 9.4 (vii) gives a simple sufficient criterion for linear dependence. In $\mathbb{R}^n$, any k vectors are linearly dependent if $k > n$. Knowing that, we could have dispensed with the detailed linear dependence check in Example 9.5.

9.5 Rank of a matrix

In Theorem 9.4 we saw that the dimension of a span $\operatorname{sp}(z_1, \ldots, z_k)$ is at most equal to k. To find its exact value, one must find the maximum number of linearly independent vectors among $z_1, \ldots, z_k$. This question is closely related to the concept of the rank of a matrix.

Definition 9.4: Rank of a matrix

Let $\boldsymbol{A}$ be an $m \times n$ matrix. For $1 \leq i \leq n$, let $\boldsymbol{a}_i$ denote the i-th column of $\boldsymbol{A}$. Then,

$$\boldsymbol{A} = [\boldsymbol{a}_1, \cdots, \boldsymbol{a}_n] .$$

The **rank** of $\boldsymbol{A}$ is the maximum number of linearly independent columns of A. We denote the rank of a matrix $\boldsymbol{A}$ by $\operatorname{rk} \boldsymbol{A}$.

Example 9.10. Consider the matrix

$$\boldsymbol{A} = \begin{bmatrix} 2 & 1 & 4 \\ 0 & 1 & 2 \end{bmatrix} .$$

Then the three columns of $\boldsymbol{A}$ are linearly dependent, since three vectors in $\mathbb{R}^2$ are always linearly dependent; it holds

$$\begin{bmatrix} 2 \\ 0 \end{bmatrix} + 2 \begin{bmatrix} 1 \\ 1 \end{bmatrix} = \begin{bmatrix} 4 \\ 2 \end{bmatrix} .$$

Since the first two columns are linearly independent, $\operatorname{rk} \boldsymbol{A} = 2$.

Example 9.11. The rank of the $n \times n$ identity matrix $\boldsymbol{I}_n$ is $\operatorname{rk} \boldsymbol{I}_n = n$. The rank of an $m \times n$ zero matrix $\boldsymbol{0}_{m\times n}$ is $\operatorname{rk} \boldsymbol{0}_{m\times n} = 0$.

Apparently, the following theorem holds:

Theorem 9.5: Rank and dimension

Let $\boldsymbol{A} = [\boldsymbol{a}_1, \cdots, \boldsymbol{a}_n]$ be an $m \times n$ matrix. Then it is

$$\operatorname{rk} \boldsymbol{A} = \dim \operatorname{sp}(\boldsymbol{a}_1, \ldots, \boldsymbol{a}_n).$$

Of course, one could also ask for the maximum number of linearly independent **rows** of a given matrix. Since the rows of $\boldsymbol{A}$ are the columns of $\boldsymbol{A}^T$, this number is equal to $\operatorname{rk} \boldsymbol{A}^T$. It is therefore natural to ask about the relationship between $\operatorname{rk} \boldsymbol{A}$ and $\operatorname{rk} \boldsymbol{A}^T$. The following theorem provides the answer.

Theorem 9.6: Rank of the transpose

Let $\boldsymbol{A}$ be an $m \times n$ matrix. Then it is

$$\operatorname{rk} \boldsymbol{A} = \operatorname{rk} \boldsymbol{A}^T.$$

Therefore, the rank of a matrix can neither be greater than the number of columns nor greater than the number of rows. In other words: If $\boldsymbol{A}$ is an $m \times n$ matrix, then

$$\operatorname{rk} \boldsymbol{A} \leq \min\{m, n\}.$$

If a matrix has the maximum possible rank, i.e., if $\operatorname{rk} \boldsymbol{A} = \min\{m, n\}$, then $\boldsymbol{A}$ is said to have **full rank**.

Is there a systematic procedure for finding the rank of a matrix $\boldsymbol{A}$? Yes, and we already know this procedure. It is the Gauss-Jordan elimination from Chapter 8. This is because elementary row operations do not change the rank of a matrix. Therefore, when applying Gauss-Jordan elimination (of course, without a right-hand side vector $\boldsymbol{b}$), the rank of the matrix remains unchanged. From the final tableau one can directly read the rank of a matrix. We illustrate this with a few examples.

Example 9.12. Consider the matrix

$$\boldsymbol{A} = \begin{bmatrix} 1 & 2 & 4 \\ 3 & 6 & 13 \\ 1 & 5 & 7 \end{bmatrix}.$$

We transform the matrix A using Gauss-Jordan elimination:

$$\begin{array}{ccc} a_1 & a_2 & a_3 \\ \hline 1 & 2 & 4 \\ 3 & 6 & 13 \\ 1 & 5 & 7 \end{array} \xrightarrow[\text{III}-\text{I}]{\text{II}-3\cdot\text{I}} \begin{array}{ccc} a_1 & a_2 & a_3 \\ \hline 1 & 2 & 4 \\ 0 & 0 & 1 \\ 0 & 3 & 3 \end{array} \xrightarrow{\text{II}\leftrightarrow\text{III}} \begin{array}{ccc} a_1 & a_2 & a_3 \\ \hline 1 & 2 & 4 \\ 0 & 3 & 3 \\ 0 & 0 & 1 \end{array}$$

$$\xrightarrow[\text{II}/3]{\text{I}-2/3\cdot\text{II}} \begin{array}{ccc} a_1 & a_2 & a_3 \\ \hline 1 & 0 & 2 \\ 0 & 1 & 1 \\ 0 & 0 & 1 \end{array} \xrightarrow[\text{II}-\text{III}]{\text{I}-2\cdot\text{III}} \begin{array}{ccc} a_1 & a_2 & a_3 \\ \hline 1 & 0 & 0 \\ 0 & 1 & 0 \\ 0 & 0 & 1 \end{array}$$

Since the rank of the 3×3 identity matrix is three, we have $\text{rk}\, \boldsymbol{A} = 3$.

When using Gauss-Jordan elimination to find the rank of $\boldsymbol{A}$, one can make the work a bit easier. In example 9.12, the computational effort can be reduced by omitting the last two tableaux. The third-to-last tableau already contains only zeros below the main diagonal and non-zero entries on the main diagonal. In this case, it is clear that Gauss-Jordan elimination will ultimately produce the identity matrix.

Example 9.13. Consider the matrix

$$\boldsymbol{A} = \begin{bmatrix} 1 & 2 & 4 & 8 \\ 3 & 6 & 13 & 25 \\ 1 & 2 & 7 & 11 \end{bmatrix}.$$

Gauss-Jordan elimination is carried out as follows:

$$\begin{array}{cccc} a_1 & a_2 & a_3 & a_4 \\ \hline 1 & 2 & 4 & 8 \\ 3 & 6 & 13 & 25 \\ 1 & 2 & 7 & 11 \end{array} \xrightarrow[\text{III}-\text{I}]{\text{II}-3\cdot\text{I}} \begin{array}{cccc} a_1 & a_2 & a_3 & a_4 \\ \hline 1 & 2 & 4 & 8 \\ 0 & 0 & 1 & 1 \\ 0 & 0 & 3 & 3 \end{array}$$

$$\xrightarrow[a_2 \leftrightarrow a_3]{} \begin{array}{cccc} a_1 & a_3 & a_2 & a_4 \\ \hline 1 & 4 & 2 & 8 \\ 0 & 1 & 0 & 1 \\ 0 & 3 & 0 & 3 \end{array} \xrightarrow[\text{III}-3\cdot\text{II}]{\text{I}-4\cdot\text{II}} \begin{array}{cccc} a_1 & a_3 & a_2 & a_4 \\ \hline 1 & 0 & 2 & 4 \\ 0 & 1 & 0 & 1 \\ 0 & 0 & 0 & 0 \end{array}$$

Here, the first two columns are linearly independent and the last two columns are linear combinations of the first two. Thus, $\text{rk}\, \boldsymbol{A} = 2$. Additionally, from the final tableau, one can see that the columns $\boldsymbol{a}_1$ and $\boldsymbol{a}_3$ of the original matrix $\boldsymbol{A}$ are linearly independent and that the columns $\boldsymbol{a}_2$ and $\boldsymbol{a}_4$ of $\boldsymbol{A}$ are linear combinations of $\boldsymbol{a}_1$ and $\boldsymbol{a}_3$, namely $\boldsymbol{a}_2 = 2\boldsymbol{a}_1$ and $\boldsymbol{a}_4 = 4\boldsymbol{a}_1 + 1\boldsymbol{a}_3$.

The Gauss-Jordan method can also be used to check whether vectors are linearly independent or not. It holds that

$$\boldsymbol{a}_1, \ldots, \boldsymbol{a}_k \text{ are linearly independent} \quad \Longleftrightarrow \quad \text{rk}[\boldsymbol{a}_1, \cdots, \boldsymbol{a}_k] = k\,.$$

The vectors are therefore linearly independent if the rank of the matrix with columns $\boldsymbol{a}_1, \ldots, \boldsymbol{a}_k$ is equal to k. They are linearly dependent if the rank is less than k.

9.6 More about linear equations

We have already dealt with the question of how to find the solution to a linear equation in Chapter 8. In this section, we analyze the structure of the general solution and establish criteria for both the solvability of the linear equation $\boldsymbol{A}\boldsymbol{x} = \boldsymbol{b}$ and the invertibility of the matrix $\boldsymbol{A}$.

Solution to a homogeneous linear equation

We first deal with the solution to a **homogeneous** linear equation. This refers to a linear equation $\boldsymbol{A}\boldsymbol{x} = \boldsymbol{0}$, where the right-hand side is the zero vector $\boldsymbol{0}$.

Theorem 9.7: Solution to a homogeneous equation is a subspace

Let $\boldsymbol{A}$ be an $m \times n$ matrix. Then the general solution to the homogeneous linear equation $\boldsymbol{A}\boldsymbol{x} = \boldsymbol{0}$, $L(\boldsymbol{A}, \boldsymbol{0}) = \{\boldsymbol{x} \in \mathbb{R}^n : \boldsymbol{A}\boldsymbol{x} = \boldsymbol{0}\}$, is a subspace of $\mathbb{R}^n$.

Proof: We have to show that $L(\boldsymbol{A}, \boldsymbol{0})$ is non-empty and closed regarding addition and scalar multiplication of vectors. Obviously, the null vector $\boldsymbol{0}$ is a particular solution, $\boldsymbol{A}\boldsymbol{0} = \boldsymbol{0}$, hence $L(\boldsymbol{A}, \boldsymbol{0})$ is not empty. If $\boldsymbol{x}$ and $\boldsymbol{y}$ are two solutions to the linear equation, then

$$\boldsymbol{A}(\boldsymbol{x} + \boldsymbol{y}) = \boldsymbol{A}\boldsymbol{x} + \boldsymbol{A}\boldsymbol{y} = \boldsymbol{0} + \boldsymbol{0} = \boldsymbol{0}\,,$$

i.e., $\boldsymbol{x} + \boldsymbol{y}$ is also a solution. If $\boldsymbol{x}$ is a solution, then for every real number λ

$$\boldsymbol{A}(\lambda\boldsymbol{x}) = \lambda(\boldsymbol{A}\boldsymbol{x}) = \lambda\boldsymbol{0} = \boldsymbol{0}\,,$$

i.e., $\lambda\boldsymbol{x}$ is also a solution. □

As $L(\boldsymbol{A}, \boldsymbol{0})$ is a subspace, two questions arise. What can be said about the dimension of the solution space and how can one find a basis of the solution space? We firstly discuss the dimension.

Theorem 9.8: Dimension formula

Let $\boldsymbol{A}$ be an $m \times n$ matrix. Then

$$n = \operatorname{rk} \boldsymbol{A} + \dim L(\boldsymbol{A}, \boldsymbol{0})\,.$$

For the dimension of the solution space, we have

$$\boxed{\dim L(\boldsymbol{A}, \boldsymbol{0}) = n - \operatorname{rk} \boldsymbol{A}\,.}$$

The solution to the homogeneous linear equation is unique if the zero vector of $\mathbb{R}^n$ is the only solution. Then $L(\boldsymbol{A}, \boldsymbol{0}) = \{\boldsymbol{0}\}$, i.e., $\dim L(\boldsymbol{A}, \boldsymbol{0}) = 0$. Because of the dimension formula, this is equivalent to $n = \operatorname{rk} \boldsymbol{A}$. We formulate this important result as a theorem.

Theorem 9.9: Uniqueness of the solution to a homogeneous equation

Let $\boldsymbol{A}$ be an $m \times n$ matrix. Then the following three statements are equivalent:

(i) The solution to the homogeneous linear equation $\boldsymbol{Ax} = \boldsymbol{0}$ is unique.

(ii) The rank of $\boldsymbol{A}$ is equal to the number of columns of $\boldsymbol{A}$, i.e., $\operatorname{rk} \boldsymbol{A} = n$.

(iii) The columns of $\boldsymbol{A}$ are linearly independent.

A basis of the general solution to a homogeneous linear equation, is easily found using Gauss-Jordan elimination as described in Chapter 8. As the right-hand side vector $\boldsymbol{b} = \boldsymbol{0}$ remains unchanged under elementary row operations, it is omitted in the tableaux. Assume, e.g., that the final tableau is

x_1	x_2	x_3	x_4	x_5
1	0	0	2	1
0	1	0	1	0
0	0	1	1	2

Then, the rank of the matrix $\boldsymbol{A}$ is three. It follows from the dimension formula that the dimension of the solution space is two. As can be easily verified, the two vectors

$$\begin{bmatrix} 2 \\ 1 \\ 1 \\ -1 \\ 0 \end{bmatrix} \quad \text{and} \quad \begin{bmatrix} 1 \\ 0 \\ 2 \\ 0 \\ -1 \end{bmatrix}$$

are solutions to the linear equation. Moreover, they are linearly independent and therefore, since $\dim L(\boldsymbol{A}, \boldsymbol{0}) = 2$, form a basis of the solution space. These basis vectors are obtained by appending a negative identity matrix below the columns in the final tableau where no elimination step could be performed:

x_1	x_2	x_3	x_4	x_5
1	0	0	2	1
0	1	0	1	0
0	0	1	1	2
			−1	0
			0	−1

Using these basis vectors, the solution space is written as

$$L(\boldsymbol{A}, \boldsymbol{0}) = \left\{ \lambda \begin{bmatrix} 2 \\ 1 \\ 1 \\ -1 \\ 0 \end{bmatrix} + \mu \begin{bmatrix} 1 \\ 0 \\ 2 \\ 0 \\ -1 \end{bmatrix} : \lambda, \mu \in \mathbb{R} \right\}.$$

This representation is called a **general solution in basis representation**.

Solution to a nonhomogeneous linear equation

A linear equation $\boldsymbol{A}\boldsymbol{x} = \boldsymbol{b}$, where the right-hand side vector $\boldsymbol{b}$ is not the zero vector, i.e., $\boldsymbol{b} \neq \boldsymbol{0}$, is called **nonhomogeneous** linear equation. We are again interested in the **general solution** $L(\boldsymbol{A}, \boldsymbol{b}) = \{\boldsymbol{x} \in \mathbb{R}^n : \boldsymbol{A}\boldsymbol{x} = \boldsymbol{b}\}$. Any element $\boldsymbol{x}^p \in L(\boldsymbol{A}, \boldsymbol{b})$ is called a **particular solution**.

Theorem 9.10: General solution to a nonhomogeneous equation

Let $\boldsymbol{A}$ be an $m \times n$ matrix, $\boldsymbol{b} \neq \boldsymbol{0}$ and $\boldsymbol{x}^p \in L(\boldsymbol{A}, \boldsymbol{b})$ a particular solution to the nonhomogeneous equation. Then the general solution to the nonhomogeneous equation is given by

$$L(\boldsymbol{A}, \boldsymbol{b}) = \{\boldsymbol{x}^p + \boldsymbol{z} : \boldsymbol{z} \in L(\boldsymbol{A}, \boldsymbol{0})\},$$

i.e., every solution to $\boldsymbol{A}\boldsymbol{x} = \boldsymbol{b}$ is obtained as the sum of the particular solution $\boldsymbol{x}^p$ and a solution $\boldsymbol{z}$ to the homogeneous linear equation $\boldsymbol{A}\boldsymbol{z} = \boldsymbol{0}$.

Proof: If $\boldsymbol{z}$ is a solution to the homogeneous equation, then

$$\boldsymbol{A}(\boldsymbol{x}^p + \boldsymbol{z}) = \boldsymbol{A}\boldsymbol{x}^p + \boldsymbol{A}\boldsymbol{z} = \boldsymbol{b} + \boldsymbol{0} = \boldsymbol{b}.$$

Hence, every vector of the form $\boldsymbol{x}^p + \boldsymbol{z}$ is a solution to the nonhomogeneous equation.

Conversely, let $\boldsymbol{x}$ be an arbitrary solution to the nonhomogeneous equation. Define $\boldsymbol{z} = \boldsymbol{x} - \boldsymbol{x}^p$. Then,

$$\boldsymbol{A}\boldsymbol{z} = \boldsymbol{A}(\boldsymbol{x} - \boldsymbol{x}^p) = \boldsymbol{A}\boldsymbol{x} - \boldsymbol{A}\boldsymbol{x}^p = \boldsymbol{b} - \boldsymbol{b} = \boldsymbol{0},$$

which shows that $\boldsymbol{z}$ is a solution to the homogeneous equation. Hence, every solution $\boldsymbol{x}$ can be expressed as $\boldsymbol{x} = \boldsymbol{x}^p + \boldsymbol{z}$, i.e., the sum of a particular solution $\boldsymbol{x}^p$ and a solution to the homogeneous equation $\boldsymbol{z}$. □

If $\boldsymbol{x}^p$ is a particular solution to the nonhomogeneous equation and $\{\boldsymbol{v}_1, \ldots, \boldsymbol{v}_k\}$ is a basis of the solution space of the homogeneous equation, then the general solution to the nonhomogeneous equation is written in the form

$$L(\boldsymbol{A}, \boldsymbol{b}) = \{\boldsymbol{x}^p + \lambda_1 \boldsymbol{v}_1 + \ldots + \lambda_k \boldsymbol{v}_k : \lambda_1, \ldots, \lambda_k \in \mathbb{R}\}.$$

As with the homogeneous equation, this representation is called a **general solution in basis representation**.

We now address the question, under which conditions a linear equation $\boldsymbol{A}\boldsymbol{x} = \boldsymbol{b}$ has a solution $\boldsymbol{x} \in \mathbb{R}^n$ and under which additional conditions this solution is unique, i.e., the solution space $L(\boldsymbol{A}, \boldsymbol{b})$ is a singleton. Denote again by $\boldsymbol{a}_1, \ldots, \boldsymbol{a}_n$ the columns of the matrix $\boldsymbol{A}$. Then, the left-hand side of the equation can be written as:

$$\boldsymbol{A}\boldsymbol{x} = x_1 \boldsymbol{a}_1 + \ldots + x_n \boldsymbol{a}_n.$$

Thus, a solution exists if and only if there exist real numbers $x_1, \ldots, x_n$ such that

$$x_1\boldsymbol{a}_1 + \ldots + x_n\boldsymbol{a}_n = \boldsymbol{b},$$

that is, $\boldsymbol{b}$ is a linear combination of the columns of $\boldsymbol{A}$. However, this is equivalent to the rank of $\boldsymbol{A}$ being equal to the rank of the augmented matrix $[\boldsymbol{A}, \boldsymbol{b}]$.

Under what conditions is the solution to a nonhomogeneous linear equation unique? According to Theorem 9.10, the solution is unique if and only if the solution to the associated homogeneous equation is unique. We have seen in Theorem 9.9, that this is the case if and only if the rank of $\boldsymbol{A}$ is equal to the number n of columns of $\boldsymbol{A}$. Summarizing, we obtain the following theorem:

Theorem 9.11: Solution criteria for nonhomogeneous linear equations

Let $\boldsymbol{A}$ be an $m \times n$ matrix and $\boldsymbol{b} \in \mathbb{R}^m$.

(i) The linear equation $\boldsymbol{A}\boldsymbol{x} = \boldsymbol{b}$ has at least one solution if and only if the rank of the augmented matrix $[\boldsymbol{A}, \boldsymbol{b}]$ is equal to the rank of the matrix $\boldsymbol{A}$:

$$L(\boldsymbol{A}, \boldsymbol{b}) \neq \emptyset \quad \Longleftrightarrow \quad \operatorname{rk} \boldsymbol{A} = \operatorname{rk}[\boldsymbol{A}, \boldsymbol{b}].$$

(ii) The linear equation $\boldsymbol{A}\boldsymbol{x} = \boldsymbol{b}$ has a **unique** solution if and only if the rank of the augmented matrix $[\boldsymbol{A}, \boldsymbol{b}]$ and the rank of $\boldsymbol{A}$ are equal to the number n of unknowns:

$$L(\boldsymbol{A}, \boldsymbol{b}) = \{\boldsymbol{x}\} \quad \Longleftrightarrow \quad \operatorname{rk} \boldsymbol{A} = \operatorname{rk}[\boldsymbol{A}, \boldsymbol{b}] = n.$$

Existence of the inverse matrix

To conclude this section, we provide a criterion to determine whether a given square matrix $\boldsymbol{A}$ is regular or singular.

Theorem 9.12: Regularity criterion

Let $\boldsymbol{A}$ be an $n \times n$ matrix. Then:

- If $\operatorname{rk} \boldsymbol{A} = n$, then $\boldsymbol{A}$ is regular.
- If $\operatorname{rk} \boldsymbol{A} < n$, then $\boldsymbol{A}$ is singular.

Proof: If $\operatorname{rk} \boldsymbol{A} = n$, then the linear equations $\boldsymbol{A}\boldsymbol{x} = \boldsymbol{e}_i$, $i = 1, \ldots, n$, and thus also the equation $\boldsymbol{A}\boldsymbol{X} = \boldsymbol{I}_n$ have unique solutions, i.e., the inverse $\boldsymbol{A}^{-1}$ exists. In other words, $\boldsymbol{A}$ is regular.

Conversely, if $\operatorname{rk} \boldsymbol{A} < n$, then the homogeneous linear equation $\boldsymbol{A}\boldsymbol{x} = \boldsymbol{0}$ has a solution $\boldsymbol{x} \neq \boldsymbol{0}$. If $\boldsymbol{A}$ had an inverse, then

$$\boldsymbol{x} = \boldsymbol{I}_n\boldsymbol{x} = (\boldsymbol{A}^{-1}\boldsymbol{A})\boldsymbol{x} = \boldsymbol{A}^{-1}(\boldsymbol{A}\boldsymbol{x}) = \boldsymbol{A}^{-1}\boldsymbol{0} = \boldsymbol{0},$$

which contradicts $\boldsymbol{x} \neq \mathbf{0}$. Thus, $\boldsymbol{A}$ cannot have an inverse, i.e., $\boldsymbol{A}$ is singular. □

9.7 Vector spaces

The elements of $\mathbb{R}^n$ are vectors. For vectors (as special cases of matrices), the rules from Theorem 2.1 apply. However, these rules do not only apply to the elements of $\mathbb{R}^n$. There are many other examples of sets on which the operations of addition and scalar multiplication are defined and for which the above rules apply. Any such set is called a **vector space**.

Definition 9.5: Vector space

A **vector space** is a set V with two operations "+": $V \times V \to \mathbb{R}$ ("addition") and "·": $\mathbb{R} \times V \to V$ ("scalar multiplication"), for which the following rules apply:

(i) $\boldsymbol{x} + \boldsymbol{y} = \boldsymbol{y} + \boldsymbol{x}$ for all $\boldsymbol{x}, \boldsymbol{y} \in V$.

(ii) $(\boldsymbol{x} + \boldsymbol{y}) + \boldsymbol{z} = \boldsymbol{x} + (\boldsymbol{y} + \boldsymbol{z})$ for all $\boldsymbol{x}, \boldsymbol{y}, \boldsymbol{z} \in V$.

(iii) There is an element $\mathbf{0} \in V$, such that $\boldsymbol{x} + \mathbf{0} = \boldsymbol{x}$ for all $\boldsymbol{x} \in V$.

(iv) For each $\boldsymbol{x} \in V$ there is $-\boldsymbol{x} \in V$, such that $\boldsymbol{x} + (-\boldsymbol{x}) = \mathbf{0}$.

(v) $\lambda(\boldsymbol{x} + \boldsymbol{y}) = \lambda\boldsymbol{x} + \lambda\boldsymbol{y}$ for all $\boldsymbol{x}, \boldsymbol{y} \in V$ and $\lambda \in \mathbb{R}$.

(vi) $(\lambda + \mu)\boldsymbol{x} = \lambda\boldsymbol{x} + \mu\boldsymbol{x}$ for all $\boldsymbol{x} \in V$ and $\lambda, \mu \in \mathbb{R}$.

(vii) $\lambda(\mu\boldsymbol{x}) = (\lambda\mu)\boldsymbol{x}$ for all $\boldsymbol{x} \in V$ and $\lambda, \mu \in \mathbb{R}$.

(viii) $1 \cdot \boldsymbol{x} = \boldsymbol{x}$ for all $\boldsymbol{x} \in V$.

The element $\mathbf{0}$ from (iii) is referred to as **zero element** or **neutral element**. The element $-\boldsymbol{x}$ is called **negative element** of $\boldsymbol{x}$. All concepts of this chapter, such as linear independence, basis, or dimension, apply in the same way to general vector spaces.

We have already seen two examples of vector spaces:

Example 9.14. The set of all vectors in $\mathbb{R}^n$, together with vector addition and scalar multiplication, forms a vector space. The zero element is the zero vector $\mathbf{0}_n$.

Example 9.15. The set of all $m \times n$ matrices, together with matrix addition and scalar multiplication, forms a vector space. The zero element is the zero matrix $\mathbf{0}_{m,n}$.

Let us now consider another important example of a vector space: Let D be a subset of $\mathbb{R}^n$ and $\mathcal{F}(D, \mathbb{R})$ the set of all real-valued functions $f : D \to \mathbb{R}$ with domain D. We have seen in Chapter 1 that two functions with a common domain can be added.

Similarly, a real-valued function can be multiplied by a real number λ:

$$(\lambda f) : D \to \mathbb{R}, \quad x \mapsto \lambda f(x)\,.$$

The zero element in the set of all functions $f : D \to \mathbb{R}$ is the **zero function**

$$\mathbf{0} : D \to \mathbb{R}, \quad x \mapsto 0\,,$$

which has at any argument has function value zero. The negative function to f is $-f : D \to \mathbb{R}, x \mapsto -f(x)$. It is easy to verify that for $\mathcal{F}(D, \mathbb{R})$ the conditions from Definition 9.5 are met.

Theorem 9.13: Function space

The set $\mathcal{F}(D, \mathbb{R})$ of all functions with domain D and codomain $\mathbb{R}$ forms a vector space.

A subset U of a vector space is called **subspace**, if U itself is a vector space. It is easy to check whether a subset U of a vector space is a subspace. U is a subspace if and only if the properties **U0**, **U1** and **U2** (cf. Section 9.2) hold:

U0: $U \neq \emptyset$.

U1: If $\boldsymbol{x}$ and $\boldsymbol{y}$ are in U, then their sum $\boldsymbol{x} + \boldsymbol{y}$ is also in U.

U2: If $\boldsymbol{x}$ is in U, then for every real number λ, $\lambda\boldsymbol{x}$ is also in U.

Important subspaces of $\mathcal{F}(D, \mathbb{R})$ are

- the set of all bounded functions from D to $\mathbb{R}$,
- the set of all continuous functions from D to $\mathbb{R}$,
- the set of all differentiable functions from D to $\mathbb{R}$.

Example 9.16 (Vector space of polynomials). The set $\mathcal{P}(\mathbb{R}, \mathbb{R})$ of all **polynomials** forms a subspace of $\mathcal{F}(\mathbb{R}, \mathbb{R})$. Unlike the function spaces discussed above, the vector space of all polynomials admits a straightforward basis, namely $\{1, x, x^2, x^3, \ldots\}$. These functions are linearly independent, as no linear combination of them results in the zero function. At the same time, every polynomial can be represented as a linear combination of these functions. The space of polynomials (and likewise all previous function spaces) is therefore an infinite-dimensional vector space, since the basis contains infinitely many elements.

In Chapters 12 and 13 we will get to know more examples of function spaces. There it will turn out that the solution sets of linear homogeneous differential equations and linear homogeneous difference equations form vector spaces.

Important terms for review

After reading this chapter, the following terms should be familiar:

- Linear combination, linear hull, linear span
- Linear subspace
- Linear dependence or independence
- Orthonormal system
- Basis, canonical basis, orthonormal basis
- Dimension
- Rank of a matrix
- Homogeneous and nonhomogeneous linear equations
- Dimension formula
- General solution in basis representation
- Vector space

Self-test

You can check your understanding of this chapter using the following true-or-false questions. For each statement, indicate whether it is true (T) or false (F). Short solutions to these questions can be found in Appendix H.

I. Mark each of the following statements about linear independence as true or false.

	T	F
One can find four vectors in $\mathbb{R}^3$ that are linearly independent.	☐	☐
$[1,0,0]^T, [1,1,0]^T, [1,1,1]^T$ are linearly independent.	☐	☐
$[1,0,1]^T, [2,0,1]^T, [1,0,2]^T$ are linearly independent.	☐	☐
If three vectors lie in a plane in $\mathbb{R}^3$, then they are linearly independent.	☐	☐

II. Decide whether the following statements about the rank of a matrix are true or false.

	T	F
If the columns of a matrix are linearly independent, then the matrix has full rank.	☐	☐
If $\boldsymbol{A}$ is not symmetric, then $\operatorname{rk} \boldsymbol{A} \neq \operatorname{rk} \boldsymbol{A}^T$.	☐	☐
An equation $\boldsymbol{A}\boldsymbol{x} = \boldsymbol{b}$ has a solution if $\boldsymbol{A}$ has the same rank as $[\boldsymbol{A}, \boldsymbol{b}]$.	☐	☐

III. Are the following statements true or false?

	T	F
The linear span of $[1,0,0]^T$ and $[0,1,0]^T$ is a plane in $\mathbb{R}^3$.	☐	☐
With respect to the multiplication of square matrices, the identity matrix is the neutral element.	☐	☐
The set of all square matrices with n rows, together with matrix addition and scalar multiplication, forms a vector space.	☐	☐

Exercises

Exercise 1
Check whether the given vectors are linearly independent and compute their span:

$$\text{a)}\quad \begin{bmatrix}1\\1\\1\\1\end{bmatrix}, \begin{bmatrix}2\\2\\2\\0\end{bmatrix}, \begin{bmatrix}3\\3\\0\\0\end{bmatrix}, \begin{bmatrix}4\\0\\0\\0\end{bmatrix}, \qquad \text{b)}\quad \begin{bmatrix}1\\0\\1\\0\end{bmatrix}, \begin{bmatrix}2\\2\\0\\0\end{bmatrix}, \begin{bmatrix}0\\3\\0\\0\end{bmatrix}.$$

Exercise 2
Consider the following vectors $\boldsymbol{x}, \boldsymbol{y}, \boldsymbol{z}$:

a) $\boldsymbol{x} = [0,2,1]^T$, $\boldsymbol{y} = [1,3,3]^T$, and $\boldsymbol{z} = [1,3,0]^T$,

b) $\boldsymbol{x} = [1,2,7]^T$, $\boldsymbol{y} = [0,2,3]^T$, and $\boldsymbol{z} = [2,0,8]^T$,

c) $\boldsymbol{x} = [1,0,1]^T$, $\boldsymbol{y} = [4,1,-2]^T$, and $\boldsymbol{z} = [9,3,-6]^T$.

Are the vectors linearly independent?

Exercise 3
For which values $a, b, c \in \mathbb{R}$ are the vectors $\boldsymbol{x}, \boldsymbol{y}, \boldsymbol{z}$ linearly independent?

a) $\boldsymbol{x} = [a,0,0]^T$, $\boldsymbol{y} = [b,b,0]^T$, and $\boldsymbol{z} = [c,c,c]^T$,

b) $\boldsymbol{x} = [a,0,-c]^T$, $\boldsymbol{y} = [a,b,0]^T$, and $\boldsymbol{z} = [0,b,c]^T$.

Exercise 4
Consider the following vectors in $\mathbb{R}^4$:

$$\boldsymbol{x} = \begin{bmatrix}1\\-2\\0\\1\end{bmatrix}, \quad \boldsymbol{y} = \begin{bmatrix}0\\0\\2\\5\end{bmatrix}, \quad \boldsymbol{z} = \begin{bmatrix}-2\\4\\2\\3\end{bmatrix}.$$

a) Compute the dimension of $\mathrm{sp}(\boldsymbol{x}, \boldsymbol{y}, \boldsymbol{z})$.

b) Find a basis of $\mathrm{sp}(\boldsymbol{x}, \boldsymbol{y}, \boldsymbol{z})$.

c) If the vectors are linearly dependent, represent one of the three vectors as a linear combination of the other vectors.

Exercise 5
Consider the following matrices:

$$A = \begin{bmatrix} 1 & 0 & 1 & -1 \\ 1 & 2 & 1 & 0 \\ 1 & 2 & 3 & -4 \\ 2 & 0 & 2 & 1 \end{bmatrix}, \qquad B = \begin{bmatrix} 1 & 0 & 3 & -1 & 2 \\ 3 & -2 & 1 & 0 & -1 \\ -1 & 2 & 5 & -2 & 5 \end{bmatrix}.$$

a) Determine the rank of both matrices.

b) Find the dimensions of the solution spaces of the homogeneous linear equations $\boldsymbol{Ax} = \boldsymbol{0}$ and $\boldsymbol{Bx} = \boldsymbol{0}$.

c) Find for each homogeneous linear equation, $\boldsymbol{Ax} = \boldsymbol{0}$ and $\boldsymbol{Bx} = \boldsymbol{0}$, a general solution in basis representation.

Chapter 10

Determinants and eigenvalues of matrices

To every square matrix there is assigned a real number, the so-called determinant. Determinants have many diverse applications, among them checking, whether a matrix has an inverse, and finding the eigenvalues and eigenvectors of a matrix. Furthermore, the sufficient conditions in constrained as well as unconstrained optimization (see Chapter 6) can be formulated in terms of determinants. Moreover, determinants are closely related to the concept of volume and therefore important in the integration of multivariate functions.

10.1 Determinants

The determinant of a 2×2 matrix $\boldsymbol{A}$ is defined as follows

$$\det \boldsymbol{A} = \det \begin{bmatrix} a_{11} & a_{12} \\ a_{21} & a_{22} \end{bmatrix} = a_{11}a_{22} - a_{12}a_{21} ,$$

that is, the determinant of a 2×2 matrix is the product of the main diagonal elements minus the product of the off-diagonal elements.

Example 10.1. The 2×2 matrix

$$\boldsymbol{A} = \begin{bmatrix} 2 & 1 \\ -1 & 3 \end{bmatrix}$$

has the determinant

$$\det \boldsymbol{A} = 2 \cdot 3 - 1 \cdot (-1) = 7 .$$

K. Mosler et al., *Mathematical Methods for Business and Economics*, Mathematics Study Resources 20, https://doi.org/10.1007/978-3-662-73831-3_10

Geometrically, the (absolute value of the) determinant can be interpreted as follows (Figure 10.1): Consider the linear function $f_{\boldsymbol{A}} : \boldsymbol{x} \mapsto \boldsymbol{A}\boldsymbol{x}$; it maps the unit square $[0,1]^2$ to the parallelogram spanned by the vectors $[2,-1]^T$ and $[1,3]^T$. The absolute value of the determinant of $\boldsymbol{A}$ is equal to the area of this parallelogram.

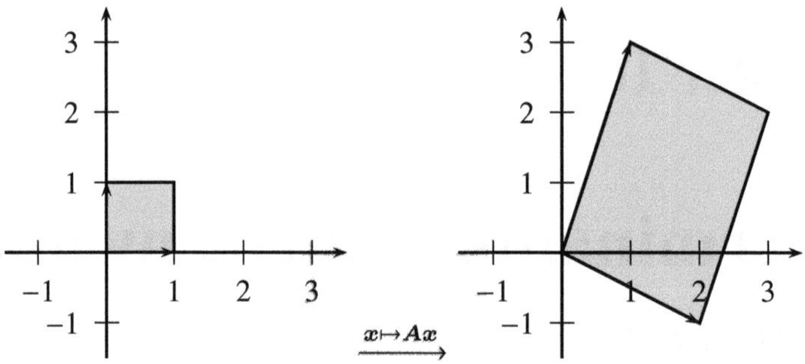

Figure 10.1: Interpretation of the determinant.

Hence, the image of the unit square under this linear transformation has seven times the area of the unit square.

It can be shown that the same holds for any subset of $\mathbb{R}^2$: If $\boldsymbol{A}$ is a 2×2 matrix and S is any subset[1] of $\mathbb{R}^2$, then the area of the image $f_{\boldsymbol{A}}(S)$ of S is equal to $|\det \boldsymbol{A}|$ times the area of S.

Analogous statements also apply to higher dimensions: If $\boldsymbol{A}$ is a 3×3 matrix and S is a subset of $\mathbb{R}^3$, then the volume of the set $f_{\boldsymbol{A}}(S)$ is equal to $|\det \boldsymbol{A}|$ times the volume of S. Also in dimension $n > 3$ the determinant can be interpreted in this way.

To define the determinant of an arbitrary $n \times n$ matrix $\boldsymbol{A}$, we need some notation. Let $\boldsymbol{A}_{ij}$ denote the matrix obtained by deleting the i-th row and the j-th column from $\boldsymbol{A}$.

Example 10.2. For

$$\boldsymbol{A} = \begin{bmatrix} 1 & 2 & 3 \\ 4 & 5 & 6 \\ 7 & 8 & 9 \end{bmatrix},$$

we obtain

$$\boldsymbol{A}_{11} = \begin{bmatrix} 5 & 6 \\ 8 & 9 \end{bmatrix}, \quad \boldsymbol{A}_{12} = \begin{bmatrix} 4 & 6 \\ 7 & 9 \end{bmatrix} \text{ and } \boldsymbol{A}_{23} = \begin{bmatrix} 1 & 2 \\ 7 & 8 \end{bmatrix}.$$

The determinant of an arbitrary square matrix is defined as follows.

[1]More precisely, one should say "measurable subset", since there are subsets of $\mathbb{R}^2$ to which no area can be assigned. However, this goes far beyond the scope of this book.

Definition 10.1: Determinant of a matrix

Let $\boldsymbol{A}$ be an $n \times n$ matrix. The **determinant** of a 1×1 matrix is given by

$$\det[a_{11}] = a_{11} .$$

For $n > 1$, the **determinant** of an $n \times n$ matrix is defined by

$$\begin{aligned}\det \boldsymbol{A} &= a_{11} \det \boldsymbol{A}_{11} - a_{12} \det \boldsymbol{A}_{12} + \ldots + (-1)^{n+1} a_{1n} \det \boldsymbol{A}_{1n} \\ &= \sum_{j=1}^{n} (-1)^{1+j} a_{1j} \det \boldsymbol{A}_{1j} .\end{aligned}$$

Note that the terms in the above formula have alternating signs.

In the following, instead of "determinant of an $n \times n$ matrix" we simply say "$n \times n$ determinant".

Our definition of the determinant is a so-called **recursive definition**. The calculation of the determinant of an $n \times n$ matrix is reduced to the calculation of $(n-1) \times (n-1)$ determinants. The calculation of these determinants can then be reduced in the same way to the calculation of $(n-2) \times (n-2)$ determinants, and so on. Ultimately, this leads to the calculation of 1×1 determinants, which we already know how to calculate.

In particular, for 2×2 determinants, the formula above is obtained,

$$\det \begin{bmatrix} a_{11} & a_{12} \\ a_{21} & a_{22} \end{bmatrix} = a_{11}a_{22} - a_{12}a_{21} .$$

For 3×3 determinants, the following formula is obtained:

$$\begin{aligned}\det &\begin{bmatrix} a_{11} & a_{12} & a_{13} \\ a_{21} & a_{22} & a_{23} \\ a_{31} & a_{32} & a_{33} \end{bmatrix} \\ &= a_{11} \cdot \det \begin{bmatrix} a_{22} & a_{23} \\ a_{32} & a_{33} \end{bmatrix} - a_{12} \cdot \det \begin{bmatrix} a_{21} & a_{23} \\ a_{31} & a_{33} \end{bmatrix} + a_{13} \cdot \det \begin{bmatrix} a_{21} & a_{22} \\ a_{31} & a_{32} \end{bmatrix} \\ &= a_{11}a_{22}a_{33} - a_{11}a_{23}a_{32} - a_{12}a_{21}a_{33} \\ &\quad + a_{12}a_{23}a_{31} + a_{13}a_{21}a_{32} - a_{13}a_{22}a_{31} .\end{aligned}$$

The determinant of a 3×3 matrix $\boldsymbol{A}$ can be conveniently calculated using the so-called **rule of Sarrus**[2]: The first two columns of the matrix are replicated at its right side:

$$\begin{array}{ccc|cc} a_{11} & a_{12} & a_{13} & a_{11} & a_{12} \\ a_{21} & a_{22} & a_{23} & a_{21} & a_{22} \\ a_{31} & a_{32} & a_{33} & a_{31} & a_{32} \end{array} .$$

[2]Pierre Frédéric Sarrus (1798–1861)

Now, for each of the six diagonals (three from top left to bottom right, three from bottom left to top right), one computes the product of the three diagonal elements. Then, the products of the three diagonals from top left to bottom right are summed up, and the three other products are subtracted from this sum, which results in the same formula as above:

$$\begin{aligned}\det \boldsymbol{A} &= a_{11}a_{22}a_{33} + a_{12}a_{23}a_{31} + a_{13}a_{21}a_{32} \\ &\quad - a_{31}a_{22}a_{13} - a_{32}a_{23}a_{11} - a_{33}a_{21}a_{12} \,.\end{aligned}$$

Example 10.3. For the matrix

$$\boldsymbol{A} = \begin{bmatrix} 1 & 0 & 5 \\ 0 & 2 & 7 \\ -1 & 2 & 3 \end{bmatrix}$$

one gets using the rule of Sarrus

$$\det \boldsymbol{A} = 1\cdot 2\cdot 3 + 0\cdot 7\cdot(-1) + 5\cdot 0\cdot 2 - (-1)\cdot 2\cdot 5 - 2\cdot 7\cdot 1 - 3\cdot 0\cdot 0 = 6 + 10 - 14 = 2\,.$$

However, note that the rule of Sarrus works *only for* 3×3 *determinants.* Determinants of larger matrices ($n > 3$) must be calculated recursively from the determinants of smaller matrices or by other methods discussed below.

The formula in Definition 10.1 is referred to as **expansion** of the determinant along the first row. Similarly, the determinant can be expanded along *any other row.*

Theorem 10.1: Expansion along the i-th row

Let $\boldsymbol{A}$ be an $n \times n$ matrix and $1 \le i \le n$. Then,

$$\det \boldsymbol{A} = \sum_{j=1}^{n} (-1)^{i+j} a_{ij} \det \boldsymbol{A}_{ij}\,.$$

Furthermore, it holds:

Theorem 10.2: Determinant of the transpose

For every square matrix $\boldsymbol{A}$ it holds

$$\det \boldsymbol{A} = \det \boldsymbol{A}^T\,.$$

It follows that a determinant can as well be expanded along a column of $\boldsymbol{A}$, i.e., a row of $\boldsymbol{A}^T$.

Theorem 10.3: Expansion along the j-th column

Let $\boldsymbol{A}$ be an $n \times n$ matrix and $1 \le j \le n$. Then,

$$\det \boldsymbol{A} = \sum_{i=1}^{n} (-1)^{i+j} a_{ij} \det \boldsymbol{A}_{ij}\,.$$

It is advisable to expand a determinant along a row or column that contains as many zeros as possible, as this simplifies the calculation. We illustrate this by an example.

Example 10.4. For

$$\boldsymbol{A} = \begin{bmatrix} 4 & 0 & 1 & 0 \\ 2 & 1 & 2 & 5 \\ 0 & 2 & 0 & 1 \\ 5 & 0 & 2 & 0 \end{bmatrix}$$

one expands the determinant along the first row and obtains

$$\det \boldsymbol{A} = 4 \cdot \det \begin{bmatrix} 1 & 2 & 5 \\ 2 & 0 & 1 \\ 0 & 2 & 0 \end{bmatrix} + 1 \cdot \det \begin{bmatrix} 2 & 1 & 5 \\ 0 & 2 & 1 \\ 5 & 0 & 0 \end{bmatrix}.$$

The two 3×3 determinants are each expanded along the third row.

$$\det \begin{bmatrix} 1 & 2 & 5 \\ 2 & 0 & 1 \\ 0 & 2 & 0 \end{bmatrix} = (-2) \cdot \det \begin{bmatrix} 1 & 5 \\ 2 & 1 \end{bmatrix} = (-2) \cdot (1 \cdot 1 - 2 \cdot 5) = 18\,,$$

$$\det \begin{bmatrix} 2 & 1 & 5 \\ 0 & 2 & 1 \\ 5 & 0 & 0 \end{bmatrix} = 5 \cdot \det \begin{bmatrix} 1 & 5 \\ 2 & 1 \end{bmatrix} = 5 \cdot (1 \cdot 1 - 2 \cdot 5) = -45\,.$$

This gives

$$\det \boldsymbol{A} = 4 \cdot 18 - 45 = 27\,.$$

In the above example, the calculation of the determinant is relatively simple, because the matrix $\boldsymbol{A}$ contained many zeros. The determinant can be calculated particularly simply if the matrix $\boldsymbol{A}$ is triangular:

Theorem 10.4: Determinant of a triangular matrix

Let $\boldsymbol{A}$ be (upper or lower) triangular. Then the determinant of $\boldsymbol{A}$ is equal to the product of the main diagonal elements,

$$\det \boldsymbol{A} = a_{11} \cdot a_{22} \cdot \ldots \cdot a_{nn} = \prod_{i=1}^{n} a_{ii}\,.$$

In particular, the above theorem applies to diagonal matrices. Specifically, $\det \boldsymbol{I}_n = 1$ and $\det \lambda \boldsymbol{I}_n = \lambda^n$.

Since the determinant of a triangular matrix is simple to calculate, it is reasonable to transform a given matrix into triangular form by elementary row (or column) operations, then calculate the determinant of the transformed matrix and finally conclude the determinant of the original matrix. However, to do so one must know how the determinant of a matrix changes with elementary row operations.

Theorem 10.5: Determinants and elementary operations

(i) If $\widetilde{\boldsymbol{A}}$ results from $\boldsymbol{A}$ by swapping two rows, then $\det \widetilde{\boldsymbol{A}} = -\det \boldsymbol{A}$.

(ii) If $\widetilde{\boldsymbol{A}}$ results from $\boldsymbol{A}$ by adding or subtracting a multiple of a row to another row, then $\det \widetilde{\boldsymbol{A}} = \det \boldsymbol{A}$.

(iii) If $\widetilde{\boldsymbol{A}}$ results from $\boldsymbol{A}$ by multiplying a row by $\lambda \in \mathbb{R}$, then $\det \widetilde{\boldsymbol{A}} = \lambda \det \boldsymbol{A}$.

The above statements are still valid when 'row' is replaced by 'column'.

This theorem enables calculating a determinant by using Gauss-Jordan elimination. Unlike standard Gauss-Jordan elimination, the goal here is to transform the matrix into triangular form rather than into the identity matrix. This is because the determinant of a triangular matrix is as easy to calculate as that of a diagonal matrix. Transformation into a triangular matrix significantly reduces the computational effort. Furthermore, it is recommended to use only the elementary row operations **ERO 1** and **ERO 2**. Using only these elementary row operations it is still possible to transform a matrix into triangular form, albeit without ones on the main diagonal. We illustrate the procedure with an example:

Example 10.5. The determinant of the following matrix is to be calculated:

$$\boldsymbol{A} = \begin{bmatrix} 2 & 2 & 4 & 6 \\ 1 & 1 & 3 & 7 \\ 4 & 6 & 2 & 9 \\ 2 & 0 & 6 & 1 \end{bmatrix}.$$

We compute the determinant using Gauss-Jordan elimination. For simplicity, zeros below the main diagonal are eliminated:

$$\begin{bmatrix} 2 & 2 & 4 & 6 \\ 1 & 1 & 3 & 7 \\ 4 & 6 & 2 & 9 \\ 2 & 0 & 6 & 1 \end{bmatrix} \xrightarrow[\substack{\text{II}-(1/2)\cdot\text{I} \\ \text{III}-2\cdot\text{I} \\ \text{IV}-\text{I}}]{} \begin{bmatrix} 2 & 2 & 4 & 6 \\ & 0 & 1 & 4 \\ & 2 & -6 & -3 \\ & -2 & 2 & -5 \end{bmatrix}$$

$$\xrightarrow[\text{II}\leftrightarrow\text{III}]{} \begin{bmatrix} 2 & 2 & 4 & 6 \\ & 2 & -6 & -3 \\ & 0 & 1 & 4 \\ & -2 & 2 & -5 \end{bmatrix}$$

$$\xrightarrow[\text{IV+II}]{} \begin{bmatrix} 2 & 2 & 4 & 6 \\ & 2 & -6 & -3 \\ & & 1 & 4 \\ & & -4 & -8 \end{bmatrix}$$

$$\xrightarrow[\text{IV+4·III}]{} \begin{bmatrix} 2 & 2 & 4 & 6 \\ & 2 & -6 & -3 \\ & & 1 & 4 \\ & & & 8 \end{bmatrix}$$

The product of the main diagonal elements in the final tableau is 32. Since *one row swap* was performed, $\det \boldsymbol{A} = (-1)^1 \cdot 32 = -32$.

If $\boldsymbol{A}$ is regular, $\boldsymbol{A}$ can be transformed into a identity matrix using Gauss-Jordan elimination. In this case, the determinant must be non-zero. Conversely, if $\boldsymbol{A}$ is singular, a zero row will appear during Gauss-Jordan elimination. The determinant of a matrix that has a zero row is zero. (Expansion along the zero row!) This yields the following criterion for the invertibility of a matrix.

Theorem 10.6: Determinant and invertibility

Let $\boldsymbol{A}$ be an $n \times n$ matrix. Then

$$\boldsymbol{A} \text{ is invertible} \quad \Longleftrightarrow \quad \operatorname{rk} \boldsymbol{A} = n \quad \Longleftrightarrow \quad \det \boldsymbol{A} \neq 0\,.$$

We now provide some additional rules for determinants.

Theorem 10.7: Rules for determinants

Let $\boldsymbol{A}, \boldsymbol{B}$ be $n \times n$ matrices and $\lambda \in \mathbb{R}$. Then

(i) $\det(\lambda \boldsymbol{A}) = \lambda^n \det \boldsymbol{A}$,

(ii) $\det(\boldsymbol{AB}) = \det \boldsymbol{A} \cdot \det \boldsymbol{B}$,

(iii) If $\boldsymbol{A}$ is invertible, then $\det \boldsymbol{A}^{-1} = (\det \boldsymbol{A})^{-1}$.

A final remark concerns orthogonal matrices. If $\boldsymbol{Q}$ is orthogonal, then $\boldsymbol{Q}^T\boldsymbol{Q} = \boldsymbol{I}_n$. From Theorem 10.7, we obtain

$$1 = \det \boldsymbol{I}_n = \det(\boldsymbol{Q}^T\boldsymbol{Q}) = \det \boldsymbol{Q}^T \cdot \det \boldsymbol{Q} = (\det \boldsymbol{Q})^2\,,$$

and thus $\det \boldsymbol{Q}$ must be either $+1$ or -1.

Theorem 10.8: Determinant of an orthogonal matrix

If $\boldsymbol{Q}$ is an orthogonal matrix, then $|\det \boldsymbol{Q}| = 1$.

Orthogonal transformations thus not only preserve lengths and angles, but also volumes. With an orthogonal 2×2 matrix $\boldsymbol{A}$ (cf. Theorem 2.11), $f_{\boldsymbol{A}}$ describes either a rotation or a reflection of $\mathbb{R}^2$, and the determinant can be used to distinguish between the two: $f_{\boldsymbol{A}}$ is a rotation, if $\det \boldsymbol{A} = 1$, while it is a reflection if $\det \boldsymbol{A} = -1$.

At the end of this section, we return to the sufficient conditions for the existence of an extremum from Chapter 6. There, sufficient conditions have been stated only for $n = 2$ and $n = 3$. We now define these conditions for general n. For this we need the concept of **definiteness** of a matrix.

Definition 10.2: Definiteness

Let $\boldsymbol{A}$ be a symmetric $n \times n$ matrix. Then $\boldsymbol{A}$ is called

- **positive definite**, if $\boldsymbol{x}^T \boldsymbol{A}\boldsymbol{x} > 0$ for all $\boldsymbol{x} \neq \boldsymbol{0}$,
- **positive semidefinite**, if $\boldsymbol{x}^T \boldsymbol{A}\boldsymbol{x} \geq 0$ for all $\boldsymbol{x} \neq \boldsymbol{0}$,
- **negative definite**, if $\boldsymbol{x}^T \boldsymbol{A}\boldsymbol{x} < 0$ for all $\boldsymbol{x} \neq \boldsymbol{0}$,
- **negative semidefinite**, if $\boldsymbol{x}^T \boldsymbol{A}\boldsymbol{x} \leq 0$ for all $\boldsymbol{x} \neq \boldsymbol{0}$,
- **indefinite**, if $\boldsymbol{x}^T \boldsymbol{A}\boldsymbol{x}$ takes both values greater than zero and values less than zero.

In terms of definiteness of Hessian matrices, the sufficient conditions for the existence of an extremum turn out as follows:

Theorem 10.9: Sufficient condition for the existence of an extremum

Let $f : A \to \mathbb{R}$, $A \subset \mathbb{R}^n$, be twice continuously differentiable, and let $\boldsymbol{a}$ be a point in the interior of A. If $\operatorname{grad} f(\boldsymbol{a}) = \boldsymbol{0}$, then the following holds:

- If the Hessian $H_f(\boldsymbol{a})$ at $\boldsymbol{a}$ is **positive definite**, then f has a **local minimum** at $\boldsymbol{a}$.
- If the Hessian $H_f(\boldsymbol{a})$ at $\boldsymbol{a}$ is **negative definite**, then f has a **local maximum** at $\boldsymbol{a}$.
- If the Hessian $H_f(\boldsymbol{a})$ at $\boldsymbol{a}$ is **indefinite**, then f has *no local extremum* at $\boldsymbol{a}$.

Likewise, Theorem 6.5, in which sufficient conditions for the convexity or concavity of a function of several variables are formulated, can be specified more simply and accurately using the concept of definiteness.

Theorem 10.10: Convexity, concavity of a differentiable function

Let $f : A \to \mathbb{R}$, $A \subset \mathbb{R}^n$, be twice continuously differentiable, and let A is open and convex. Then the following statements hold:

- f is convex if and only if $H_f(\boldsymbol{x})$ is positive semidefinite at every $\boldsymbol{x} \in A$.
- f is concave if and only if $H_f(\boldsymbol{x})$ is negative semidefinite at every $\boldsymbol{x} \in A$.

The two theorems above give general criteria for the existence of local extrema and for the convexity or concavity of a function. However, the problem is how to check these criteria, specifically determining the definiteness of a symmetric matrix. For diagonal matrices, this is very simple. If $\boldsymbol{A}$ is a diagonal matrix, then

$$\begin{aligned}
\boldsymbol{x}^T \boldsymbol{A}\boldsymbol{x} &= [x_1, x_2, \ldots, x_n] \begin{bmatrix} a_{11} & 0 & \ldots & 0 \\ 0 & a_{22} & \ldots & 0 \\ \vdots & \vdots & \ddots & \vdots \\ 0 & 0 & \ldots & a_{nn} \end{bmatrix} \begin{bmatrix} x_1 \\ x_2 \\ \vdots \\ x_n \end{bmatrix} \\
&= [x_1, x_2, \ldots, x_n] \begin{bmatrix} a_{11}x_1 \\ a_{22}x_2 \\ \vdots \\ a_{nn}x_n \end{bmatrix} \\
&= a_{11}x_1^2 + a_{22}x_2^2 + \cdots + a_{nn}x_n^2 = \sum_{i=1}^{n} a_{ii}x_i^2 .
\end{aligned}$$

In particular, for $\boldsymbol{x} = \boldsymbol{e}_i$, we get

$$\boldsymbol{e}_i^T \boldsymbol{A} \boldsymbol{e}_i = a_{ii} .$$

So for $\boldsymbol{A}$ to be positive definite, all diagonal elements a_{ii} must be positive. Obviously, this is also sufficient for $\boldsymbol{A}$ to be positive definite. Similar considerations show that the following theorem holds.

Theorem 10.11: Definiteness for diagonal matrices

A **diagonal matrix $\boldsymbol{A}$** is

- positive definite, if and only if $a_{ii} > 0$ for all i,
- positive semidefinite, if and only if $a_{ii} \geq 0$ for all i,
- negative definite, if and only if $a_{ii} < 0$ for all i,
- negative semidefinite, if and only if $a_{ii} \leq 0$ for all i,
- indefinite, if and only if the diagonal contains both values less than zero and values greater than zero.

For arbitrary symmetric matrices, it is more difficult to decide whether a given matrix is positive (or negative) definite or not. An easy-to-apply criterion is **Sylvester's**[3]

[3]James Joseph Sylvester (1814–1897)

criterion. For $k = 1, \dots, n$, the k-th **principal minor** Δ_k of a symmetric $n \times n$ matrix is defined as

$$\Delta_k = \det \begin{bmatrix} a_{11} & \cdots & a_{1k} \\ \vdots & \ddots & \vdots \\ a_{k1} & \cdots & a_{kk} \end{bmatrix}.$$

The k-th principal minor [4] is thus the determinant of the upper left $k \times k$ submatrix of $\boldsymbol{A}$. The definiteness of a symmetric matrix depends can be checked by the signs of its principal minors. The sufficient conditions for the existence of an extremum in Chapter 6 are deduced from Theorem 10.9 and the following criterion of Sylvester:

Theorem 10.12: Sylvester's criterion

Let $\boldsymbol{A}$ be a symmetric $n \times n$ matrix. Then the following statements hold:

(1) $\boldsymbol{A}$ **positive definite** $\iff$ $\Delta_i > 0$ for $i = 1, \dots, n$,

(2) $\boldsymbol{A}$ **negative definite** $\iff$ $(-1)^i \Delta_i > 0$ for $i = 1, \dots, n$,

(3) $\boldsymbol{A}$ **positive semidefinite** $\implies$ $\Delta_i \geq 0$ for $i = 1, \dots, n$,

(4) $\boldsymbol{A}$ **negative semidefinite** $\implies$ $(-1)^i \Delta_i \geq 0$ for $i = 1, \dots, n$.

The condition $(-1)^i \Delta_i > 0$ states that the *odd* minors $\Delta_1, \Delta_3, \dots$ are *less* than zero, and the *even* minors $\Delta_2, \Delta_4, \dots$ are *greater* than zero.

It should be kept in mind that in statements (3) and (4) the reverse implication does not hold. Even if all principal minors are non-negative, the matrix $\boldsymbol{A}$ does not have to be positive semidefinite. This can be easily seen from the matrix

$$\boldsymbol{A} = \begin{bmatrix} 1 & 0 & 0 \\ 0 & 0 & 0 \\ 0 & 0 & -1 \end{bmatrix}.$$

The principal minors of $\boldsymbol{A}$ are $\Delta_1 = 1$, $\Delta_2 = 0$ and $\Delta_3 = 0$. Hence, all the principal minors are non-negative. Nevertheless, according to Theorem 10.11, $\boldsymbol{A}$ is not positive semidefinite but indefinite, since there are positive as well as negative entries on the diagonal.

In the cases $n = 2$ and $n = 3$, this theorem in combination with Theorem 10.9 provides the sufficient criterion for a local extremum (Theorem 6.2) from Chapter 6.

From Sylvester's criterion (Theorem 10.12), a simple sufficient criterion for indefiniteness can be derived:

[4] More precisely, these principal minors are called **leading principal minors**, as principal minors are usually defined more generally than here.

Theorem 10.13: Indefiniteness criterion

Let A be a symmetric $n \times n$ matrix. Then the following statements hold:

- If there is an **even** number i with $\Delta_i < 0$, then A is indefinite.
- If there are **odd** numbers i and j with $\Delta_i < 0$ and $\Delta_j > 0$, then A is indefinite.

10.2 Eigenvalues and eigenvectors

In this section we deal with the so-called **eigenvalues** and **eigenvectors** of a matrix. Eigenvalues and eigenvectors have a wide range of applications. They play important roles in the analysis of input-output models (see Example 10.6 below), in multivariate statistics (principal component or factor analysis), in time series analysis (vector autoregressive models), and in the study of systems of linear differential and difference equations (see Sections 12.4 and 13.4).

Consider the diagonal matrix

$$A = \begin{bmatrix} 2 & 0 & 0 \\ 0 & \frac{1}{2} & 0 \\ 0 & 0 & -2 \end{bmatrix}.$$

For this matrix, we have

$$Ae_1 = 2e_1\,, \quad Ae_2 = \frac{1}{2}e_2 \quad \text{and} \quad Ae_3 = -2e_3\,.$$

The multiplication with the diagonal matrix A only causes a **scaling** of the three canonical unit vectors. The directions of these vectors remain unchanged (unless the scaling factor is negative, in which case the direction is reversed). In contrast, for $x = [1, 1, 0]^T$,

$$Ax = \left[\, 2, \tfrac{1}{2}, 0 \,\right]^T,$$

i.e., Ax is not a multiple of x. When examining a given matrix A, those vectors are of particular interest whose multiplication with A just causes a scaling of the vector. They are called **eigenvectors** of A.

Definition 10.3: Eigenvalue, eigenvector, eigenspace

Let A be a $n \times n$ matrix. If for $x \neq 0$

$$Ax = \lambda x$$

for some $\lambda \in \mathbb{R}$, then x is called **eigenvector** of A corresponding to the **eigenvalue** λ.

The set

$$V(\lambda) = \{x : Ax = \lambda x\}$$

is called **eigenspace** corresponding to the eigenvalue λ of $\boldsymbol{A}$.

Example 10.6 (Leontief's closed input-output model). If Leontief's open input-output model (see Example 2.2) has no exogenous demand, it is called **closed** input-output model. In that model the output x_i of the i-th sector must be at least as large as the amount of the good that is used as input for all sectors, the so-called endogenous demand. With the notations from Example 2.2,

$$a_{i1}x_1 + a_{i2}x_2 + \ldots + a_{in}x_n \leq x_i \qquad \text{for all } i$$

must hold, that is in matrix notation

$$\boldsymbol{A}\boldsymbol{x} \leq \boldsymbol{x}\,,$$

where the inequality sign is to be understood componentwise. An output vector $\boldsymbol{x}$ that satisfies the above condition is called **feasible**. If $\boldsymbol{x}$ satisfies the stronger condition

$$\boldsymbol{A}\boldsymbol{x} = \boldsymbol{x}\,,$$

it is called an **equilibrium output**.

An equilibrium output is therefore an eigenvector of the production matrix $\boldsymbol{A}$ corresponding to the eigenvalue one.

Remark: The equation $\boldsymbol{A}\boldsymbol{x} = \lambda\boldsymbol{x}$ can also be written

$$\boxed{(\boldsymbol{A} - \lambda\boldsymbol{I}_n)\boldsymbol{x} = \boldsymbol{0}\,.}$$

Thus, the eigenspace to a given eigenvalue λ is the solution of a homogeneous linear equation, hence a subspace. It consists of all eigenvectors corresponding to λ as well as the zero vector. If, given some $\lambda \in \mathbb{R}$, $\text{rk}(\boldsymbol{A} - \lambda\boldsymbol{I}_n) = n$, i.e., if $\boldsymbol{A} - \lambda\boldsymbol{I}_n$ is regular, then $\boldsymbol{0}$ is the only solution of this equation, thus λ is no eigenvalue. Conversely, if $\text{rk}(\boldsymbol{A} - \lambda\boldsymbol{I}_n) < n$, i.e., if $\boldsymbol{A} - \lambda\boldsymbol{I}_n$ is singular, there exist solutions $\boldsymbol{x} \neq \boldsymbol{0}$ and λ is an eigenvalue of $\boldsymbol{A}$. In other words:

$$\lambda \text{ is an eigenvalue of } \boldsymbol{A} \quad \Longleftrightarrow \quad \boldsymbol{A} - \lambda\boldsymbol{I}_n \text{ is singular.}$$

According to Theorem 10.6, $\boldsymbol{A} - \lambda\boldsymbol{I}_n$ is singular if and only if $\det(\boldsymbol{A} - \lambda\boldsymbol{I}_n) = 0$. It can be shown that $\det(\boldsymbol{A} - \lambda\boldsymbol{I}_n)$ is a polynomial of degree n in λ. This polynomial is of fundamental importance in determining the eigenvalues of a matrix.

Theorem 10.14: Characteristic polynomial, characteristic equation

If $\boldsymbol{A}$ is an $n \times n$ matrix, then $\det(\boldsymbol{A} - \lambda\boldsymbol{I}_n)$ is a polynomial of degree n in the variable λ. It is called the **characteristic polynomial** of $\boldsymbol{A}$.

The equation $\det(\boldsymbol{A} - \lambda \boldsymbol{I}_n) = 0$ is called the **characteristic equation**. Its solutions λ_i are the eigenvalues of $\boldsymbol{A}$.

The eigenspace corresponding to the eigenvalue λ_i is $V(\lambda_i) = L(\boldsymbol{A} - \lambda_i \boldsymbol{I}_n, \boldsymbol{0})$.

Example 10.7. The eigenvalues of the matrix

$$\boldsymbol{A} = \begin{bmatrix} 1 & -2 \\ 1 & 4 \end{bmatrix}$$

are to be found. The characteristic polynomial is

$$\det(\boldsymbol{A} - \lambda \boldsymbol{I}_2) = \det \begin{bmatrix} 1-\lambda & -2 \\ 1 & 4-\lambda \end{bmatrix} = (1-\lambda)(4-\lambda) + 2 = \lambda^2 - 5\lambda + 6 .$$

The characteristic equation is thus $\lambda^2 - 5\lambda + 6 = 0$. Its solutions are obtained with the quadratic formula:

$$\lambda_{1,2} = \frac{5}{2} \pm \sqrt{\frac{25}{4} - 6} = \frac{5}{2} \pm \sqrt{\frac{1}{4}} = \frac{5}{2} \pm \frac{1}{2} .$$

Thus, $\boldsymbol{A}$ has the two eigenvalues $\lambda_1 = 3$ and $\lambda_2 = 2$.

To determine the eigenspace associated with the eigenvalue $\lambda_1 = 3$, solve the equation $(\boldsymbol{A} - 3\boldsymbol{I}_2)\boldsymbol{x} = \boldsymbol{0}$. As

$$\boldsymbol{A} - 3\boldsymbol{I}_2 = \begin{bmatrix} -2 & -2 \\ 1 & 1 \end{bmatrix}$$

the eigenspace is

$$V(\lambda_1) = V(3) = \left\{ \alpha \begin{bmatrix} 1 \\ -1 \end{bmatrix} : \alpha \in \mathbb{R} \right\} .$$

Similarly, one obtains

$$V(\lambda_2) = V(2) = \left\{ \alpha \begin{bmatrix} 2 \\ -1 \end{bmatrix} : \alpha \in \mathbb{R} \right\} .$$

It is now useful to recall some facts about the roots (i.e., zeros) of polynomials. The most important result for our purposes is:

Mnemonic 10.1: Roots of polynomials

Every polynomial of degree n with real coefficients has at most n different roots in $\mathbb{R}$. If n is odd, there is at least one root in $\mathbb{R}$.

This is a direct consequence of the fundamental theorem of algebra (Theorem E.2 in the Appendix).

Some remarks about the multiplicity of a root are in order here. If λ is a root of a polynomial $p_n(x)$, then one can represent $p_n(x)$ in the form

$$p_n(x) = (x - \lambda)p_{n-1}(x),$$

where $p_{n-1}(x)$ is a polynomial of degree $(n-1)$. $p_{n-1}(x)$ is obtained by polynomial long division $p_n(x)/(x-\lambda)$. If λ is as well a root of $p_{n-1}(x)$, then the procedure can be continued and one obtains

$$p_n(x) = (x - \lambda)[(x - \lambda)p_{n-2}(x)] = (x - \lambda)^2 p_{n-2}(x).$$

This procedure can be continued until λ is no longer a root of $p_{n-k}(x)$. One then obtains

$$p_n(x) = (x - \lambda)^k p_{n-k}(x) \quad \text{and } p_{n-k}(\lambda) \neq 0.$$

In this case, λ is called **root** (or **zero**) **of multiplicity** k of $p_n(x)$. A root of multiplicity one is called **simple root**, a root of multiplicity two **double root**.

Theorem 10.15: Properties of eigenvalues and eigenvectors

Let $\boldsymbol{A}$ be an $n \times n$ matrix. Then the following holds:

(i) $\boldsymbol{A}$ has at most n different real eigenvalues.

(ii) If the eigenvalue $\lambda \in \mathbb{R}$ is a root of multiplicity k of the characteristic polynomial, then the dimension of the associated eigenspace $V(\lambda)$ is at most k, i.e., $\dim V(\lambda) \leq k$.

(iii) If $\lambda_1, \ldots, \lambda_m$ are different eigenvalues of $\boldsymbol{A}$ and $\boldsymbol{v}_1, \ldots, \boldsymbol{v}_m$ are corresponding eigenvectors, then $\boldsymbol{v}_1, \ldots, \boldsymbol{v}_m$ are linearly independent.

Remarks:

1. If the eigenvalue λ is a root of multiplicity k of the characteristic polynomial, then λ is referred to as an **eigenvalue of multiplicity** k and k as the **algebraic multiplicity** of λ.

2. The dimension of the eigenspace $V(\lambda)$ is referred to as the **geometric multiplicity** of the eigenvalue. Statement (ii) of the above theorem can thus be formulated as follows: The geometric multiplicity of an eigenvalue is *at most* as large as its algebraic multiplicity.

3. In **complex numbers** (see Appendix E), a polynomial of degree n always has exactly n roots (roots being counted with their multiplicity). If λ is a complex root of the characteristic polynomial with a non-zero imaginary part, then λ is said to be a complex eigenvalue. Complex eigenvalues are particularly important in the context of systems of differential or difference equations (see Sections 12.4 and 13.4). We will return to complex eigenvalues in Section 10.4.

A matrix does not need to have real eigenvalues, as the following example shows:

Example 10.8. The matrix

$$A = \begin{bmatrix} 0 & 1 \\ -1 & 0 \end{bmatrix}$$

has the characteristic polynomial $\lambda^2 + 1$, which has no real root. $\boldsymbol{A}$ has a complex eigenvalue of multiplicity 2 at $\lambda = i$, but no real eigenvalue.

The next example shows that the dimension of an eigenspace can be smaller than the multiplicity of the corresponding eigenvalue.

Example 10.9. For the matrix

$$\boldsymbol{A} = \begin{bmatrix} 4 & 1 \\ 0 & 4 \end{bmatrix}$$

the characteristic polynomial is $(4 - \lambda)^2$, i.e., $\lambda = 4$ is a double root. For $\lambda = 4$ we have

$$(\boldsymbol{A} - 4\boldsymbol{I}_2) = \begin{bmatrix} 0 & 1 \\ 0 & 0 \end{bmatrix}.$$

Obviously, the eigenspace is given by $V(4) = \{\alpha[1,0]^T : \alpha \in \mathbb{R}\}$ and thus has dimension one. Here, the geometric multiplicity of $\lambda = 4$ is smaller than its algebraic multiplicity.

We illustrate the calculation of eigenvalues and eigenspaces with another example.

Example 10.10. The eigenvalues and eigenspaces of the matrix

$$\boldsymbol{A} = \begin{bmatrix} 7 & 1 & 1 & 0 \\ 0 & 2 & 1 & 3 \\ 0 & 0 & -1 & 0 \\ 0 & 0 & 0 & -1 \end{bmatrix}$$

are to be determined. We have

$$\boldsymbol{A} - \lambda\boldsymbol{I}_4 = \begin{bmatrix} 7-\lambda & 1 & 1 & 0 \\ 0 & 2-\lambda & 1 & 3 \\ 0 & 0 & -1-\lambda & 0 \\ 0 & 0 & 0 & -1-\lambda \end{bmatrix}.$$

Consequently, the characteristic equation is

$$\det(\boldsymbol{A} - \lambda\boldsymbol{I}_4) = (7-\lambda)(2-\lambda)(-1-\lambda)^2 = 0.$$

The solutions of the characteristic equation and thus the eigenvalues are directly obtained as $\lambda_1 = 7$, $\lambda_2 = 2$ and $\lambda_{3,4} = -1$. $\lambda_{3,4}$ is a double eigenvalue, as $\lambda_{3,4} = -1$ is a double root of the characteristic polynomial.

We now calculate the eigenspaces. For $\lambda_1 = 7$ we have

$$A - 7I_4 = \begin{bmatrix} 0 & 1 & 1 & 0 \\ 0 & -5 & 1 & 3 \\ 0 & 0 & -8 & 0 \\ 0 & 0 & 0 & -8 \end{bmatrix}.$$

Clearly, x_1 can be chosen freely, while x_2, x_3 and x_4 must be zero. Thus,

$$V(\lambda_1) = V(7) = \left\{ \alpha \begin{bmatrix} 1 \\ 0 \\ 0 \\ 0 \end{bmatrix} : \alpha \in \mathbb{R} \right\}.$$

Similarly, for $\lambda_2 = 2$ we get

$$A - 2I_4 = \begin{bmatrix} 5 & 1 & 1 & 0 \\ 0 & 0 & 1 & 3 \\ 0 & 0 & -3 & 0 \\ 0 & 0 & 0 & -3 \end{bmatrix}.$$

We have $x_3 = x_4 = 0$ and $x_2 = -5x_1$. The eigenspace for $\lambda_2 = 2$ is therefore

$$V(\lambda_2) = V(2) = \left\{ \alpha \begin{bmatrix} 1 \\ -5 \\ 0 \\ 0 \end{bmatrix} : \alpha \in \mathbb{R} \right\}.$$

For $\lambda_3 = -1$ we have

$$A - (-1)I_4 = A + I_4 = \begin{bmatrix} 8 & 1 & 1 & 0 \\ 0 & 3 & 1 & 3 \\ 0 & 0 & 0 & 0 \\ 0 & 0 & 0 & 0 \end{bmatrix}.$$

Here, the solution of the linear equation is not so obvious. We therefore determine the eigenspace using Gauss-Jordan elimination. First, we remove the two zero rows and obtain

$$\begin{array}{cccc} x_1 & x_2 & x_3 & x_4 \\ \hline 8 & 1 & 1 & 0 \\ 0 & 3 & 1 & 3 \end{array} \quad \xrightarrow[\text{II}/3]{\text{I}/8} \quad \begin{array}{cccc} x_1 & x_2 & x_3 & x_4 \\ \hline 1 & \frac{1}{8} & \frac{1}{8} & 0 \\ 0 & 1 & \frac{1}{3} & 1 \end{array} \quad \xrightarrow[\text{I}-\frac{1}{8}\cdot\text{II}]{} \quad \begin{array}{cccc} x_1 & x_2 & x_3 & x_4 \\ \hline 1 & 0 & \frac{1}{12} & -\frac{1}{8} \\ 0 & 1 & \frac{1}{3} & 1 \end{array}$$

The eigenspace $V(\lambda_3)$ thus has dimension two. It is

$$V(\lambda_3) = V(-1) = \left\{ \alpha \begin{bmatrix} \frac{1}{12} \\ \frac{1}{3} \\ -1 \\ 0 \end{bmatrix} + \beta \begin{bmatrix} -\frac{1}{8} \\ 1 \\ 0 \\ -1 \end{bmatrix} : \alpha, \beta \in \mathbb{R} \right\}.$$

Another view on the eigenvalues of a matrix and their determination is provided by the important concept of **diagonalizability**.

Definition 10.4: Diagonalizability

An $n \times n$ matrix $\boldsymbol{A}$ is called (real) **diagonalizable**, if there is a regular matrix $\boldsymbol{S}$ and a diagonal matrix $\boldsymbol{D}$ such that

$$\boldsymbol{S}^{-1}\boldsymbol{A}\boldsymbol{S} = \boldsymbol{D}\,.$$

This definition naturally raises the question under which conditions a matrix is diagonalizable. The condition $\boldsymbol{S}^{-1}\boldsymbol{A}\boldsymbol{S} = \boldsymbol{D}$ is equivalent to $\boldsymbol{A}\boldsymbol{S} = \boldsymbol{S}\boldsymbol{D}$. If we denote the i-th column of $\boldsymbol{S}$ by $\boldsymbol{s}_i$ and the i-th diagonal element of $\boldsymbol{D}$ by d_i, then

$$\boldsymbol{A}\boldsymbol{S} = \boldsymbol{A}[\boldsymbol{s}_1, \ldots, \boldsymbol{s}_n] = [\boldsymbol{A}\boldsymbol{s}_1, \ldots, \boldsymbol{A}\boldsymbol{s}_n]$$

and

$$\boldsymbol{S}\boldsymbol{D} = [\boldsymbol{s}_1, \ldots, \boldsymbol{s}_n] \begin{bmatrix} d_1 & & \\ & \ddots & \\ & & d_n \end{bmatrix} = [d_1\boldsymbol{s}_1, \ldots, d_n\boldsymbol{s}_n]\,.$$

Thus, $\boldsymbol{A}\boldsymbol{S} = \boldsymbol{S}\boldsymbol{D}$ if and only if

$$\boldsymbol{A}\boldsymbol{s}_i = d_i\boldsymbol{s}_i\,, \quad i = 1, \ldots, n.$$

So, if a matrix $\boldsymbol{A}$ is diagonalizable, the diagonal elements of $\boldsymbol{D}$ are eigenvalues of $\boldsymbol{A}$ and the columns of $\boldsymbol{S}$ are corresponding eigenvectors. From this it follows directly that a matrix $\boldsymbol{A}$ is diagonalizable if and only if there are n linearly independent eigenvectors of $\boldsymbol{A}$. Since n linearly independent vectors form a basis of $\mathbb{R}^n$, this result is formulated briefly and succinctly as follows:

Mnemonic 10.2: Diagonalizability

A matrix $\boldsymbol{A}$ is **diagonalizable** if and only if there is a basis of eigenvectors.

From Theorem 10.15 the following criterion follows.

Theorem 10.16: Diagonalizability of matrices

An $n \times n$ matrix $\boldsymbol{A}$ is (real) diagonalizable if and only if it

1. has n real eigenvalues, where multiple eigenvalues are to be counted with their algebraic multiplicity, and
2. for each eigenvalue algebraic and geometric multiplicity coincide.

The previous theorem immediately yields the following easy-to-apply criterion for diagonalizability.

Theorem 10.17: Diagonalizability of matrices

An $n \times n$ matrix $\boldsymbol{A}$ is (real) diagonalizable if and only if the sum of the geometric multiplicities of the eigenvalues is equal to n.

Example 10.11. The 2×2 matrix $\boldsymbol{A}$ from Example 10.7 is diagonalizable, since it has two simple eigenvalues and the associated eigenspaces each have dimension one. With the matrices

$$\boldsymbol{S} = \begin{bmatrix} 1 & 2 \\ -1 & -1 \end{bmatrix} \quad \text{and} \quad \boldsymbol{D} = \begin{bmatrix} 3 & 0 \\ 0 & 2 \end{bmatrix}$$

we have $\boldsymbol{S}^{-1}\boldsymbol{A}\boldsymbol{S} = \boldsymbol{D}$.

The 2×2 matrix from Example 10.8 is not diagonalizable, as it has no eigenvalues.

The 2×2 matrix $\boldsymbol{A}$ from Example 10.9 is not diagonalizable, since the algebraic multiplicity of the eigenvalue $\lambda = 4$ is two, but its geometric multiplicity is one.

The 4×4 matrix $\boldsymbol{A}$ from Example 10.10 is diagonalizable, since it has two simple eigenvalues 7 and 2 as well as a double eigenvalue -1 and the associated eigenspaces have dimensions 1, 1 and 2. The matrix S is formed from the basis vectors of the eigenspaces and the diagonal matrix $\boldsymbol{D}$ from the eigenvalues,

$$\boldsymbol{S} = \begin{bmatrix} 1 & 1 & \frac{1}{12} & -\frac{1}{8} \\ 0 & -5 & \frac{1}{3} & 1 \\ 0 & 0 & -1 & 0 \\ 0 & 0 & 0 & -1 \end{bmatrix} \quad \text{and} \quad \boldsymbol{D} = \begin{bmatrix} 7 & 0 & 0 & 0 \\ 0 & 2 & 0 & 0 \\ 0 & 0 & -1 & 0 \\ 0 & 0 & 0 & -1 \end{bmatrix}.$$

It is easily verified that $\boldsymbol{S}^{-1}\boldsymbol{A}\boldsymbol{S} = \boldsymbol{D}$.

Since the determinant of a triangular matrix (particularly of a diagonal one) is simply the product of all diagonal elements, the eigenvalues can be very simply determined in this case, as we have already seen in the previous Example 10.10.

Theorem 10.18: Eigenvalues of triangular matrices

Let $\boldsymbol{A}$ be (upper or lower) triangular. Then the eigenvalues of $\boldsymbol{A}$ are the diagonal elements of $\boldsymbol{A}$.

Proof: The statement follows directly from the fact that $\boldsymbol{A} - \lambda\boldsymbol{I}_n$ is triangular as well and that according to Theorem 10.4 we have

$$\det(\boldsymbol{A} - \lambda\boldsymbol{I}_n) = \prod_{i=1}^{n}(a_{ii} - \lambda)\,.$$

□

Remarks:

1. In light of this theorem, one might be tempted to determine the eigenvalues of a matrix by transforming it into a triangular matrix through elementary row or column operations, and then simply deduce the eigenvalues of the original matrix from the diagonal elements of the final triangular matrix. However, this approach *does not work!* Elementary operations alter the eigenvalues, and it is not straightforward to deduce the eigenvalues of the original matrix from those of the transformed matrix.

2. In contrast to the eigenvalues, the eigenvectors of triangular matrices cannot simply be read off. Always, a homogeneous linear equation must be solved. The only exception is the case of diagonal matrices, where the eigenvectors are given by the canonical unit vectors.

If an orthogonal matrix has real eigenvalues, then they must be +1 or −1. This follows from the fact, that for an orthogonal matrix we have $||\boldsymbol{Qx}|| = ||\boldsymbol{x}||$. Thus, from the equation $\boldsymbol{Qx} = \lambda\boldsymbol{x}$ it follows

$$||\boldsymbol{x}|| = ||\boldsymbol{Qx}|| = ||\lambda\boldsymbol{x}|| = |\lambda| \cdot ||\boldsymbol{x}||$$

and hence $|\lambda| = 1$, i.e., $\lambda = 1$ or $\lambda = -1$. We formulate this result as a theorem.

Theorem 10.19: Eigenvalues of orthogonal matrices

Let $\boldsymbol{Q}$ be an orthogonal $n \times n$ matrix. If λ is a (real) eigenvalue of $\boldsymbol{Q}$, then $\lambda = 1$ or $\lambda = -1$.

10.3 Eigenvalues of symmetric matrices

As symmetric matrices arise in numerous applications, the study of the eigenvalues of symmetric matrices is particularly interesting. For instance, the Hessian matrix, which appears in the formulation of sufficient conditions for the existence of extrema (see Chapter 6), is symmetric. In statistics, the covariance matrix of a random vector is also symmetric.

In Theorem 10.15 we saw that eigenvectors to *different* eigenvalues are linearly independent. For symmetric matrices a much stronger result holds, as stated in the following theorem.

Theorem 10.20: Orthogonality of eigenvectors of symmetric matrices

Let $\boldsymbol{A}$ be a symmetric $n \times n$ matrix. If λ_1 and λ_2 are different eigenvalues and $\boldsymbol{v}_1$ and $\boldsymbol{v}_2$ are associated eigenvectors, then $\boldsymbol{v}_1$ and $\boldsymbol{v}_2$ are orthogonal.

Proof: We have to show that $\boldsymbol{v}_1^T\boldsymbol{v}_2 = 0$. Using the fact that $\boldsymbol{A}^T = \boldsymbol{A}$, it follows

$$\lambda_1(\boldsymbol{v}_1^T\boldsymbol{v}_2) = (\lambda_1\boldsymbol{v}_1)^T\boldsymbol{v}_2 = (\boldsymbol{A}\boldsymbol{v}_1)^T\boldsymbol{v}_2 = (\boldsymbol{v}_1^T\boldsymbol{A}^T)\boldsymbol{v}_2$$

$$= v_1^T(Av_2) = v_1^T(\lambda_2 v_2) = \lambda_2(v_1^T v_2)\,.$$

Since $\lambda_1 \neq \lambda_2$, this equation can be satisfied only if $v_1^T v_2 = 0$, as claimed. □

The following theorem strengthens Theorem 10.15 for symmetric matrices.

Theorem 10.21: Eigenvalues and eigenvectors of symmetric matrices

Let A be a symmetric $n \times n$ matrix. Then the following holds:

(i) A has exactly n real eigenvalues, where multiple eigenvalues are counted according to their multiplicity.

(ii) If λ is an eigenvalue of multiplicity k, then the dimension of the associated eigenspace $V(\lambda)$ is k, $\dim V(\lambda) = k$.

(iii) If $\lambda_1, \ldots, \lambda_n$ are the eigenvalues of A (with multiple eigenvalues appearing according to their multiplicity), then for each eigenvalue λ_i, one can choose an eigenvector v_i such that the vectors $v_1, \ldots, v_n$ form an orthonormal system (and thus also a basis of $\mathbb{R}^n$).

In particular, a symmetric matrix has always *exactly* n real eigenvalues and algebraic and geometric multiplicity of each eigenvalue coincide. Hence, the conditions for (real) diagonalizability from Theorem 10.16 are met and thus, symmetric matrices are diagonalizable. In fact, an even stronger result holds. Part (iii) of the above theorem can – somewhat imprecisely – be formulated as follows:

> If A is a symmetric $n \times n$ matrix, then there is an orthonormal basis of $\mathbb{R}^n$ consisting of eigenvectors of A.

From this it follows that for symmetric matrices the transformation matrix S can always be chosen as an orthogonal matrix. We formulate this result as a theorem.

Theorem 10.22: Diagonalizability of symmetric matrices

Let A be a symmetric $n \times n$ matrix. Then there is a diagonal matrix Λ and an orthogonal matrix Q, such that

$$Q^T A Q = \Lambda\,.$$

In particular, A is diagonalizable.

Proof: If one chooses $\Lambda = \mathrm{diag}(\lambda_1, \ldots, \lambda_n)$ as the diagonal matrix of eigenvalues of A and Q as the matrix whose columns are the eigenvectors $v_1, \ldots, v_n$ from statement (iii) of Theorem 10.21, then

$$AQ = A[v_1, \cdots, v_n] = [\lambda_1 v_1, \cdots, \lambda_n v_n] = [v_1, \cdots, v_n]\Lambda = Q\Lambda\,.$$

Since Q is orthogonal, $Q^{-1} = Q^T$ and multiplication from the left with Q^T yields the statement. □

At the end of this section, we return to the concept of definiteness and provide another criterion for the definiteness of a matrix. According to Theorem 10.22, for every symmetric matrix $\boldsymbol{A}$, there is an orthogonal matrix $\boldsymbol{Q}$ and a diagonal matrix $\boldsymbol{\Lambda}$ such that

$$\boldsymbol{A} = \boldsymbol{Q}\boldsymbol{\Lambda}\boldsymbol{Q}^T .$$

Then for $\boldsymbol{x} \neq \boldsymbol{0}$ we get

$$\boldsymbol{x}^T\boldsymbol{A}\boldsymbol{x} = \boldsymbol{x}^T\boldsymbol{Q}\boldsymbol{\Lambda}\boldsymbol{Q}^T\boldsymbol{x} = (\boldsymbol{Q}^T\boldsymbol{x})^T\boldsymbol{\Lambda}(\boldsymbol{Q}^T\boldsymbol{x}) = \boldsymbol{y}^T\boldsymbol{\Lambda}\boldsymbol{y} ,$$

where, in the final step, the substitution $\boldsymbol{y} = \boldsymbol{Q}^T\boldsymbol{x}$ was made. From this it follows that $\boldsymbol{A}$ is positive definite if and only if $\boldsymbol{\Lambda}$ is positive definite. Since $\boldsymbol{\Lambda}$ is a diagonal matrix with the eigenvalues of $\boldsymbol{A}$ on the diagonal, Theorem 10.11 yields the following criterion, which characterizes the definiteness of $\boldsymbol{A}$ by its eigenvalues.

Theorem 10.23: Definiteness and eigenvalues

Let $\boldsymbol{A}$ be a symmetric $n \times n$ matrix. Then $\boldsymbol{A}$ is

- **positive definite**, if and only if all eigenvalues of $\boldsymbol{A}$ are positive,
- **positive semidefinite**, if and only if all eigenvalues of $\boldsymbol{A}$ are non-negative,
- **negative definite**, if and only if all eigenvalues of $\boldsymbol{A}$ are negative,
- **negative semidefinite**, if and only if all eigenvalues of $\boldsymbol{A}$ are non-positive,
- **indefinite**, if $\boldsymbol{A}$ has both positive and negative eigenvalues.

This theorem is primarily of theoretical interest, as it demonstrates that definiteness is a property of a matrix's eigenvalues. However, if the eigenvalues are not known, it is of limited practical value, as computing eigenvalues is generally quite involved. For this reason, definiteness is usually checked more easily using Sylvester's criterion.

10.4 Complex eigenvalues

This section requires a basic knowledge of complex numbers. A brief introduction to complex numbers and their most important properties is found in Section E of the Appendix.

The fundamental theorem of algebra (Theorem E.2) states that every polynomial of degree n has exactly n zeros in the complex numbers, where zeros are counted according to their multiplicity. The characteristic polynomial of a matrix in general can also have complex zeros[5]. Accordingly, a matrix may possess not only real but also complex eigenvalues. We now discuss how the theory changes when complex eigenvalues are also taken into consideration.

[5]When we speak of complex zeros or complex eigenvalues, we implicitly assume that the imaginary part of the zero or eigenvalue is non-zero.

Theorem 10.24: Complex eigenvalues

Let $\boldsymbol{A}$ be a real $n \times n$ matrix. Then the following holds:

(i) $\boldsymbol{A}$ has exactly n eigenvalues in the complex numbers.

(ii) If $\lambda = a + bi$ with $b \neq 0$ is a complex eigenvalue, then $\overline{\lambda} = a - bi$ is also a complex eigenvalue. If $\boldsymbol{u} + i\boldsymbol{v}$ is an eigenvector corresponding to the eigenvalue λ, then $\boldsymbol{u} - i\boldsymbol{v}$ is an eigenvector corresponding to $\overline{\lambda}$.

Example 10.12 (Continuation of Example 10.8). The matrix

$$\boldsymbol{A} = \begin{bmatrix} 0 & 1 \\ -1 & 0 \end{bmatrix}$$

has the characteristic polynomial $\lambda^2 + 1$, which has the complex zeros $\lambda_1 = i$ and $\lambda_2 = -i$.

To calculate the eigenspace for $\lambda_1 = i$, we form

$$\boldsymbol{A} - i\boldsymbol{I} = \begin{bmatrix} -i & 1 \\ -1 & -i \end{bmatrix}.$$

The solution of the linear equation $(\boldsymbol{A} - i\boldsymbol{I})\boldsymbol{x} = \boldsymbol{0}$ is given by

$$V(i) = \left\{\alpha \begin{bmatrix} 1 \\ i \end{bmatrix} : \alpha \in \mathbb{C}\right\}.$$

Correspondingly,

$$V(-i) = \left\{\alpha \begin{bmatrix} 1 \\ -i \end{bmatrix} : \alpha \in \mathbb{C}\right\}.$$

If we allow for complex eigenvalues, matrices that were not diagonalizable before may become diagonalizable. However, then the transformation matrix $\boldsymbol{S}$ as well as the diagonal matrix $\boldsymbol{D}$ may include complex components.

Definition 10.5: Complex diagonalizability

An $n \times n$ matrix $\boldsymbol{A}$ is called **complex diagonalizable**, if there is a regular matrix $\boldsymbol{S}$ and a diagonal matrix $\boldsymbol{D}$, both with possibly complex components, such that

$$\boldsymbol{S}^{-1}\boldsymbol{A}\boldsymbol{S} = \boldsymbol{D}.$$

Obviously, every real diagonalizable matrix is also complex diagonalizable. However, the reverse is not true.

Since a polynomial of degree n has exactly n roots (real or complex) when roots are counted with their multiplicities, every $n \times n$ matrix $\boldsymbol{A}$ consequently possesses exactly n eigenvalues (real or complex) when accounting for their algebraic multiplicities. Thus, condition 1) of Theorem 10.16 is always satisfied when considering complex eigenvalues. Therefore, we obtain the following criterion for complex diagonalizability.

Theorem 10.25: Complex diagonalizability of matrices

An $n \times n$ matrix $\boldsymbol{A}$ is complex diagonalizable if and only if for every (real or complex) eigenvalue the algebraic and geometric multiplicity coincide.

Example 10.13 (Continuation of Examples 10.8 and 10.12). The matrix

$$\boldsymbol{A} = \begin{bmatrix} 0 & 1 \\ -1 & 0 \end{bmatrix}$$

has the eigenvalues i and $-i$ with associated eigenvectors $[1, i]^T$ and $[1, -i]^T$, respectively.

The matrix $\boldsymbol{A}$ is not real diagonalizable, as it has no real eigenvalues. On the other hand, $\boldsymbol{A}$ is complex diagonalizable with

$$\boldsymbol{S} = \begin{bmatrix} 1 & 1 \\ i & -i \end{bmatrix} \quad \text{and} \quad \boldsymbol{D} = \begin{bmatrix} i & 0 \\ 0 & -i \end{bmatrix}.$$

Powers of a square matrix $\boldsymbol{A}$ are defined for natural exponents in the usual way, that is, $\boldsymbol{A}^2 = \boldsymbol{A} \cdot \boldsymbol{A}$, $\boldsymbol{A}^3 = \boldsymbol{A} \cdot \boldsymbol{A} \cdot \boldsymbol{A}$ and in general $\boldsymbol{A}^n = \boldsymbol{A} \cdot \ldots \cdot \boldsymbol{A}$, where the product consists of n identical factors $\boldsymbol{A}$. Additionally, we define $\boldsymbol{A}^0 = \boldsymbol{I}$ and $\boldsymbol{A}^1 = \boldsymbol{A}$. Correspondingly, a matrix polynomial of degree m has the form $p_m(\boldsymbol{A}) = a_m \boldsymbol{A}^m + a_{m-1}\boldsymbol{A}^{m-1} + \cdots + a_1 \boldsymbol{A} + a_0 \boldsymbol{I}$, where $a_0, \ldots, a_m \in \mathbb{R}$, $a_m \neq 0$.

Theorem 10.26: Cayley-Hamilton

Let $\boldsymbol{A}$ be an $n \times n$ matrix and $p_{\boldsymbol{A}}(x) = (-1)^n x^n + a_{n-1}x^{n-1} + \cdots + a_1 x + a_0$ its characteristic polynomial. Then $\boldsymbol{A}$ is a root of $p_{\boldsymbol{A}}$, that is,

$$p_{\boldsymbol{A}}(\boldsymbol{A}) = (-1)^n \boldsymbol{A}^n + a_{n-1}\boldsymbol{A}^{n-1} + \cdots + a_1 \boldsymbol{A} + a_0 \boldsymbol{I} = \boldsymbol{0}.$$

In short, the statement of the Cayley[6]-Hamilton[7] theorem is as follows: "Every square matrix is a root of its characteristic polynomial."

Example 10.14. The matrix

$$\boldsymbol{A} = \begin{bmatrix} 3 & 2 & -1 \\ -1 & 0 & 2 \\ 1 & 2 & -1 \end{bmatrix}$$

[6] Arthur Cayley (1821–1895)
[7] William Rowan Hamilton (1805–1865)

has the characteristic polynomial $p_{\boldsymbol{A}}(\lambda) = -\lambda^3 + 2\lambda^2 + 4\lambda - 8$. According to the Cayley-Hamilton theorem, we have

$$-\boldsymbol{A}^3 + 2\boldsymbol{A}^2 + 4\boldsymbol{A} - 8\boldsymbol{I} = \boldsymbol{0}\,.$$

In particular, the Cayley-Hamilton theorem implies that the matrix power $\boldsymbol{A}^n$ can be represented as a linear combination of the matrix powers $\boldsymbol{I}, \boldsymbol{A}, \boldsymbol{A}^2, \ldots, \boldsymbol{A}^{n-1}$. By mathematical induction, it follows that every power $\boldsymbol{A}^k$ with $k \geq n$ can be represented as a linear combination of the matrix powers $\boldsymbol{I}, \boldsymbol{A}, \boldsymbol{A}^2, \ldots, \boldsymbol{A}^{n-1}$. The same holds for any matrix polynomial.

Theorem 10.27: Matrix polynomials

Let $\boldsymbol{A}$ be an $n \times n$ matrix and $p(\boldsymbol{A})$ a polynomial in $\boldsymbol{A}$. Then there are constants $a_0, \ldots, a_{n-1} \in \mathbb{R}$, such that

$$p(\boldsymbol{A}) = a_0\boldsymbol{I} + a_1\boldsymbol{A} + a_2\boldsymbol{A}^2 + \cdots + a_{n-1}\boldsymbol{A}^{n-1}\,.$$

We now address the question of how to find the coefficients $a_0, \ldots, a_{n-1}$ for a given polynomial $p(\boldsymbol{A})$. If the polynomial p is divided by the characteristic polynomial $p_{\boldsymbol{A}}$ of $\boldsymbol{A}$ using polynomial long division, one obtains

$$\frac{p(x)}{p_{\boldsymbol{A}}(x)} = q(x) + \frac{r(x)}{p_{\boldsymbol{A}}(x)}$$

or equivalently

$$p(x) = q(x)p_{\boldsymbol{A}}(x) + r(x)\,, \tag{10.1}$$

where $q(x)$ is the quotient polynomial and $r(x)$ the remainder polynomial with a degree less than n. Because of the Cayley-Hamilton theorem, $\boldsymbol{A}$ is a root of its characteristic polynomial and one obtains

$$p(\boldsymbol{A}) - q(\boldsymbol{A})p_{\boldsymbol{A}}(\boldsymbol{A}) + r(\boldsymbol{A}) = r(\boldsymbol{A})\,,$$

Thus, the coefficients of the polynomial $r(x)$ are the desired coefficients $a_0, \ldots, a_{n-1}$.

Another approach for determining the coefficients $a_0, \ldots, a_{n-1}$ is provided by the following considerations. Since the eigenvalues of the matrix $\boldsymbol{A}$ are zeros of the characteristic polynomial $p_{\boldsymbol{A}}(x)$, it holds that

$$p(\lambda) = q(\lambda)p_{\boldsymbol{A}}(\lambda) + r(\lambda) = r(\lambda)$$

for every (real or complex) eigenvalue λ of $\boldsymbol{A}$. So, if there are n different eigenvalues, this gives n equations

$$p(\lambda_i) = a_0 + a_1\lambda_i + a_2\lambda_i^2 + \cdots + a_{n-1}\lambda_i^{n-1}\,, \qquad i = 1, \ldots, n,$$

from which the coefficients $a_0, \ldots, a_{n-1}$ can be uniquely determined.

If multiple eigenvalues occur, this approach yields fewer than n equations for determining the coefficients. However, if λ_i is an eigenvalue of multiplicity m, then λ_i is also a root of the polynomials $p'_{\boldsymbol{A}}, p''_{\boldsymbol{A}}, \ldots, p^{(m-1)}_{\boldsymbol{A}}$. Differentiating Equation (10.1), we obtain

$$p'(x) = q'(x)p_{\boldsymbol{A}}(x) + q(x)p'_{\boldsymbol{A}}(x) + r'(x)\,.$$

Since a double eigenvalue λ_i is both a root of $p_{\boldsymbol{A}}$ and $p'_{\boldsymbol{A}}$, by substituting λ_i we get

$$p'(\lambda_i) = q'(\lambda_i)p_{\boldsymbol{A}}(\lambda_i) + q(\lambda_i)p'_{\boldsymbol{A}}(\lambda_i) + r'(\lambda_i) = r'(\lambda_i)\,.$$

A double eigenvalue λ_i thus provides a second equation

$$p'(\lambda_i) = a_1 + 2a_2\lambda_i + \cdots + (n-1)a_{n-1}\lambda_i^{n-2}\,.$$

Similarly, for an eigenvalue λ_i of multiplicity m we get the m equations

$$p(\lambda_i) = r(\lambda_i), \quad p'(\lambda_i) = r'(\lambda_i), \quad \ldots, \quad p^{(m-1)}(\lambda_i) = r^{(m-1)}(\lambda_i)\,.$$

Hence, it is always possible to find n equations, from which the coefficients $a_0, \ldots,$ a_{n-1} can be uniquely determined.

Example 10.15 (Continuation of Example 10.14). The matrix

$$\boldsymbol{A} = \begin{bmatrix} 3 & 2 & -1 \\ -1 & 0 & 2 \\ 1 & 2 & -1 \end{bmatrix}$$

has the characteristic polynomial $p_{\boldsymbol{A}}(\lambda) = \lambda^3 + 2\lambda^2 + 4\lambda - 8$. Thus, $\lambda_1 = -2$ is a simple and $\lambda_{2,3} = 2$ is a double eigenvalue. We want to represent $\boldsymbol{A}^5$ as a second degree polynomial in $\boldsymbol{A}$. To determine the coefficients a_0, a_1, a_2 we get the system of equations.

$$\begin{array}{rclclcl} (-2)^5 & = & a_0 & + & a_1 \cdot (-2) & + & a_2 \cdot (-2)^2 \\ 2^5 & = & a_0 & + & a_1 \cdot 2 & + & a_2 \cdot 2^2 \\ 5 \cdot 2^4 & = & & & a_1 & + & 2a_2 \cdot 2 \end{array}$$

Solving the system of equations using Gauss-Jordan elimination yields $a_0 = -64$, $a_1 = 16$ and $a_2 = 16$. Thus,

$$\boldsymbol{A}^5 = -64\boldsymbol{I} + 16\boldsymbol{A} + 16\boldsymbol{A}^2\,.$$

This method can also be used to represent $\boldsymbol{A}^n$ for arbitrary n as a quadratic polynomial in $\boldsymbol{A}$. This yields the following system of equations:

$$\begin{array}{rclclcl} (-2)^n & = & a_0 & + & a_1 \cdot (-2) & + & a_2 \cdot (-2)^2 \\ 2^n & = & a_0 & + & a_1 \cdot 2 & + & a_2 \cdot 2^2 \\ n \cdot 2^{n-1} & = & & & a_1 & + & 2a_2 \cdot 2 \end{array}$$

The solution is

$$\begin{aligned} a_0 &= 3 \cdot 2^{n-2} + (-2)^{n-2} - n \cdot 2^{n-1}, \\ a_1 &= 2^{n-2} - (-2)^{n-2}, \\ a_2 &= -2^{n-4} + (-2)^{n-4} + n \cdot 2^{n-3}. \end{aligned}$$

With these coefficients, we have

$$\boldsymbol{A}^n = a_o \boldsymbol{I} + a_1 \boldsymbol{A} + a_2 \boldsymbol{A}^2 .$$

Important terms for review

After reading this chapter, the following terms should be familiar:

- Determinant
- Expansion of a determinant along a row or a column
- Rule of Sarrus
- Eigenvalue and eigenvector
- Characteristic polynomial, characteristic equation
- Eigenvalue of multiplicity k, root of multiplicity k, zero of multiplicity k
- Algebraic and geometric multiplicity
- Positive and negative (semi-)definite matrix
- Principal minor
- Diagonalizability

Self-test

You can check your understanding of this chapter using the following true-or-false questions. For each statement, indicate whether it is true (T) or false (F). Short solutions to these questions can be found in Appendix H.

I. Are the following statements about determinants true or false?

	T	F
The determinant of a square matrix is a number.	☐	☐
The determinant is zero if the matrix is invertible.	☐	☐
The determinant of a triangular matrix is equal to the sum of the main diagonal elements.	☐	☐
$\det(\boldsymbol{AB})^{-1} = \frac{1}{\det \boldsymbol{A} \det \boldsymbol{B}}$, if $\boldsymbol{AB}$ is invertible.	☐	☐
If one expands the determinant of the identity matrix $\boldsymbol{I}_n$ ($n \geq 2$) along any row or column, one gets the determinant of $\boldsymbol{I}_{n-1}$.	☐	☐

II. Mark each of the following statements about eigenvalues and eigenvectors as true or false.

	T	F
If one multiplies a matrix from the right by one of its eigenvectors, the result is a multiple of that eigenvector.	☐	☐
The only eigenvalue of the identity matrix is 1.	☐	☐
Every matrix has exactly as many real eigenvalues as rows.	☐	☐
A matrix that has no real eigenvalues is singular.	☐	☐

III. Decide whether the following statements about definiteness of matrices are true or false.

	T	F
If a matrix is positive semidefinite, then it is positive definite.	☐	☐
The identity matrix is positive definite.	☐	☐
A diagonal matrix with all diagonal elements negative is negative definite.	☐	☐

Exercises

Exercise 1
Find the determinants of the following matrices:

$$A = \begin{bmatrix} 1 & 2 & 3 \\ 8 & 9 & 4 \\ 7 & 6 & 5 \end{bmatrix}, \quad B = \begin{bmatrix} 1 & 4 & 7 \\ -2 & 3 & -5 \\ 1 & 0 & 1 \end{bmatrix}, \quad C = \begin{bmatrix} 4 & 3 & 9 \\ 4 & 2 & 1 \\ 1 & 2 & -1 \end{bmatrix}.$$

Exercise 2
Consider again the matrices from Exercise 1. Solve the following tasks using your results from Exercise 1.

a) Find the determinant of ABC.

b) Are the matrices invertible? If yes, find the determinant of $A^{-1}BC^{-1}$.

c) Find the determinant of $-\frac{1}{2}A^T$.

Exercise 3
Compute the determinants of the matrices

$$A = \begin{bmatrix} 1 & 0 & 0 & 2 \\ 0 & 2 & 1 & 0 \\ 0 & -1 & 4 & 1 \\ 3 & 0 & 1 & 1 \end{bmatrix} \quad \text{and} \quad B = \begin{bmatrix} 2 & 0 & 0 & 4 & 9 \\ 4 & -1 & 0 & 3 & -2 \\ 4 & 6 & 2 & -1 & -3 \\ 0 & 0 & 0 & 0 & 4 \\ 1 & 0 & 0 & 3 & -3 \end{bmatrix}.$$

Exercise 4
Find the eigenvalues and corresponding eigenspaces of the matrices

$$\boldsymbol{A} = \begin{bmatrix} 4 & -1 & 0 \\ 1 & 2 & 0 \\ 1 & -1 & 3 \end{bmatrix}, \quad \boldsymbol{B} = \begin{bmatrix} 5 & 12 & -12 \\ 0 & 2 & 0 \\ 1 & 4 & -2 \end{bmatrix} \quad \text{and} \quad \boldsymbol{C} = \begin{bmatrix} 9 & -5 & -5 \\ 4 & -3 & -4 \\ 6 & -3 & -2 \end{bmatrix}.$$

Exercise 5
Calculate all principal minors of the following matrices. Which of the matrices are positive, which are negative definite?

$$\boldsymbol{A} = \begin{bmatrix} -9 & -2 & -1 \\ -2 & -4 & -2 \\ -1 & -2 & -5 \end{bmatrix}, \quad \boldsymbol{B} = \begin{bmatrix} 4 & 1 & 1 \\ 1 & 4 & 2 \\ 1 & 2 & 4 \end{bmatrix}, \quad \boldsymbol{C} = \begin{bmatrix} -2 & 1 & -4 \\ 1 & -3 & 2 \\ -4 & 2 & 1 \end{bmatrix}.$$

Exercise 6
Consider the matrix

$$\boldsymbol{A} = \begin{bmatrix} c & 0 & d \\ 0 & c & 0 \\ d & 0 & c \end{bmatrix},$$

where $c, d \in \mathbb{R}$. For which values of c and d is $\boldsymbol{A}$ positive definite?

Exercise 7
Find a 2×2 matrix $\boldsymbol{A}$ with the following properties: $\boldsymbol{A}$ has the eigenvalues 4 and -2; the corresponding eigenvectors are $[2,4]^T$ and $[2,5]^T$.

Exercise 8
Which of the matrices of Exercise 4 are (real) diagonalizable? For each diagonalizable matrix, find a matrix $\boldsymbol{S}$, such that $\boldsymbol{S}^{-1}\boldsymbol{A}\boldsymbol{S}$ is a diagonal matrix.

Chapter 11

Linear optimization

We begin this chapter with an example.

Example 11.1 (Production planning). Consider a company that manufactures two products, a milk drink and ice cream, using three raw materials: cream, milk powder, and strawberry flavor. Producing one unit of the milk drink requires 5 units of cream, 3 units of milk powder, and 3 units of strawberry flavor. Similarly, each unit of ice cream demands 10 units of cream, 3 units of milk powder, and 1 unit of strawberry flavor. The available quantities of the inputs are limited to 100 units of cream, 36 units of milk powder, and 30 units of strawberry flavor. The contribution margins per unit are 50 monetary units for the milk drink and 33 monetary units for ice cream. The objective is to determine the optimal production plan that maximizes the total contribution margin.

In the following, we denote the products by P1 and P2 and the inputs by I1, I2, and I3. Let x_1 and x_2 represent the quantities produced of P1 and P2, respectively; then the total contribution margin is given by $50x_1 + 33x_2$. The required amount of input I1 is expressed as $5x_1 + 10x_2$; accordingly, those of I2 and I3 are $3x_1 + 3x_2$ and $3x_1 + x_2$, respectively. Given the limited availability of inputs, the **capacity constraints** are $5x_1 + 10x_2 \leq 100$ (input I1), $3x_1 + 3x_2 \leq 36$ (input I2), and $3x_1 + x_2 \leq 30$ (input I3). Since production quantities cannot be negative, we also impose the non-negativity constraints $x_1, x_2 \geq 0$. Thus, the production planning

K. Mosler et al., *Mathematical Methods for Business and Economics*, Mathematics Study Resources 20, https://doi.org/10.1007/978-3-662-73831-3_11

problem is formulated as the following optimization problem:

$$
\boxed{
\begin{array}{ll}
\text{Maximize} & 50x_1 + 33x_2 \\
\text{subject to the constraints} & 5x_1 + 10x_2 \le 100\,, \\
 & 3x_1 + 3x_2 \le 36\,, \\
 & 3x_1 + x_2 \le 30\,, \\
 & x_1, x_2 \ge 0\,.
\end{array}
}
\tag{11.1}
$$

Instead of "Maximize" and "subject to the constraints", one often writes briefly "max" and "s.t.".

In the present optimization problem both the objective function and the constraints are described by **linear functions**. Such an optimization problem is referred to as a **linear optimization problem** or **linear program**, which is usually abbreviated as **LP**. Linear programs can appear in many variants.

- In the above example, the **objective function** $50x_1 + 33x_2$ is to be maximized. In other applications, the objective function is to be minimized.
- The constraints (besides non-negativity) in the example come as less-than-or-equal-to inequalities. In general, the constraints can also be greater-than-or-equal-to inequalities or equations.
- In general, variables can also be subject to nonpositivity constraints as well as have no sign constraints.

However, the various variants of LPs can be transformed into each other:

- Since
$$\min \sum_{j=1}^{n} c_j x_j = -\max\left(-\sum_{j=1}^{n} c_j x_j\right),$$
a minimization problem can be transformed into a maximization problem by multiplying the objective function by -1.
- Consider a constraint of the form $\sum_{j=1}^{n} a_{ij}x_j \le b_i$. By introducing a so-called **slack variable** s_i, this constraint can be transformed into an equation,
$$\sum_{j=1}^{n} a_{ij}x_j + s_i = b_i\,, \quad \text{with } s_i \ge 0\,.$$
- The constraint $\sum_{j=1}^{n} a_{ij}x_j \ge b_i$ can be transformed into an equivalent equation $\sum_{j=1}^{n} a_{ij}x_j - s_i = b_i$ by introducing a slack variable $s_i \ge 0$.
- The constraint $\sum_{j=1}^{n} a_{ij}x_j = b_i$ is equivalent to the inequalities $\sum_{j=1}^{n} a_{ij}x_j \le b_i$ and $\sum_{j=1}^{n} a_{ij}x_j \ge b_i$ holding simultaneously.

- The inequality $\sum_{j=1}^{n} a_{ij}x_j \leq b_i$ is equivalent to $\sum_{j=1}^{n}(-a_{ij})x_j \geq -b_i$.
- If a variable x_j is unconstrained, it can be replaced by $x'_j - x''_j$, where $x'_j, x''_j \geq 0$.

When considering LPs, we therefore focus on two basic forms, the so-called **standard form** and the **canonical form**.

An LP is in **standard form** if all constraints are equations and all variables are nonnegative. A maximization problem is in **canonical form** if all constraints are $\leq$-inequalities and all variables are nonnegative. Similarly, a minimization problem is in **canonical form** if all constraints are $\geq$-inequalities and all variables are nonnegative.

Maximization problem in standard form	Maximization problem in canonical form
$\max \sum_{j=1}^{n} c_j x_j$ s.t. $\sum_{j=1}^{n} a_{ij}x_j = b_i,\ i = 1,\dots,m$; $x_j \geq 0,\ j = 1,\dots,n$	$\max \sum_{j=1}^{n} c_j x_j$ s.t. $\sum_{j=1}^{n} a_{ij}x_j \leq b_i,\ i = 1,\dots,m$; $x_j \geq 0,\ j = 1,\dots,n$

Minimization problem in standard form	Minimization problem in canonical form
$\min \sum_{j=1}^{n} c_j x_j$ s.t. $\sum_{j=1}^{n} a_{ij}x_j = b_i,\ i = 1,\dots,m$; $x_j \geq 0,\ j = 1,\dots,n$	$\min \sum_{j=1}^{n} c_j x_j$ s.t. $\sum_{j=1}^{n} a_{ij}x_j \geq b_i,\ i = 1,\dots,m$; $x_j \geq 0,\ j = 1,\dots,n$

Example 11.2 (Continuation of Example 11.1). The LP in Example 11.1 is a maximization problem in canonical form. To transform this LP into standard form, we need to introduce slack variables s_1, s_2, s_3 for the three capacity constraints. This gives:

$$\begin{array}{lrcl} \max & 50x_1 + 33x_2 & & \\ \text{s.t.} & 5x_1 + 10x_2 + s_1 & = & 100 \\ & 3x_1 + 3x_2 + s_2 & = & 36 \\ & 3x_1 + x_2 + s_3 & = & 30 \\ & x_1, x_2, s_1, s_2, s_3 & \geq & 0 \end{array} \tag{11.2}$$

The i-th slack variable s_i indicates the free capacity with respect to the i-th input.

11.1 Graphical solution

If an LP has only two variables, a graphical solution is possible. We illustrate this with our production planning example. First, draw the three lines $5x_1 + 10x_2 = 100$, $3x_1 + 3x_2 = 36$ and $3x_1 + x_2 = 30$ in a coordinate system. These lines represent production plans that exactly exhaust the available input capacities. In Figure 11.1, these lines are shown in black. The set of production plans in which the capacity of a given input is not fully utilized is given by the halfplane bounded on the upper right by the corresponding straight line. The set of feasible production plans is therefore given by the intersection of the corresponding halfplanes. Due to the nonnegativity constraints, all feasible production plans must lie in the first quadrant. In general, the set of all points that satisfy the constraints is called the **feasible region**. Each point in the feasible region is called a **feasible solution**.

The set of feasible production plans is shown in light gray in Figure 11.2. The black lines represent the isoquants of the total contribution margin. The further "up and to the right" an isoquant lies, the greater the total contribution margin. The leftmost line in Figure 11.2 still intersects the feasible region, indicating that the total contribution margin can be increased by shifting the line parallel to the right. The rightmost line does not intersect the feasible region, meaning that none of the production plans on this line can be realized given the existing capacities.

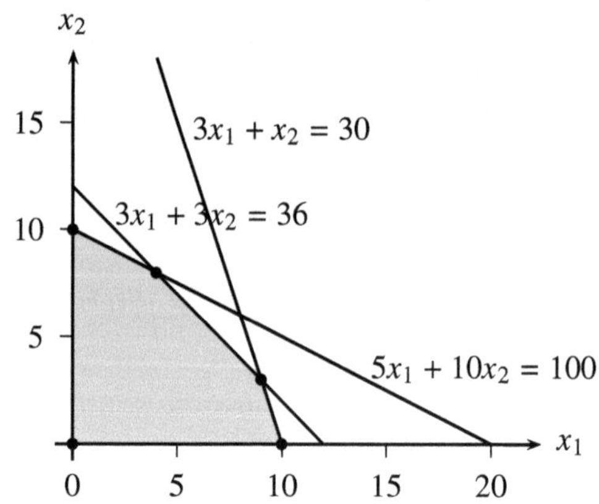

Figure 11.1: The feasible region from Example 11.1.

To find an optimal production plan, move the left isoquant upward and to the right as long as it still intersects the feasible region at least at one point. Alternatively, you can move the right isoquant downward and to the left until it first intersects the region of feasible production plans at at least one point. The resulting line is highlighted in bold. All intersections of this line with the feasible region represent optimal production plans that maximize the contribution margin. In our example, there is a single intersection with the feasible region at the point $(x_1, x_2) = (9, 3)$. This production plan yields the maximum contribution margin of 549 monetary units.

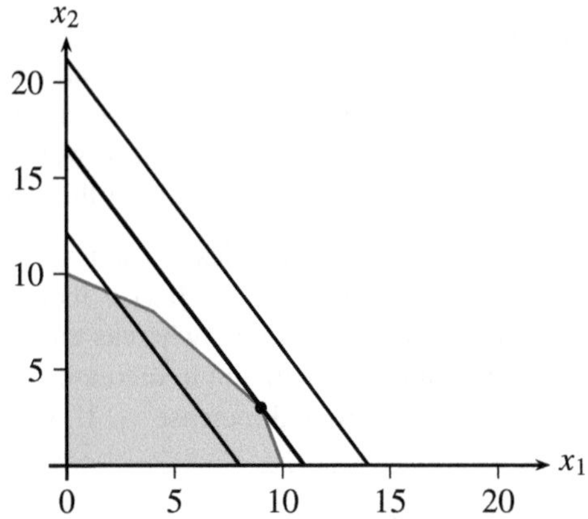

Figure 11.2: Finding the optimal production plan in Example 11.1.

11.2 The simplex method

A graphical solution to an LP is only possible for problems with two variables and not too many constraints. Therefore, it is essential to have a method capable of solving linear programs with many variables and constraints. Such a method is the **simplex method**, developed by Dantzig[1]. The basic idea of the simplex method is as follows.

Let us assume that the feasible region is not empty, as otherwise no solution to the LP exists. From the considerations of the graphical solution, it is evident that the optimum value is achieved at a vertex (corner point) of the feasible region. With two variables ($n = 2$), the optimum is at a vertex, but it can also occur along an entire edge, when the function's level curves are parallel to that edge. Also in higher dimensions, the optimum may occur not only at a vertex but at a whole face of the feasible region. However, there is always at least one vertex where the objective value is optimal.

The simplex method exploits this property. Starting at a vertex of the feasible region, it moves through a sequence of adjacent vertices, choosing each subsequent vertex so that the objective value does not decrease. After a finite number of steps, the method typically reaches a vertex that yields the optimal value of the objective. We will now illustrate this approach computationally using our production example.

For the computational solution of the LP from Example 11.1 we start with the standard form (11.2). First, we solve the constraints of the LP for the slack variables s_1, s_2, s_3:

$$\boxed{\begin{aligned} s_1 &= 100 - 5x_1 - 10x_2 \\ s_2 &= 36 - 3x_1 - 3x_2 \\ s_3 &= 30 - 3x_1 - x_2 \end{aligned}} \tag{11.3}$$

[1]George Bernard Dantzig (1914–2005)

If the values of the variables x_1 and x_2 are known, the variables s_1, s_2 and s_3 are uniquely determined. For example, for $x_1 = x_2 = 0$ the values $s_1 = 100$, $s_2 = 36$ and $s_3 = 30$ are obtained. This solution obviously corresponds to a feasible (though not very purposive) production program: all constraints as well as the non-negativity conditions are satisfied. The total contribution margin of this solution is zero.

What happens now if we "adjust" the variables x_1 and x_2? Let us consider the objective function to be maximized: $50x_1 + 33x_2$. If x_1 is increased by one unit, the total contribution margin increases by 50 units, whereas an increase of x_2 by one unit raises it by only 33 units. Thus, one will aim to increase x_1 as much as possible first. We therefore keep x_2 fixed at zero and increase x_1. If we set $x_2 = 0$ in (11.3), we obtain:

$$\begin{aligned} s_1 &= 100 - 5x_1 \\ s_2 &= \ \ 36 - 3x_1 \\ s_3 &= \ \ 30 - 3x_1 \end{aligned}$$

By increasing x_1, the variables s_1, s_2, s_3 all become smaller. However, since all variables must be nonnegative, we cannot increase x_1 arbitrarily. From $x_1 = 20$ onward, s_1 becomes negative, from $x_1 = 12$ onward, s_2 becomes negative, and already from $x_1 = 10$ onward, s_3 becomes negative. Since all variables should remain nonnegative, we can increase x_1 only to a maximum value of 10. Therefore, the third constraint is called the **blocking constraint**, and the variable s_3 that would become negative is called the **blocking variable**. This leads to the following solution of the system: $x_1 = 10$, $x_2 = 0$, $s_1 = 50$, $s_2 = 6$ and $s_3 = 0$.

If we now solve the third equation in (11.3) for x_1 and substitute this in the first and second equation in (11.3), we get:

$$\begin{aligned} s_1 &= 50 + \frac{5}{3}s_3 - \frac{25}{3}x_2 \\ s_2 &= \ \ 6 + \ \ s_3 - \ \ 2x_2 \\ x_1 &= 10 - \frac{1}{3}s_3 - \ \ \frac{1}{3}x_2 \end{aligned} \tag{11.4}$$

We now have s_1, s_2 and x_1 expressed in terms of s_3 and x_2. Next, we express the objective function in terms of s_3 and x_2 and get

$$500 - \frac{50}{3}s_3 + \frac{49}{3}x_2 \,.$$

At the current solution $x_1 = 10$, $x_2 = 0$, $s_1 = 50$, $s_2 = 6$ and $s_3 = 0$ the value of the objective function is 500. Now, what is the effect of an increase in one of the two zero variables x_2 and s_3? An increase in s_3 would mean a reduction of the objective value, as the corresponding coefficient is negative. On the other hand, an increase in x_2 results in a further improvement of the objective value. So we leave s_3 at the value zero and check how far we can increase x_2 without violating the constraints. To do this, we set $s_3 = 0$ in (11.4) and get:

$$\begin{aligned} s_1 &= 50 - \frac{25}{3}x_2 \\ s_2 &= 6 - 2x_2 \\ x_1 &= 10 - \frac{1}{3}x_2 \end{aligned}$$

The variable s_1 becomes negative from $x_2 = 6$ onward, the variable s_2 from $x_2 = 3$, and the variable x_1 from $x_2 = 30$. So, the second constraint is blocking and we can increase x_2 to a maximum value of 3, without violating the nonnegativity constraints. This leads to the solution $x_1 = 9$, $x_2 = 3$, $s_1 = 25$ and $s_2 = s_3 = 0$. If we express – analogous to above – the variables x_1, x_2 and s_1 with s_2 and s_3, we obtain from (11.4):

$$\begin{aligned} s_1 &= 25 + \frac{25}{6}s_2 - \frac{5}{2}s_3 \\ x_2 &= 3 - \frac{1}{2}s_2 + \frac{1}{2}s_3 \\ x_1 &= 9 + \frac{1}{6}s_2 - \frac{1}{2}s_3 \end{aligned}$$

Now, if we express the objective function in terms of the variables s_2 and s_3, we get

$$549 - \frac{49}{6}s_2 - \frac{17}{2}s_3\,.$$

Here, it is immediately clear that the current solution $x_1 = 9$, $x_2 = 3$, $s_1 = 25$ and $s_2 = s_3 = 0$ is indeed the optimal solution of our LP. For this solution, the objective value 549. An increase in any of the free variables s_2 and s_3 would now lead to a decrease of the objective value.

Therefore, the maximum of the objective function is achieved when nine units of product P1 and three units of product P2 are produced. This yields a total contribution margin of 549. At the same time, the values of the slack variables at the optimum indicate that the inputs I2 and I3 are completely consumed ($s_2 = s_3 = 0$), while a residual quantity of 25 units remains from input I1 ($s_1 = 25$).

The above calculations are exactly the steps that are carried out when solving the LP with the simplex method. We reached the solution of the LP by moving through a sequence of vertices. We started at the vertex $(0, 0)$, moved to the vertex $(10, 0)$ in the first step, and finally arrived at the optimal solution $(9, 3)$.

11.3 The mathematics of the simplex method

In the following, we consider a maximization problem in standard form. We denote by $\boldsymbol{A} = [a_{ij}]$ the $m \times n$ matrix of the constraint coefficients and by $\boldsymbol{b} = [b_i]$ the column vector of the right-hand sides of the constraints. Furthermore, $\boldsymbol{c} = [c_j]$ is the row vector of coefficients of the objective function and $\boldsymbol{x}$ is the column vector of the variables $x_1, \ldots, x_n$. Thus, the maximization problem in standard form is written as:

$$\begin{array}{ll} \max & \boldsymbol{c}\boldsymbol{x} \\ \text{s.t.} & \boldsymbol{A}\boldsymbol{x} = \boldsymbol{b} \\ & \boldsymbol{x} \geq \boldsymbol{0} \end{array}$$

In the above example, we have

$$\boldsymbol{A} = \begin{bmatrix} 5 & 10 & 1 & 0 & 0 \\ 3 & 3 & 0 & 1 & 0 \\ 3 & 1 & 0 & 0 & 1 \end{bmatrix}, \quad \boldsymbol{b} = \begin{bmatrix} 100 \\ 36 \\ 30 \end{bmatrix}, \quad \text{and} \quad \boldsymbol{c} = \begin{bmatrix} 50 & 33 & 0 & 0 & 0 \end{bmatrix}.$$

The set of all $\boldsymbol{x} \in \mathbb{R}^n$ that satisfy the constraints $\boldsymbol{A}\boldsymbol{x} = \boldsymbol{b}$ and the nonnegativity constraints $\boldsymbol{x} \geq \boldsymbol{0}$ is the feasible region. It can be shown that the feasible region of an LP is a so-called **convex polyhedron**[2]. A convex polyhedron has a finite number of vertices.

The feasible region of an LP can be empty. In this case, there are no points that satisfy all constraints and the LP obviously has no solution.

To understand the simplex method, we need to know how the vertices of the feasible region can be characterized. If we look at the feasible region of our initial example (see Figure 11.3), we recognize the following two properties:

- At each vertex, exactly two of the five variables are equal to zero.
- At two adjacent vertices, the same variables are zero except for one.

These two properties apply not only in our example, but in general: Consider an LP in standard form, having constraints $\boldsymbol{A}\boldsymbol{x} = \boldsymbol{b}$, $\boldsymbol{x} \geq \boldsymbol{0}$. If $\boldsymbol{x}_0$ is a vertex of the feasible region, then $n-m$ of the n variables $x_1, \ldots, x_n$ are zero. Therefore, to obtain a vertex of the feasible region, one can set $n-m$ variables equal to zero and solve the equation $\boldsymbol{A}\boldsymbol{x} = \boldsymbol{b}$ for the remaining m variables. If the values of these m variables are all nonnegative, such a solution represents a vertex of the feasible region and is referred to as a **basic feasible solution**. We will clarify this term in the following definition.

[2] A convex polyhedron is the intersection of finitely many **halfspaces**. A halfspace is a subset of $\mathbb{R}^n$ described by an inequality of the form $\boldsymbol{a}^T\boldsymbol{x} \leq b$, for given $\boldsymbol{a} \in \mathbb{R}^n$, $b \in \mathbb{R}$.

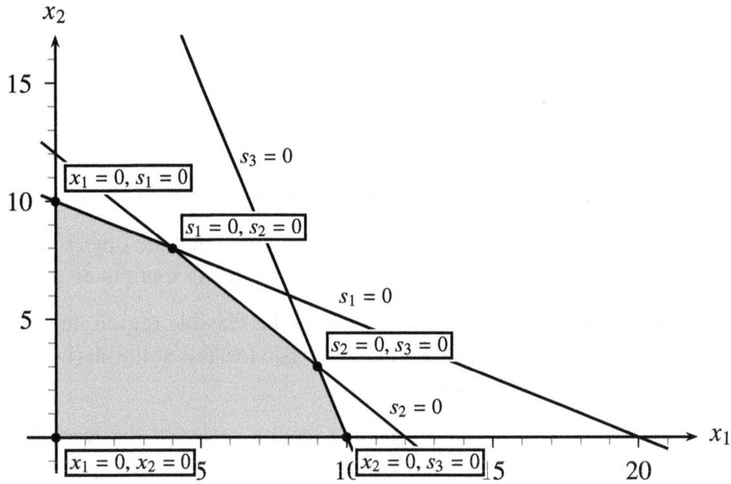

Figure 11.3: Vertices of the feasible region from Example 11.1.

Definition 11.1: Basic feasible solution, basis matrix

We consider the system $\boldsymbol{A}\boldsymbol{x} = \boldsymbol{b}$, $\boldsymbol{x} \geq \boldsymbol{0}$, where $\boldsymbol{A}$ is an $m \times n$ matrix with $\operatorname{rk} \boldsymbol{A} = m$. After possibly rearranging the columns of $\boldsymbol{A}$, one can write $\boldsymbol{A}$ in the form $\boldsymbol{A} = [\boldsymbol{B}, \boldsymbol{N}]$, where $\boldsymbol{B}$ is an invertible $m \times m$ matrix and $\boldsymbol{N}$ is an $m \times (n-m)$ matrix. $\boldsymbol{B}$ is called **basis matrix**, $\boldsymbol{N}$ is called **nonbasic matrix**. The variables corresponding to the columns of $\boldsymbol{B}$ are referred to as **dependent variables** or **basic variables**, the remaining variables as **independent variables** or **nonbasic variables**. The set of basic variables is referred to as **basis**. Let $\boldsymbol{x}_B$ be the vector of basic variables and $\boldsymbol{x}_N$ the vector of nonbasic variables. Obviously, for

$$\boldsymbol{x}_B = \boldsymbol{B}^{-1}\boldsymbol{b} \quad \text{and} \quad \boldsymbol{x}_N = \boldsymbol{0}$$

one obtains a solution of the system $\boldsymbol{A}\boldsymbol{x} = \boldsymbol{b}$. Such a solution is called **basic solution**. If all components of $\boldsymbol{x}_B = \boldsymbol{B}^{-1}\boldsymbol{b}$ are nonnegative, the solution is called **basic feasible solution**. If all components of $\boldsymbol{x}_B$ are positive, it is called a **nondegenerate** basic feasible solution, otherwise it is called **degenerate**.

As already mentioned, the notion of a basic feasible solution is of such importance because the basic feasible solutions are precisely the vertices of the feasible region. This is the statement of the following theorem:

Theorem 11.1: Vertices and basic feasible solutions

If $\boldsymbol{x}$ is a vertex of the feasible region, there is (at least) one basis such that $\boldsymbol{x}$ is the corresponding basic feasible solution. Conversely, if $\boldsymbol{x}$ is a basic feasible solution, then $\boldsymbol{x}$ is a vertex of the feasible region.

What can be stated about the solvability of an LP? Essentially, there are three possible scenarios in a maximization problem:

- The feasible region of an LP may be empty. In this case, the LP has no solution.
- If the feasible region is not empty, it may still happen that the objective function is unbounded over the feasible region. In this case, again there is no solution.
- If the objective function is bounded above on the feasible region, the LP has at least one solution. In this case, there exists a basic feasible solution (i.e., a vertex) at which the maximum is attained.

The simplex method checks whether, starting from a vertex of the feasible region, the objective value can be further improved at an adjacent vertex. An adjacent vertex is characterized by the fact that the set of basic variables (and thus also the set of nonbasic variables) differs at exactly one position from the set of current basic variables. The transition to an adjacent vertex (basic feasible solution) thus means that one of the previous nonbasic variables becomes a basic variable and one of the previous basic variables leaves the basis.

Selection of the nonbasic variable that enters the basis

Assume we have a basic feasible solution with basis matrix $\boldsymbol{B}$ and nonbasic matrix $\boldsymbol{N}$. Then, $\boldsymbol{B}\boldsymbol{x}_B + \boldsymbol{N}\boldsymbol{x}_N = \boldsymbol{b}$ holds. Multiplying this equation from the left with the inverse of $\boldsymbol{B}$, we get

$$\boldsymbol{x}_B = \boldsymbol{B}^{-1}\boldsymbol{b} - \boldsymbol{B}^{-1}\boldsymbol{N}\boldsymbol{x}_N\,.$$

With the notation $\overline{\boldsymbol{b}} = \boldsymbol{B}^{-1}\boldsymbol{b}$ we have

$$\boxed{\boldsymbol{x}_B = \overline{\boldsymbol{b}} - \boldsymbol{B}^{-1}\boldsymbol{N}\boldsymbol{x}_N\,.} \tag{11.5}$$

The associated basic feasible solution is $\boldsymbol{x}_B = \overline{\boldsymbol{b}}$, $\boldsymbol{x}_N = \boldsymbol{0}$. Replacing $\boldsymbol{x}_B$ in the objective function with the corresponding expression from (11.5) gives

$$\begin{aligned}\boldsymbol{c}\boldsymbol{x} &= \boldsymbol{c}_B\boldsymbol{x}_B + \boldsymbol{c}_N\boldsymbol{x}_N \\ &= \boldsymbol{c}_B(\overline{\boldsymbol{b}} - \boldsymbol{B}^{-1}\boldsymbol{N}\boldsymbol{x}_N) + \boldsymbol{c}_N\boldsymbol{x}_N \\ &= \boldsymbol{c}_B\overline{\boldsymbol{b}} + (\boldsymbol{c}_N - \boldsymbol{c}_B\boldsymbol{B}^{-1}\boldsymbol{N})\boldsymbol{x}_N\,.\end{aligned} \tag{11.6}$$

For short, we write z_0 for $\boldsymbol{c}_B\overline{\boldsymbol{b}}$ and z_j for $\boldsymbol{c}_B\boldsymbol{B}^{-1}\boldsymbol{a}_j$, where $\boldsymbol{a}_j$ is the j-th column of $\boldsymbol{A}$. Thus, for the objective function, we get

$$\boxed{\boldsymbol{c}\boldsymbol{x} = z_0 + \sum_{j\in N}(c_j - z_j)x_j\,,}$$

where the sum extends over all indices j of nonbasic variables. We now have expressed the objective function in terms of the nonbasic variables. If all nonbasic variables are zero, the objective value is z_0. It is also clear that an increase in the nonbasic variable x_j by one unit results in an increase of the objective value by $c_j - z_j$.

What is interesting at this point is the interpretation of the quantities $c_j - z_j$, which we would like to discuss briefly here. The quantity c_j is the coefficient of x_j in the objective function and thus indicates the change in the objective value when x_j is increased by one unit. Since the basic variables depend on the nonbasic variables via the relation (11.5), an increase in a nonbasic variable also results in a change in the basic variables, which again changes the objective value. An increase in a nonbasic variable thus has a *direct influence* on the objective value, which is expressed by the coefficient c_j, as well as an *indirect influence* via the change in the basic variables. This indirect influence is described by the coefficient z_j. An increase in a nonbasic variable x_j causes a change in the basic variables, resulting in a reduction in the objective value by z_j. The value $c_j - z_j$ thus represents the direct influence of the nonbasic variable x_j reduced by the indirect influence of this variable. Therefore, depending on the context the quantity $c_j - z_j$ is referred to as **reduced contribution margin** (as in our introductory example) or more common as **reduced costs** (when, for example, a linear cost function is to be minimized). $c_j - z_j$ represents the **net contribution** of x_j to the objective function. Throughout the remainder of this chapter, we will refer to $c_j - z_j$ as **reduced cost**, regardless of the context.

If x_j is a basic variable, then obviously $c_j - z_j = c_j - \boldsymbol{c}_B \boldsymbol{B}^{-1} \boldsymbol{a}_j = c_j - c_j = 0$, i.e., the reduced costs of the basic variables are equal to zero.

What does this mean for the solution of an LP? In general, two cases can occur:

1. If all reduced costs $c_j - z_j$ of nonbasic variables are nonpositive, then no further improvement of the objective value is possible. The current basic feasible solution is optimal. The optimal objective value is z_0.

2. If the reduced cost of some nonbasic variable is still positive, then the objective value can be further improved by increasing this variable. Typically, one tries to increase that nonbasic variable, whose reduced cost is maximum among all positive reduced costs. So if

$$c_{j^*} - z_{j^*} = \max\left\{c_j - z_j \; : \; c_j - z_j > 0\right\},$$

then the nonbasic variable x_{j^*} will enter the basis in the next step.

Selection of the basic variable that leaves the basis

If a nonbasic variable x_{j^*} having positive reduced cost enters the basis, one of the previous basic variables must leave the basis. Increasing x_{j^*} and keeping all other nonbasic variables fixed at zero gives

$$\boldsymbol{x}_B = \overline{\boldsymbol{b}} - \boldsymbol{B}^{-1} \boldsymbol{a}_{j^*} x_{j^*} = \overline{\boldsymbol{b}} - \boldsymbol{y}_{j^*} x_{j^*},$$

thus

$$\begin{bmatrix} x_{B_1} \\ x_{B_2} \\ \vdots \\ x_{B_m} \end{bmatrix} = \begin{bmatrix} \overline{b}_1 \\ \overline{b}_2 \\ \vdots \\ \overline{b}_m \end{bmatrix} - x_{j^*} \begin{bmatrix} y_{1j^*} \\ y_{2j^*} \\ \vdots \\ y_{mj^*} \end{bmatrix},$$

where for simplicity the notation $\boldsymbol{B}^{-1}\boldsymbol{a}_{j^*} = \boldsymbol{y}_{j^*} = [y_{1j^*} \cdots y_{mj^*}]^T$ was used. The i-th basic variable x_{B_i} thus changes to

$$x_{B_i} = \overline{b}_i - x_{j^*} y_{ij^*} .$$

If y_{ij^*} is nonpositive, x_{j^*} can be increased arbitrarily without x_{B_i} becoming zero. However, if y_{ij^*} is positive, increasing x_{j^*} decreases x_{B_i}. For $x_{j^*} = \overline{b}_i / y_{ij^*}$, the basic variable x_{B_i} becomes zero. Since all variables must be nonnegative, x_{j^*} can only be increased until the first basic variable reaches zero. This variable is called the **blocking variable**. To find the blocking variable, we form the quotients $\overline{b}_i / y_{ij^*}$ for all positive y_{ij^*}. The smallest of these quotients gives the maximum amount up to which x_{j^*} can be increased and thus determines the blocking variable. Overall, two situations can occur:

1. If all y_{ij^*} are nonpositive, x_{j^*} can be increased arbitrarily without violating the constraints. Since the reduced cost of x_{j^*} is positive, the objective value can be made arbitrarily large. Hence, the objective function is unbounded over the feasible region and there is no solution to the LP.

2. If at least one of the values y_{ij^*} is positive, we determine

$$\frac{\overline{b}_{i^*}}{y_{i^*j^*}} = \min_i \left\{ \frac{\overline{b}_i}{y_{ij^*}} : y_{ij^*} > 0 \right\} .$$

   This is the value to which x_{j^*} can be increased. The variable $x_{B_{i^*}}$ then falls to zero, i.e., it leaves the basis and the variable x_{j^*} enters the basis. Thus, a transition to a new basis is performed by exchanging the variables $x_{B_{i^*}}$ and x_{j^*}.

In this way, the simplex method determines a new basis for which the objective value does not decrease. Starting from this new basis, we examine whether the solution can be further improved. The procedure repeats until an optimum is attained or it is established that the objective function is unbounded over the feasible region.

11.4 The simplex method in tableau form

The simplex method in tableau form can be executed either with a **full tableau** or with a **shortened tableau**, where columns corresponding to basic variables are omitted. In this section, we first describe the full tableau and subsequently show how the algorithm can be performed more compactly using the shortened tableau.

Simplex method with full tableau

The simplex method in tableau form starts from a maximization problem in standard form. In addition, the variable $z = \boldsymbol{cx}$ is introduced, which represents the objective value. If a basic feasible solution is known, the LP can be written as follows:

$$
\begin{aligned}
\max\quad & z \\
\text{s.t.}\quad & z - \boldsymbol{c_B x_B} - \boldsymbol{c_N x_N} = 0 \\
& \boldsymbol{B x_B} + \boldsymbol{N x_N} = \boldsymbol{b} \\
& \boldsymbol{x_B},\quad \boldsymbol{x_N} \geq \boldsymbol{0}
\end{aligned}
$$

Solving the constraints for the basic variables and expressing the objective function completely in terms of nonbasic variables gives

$$
\begin{aligned}
\max\quad & z \\
\text{s.t.}\quad & z \qquad\quad + (\boldsymbol{c_B B^{-1} N} - \boldsymbol{c_N})\boldsymbol{x_N} = \boldsymbol{c_B \bar{b}} \\
& \quad \boldsymbol{x_B} + \qquad\qquad \boldsymbol{B^{-1} N x_N} = \boldsymbol{\bar{b}} \\
& \quad \boldsymbol{x_B}, \qquad\qquad\qquad\quad \boldsymbol{x_N} \geq \boldsymbol{0}
\end{aligned}
$$

The constraints represent a linear system of equations with m equations and $n + 1$ variables, which we write in tableau form like in the Gauß algorithm:

	z	$\boldsymbol{x_B}$	$\boldsymbol{x_N}$	
z	1	$\boldsymbol{0}$	$\boldsymbol{c_B B^{-1} N} - \boldsymbol{c_N}$	$\boldsymbol{c_B \bar{b}}$
$\boldsymbol{x_B}$	$\boldsymbol{0}$	$\boldsymbol{I}$	$\boldsymbol{B^{-1} N}$	$\boldsymbol{\bar{b}}$

For better clarity, the variables are noted at the top and left in the tableau. These include the objective function variable, the basic variables, as well as the nonbasic variables. Below the first $m + 1$ variables (objective function variable and basic variables) is an identity matrix. Below the following $n - m$ variables (nonbasic variables) the objective function row contains the negative of the reduced costs, and the subsequent rows contain the matrix $\boldsymbol{B}^{-1}\boldsymbol{N}$. In the last column are the entries of the right-hand side, i.e., the objective value corresponding to the basic solution, and in the following m rows the values of the basic variables of the current basic feasible solution. It makes sense to list the objective function variable and the basic variables also to the left of the tableau.

Remark: In the objective function row, not the reduced costs themselves are shown, but their negative values. Therefore, when determining the variable that enters the basis, one has to find the minimum entry in the objective function row of the tableau, rather than the maximum.

To solve our example 11.1 with the simplex method we therefore start from the following LP:

$$
\begin{array}{llr}
\max & z & \\
\text{s.t.} & z - 50x_1 - 33x_2 & = \ \ 0 \\
& \quad 5x_1 + 10x_2 + s_1 & = 100 \\
& \quad 3x_1 + \ 3x_2 \qquad + s_2 & = \ \ 36 \\
& \quad 3x_1 + \quad x_2 \qquad\qquad + s_3 & = \ \ 30 \\
& \qquad\qquad x_1, x_2, s_1, s_2, s_3 & \geq \ \ 0
\end{array}
\tag{11.7}
$$

To create a simplex tableau, a basic feasible solution must be known. In our example, the columns corresponding to the slack variables already form a basis, the corresponding basis matrix $\boldsymbol{B}$ is then the identity matrix $\boldsymbol{I}$. Thus, the initial tableau of our LP is written as follows. In the first row of the tableau are the *negative* coefficients of the objective function and on the right the current value of the objective function. Below these entries are the coefficient matrix $\boldsymbol{A}$ and in the last column the values of the basic variables:

	z	s_1	s_2	s_3	x_1	x_2	
z	1				−50	−33	0
s_1		1			5	10	100
s_2			1		3	3	36
s_3				1	$\boxed{3}$	1	30

This corresponds to the initial situation of the above example: The basic variables are s_1, s_2 and s_3, their corresponding values 100, 36 and 30. The nonbasic variables are x_1 and x_2. The value of the objective function is zero.

In the next step, we search the smallest entry in the objective function row, which corresponds to the largest coefficient of the objective function. In our example, this is −50, the corresponding variable is x_1. The corresponding column in the simplex tableau is referred to as the **pivot column**. The variable x_1 will enter the basis in the next step. Thus, one of the previous basic variables must leave the basis. This variable is determined by the **minimum ratio test**, that is, we must find the minimum of the ratios $\overline{b}_i/y_{ij^*}$, for which $y_{ij^*} > 0$. The values y_{ij^*} are just the entries in the pivot column. By forming the quotients $100/5 = 20$, $36/3 = 12$ and $30/3 = 10$ we see that s_3 is the blocking variable. The corresponding row is referred to as the **pivot row**, the element at the intersection of pivot row and column as the **pivot element**. In the following simplex step, s_3 leaves the basis and x_1 enters the basis. This basis exchange can be carried out by elementary row transformations as in the Gauß-Jordan algorithm. The tableau is transformed in such a way, that in the pivot column (the column of the entering variable) there is only a single one in the pivot row (the row of the leaving variable) and otherwise zeros. Therefore, the one in the updated tableau is at the position of the pivot element. So, in our example, the third row must be divided by three and the resulting third row added 50 times to the objective function row and subtracted five and three times from the first and second row, respectively. This gives the updated tableau:

	z	s_1	s_2	s_3	x_1	x_2	
z	1			$\frac{50}{3}$		$-\frac{49}{3}$	500
s_1		1		$-\frac{5}{3}$		$\frac{25}{3}$	50
s_2			1	-1		$\boxed{2}$	6
x_1				$\frac{1}{3}$	1	$\frac{1}{3}$	10

The new basic variables are now s_1, s_2 and x_1 with corresponding values 50, 6 and 10. The nonbasic variables are x_2 and s_3, the objective value is 500. Now in the objective function row, only one entry is negative, namely the one corresponding to x_2. To further improve the objective value, x_2 must enter the basis. The variable that leaves the basis is again determined by the minimum ratio test, i.e., by forming the ratios $50/\frac{25}{3} = 6$, $6/2 = 3$ and $10/\frac{1}{3} = 30$. The minimum ratio is obtained in the row corresponding to s_2. So, s_2 leaves the basis and x_2 enters the basis. As before, this is achieved by elementary row transformations, transforming the tableau so that the pivot element (at the intersection of the pivot row and pivot column) becomes one, and all other entries in the pivot column become zero.

	z	s_1	s_2	s_3	x_1	x_2	
z	1		$\frac{49}{6}$	$\frac{17}{2}$			549
s_1		1	$-\frac{25}{6}$	$\frac{5}{2}$			25
x_2			$\frac{1}{2}$	$-\frac{1}{2}$		1	3
x_1			$-\frac{1}{6}$	$\frac{1}{2}$	1		9

This tableau now corresponds to the basic feasible solution, where the basic variables s_1, x_2 and x_1 have the values 25, 3 and 9. The nonbasic variables are s_2 and s_3, the value of the objective function is 549. Since the entries in the objective function row are all nonnegative, a further increase in the objective value by increasing the nonbasic variables is not possible. The optimal value of our LP is therefore 549. This solution is obtained for $x_1 = 9$ and $x_2 = 3$.

Simplex method with shortened tableau

All tableaux of the previous section have one thing in common: four of the columns are equal to the canonical unit vectors. These are exactly the columns that correspond to the objective function variable and the three basic variables. Therefore, the tableau can also be represented in a shortened form by simply omitting these columns. No information is lost as long as you keep track of which nonbasic variables correspond to which columns, and which basic variables correspond to which rows. Thus, the initial tableau in shortened form for the example above is as follows:

	z	s_1	s_2	s_3	x_1	x_2	
z	1				−50	−33	0
s_1		1			5	10	100
s_2			1		3	3	36
s_3				1	3	1	30

$\xrightarrow[\text{form}]{\text{shortened}}$

	x_1	x_2	
z	−50	−33	0
s_1	5	10	100
s_2	3	3	36
s_3	3	1	30

In the next simplex step, x_1 entered the basis, while s_3 left the basis. When updating the full tableau, the x_1 column is transformed into a canonical unit vector. Since in the shortened tableau we do not show columns that correspond to basic variables, the s_3 column (new nonbasic variable) takes the place of the x_1 column. We illustrate this again by juxtaposing the corresponding tableaux:

	z	s_1	s_2	s_3	x_1	x_2	
z	1			$\frac{50}{3}$		$-\frac{49}{3}$	500
s_1		1		$-\frac{5}{3}$		$\frac{25}{3}$	50
s_2			1	−1		2	6
x_1				$\frac{1}{3}$	1	$\frac{1}{3}$	10

$\xrightarrow[\text{form}]{\text{shortened}}$

	s_3	x_2	
z	$\frac{50}{3}$	$-\frac{49}{3}$	500
s_1	$-\frac{5}{3}$	$\frac{25}{3}$	50
s_2	−1	2	6
x_1	$\frac{1}{3}$	$\frac{1}{3}$	10

In the next simplex step, x_2 enters the basis, while s_2 leaves the basis. In the shortened tableau, the new s_2 column takes the place of the x_2 column:

	z	s_1	s_2	s_3	x_1	x_2	
z	1		$\frac{49}{6}$	$\frac{17}{2}$			549
s_1		1	$-\frac{25}{6}$	$\frac{5}{2}$			25
x_2			$\frac{1}{2}$	$-\frac{1}{2}$		1	3
x_1			$-\frac{1}{6}$	$\frac{1}{2}$	1		9

$\xrightarrow[\text{form}]{\text{shortened}}$

	s_3	s_2	
z	$\frac{17}{2}$	$\frac{49}{6}$	549
s_1	$\frac{5}{2}$	$-\frac{25}{6}$	25
x_2	$-\frac{1}{2}$	$\frac{1}{2}$	3
x_1	$\frac{1}{2}$	$-\frac{1}{6}$	9

In general, when using the shortened tableau, proceed as follows: If there is a basic feasible solution with basis matrix $\boldsymbol{B}$, the shortened tableau is initialized as follows:

	$\boldsymbol{x}_N$	
z	$\boldsymbol{c}_B\boldsymbol{B}^{-1}\boldsymbol{N} - \boldsymbol{c}_N$	$\boldsymbol{c}_B\overline{\boldsymbol{b}}$
$\boldsymbol{x}_B$	$\boldsymbol{B}^{-1}\boldsymbol{N}$	$\overline{\boldsymbol{b}}$

For simplicity, we denote the entries of the tableau as follows, deviating from our previous notation:

	x_1	$\dots$	x_n	
z	$-c_1$	$\dots$	$-c_n$	z_0
s_1	a_{11}	$\dots$	a_{1n}	b_1
$\vdots$	$\vdots$		$\vdots$	$\vdots$
s_m	a_{m1}	$\dots$	a_{mn}	b_m

In a simplex step, a basic variable is exchanged for a nonbasic variable. This corresponds to the transition to an adjacent vertex of the feasible region. A simplex step proceeds as follows:

Selecting the variable that enters the basis. Find the minimum

$$-c_{j^*} = \min\{-c_1, \dots, -c_n\}.$$

If the minimum is nonnegative, no further improvement of the objective value is possible; the current solution is optimal.

If the minimum is negative, the corresponding variable x_{j^*} enters the basis. The corresponding column is called **pivot column**.

Selecting the variable that leaves the basis. If all entries a_{ij^*} in the pivot column are nonpositive, the objective function is unbounded over the feasible region; the LP has no solution.

Otherwise, determine

$$\frac{b_{i^*}}{a_{i^*j^*}} = \min_i \left\{ \frac{b_i}{a_{ij^*}} \,\middle|\, a_{ij^*} > 0 \right\}.$$

The variable s_{i^*} leaves the basis. The corresponding row is called **pivot row**.

Updating the tableau. The element at the intersection of pivot row and column is called **pivot element**. Update the tableau as follows:

	$\dots$	x_j	$\dots$	x_{j^*}	$\dots$	
z	$\dots$	k_j	$\dots$	k_{j^*}	$\dots$	z_0
$\vdots$		$\vdots$		$\vdots$		$\vdots$
s_i	$\dots$	a_{ij}	$\dots$	a_{ij^*}	$\dots$	b_i
$\vdots$		$\vdots$		$\vdots$		$\vdots$
s_{i^*}	$\dots$	a_{i^*j}	$\dots$	$\boxed{a_{i^*j^*}}$	$\dots$	b_{i^*}
$\vdots$		$\vdots$		$\vdots$		$\vdots$

$\longrightarrow$

	$\dots$	x_j	$\dots$	s_{i^*}	$\dots$	
z	$\dots$	$\hat{k}_j$	$\dots$	$\hat{k}_{j^*}$	$\dots$	$\hat{z}_0$
$\vdots$		$\vdots$		$\vdots$		$\vdots$
s_i	$\dots$	$\hat{a}_{ij}$	$\dots$	$\hat{a}_{ij^*}$	$\dots$	$\hat{b}_i$
$\vdots$		$\vdots$		$\vdots$		$\vdots$
x_{j^*}	$\dots$	$\hat{a}_{i^*j}$	$\dots$	$\boxed{\hat{a}_{i^*j^*}}$	$\dots$	$\hat{b}_{i^*}$
$\vdots$		$\vdots$		$\vdots$		$\vdots$

Here, for simplicity, we have set $k_j = -c_j$. In the updated tableau, the variables x_{j^*} and s_{i^*} are swapped. The remaining entries are updated as follows:

a) $\hat{a}_{i^*j^*} = \dfrac{1}{a_{i^*j^*}}$,

b) $\hat{a}_{i^*j} = \dfrac{a_{i^*j}}{a_{i^*j^*}}$, $j \neq j^*$, and $\hat{b}_{i^*} = \dfrac{b_{i^*}}{a_{i^*j^*}}$,

c) $\hat{a}_{ij^*} = -\dfrac{a_{ij^*}}{a_{i^*j^*}}$, $i \neq i^*$, and $\hat{k}_{j^*} = -\dfrac{k_{j^*}}{a_{i^*j^*}}$,

d) $\hat{a}_{ij} = a_{ij} - \dfrac{a_{ij^*}a_{i^*j}}{a_{i^*j^*}}, i \neq i^*, j \neq j^*,$

$$\hat{b}_i = b_i - \frac{a_{ij^*}b_{i^*}}{a_{i^*j^*}}, i \neq i^*,$$

$$\hat{k}_j = k_j - \frac{k_{j^*}a_{i^*j}}{a_{i^*j^*}}, j \neq j^*,$$

$$\hat{z}_0 = z_0 - \frac{b_{i^*}k_{j^*}}{a_{i^*j^*}}.$$

After updating the tableau, continue with the next simplex step.

It should be noted that both the right-hand side column and the objective function row are updated in the same way as the coefficient matrix. The transformations for updating the tableau can be remembered as follows:

- Replace the pivot element with its reciprocal.
- Divide the remaining entries in the pivot row by the pivot element.
- Divide the remaining entries in the pivot column by the negative of the pivot element.
- From all other entries, subtract the product of the corresponding values in the pivot row and column, divided by the pivot element.

11.5 The two-phase method for obtaining an initial basic feasible solution

If a maximization problem is in canonical form and the right-hand side values $b_1, \dots, b_m$ are all nonnegative, the simplex method with shortened tableau can be started directly. The maximization problem in canonical form is given by

$$\begin{array}{rl} \max & \boldsymbol{cx} \\ \text{s.t.} & \boldsymbol{Ax} \leq \boldsymbol{b} \\ & \boldsymbol{x} \geq \boldsymbol{0} \end{array}$$

and one starts with the following tableau:

	x_1	$\cdots$	x_n	
z	$-c_1$	$\cdots$	$-c_n$	0
s_1	a_{11}	$\cdots$	a_{1n}	b_1
$\vdots$	$\vdots$		$\vdots$	$\vdots$
s_m	a_{m1}	$\cdots$	a_{mn}	b_m

This corresponds to the basic feasible solution where the slack variables are the basic variables and the original model variables are the nonbasic variables.

The simplex method assumes that a basic feasible solution is known. For LPs whose constraints have the form $\boldsymbol{Ax} \leq \boldsymbol{b}$ with $\boldsymbol{b}$ nonnegative, it is straightforward to construct a basic feasible solution. By introducing slack variables, we obtain the system $\boldsymbol{Ax} + \boldsymbol{s} = \boldsymbol{b}$, in which the slack variables form a basis with the identity matrix as the basis matrix. In this case, the simplex method can be started directly.

However, for general constraints that are already in the form $\boldsymbol{Ax} = \boldsymbol{b}$, finding a basic feasible solution is not so straightforward. Although one can try to find a basic feasible solution "by hand", this can be time-consuming. Moreover, this method fails when the feasible region is empty, i.e., no basic feasible solution exists at all. We therefore need a systematic method that provides us with an initial basic feasible solution or signals that no basic feasible solution exists.

There are several systematic methods for obtaining an initial basic feasible solution. The most well-known are the so-called **two-phase method** and the big-M method. We will only discuss the two-phase method here.

In the two-phase method, one obtains an initial basic feasible solution by introducing so-called **artificial variables**. In **Phase I** the aim is to remove these artificial variables from the basis, so that at the end of Phase I a basic feasible solution exists that contains no artificial variables at all. In **Phase II**, starting from this basic feasible solution, the LP is solved. In the following, we consider a maximization problem in standard form:

$$\begin{array}{rl} \max & \boldsymbol{cx} \\ \text{s.t.} & \boldsymbol{Ax} = \boldsymbol{b} \\ & \boldsymbol{x} \geq \boldsymbol{0} \end{array}$$

In Phase I, we introduce m artificial variables $\boldsymbol{x}_a = [x_{a1}, \cdots, x_{am}]^T$ and consider the following auxiliary LP, in which the sum of the artificial variables is used as the objective function[3]:

$$\begin{array}{ll} \min & \mathbf{1}^T \boldsymbol{x}_a \\ \text{s.t.} & \boldsymbol{A}\boldsymbol{x} + \boldsymbol{x}_a = \boldsymbol{b} \\ & \boldsymbol{x}, \boldsymbol{x}_a \geq \mathbf{0} \end{array}$$

Now we are in the pleasant situation that the artificial variables form a basis with the identity matrix as the basis matrix. The corresponding basic feasible solution is $\boldsymbol{x}_a = \boldsymbol{b}$, $\boldsymbol{x} = \mathbf{0}$. We can therefore immediately write down an initial tableau for the simplex method. One must merely keep in mind that we are now dealing with a minimization problem. By multiplying the objective function by -1, however, we can convert it into a maximization problem.

If the original LP has a basic feasible solution, then there also exists a basic feasible solution of the LP of Phase I, in which all artificial variables are nonbasic variables and thus have the value zero. The minimum value of the objective function of the auxiliary problem is then also zero.

In the following, we denote by z_{opt} the optimal objective value of the LP of Phase I. Once z_{opt} has been determined, the following cases can occur:

Case 1: $z_{opt} > 0$**.** As already explained, in this case the original LP has no basic feasible solution, so the feasible region is empty.

Case 2: $z_{opt} = 0$**.** Here we distinguish two further cases:

Case 2a: All artificial variables have left the basis. In this case, delete the columns corresponding to the artificial variables to obtain a basic feasible solution of the original LP. Then proceed directly to Phase II (see below).

Case 2b: There are still artificial variables in the basis. For each artificial variable still in the basis, perform simplex steps using nonbasic original variables to remove it from the basis. If this is not possible because all entries in the corresponding row are zero, delete that row. In this way, all artificial variables are removed from the basis and Case 2a applies

When actually performing this procedure, artificial variables that leave the basis are immediately removed from the problem. This is achieved by deleting the corresponding column in the tableau.

In order to avoid having to recalculate the objective function row at the end of Phase I, two objective function rows are used in Phase I. The simplex steps in Phase I are based solely on the objective function of the auxiliary LP. The actual objective function row is updated like any other row in the simplex tableau. Once all artificial variables have left the basis, the objective function row of the auxiliary LP is also deleted. The simplex steps in Phase II are then based solely on the actual objective function.

[3] $\mathbf{1}$ denotes the column vector, all of whose entries are equal to one, i.e., $\mathbf{1} = [1, \cdots, 1]^T$.

Example 11.3. A company produces three products using two production factors. Let x_1 be the amount of factor 1 employed and x_2 the amount of factor 2 employed. The cost function is given by $c(x_1, x_2) = 5x_1 + 8x_2$, and the production function is

$$q\begin{pmatrix} x_1 \\ x_2 \end{pmatrix} = \begin{pmatrix} 2x_1 + 4x_2 \\ 6x_1 + 2x_2 \\ 5x_1 + 2x_2 \end{pmatrix}.$$

At least 100 units of product 1 and product 3, and at least 200 units of product 2 must be produced. For which amounts of factor inputs are the production costs minimized?

This leads to the following LP:

$$\begin{array}{ll} \min & 5x_1 + 8x_2 \\ \text{s.t.} & 2x_1 + 4x_2 \geq 100 \\ & 6x_1 + 2x_2 \geq 200 \\ & 5x_1 + 2x_2 \geq 100 \\ & \quad\; x_1, x_2 \geq \;\; 0 \end{array}$$

Since the simplex method is formulated for maximization problems, we convert the minimization problem into a maximization problem by multiplying the objective function by -1. After introducing slack variables, we get:

$$\begin{array}{lllllll} \max & -5x_1 - 8x_2 & & & & & \\ \text{s.t.} & 2x_1 + 4x_2 & - s_1 & & & = & 100 \\ & 6x_1 + 2x_2 & & - s_2 & & = & 200 \\ & 5x_1 + 2x_2 & & & - s_3 & = & 100 \\ & & & & x_1, x_2, s_1, s_2, s_3 & \geq & 0 \end{array}$$

Although the slack variables form a basis, the corresponding basic solution is not feasible since all slack variables would be negative ($s_1 = -100$, $s_2 = -200$, $s_3 = -100$). Therefore, to obtain an initial basic feasible solution, we introduce artificial variables a_1, a_2 and a_3. In Phase I, the sum of the artificial variables is minimized, which is equivalent to maximizing the negative of their sum:

$$\begin{array}{llllllllll} \max & & & & & - a_1 & - a_2 & - a_3 & & \\ \text{s.t.} & 2x_1 + 4x_2 & - s_1 & & & + a_1 & & & = & 100 \\ & 6x_1 + 2x_2 & & - s_2 & & & + a_2 & & = & 200 \\ & 5x_1 + 2x_2 & & & - s_3 & & & + a_3 & = & 100 \\ & & & & & & x_1, x_2, s_1, s_2, s_3, a_1, a_2, a_3 & & \geq & 0 \end{array}$$

The artificial variables are now the basic variables and yield a basic feasible solution. To begin with the simplex method, we still need to express the objective function in terms of the nonbasic variables. This is achieved by adding each constraint row once to the objective function row. This results in the following LP for Phase I:

$$\begin{array}{llr}
\max & 13x_1 + 8x_2 - s_1 - s_2 - s_3 & -400 \\
\text{s.t.} & 2x_1 + 4x_2 - s_1 + a_1 & = 100 \\
& 6x_1 + 2x_2 - s_2 + a_2 & = 200 \\
& 5x_1 + 2x_2 - s_3 + a_3 & = 100 \\
& x_1, x_2, s_1, s_2, s_3, a_1, a_2, a_3 & \geq 0
\end{array}$$

The initial tableau is therefore:

	x_1	x_2	s_1	s_2	s_3	
z_a	−13	−8	1	1	1	−400
z	5	8	0	0	0	0
a_1	2	4	−1	0	0	100
a_2	6	2	0	−1	0	200
a_3	[5]	2	0	0	−1	100

In the above tableau, the objective function for the original LP is already included as the second objective function row. As always, the pivot element for the first simplex step is marked by a frame. In the first simplex step, x_1 enters the basis and a_3 leaves it. The a_3 column in the updated tableau can be deleted. In fact, the column does not need to be calculated at all.

	a_3	x_2	s_1	s_2	s_3	
z_a	2.6	−2.8	1	1	−1.6	−140
z	−1	6	0	0	1	−100
a_1	−0.4	[3.2]	−1	0	0.4	60
a_2	−1.2	−0.4	0	−1	1.2	80
x_1	0.2	0.4	0	0	−0.2	20

In the second step, x_2 enters and a_1 leaves the basis. Again, the a_1 column in the updated tableau can be deleted.

	a_1	s_1	s_2	s_3	
z_a	0.875	0.125	1	−1.25	−87.5
z	−1.875	1.875	0	0.25	−212.5
x_2	0.3125	−0.3125	0	0.125	18.75
a_2	0.125	−0.125	−1	**[1.25]**	87.5
x_1	−0.125	0.125	0	−0.25	12.5

In the third simplex step, the last artificial variable a_2 leaves the basis and is replaced by s_3:

	s_1	s_2	a_2	
z_a	0	0	1	0
z	1.9	0.2	−0.2	−230
x_2	−0.3	0.1	−0.1	10
s_3	−0.1	−0.8	0.8	70
x_1	0.1	−0.2	0.2	30

The value of the objective function of the Phase I problem is zero and all artificial variables have been removed from the basis. We remove the column corresponding to a_2 as well as the objective function row of the auxiliary LP and obtain our initial tableau for Phase II.

	s_1	s_2	
z	1.9	0.2	−230
x_2	−0.3	0.1	10
s_3	−0.1	−0.8	70
x_1	0.1	−0.2	30

At this point, the optimum has already been reached, as both entries in the objective function row are positive. For cost-optimal production, 30 units of the first factor and 10 units of the second factor are required. The required amounts of the first and second product are produced exactly, while 70 units more than needed are produced of the third product. The costs of this production plan amount to 230 units.

Remark: In the example, during Phase I, an artificial variable left the basis in each step, so Phase I was completed after three steps. In general, however, more simplex steps may be necessary until all artificial variables have left the basis. Furthermore, an optimal tableau was already reached at the beginning of Phase II, so no further simplex steps were required. In general, though, Phase II does not start with an optimal tableau; further simplex steps are typically required to reach the optimum.

11.6 Duality

Let us consider again the production problem (Example 11.1). Assume our producer purposes ceasing the production and selling the three inputs in stock. Which prices should he demand for these inputs? Of course, the prices should be such that he is not worse off by selling the inputs than if he were continuing the production with these inputs. Since the production of one unit of product P1 requires five units of input I1 and three units each of inputs I2 and I3, and since the contribution margin of P1 is 50, it must hold $5p_1 + 3p_2 + 3p_3 \geq 50$; otherwise, it would be cheaper to produce product P1 with the existing inputs. Accordingly, $10p_1 + 3p_2 + p_3 \geq 33$ must also hold, as it otherwise would be cheaper to produce the product P2 with the existing inputs. Thus, for the producer, only prices are acceptable for which the inequalities

$$\begin{aligned} 5p_1 + 3p_2 + 3p_3 &\geq 50\,, \\ 10p_1 + 3p_2 + p_3 &\geq 33 \end{aligned}$$

are satisfied. A potential buyer wants to pay as little as possible for the three inputs. So what prices should he offer the producer to make the deal and pay as little as possible? To do this, he must solve the following LP:

$$\begin{array}{ll} \min & 100\,p_1 + 36\,p_2 + 30\,p_3 \\ \text{s.t.} & 5\,p_1 + 3\,p_2 + 3\,p_3 \geq 50 \\ & 10\,p_1 + 3\,p_2 + p_3 \geq 33 \\ & \qquad\qquad p_1, p_2, p_3 \geq 0 \end{array}$$

To solve this LP with the simplex method, we first need to introduce slack variables and convert the minimization problem into a maximization problem:

$$\begin{array}{ll} \max & -100p_1 - 36p_2 - 30p_3 \\ \text{s.t.} & 5p_1 + 3p_2 + 3p_3 - s_1 = 50 \\ & 10p_1 + 3p_2 + p_3 - s_2 = 33 \\ & \qquad\qquad p_1, p_2, p_3, s_1, s_2 \geq 0 \end{array}$$

We solve this LP using the two-phase method. The initial tableau is:

	p_1	p_2	p_3	s_1	s_2	
z_a	−15	−6	−4	1	1	−83
z	100	36	30	0	0	0
a_1	5	3	3	−1	0	50
a_2	[10]	3	1	0	−1	33

After one step, the artificial variable a_2 has left the basis. The corresponding column can be deleted:

	a_2	p_2	p_3	s_1	s_2	
z_a	1.5	−1.5	−2.5	1	−0.5	−33.5
z	−10	6	20	0	10	−330
a_1	−0.5	1.5	[2.5]	−1	0.5	33.5
p_1	0.1	0.3	0.1	0	−0.1	3.3

After the second step, both artificial variables have left the basis. Hence, we have found an initial basic feasible solution for Phase II. After deleting the column corresponding to the last artificial variable and the objective function row for the auxiliary LP, we proceed to Phase II:

	p_2	a_1	s_1	s_2	
z_a	0	1	0	0	0
z	−6	−8	8	6	−598
p_3	0.6	0.4	−0.4	0.2	13.4
p_1	[0.24]	−0.04	0.04	−0.12	1.96

After one step in Phase II, the optimum solution of the LP is obtained:

	p_1	s_1	s_2	
z	25	9	3	−549
p_3	−2.5	−0.5	0.5	8.5
p_2	4.1667	0.1667	−0.5	8.1667

The desired solution is read from the final tableau: $p_1 = 0$, $p_2 = 8.1667$ and $p_3 = 8.5$. The slack variables s_1 and s_2 are both zero. Therefore, the producer receives the same amount for his three inputs as he would receive if he were to produce with the optimal production plan himself. Examining the final tableau, we see that it also contains the optimal production plan. The reduced costs of the slack variables s_1 and s_2 correspond exactly to the optimal production quantities in our initial example; the reduced cost of p_1 reflects the available free capacity for the first input.

The prices p_1, p_2 and p_3 are referred to as **shadow prices** of the three inputs.

Perhaps even more interesting is the fact that the final tableau of our production example (see page 326) already contains the optimal prices for the three inputs. These correspond exactly to the reduced costs of the slack variables s_1, s_2, s_3. As we will see below, this is not a coincidence: either of these two LPs can be solved just as well as the other

Each LP is associated with a second LP, which is referred to as the **dual LP** of the original LP. If we start from the maximization problem in canonical form

$$\boxed{\begin{array}{rr} \max & \boldsymbol{c}\boldsymbol{x} \\ \text{s.t.} & \boldsymbol{A}\boldsymbol{x} \leq \boldsymbol{b} \\ & \boldsymbol{x} \geq \boldsymbol{0} \end{array}} \tag{P}$$

then the dual LP to this is the minimization problem in canonical form:

$$\boxed{\begin{array}{rr} \min & \boldsymbol{b}^T\boldsymbol{y} \\ \text{s.t.} & \boldsymbol{A}^T\boldsymbol{y} \geq \boldsymbol{c}^T \\ & \boldsymbol{y} \geq \boldsymbol{0} \end{array}}$$

Introducing $\boldsymbol{y}$ as a row vector of the variables $y_1, \ldots, y_m$, this can be written as:

$$\boxed{\begin{array}{rr} \min & \boldsymbol{y}\boldsymbol{b} \\ \text{s.t.} & \boldsymbol{y}\boldsymbol{A} \geq \boldsymbol{c} \\ & \boldsymbol{y} \geq \boldsymbol{0} \end{array}} \tag{D}$$

In this context, the original LP (P) is referred to as the **primal LP**.

To find the dual LP of (D), we write (D) as a maximization problem in canonical form. This is achieved by multiplying the objective function as well as the constraints by -1. This gives:

$$\boxed{\begin{array}{rr} \max & (-\boldsymbol{b}^T)\boldsymbol{y} \\ \text{s.t.} & (-\boldsymbol{A}^T)\boldsymbol{y} \leq (-\boldsymbol{c}^T) \\ & \boldsymbol{y} \geq \boldsymbol{0} \end{array}}$$

According to the above, the dual of this LP is given by

$$\boxed{\begin{array}{rr} \min & (-\boldsymbol{c})\boldsymbol{z} \\ \text{s.t.} & (-\boldsymbol{A})\boldsymbol{z} \geq (-\boldsymbol{b}) \\ & \boldsymbol{z} \geq \boldsymbol{0} \end{array}} \quad \text{thus} \quad \boxed{\begin{array}{rr} \max & \boldsymbol{c}\boldsymbol{z} \\ \text{s.t.} & \boldsymbol{A}\boldsymbol{z} \leq \boldsymbol{b} \\ & \boldsymbol{z} \geq \boldsymbol{0} \end{array}},$$

i.e., the same as the original LP (P). We summarize this in a theorem.

Theorem 11.2: Dualizing twice

The dual LP of the dual LP is the primal LP.

Since any LP can be written as a maximization problem in canonical form, it is possible to specify the dual LP for any given LP. For the maximization problem in standard form, for example, one obtains

$$\begin{array}{llr} \text{(P): max} & \boldsymbol{cx} \\ \text{s.t.} & \boldsymbol{Ax} = \boldsymbol{b}\,, \\ & \boldsymbol{x} \geq \boldsymbol{0}\,, \end{array} \quad\longleftrightarrow\quad \begin{array}{ll} \text{(D): min} & \boldsymbol{yb} \\ \text{s.t.} & \boldsymbol{yA} \geq \boldsymbol{c}\,, \\ & \boldsymbol{y} \quad \text{unbounded}\,. \end{array}$$

In a general LP, the constraints can be inequalities (in $\leq$ or $\geq$ form) as well as equations. Furthermore, the variables can be sign-restricted (≤ 0 or ≥ 0) or unrestricted. The general form of the maximization problem is:

$$\begin{array}{ll} \text{(P): max} & \boldsymbol{c}_1\boldsymbol{x}_1 + \boldsymbol{c}_2\boldsymbol{x}_2 + \boldsymbol{c}_3\boldsymbol{x}_3 \\ \text{s.t.} & \boldsymbol{A}_{11}\boldsymbol{x}_1 + \boldsymbol{A}_{12}\boldsymbol{x}_2 + \boldsymbol{A}_{13}\boldsymbol{x}_3 \leq \boldsymbol{b}_1\,, \\ & \boldsymbol{A}_{21}\boldsymbol{x}_1 + \boldsymbol{A}_{22}\boldsymbol{x}_2 + \boldsymbol{A}_{23}\boldsymbol{x}_3 \geq \boldsymbol{b}_2\,, \\ & \boldsymbol{A}_{31}\boldsymbol{x}_1 + \boldsymbol{A}_{32}\boldsymbol{x}_2 + \boldsymbol{A}_{33}\boldsymbol{x}_3 = \boldsymbol{b}_3\,, \\ & \boldsymbol{x}_1 \geq \boldsymbol{0}\,,\ \boldsymbol{x}_2 \leq \boldsymbol{0}\,,\ \boldsymbol{x}_3 \text{ unrestricted.} \end{array}$$

It has the following dual LP:

$$\begin{array}{ll} \text{(D): min} & \boldsymbol{y}_1\boldsymbol{b}_1 + \boldsymbol{y}_2\boldsymbol{b}_2 + \boldsymbol{y}_3\boldsymbol{b}_3 \\ \text{s.t.} & \boldsymbol{y}_1\boldsymbol{A}_{11} + \boldsymbol{y}_2\boldsymbol{A}_{21} + \boldsymbol{y}_3\boldsymbol{A}_{31} \geq \boldsymbol{c}_1\,, \\ & \boldsymbol{y}_1\boldsymbol{A}_{12} + \boldsymbol{y}_2\boldsymbol{A}_{22} + \boldsymbol{y}_3\boldsymbol{A}_{32} \leq \boldsymbol{c}_2\,, \\ & \boldsymbol{y}_1\boldsymbol{A}_{13} + \boldsymbol{y}_2\boldsymbol{A}_{23} + \boldsymbol{y}_3\boldsymbol{A}_{33} = \boldsymbol{c}_3\,, \\ & \boldsymbol{y}_1 \geq \boldsymbol{0}\,,\ \boldsymbol{y}_2 \leq \boldsymbol{0}\,,\ \boldsymbol{y}_3 \text{ unrestricted.} \end{array}$$

The following table provides an overview of how to form the dual LP:

objective	max	$\longleftrightarrow$	min	objective
constraints	$\leq$	$\longleftrightarrow$	$\geq$	variables
	$\geq$	$\longleftrightarrow$	$\leq$	
	$=$	$\longleftrightarrow$	unrestricted	
variables	$\geq$	$\longleftrightarrow$	$\geq$	constraints
	$\leq$	$\longleftrightarrow$	$\leq$	
	unrestricted	$\longleftrightarrow$	$=$	

Relationship between the solutions of (P) and (D)

In this section, we want to investigate the relationship between the solutions to primal and dual LP. For simplicity, we restrict ourselves to the canonical form of duality, i.e., we start from the following LPs:

$$\boxed{\begin{array}{lr} \text{(P): max} & \boldsymbol{c}\boldsymbol{x} \\ \text{s.t.} & \boldsymbol{A}\boldsymbol{x} \leq \boldsymbol{b} \\ & \boldsymbol{x} \geq \boldsymbol{0} \end{array}} \quad \longleftrightarrow \quad \boxed{\begin{array}{lr} \text{(D): min} & \boldsymbol{y}\boldsymbol{b} \\ \text{s.t.} & \boldsymbol{y}\boldsymbol{A} \geq \boldsymbol{c} \\ & \boldsymbol{y} \geq \boldsymbol{0} \end{array}}$$

If $\boldsymbol{x}$ and $\boldsymbol{y}$ are feasible in their respective LPs, then

$$\boldsymbol{c}\boldsymbol{x} \leq (\boldsymbol{y}\boldsymbol{A})\boldsymbol{x} = \boldsymbol{y}(\boldsymbol{A}\boldsymbol{x}) \leq \boldsymbol{y}\boldsymbol{b},$$

i.e., for any two feasible solutions of the respective LPs, the objective value of the dual LP is at least as large as the objective value of the primal LP. This result is known as the **weak duality theorem**.

As a simple consequence of the weak duality theorem, we obtain:

Theorem 11.3: Consequence of the weak duality theorem

- If $\boldsymbol{x}$ and $\boldsymbol{y}$ are two feasible solutions of (P) and (D) and $\boldsymbol{c}\boldsymbol{x} = \boldsymbol{y}\boldsymbol{b}$, then both (P) and (D) have a solution and the optimal values of the objective functions are equal. $\boldsymbol{x}$ and $\boldsymbol{y}$ are then optimal solutions of the LPs (P) and (D).
- If the objective function of one of the LPs is unbounded over the feasible region, then the feasible region of the other LP is empty.

This raises the question of whether the fact that the feasible region of one of the LPs is empty implies that the objective function of the other LP is unbounded. However, this is not the case. It may occur that the feasible regions of both linear programs are empty.

To obtain a stronger statement, we consider the KKT conditions (cf. Section 6.7) for our LP (P). To obtain a minimization problem, we multiply the objective function by -1 and obtain the LP:

$$\boxed{\begin{array}{lr} \text{min} & (-\boldsymbol{c})\boldsymbol{x} \\ \text{s.t.} & \boldsymbol{A}\boldsymbol{x} \leq \boldsymbol{b} \\ & \boldsymbol{x} \geq \boldsymbol{0} \end{array}}$$

The Lagrangian for this LP is

$$L(\boldsymbol{x}, \boldsymbol{y}, \boldsymbol{v}) = -\boldsymbol{c}\boldsymbol{x} + \boldsymbol{y}(\boldsymbol{A}\boldsymbol{x} - \boldsymbol{b}) - \boldsymbol{v}\boldsymbol{x}.$$

Here, $\boldsymbol{y}$ denotes the row vector of Lagrange multipliers corresponding to the constraints $\boldsymbol{Ax} - \boldsymbol{b} \leq 0$, and $\boldsymbol{v}$ denotes the row vector of Lagrange multipliers corresponding to the constraints $-\boldsymbol{x} \leq \boldsymbol{0}$. As will become clearer below, $\boldsymbol{y}$ and $\boldsymbol{v}$ are called **dual variables** and **dual slack variables**, respectively. The KKT conditions for a minimum are:

Stationarity	$-\boldsymbol{c} + \boldsymbol{yA} - \boldsymbol{v} = \boldsymbol{0}\,,$
primal feasibility	$\boldsymbol{Ax} - \boldsymbol{b} \leq \boldsymbol{0}\,, -\boldsymbol{x} \leq \boldsymbol{0}\,,$
dual feasibility	$\boldsymbol{y} \geq \boldsymbol{0}\,, \boldsymbol{v} \geq \boldsymbol{0}\,,$
complementary slackness	$\boldsymbol{y}(\boldsymbol{Ax} - \boldsymbol{b}) = \boldsymbol{0}\,, \boldsymbol{vx} = \boldsymbol{0}\,.$

From stationarity, we get $\boldsymbol{v} = \boldsymbol{yA} - \boldsymbol{c}$. If we substitute this into dual feasibility and complementary slackness, we get

$$\boldsymbol{yA} - \boldsymbol{c} \geq \boldsymbol{0} \iff \boldsymbol{yA} \geq \boldsymbol{c} \quad \text{and}$$
$$(\boldsymbol{yA} - \boldsymbol{c})\boldsymbol{x} = \boldsymbol{0}\,.$$

This simplifies the KKT conditions to:

Primal feasibility	$\boldsymbol{Ax} \leq \boldsymbol{b}\,, \boldsymbol{x} \geq \boldsymbol{0}\,,$
dual feasibility	$\boldsymbol{yA} \geq \boldsymbol{c}\,, \boldsymbol{y} \geq \boldsymbol{0}\,,$
complementary slackness	$\boldsymbol{y}(\boldsymbol{Ax} - \boldsymbol{b}) = \boldsymbol{0}\,, (\boldsymbol{yA} - \boldsymbol{c})\boldsymbol{x} = \boldsymbol{0}\,.$

Primal feasibility means that $\boldsymbol{x}$ is feasible in LP (P). Dual feasibility means that $\boldsymbol{y}$ is feasible in LP (D). Since the KKT conditions are necessary and sufficient for the existence of a global extremum in linear programs (see Theorem 6.16), it follows: A vector $\boldsymbol{x}$ is an optimal solution of (P) if and only if there exists a vector $\boldsymbol{y}$ such that the above KKT conditions are satisfied.

If $\boldsymbol{x}$ is an optimal solution of (P), then $\boldsymbol{y}$ is obviously feasible in the dual LP (D). From complementary slackness, it follows that

$$\boldsymbol{yb} = \boldsymbol{y}(\boldsymbol{Ax}) = (\boldsymbol{yA})\boldsymbol{x} = \boldsymbol{cx}\,.$$

From the weak duality theorem follows that $\boldsymbol{y}$ is an optimal solution of the dual LP (D). In summary: If there is a solution to the primal problem (P), then (D) also has a solution and the optimal values of the two objective functions are equal. Since dualizing twice yields the primal problem, the previous theorem also applies when (P) and (D) are exchanged.

Theorem 11.4: Strong duality theorem

If one of the LPs (P) and (D) has an optimal solution, then the other LP has an optimal solution as well and the optimal values of the objective functions are equal.

If we summarize the above statements, we obtain the following theorem, which completely answers the question of solvability of the two LPs.

Theorem 11.5: Fundamental duality theorem

For the solutions of the LPs (P) and (D), exactly one of the following three statements applies:

1. Both LPs have optimal solutions and the optimal values of the objective functions are equal.
2. The objective function of one of the LPs is unbounded and the feasible region of the other LP is empty.
3. The feasible regions of both LPs are empty.

Identifying the dual solution in the final tableau

We now examine how to obtain an optimal solution $\boldsymbol{y}$ of the dual problem (D) from the final tableau of the simplex method applied to the primal problem (P).

To solve the primal maximization problem (P) in canonical form, we introduce slack variables $s_1, \ldots, s_m$. This yields the extended coefficient matrix $\widetilde{\boldsymbol{A}} = [\boldsymbol{A}, \boldsymbol{I}]$ and the extended objective vector $\widetilde{\boldsymbol{c}} = [\boldsymbol{c}, \boldsymbol{0}]$ for the problem in standard form. The columns of $\widetilde{\boldsymbol{A}}$ corresponding to slack variables are just the unit vectors $\boldsymbol{e}_1, \ldots, \boldsymbol{e}_m$, and the coefficients of the objective function corresponding to slack variables are zero.

Let $(\boldsymbol{x}, \boldsymbol{s})$ be an optimal basic feasible solution with basis matrix $\boldsymbol{B}$. We want to understand why $\boldsymbol{w} = \boldsymbol{c}_B\boldsymbol{B}^{-1}$ gives an optimal solution to the dual LP (D). For this, we will show that $\boldsymbol{w} = \boldsymbol{c}_B\boldsymbol{B}^{-1}$ satisfies the KKT conditions.

The $n + m$ columns of $\widetilde{\boldsymbol{A}}$ are $\boldsymbol{a}_1, \ldots, \boldsymbol{a}_n$ and $\boldsymbol{e}_1, \ldots, \boldsymbol{e}_m$. Therefore, we obtain

$$\boldsymbol{w}\boldsymbol{a}_j = \boldsymbol{c}_B\boldsymbol{B}^{-1}\boldsymbol{a}_j = z_j \quad \text{and} \quad \boldsymbol{w}\boldsymbol{e}_j = \boldsymbol{c}_B\boldsymbol{B}^{-1}\boldsymbol{e}_j = z_{n+j}\,.$$

As noted on page 321, for the basic variables, $z_j = c_j$ holds.

Since the columns corresponding to slack variables are the unit vectors $\boldsymbol{e}_1, \ldots, \boldsymbol{e}_m$ and since the corresponding coefficients of the objective function are $c_{n+1} = \ldots = c_{n+m} = 0$, it also holds

$$w_i = \boldsymbol{w}\boldsymbol{e}_i = z_{n+i} = z_{n+i} - c_{n+i}\,.$$

Primal feasibility is obviously satisfied when $\boldsymbol{x}$ is an optimal solution to (P).

Dual feasibility: It is $\boldsymbol{w}\boldsymbol{a}_j = z_j$. At the optimum, for all nonbasic variables $z_j - c_j \geq 0$, i.e., $z_j \geq c_j$. For basic variables x_j, $z_j = c_j$. Thus, $\boldsymbol{w}\boldsymbol{a}_j \geq c_j$ for all $j = 1, \ldots, n$, i.e., $\boldsymbol{w}\boldsymbol{A} \geq \boldsymbol{c}$. Furthermore, $w_i = z_{n+i} - c_{n+i} \geq 0$, i.e., $\boldsymbol{w} \geq \boldsymbol{0}$.

Complementary slackness: Clearly, $\boldsymbol{s} = \boldsymbol{b} - \boldsymbol{A}\boldsymbol{x}$. Therefore, $\boldsymbol{w}(\boldsymbol{b} - \boldsymbol{A}\boldsymbol{x}) = \boldsymbol{w}\boldsymbol{s}$. If s_i is a nonbasic variable, then $s_i = 0$. Conversely, if s_i is a basic variable, then $w_i = z_{n+i} - c_{n+i} = 0$, since the reduced costs of basic variables are zero. Thus, either $s_i = 0$ or $w_i = 0$ and therefore, $\boldsymbol{w}(\boldsymbol{b} - \boldsymbol{A}\boldsymbol{x}) = \boldsymbol{w}\boldsymbol{s} = 0$.

If x_j is a basic variable, then $\boldsymbol{wa}_j - c_j = z_j - c_j = 0$, since the reduced costs of basic variables are zero. Conversely, if x_j is a nonbasic variable, then $x_j = 0$. Thus, $(\boldsymbol{wA} - \boldsymbol{c})\boldsymbol{x} = 0$.

The previous explanations show that the vector $\boldsymbol{w} = \boldsymbol{c_B}\boldsymbol{B}^{-1}$ satisfies the KKT conditions. Furthermore, it can be seen that the value of the i-th dual variable w_i at the optimum is equal to the negative of the reduced cost of the i-th slack variable. If s_i is a basic variable, then this value is zero. The value of the j-th dual slack variable is equal to the negative of the reduced cost of the variable x_j. If x_j is a basic variable, this value is zero.

The final tableau of the simplex method therefore contains in the first row the optimal values of the dual variables. The reduced costs of the variables x_i correspond to the values of the dual slack variables, and the reduced costs of the slack variables correspond to the values of the dual variables y_j.

Example 11.4. We consider the final tableau of our initial example:

	s_3	s_2	
z	$\frac{17}{2}$	$\frac{49}{6}$	549
s_1	$\frac{5}{2}$	$-\frac{25}{6}$	25
x_2	$-\frac{1}{2}$	$\frac{1}{2}$	3
x_1	$\frac{1}{2}$	$-\frac{1}{6}$	9

The slack variable s_1 is a basic variable. Hence, the optimal value of the corresponding dual variable is $y_1 = 0$. The reduced costs of the slack variables s_2 and s_3 are $-49/6 = -8.1667$ and $-17/2 = -8.5$, respectively. Consequently, the optimal values of the corresponding dual variables are $y_2 = 8.1667$ and $y_3 = 8.5$. These values coincide exactly with those obtained in the example at the beginning of this section. Moreover, the dual slack variables are zero, since the corresponding primal variables are basic variables.

Another important interpretation of the dual variables $\boldsymbol{w}$ is provided by the following observation: Let $\boldsymbol{B}$ be the basis matrix of the optimal solution. The optimal objective value is then given by $z_0 = \boldsymbol{c_B}\boldsymbol{B}^{-1}\boldsymbol{b}$. Since $\boldsymbol{w} = \boldsymbol{c_B}\boldsymbol{B}^{-1}$, this can be written as

$$z_0 = \boldsymbol{wb} = \sum_{i=1}^{m} w_i b_i \,.$$

If the constraints $\boldsymbol{Ax} \leq \boldsymbol{b}$ are interpreted as capacity restrictions, this equation shows how an expansion of capacities affects the optimal value of the objective function. For example, if b_1 is increased by an amount Δb_1, the objective value changes by the amount $w_1 \Delta b_1$. The same applies for the other capacities. Of course, this holds only if the changes Δb_i are small enough so that the optimal basis remains unchanged. We state this as a theorem:

Theorem 11.6: Interpretation of dual variables

If the capacity b_i of the i-th constraint is increased by a sufficiently small amount Δb_i, the optimal value of the objective function changes by $w_i \Delta b_i$, where w_i is the optimal value of the i-th dual variable.

We illustrate this with an example.

Example 11.5. We revisit Example 11.1. Suppose the producer wants to expand his output by increasing the capacities, i.e., the maximum available inputs of cream (I1), milk powder (I2) and strawberry flavor (I3). What is the highest unit price each of these inputs can have for such an expansion to remain profitable? In the previous example, the optimal values of the dual variables were found to be $w_1 = 0$, $w_2 = 8.1667$, and $w_3 = 8.5$. Therefore, increasing the capacity of I1 is not profitable, since the objective value does not change – this is because, in the optimal production plan, the capacity of I1 was not fully utilized. In contrast, increasing the capacity of input I2 increases the total contribution margin by 8.1667 per unit, and similarly, increasing the capacity of I3 increases it by 8.5 per unit. Therefore, to make expanding capacity profitable, the price per unit of input I2 must not exceed 8.1667, and the price of input I3 must not exceed 8.5.

Hence, the optimal values of the dual variables are called **shadow prices**, as they indicate the marginal value of increasing the corresponding resource capacity.

Important terms for review

After reading this chapter, the following terms should be familiar:

- Linear optimization problem, linear program, LP
- Standard form of an LP, canonical form of an LP
- Slack variable
- Simplex method
- Feasible region
- Basic variable, nonbasic variable
- Basic feasible solution
- Full tableau, shortened tableau
- Pivot element, pivot column, pivot row
- Two-phase method
- Artificial variables
- Duality, dual LP
- Primal and dual feasibility
- Complementary slackness

Self-test

You can check your understanding of this chapter using the following true-or-false questions. For each statement, indicate whether it is true (T) or false (F). Short solutions to these questions can be found in Appendix H.

I. Are the following statements regarding the simplex method true or false?

	T	F
If all entries in the objective function row are negative, the optimum of the LP has been found.	☐	☐
In the simplex tableau in shortened form, the columns corresponding to basic variables are omitted.	☐	☐
The pivot element is found in the row of the leaving variable and the column of the entering variable.	☐	☐
The pivot element is found in the column of the leaving variable and the row of the entering variable.	☐	☐
If no basic feasible solution is available at the outset of the simplex method, one can determine such a solution using Phase I of the two-phase method.	☐	☐

II. Assess the following statements about LPs.

	T	F
In the standard form of an LP, all constraints must be written as equations.	☐	☐
By introducing a slack variable, an inequality constraint can be transformed into an equality constraint.	☐	☐
Every optimal solution of an LP is a vertex of the feasible region.	☐	☐

III. Are the following statements regarding the dual LP true or false?

	T	F
If an LP has only equality constraints, it is identical to its dual LP.	☐	☐
The dual LP of a minimization problem is a maximization problem.	☐	☐
If an LP contains only equality constraints, all variables in the dual problem are unrestricted.	☐	☐
If the primal problem has a solution, the objective function of the dual problem is unbounded.	☐	☐

Exercises

Exercise 1
Consider the following optimization problem

$$\begin{array}{llcrcl} \max & x_1 & + & x_2 & & \\ \text{s.t.} & \frac{1}{2}x_1 & + & x_2 & \leq & 4 \\ & x_1 & + & \frac{2}{5}x_2 & \leq & 4 \\ & & & x_2 & \leq & 3 \\ & & & x_1, x_2 & \geq & 0 \end{array}$$

Solve the problem graphically.

Exercise 2
The Minister of Economy in a coalition government of three parties (A, B, and C) wants to allocate subsidies to three sectors. Let x_1, x_2, and x_3 denote the amounts (in billions of monetary units) allocated to sectors 1, 2, and 3, respectively, where $x_1, x_2, x_3 \geq 0$. Each party imposes certain limits on the subsidies. Party A wants a maximum of 2 billion monetary units to be spent on sectors 1 and 2. Party B wants the same, but for sectors 1 and 3. Party C wants a maximum of 2 billion monetary units to be spent on sectors 2 and 3, but a maximum of 1 billion monetary units for sector 3. Additionally, in sector 3, there are administrative fees equal to the subsidy amount. In sectors 1 and 2, the subsidy is paid directly to the respective monopolist without extra fees. The Minister wishes to maximize the total spending (including administrative fees), subject to the wishes of all coalition partners. Using the simplex method, determine the optimal subsidy allocation. How much money does the Minister need from the Treasurer?

Exercise 3
A company produces a product using three raw materials. Let x_1, x_2 and x_3 denote the daily quantities (in tons) of these raw materials. The daily production function (in hectoliters) is given by $f(x_1, x_2, x_3) = 2x_1 + x_2 + x_3$. Suppliers can deliver a maximum of 3, 4, and 3 tons of raw materials 1, 2, and 3 per day, respectively, before production begins. These raw materials are stored until production starts, but storage capacity is limited by $x_1 + 2x_2 + 2x_3 \leq 14$. All raw materials delivered must be completely processed on the day of delivery.

What daily order quantities for each raw material will maximize the company's output? What is the maximum attainable output?

Exercise 4
A company plans to produce a new product. In the manufacturing process, quantities x_1 and x_2 (each in tons) of two greenhouse gases are generated. The company has consulted two independent experts to forecast pollution. Their predictions are: $3x_1 + x_2 \geq 8$, and $3x_1 + 9x_2 \geq 24$. The regulatory authority aims to permit the company only the minimum pollution levels forecast by the experts. For regulatory purposes, one ton of greenhouse gas 2 is considered twice as harmful as one ton of

greenhouse gas 1. What emission limits will the authority set?

a) Formulate and solve the problem using the simplex method, applying the two-phase method to determine an initial basic feasible solution.

b) Solve the problem by applying the simplex method to the dual of the given LP.

Exercise 5
A large corporation is establishing a new division and plans to hire employees with two types of qualifications. Let x_1 and x_2 denote the numbers of employees hired with qualifications 1 and 2, respectively. Three board members have differing estimates of staff requirements, namely $x_1 + 2x_2 \geq 50$, $2x_1 + x_2 \geq 40$, and $5x_1 + 2x_2 \geq 10$. The personnel manager must decide how many employees to hire, given that the cost for each employee with qualification 1 is twice as high as that for qualification 2. Assume all employees of the same qualification level are paid equally.

How many employees of each type should be hired to minimize total personnel costs while meeting all requirements?

a) Solve the problem with the simplex method, starting with the two-phase method to find an initial basic feasible solution.

b) Solve the problem by applying the simplex method to the dual of the given LP.

Exercise 6
Consider the optimization problem

$$\begin{array}{rl} \max & x_1 + 7x_2 \\ \text{s.t.} & x_1 + x_2 \geq 1 \\ & x_1 + x_2 \geq 1 \\ & x_1 \leq 0,\ x_2 \geq 0 \end{array}$$

a) Formulate the dual LP.

b) Sketch the constraints of the dual problem. What can be said about the feasible region?

Exercise 7
Consider the optimization problem

$$\begin{array}{rrcl} \max & 3x_1 + 4x_2 + 5x_3 + x_4 & & \\ \text{NB} & 5x_1 + 7x_2 + 3x_3 + 2x_4 & \leq & 12 \\ & x_1 + x_2 + x_3 + x_4 & \geq & 1 \\ & x_1 + 2x_2 + 3x_3 + x_4 & \leq & 5 \\ & x_1 + 2x_2 + x_3 + x_4 & = & 3 \\ & \multicolumn{3}{l}{x_1, x_2, x_3 \geq 0,\ x_4 \text{ unrestricted}} \end{array}$$

Formulate the dual LP.

Chapter 12

Differential equations

Differential and difference equations play a central role in many economic models, especially in those where the temporal development of economic variables is modeled, such as in growth theory or in the description of market processes. These models are referred to as **dynamic economic models**. Dynamic models are classified into two types, depending on whether the relevant economic variables are considered only at discrete points (or intervals) in time, or considered continuously at any point in time. In the first case, they are called **discrete-time models**, in the second case **continuous-time models**. Discrete-time models are formulated as **difference equations**, continuous-time models as **differential equations**.

Often economic functions, such as demand or production functions, are characterized by their growth behavior. Specifically, the elasticity of a demand function may be given instead of the function itself. The goal then is to find a demand function that possesses the given elasticity. This chapter provides a discussion of differential equations, while difference equations are postponed to the following one.

We start with some examples that underline the relevance of differential equations in economic models.

Example 12.1 (Demand function). For a demand function $q = f(p)$, it is known that the price elasticity is constant, i.e., $\varepsilon_q(p) = -\delta$. Since a higher price usually results in reduced demand, we assume that $\delta > 0$. Thus, we are looking for a demand function for which

$$\varepsilon_q(p) = \frac{p}{q}q' = -\delta.$$

K. Mosler et al., *Mathematical Methods for Business and Economics*, Mathematics Study Resources 20, https://doi.org/10.1007/978-3-662-73831-3_12

If we solve the equation for q', we get[1]

$$q' = -\frac{\delta}{p} q .$$

Hence, we must solve an equation involving the unknown function and its first derivative.

Example 12.2 (Evans[2] price adjustment model). We consider a single good on a market, being interested in its development regarding demand, price and supply during an interval T of time. At $t \in T$, let $d(t)$ denote the demand for the good, $s(t)$ the supply of it, and $p(t)$ its price. We assume that both demand and supply depend linearly on the price,

$$d(t) = \alpha - \beta p(t) ,$$
$$s(t) = \gamma + \delta p(t) .$$

Moreover, we assume that the change in price is proportional to the excess demand,

$$p'(t) = \lambda(d(t) - s(t)) .$$

Let β and δ be positive, as generally an increase in price reduces demand and enlarges supply. Further let λ be positive, so that an excess demand leads to an increase in price, while an excess supply leads to a decrease. The goal here is to find a function $p(t)$ that describes the adjustment of the price over time.

Inserting $d(t)$ and $s(t)$ into the last equation yields

$$p'(t) = \lambda\Big((\alpha - \beta p(t)) - (\gamma + \delta p(t))\Big),$$

i.e.,

$$p'(t) = \lambda(\alpha - \gamma) - \lambda(\beta + \delta)p(t) ,$$

which with $a = \lambda(\alpha - \gamma)$ and $b = \lambda(\beta + \delta)$ is written as

$$p'(t) = a - bp(t) .$$

As in the previous example, we obtain an equation involving an unknown function (here $p(t)$) and its first derivative.

[1] In differential equations, we use the abbreviations y and y' for the functions $f(x)$ and $f'(x)$. So, here we write q and q' for $f(p)$ and $f'(p)$.

[2] Griffith Conrad Evans (1887–1973)

Example 12.3 (Harrod[3]-Domar[4] model). The Harrod-Domar model describes the growth of an economy over time. We use the following notations: For t in a given time interval, let

$K(t)$	be the capital stock at time t,
$Y(t)$	the national income at time t,
$S(t)$	the savings at time t,
$I(t)$	the net investment at time t.

The question of interest is how the economic variables under consideration evolve over time. The Harrod-Domar model is based on the following assumptions:

- By definition, net investment is equal to the change in the capital stock:

$$I(t) = K'(t) .$$

- The national income is proportional to the level of the capital stock:

$$Y(t) = \frac{1}{v} K(t) ,$$

  where the constant v is referred to as the **capital coefficient**, and its reciprocal $1/v$ as **capital productivity**.

- Savings are proportional to the national income:

$$S(t) = sY(t) .$$

  The constant s is referred to as the **savings rate**.

- Savings are equal to net investment:

$$I(t) = S(t) .$$

From the first two equations, it follows that

$$I(t) = vY'(t) .$$

This is referred to as the **capacity effect** of investment. From the third and fourth equation above, we obtain the so-called **income effect** of investment,

$$I(t) = sY(t) .$$

Equating the last two equations and solving for $Y'(t)$ finally yields

$$\boxed{Y'(t) = \frac{s}{v} Y(t) .}$$

As in the two preceding examples, the Harrod-Domar model leads to an equation involving an unknown function (here $Y(t)$) and its first derivative.

[3] Sir Roy Forbes Harrod (1900–1978)
[4] Evsey David Domar (1914–1997)

In each of the previous economic applications, we need to solve an equation involving an unknown function of an economic variable, its first derivative and possibly also the variable itself (as in Example 12.1). Such an equation is called a **differential equation**.

Definition 12.1: Differential equation

An **(ordinary) differential equation (ODE)** is an equation between an independent variable, a function of this variable, and at least one derivative of this function. If the highest occurring derivative is of order m, the ODE is called an **m-th order differential equation**; it has the form

$$F(x, y, y', \ldots, y^{(m)}) = 0,$$

where F is a function of several variables and y is a function of x.

A function $f : A \to \mathbb{R}$ defined on an interval A is called a **solution** of the differential equation if

$$F(x, f(x), f'(x), \ldots, f^{(m)}(x)) = 0 \text{ for all } x \in A.$$

The set of all solutions of a differential equation is called the **general** or **complete solution**.

Remark: Solutions of differential equations are functions that, when substituted into the equation, satisfy it for all values of the variable within an interval A. This contrasts significantly with solving linear equations or finding stationary points, where the goal is to identify individual values of the variables that make the equations true when substituted.

12.1 First-order differential equations

In this section, we deal with first-order differential equations. Those equations can often be solved explicitly for y' and then written in the form

$$\boxed{y' = g(x, y).}$$

Geometric interpretation of a differential equation

A differential equation of the form $y' = g(x, y)$ can be visualized geometrically as follows. The differential equation $y' = g(x, y)$ assigns a slope to points in the x-y plane. If one draws a small line segment at each point in the x-y plane (or better for each point of a fine grid) whose slope is the slope given by the differential equation, one obtains a so-called **slope field**. The solution to the differential equation consists in finding all functions $y = f(x)$ whose slope at each point matches the slope given by the slope field of the differential equation. Figure 12.1 shows the slope field of the differential equation $y' = -\frac{x}{y}$, $y \neq 0$.

As can be seen by substituting into the differential equation, the function $f : [-3, 3] \to \mathbb{R}$, defined by $x \mapsto \sqrt{9 - x^2}$, is a solution to the differential equation. It is shown in Figure 12.1.

In the following, we provide simple examples of differential equations and their solutions. In these examples, the solution function can be conveniently verified by substituting it and its derivatives into the differential equation.

Example 12.4. The differential equation

$$y' = y$$

is solved by the exponential function $y = e^x$, $x \in \mathbb{R}$. However, also $y = 5e^x$, or more generally $y = \gamma e^x$ with an arbitrary constant $\gamma \in \mathbb{R}$ is a solution. Figure 12.2 shows the slope field of the differential equation $y' = y$ as well as a solution $y = \frac{1}{4}e^x$.

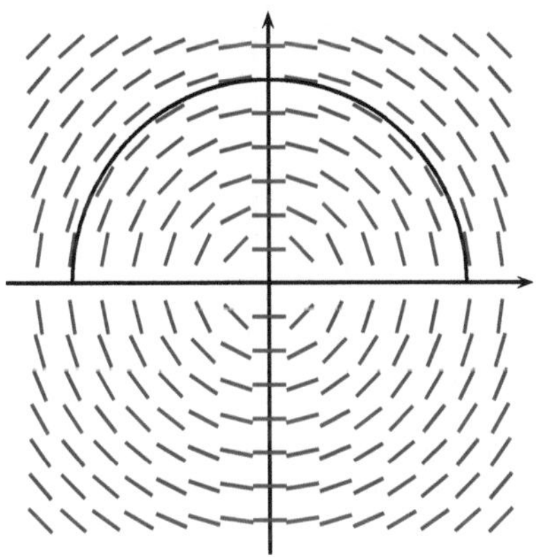

Figure 12.1: Slope field of the differential equation $y' = -\frac{x}{y}$.

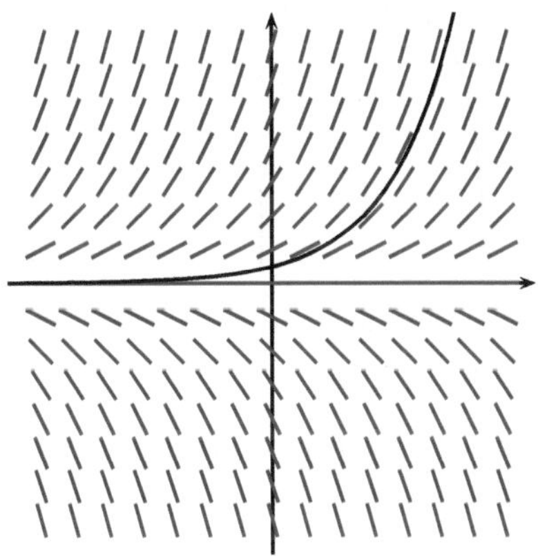

Figure 12.2: Slope field of the differential equation $y' = y$.

Example 12.5. For $\lambda \in \mathbb{R}$, the differential equation

$$y' = \lambda y$$

has the function $y = e^{\lambda x}$, $x \in \mathbb{R}$, as a solution. As in the previous example, the solution is not unique here either. Any function $y = \gamma e^{\lambda x}$, $\gamma \in \mathbb{R}$, is a solution to the differential equation.

Example 12.6. Consider the differential equation

$$y' = xy\,.$$

Every function $y = \gamma e^{\frac{1}{2}x^2}$, $x \in \mathbb{R}$, is a solution to this differential equation. Here again, γ denotes an arbitrary real constant. The slope field of this differential equation and a particular solution $y = -\frac{1}{100}e^{\frac{1}{2}x^2}$ are shown in Figure 12.3.

Example 12.7. The differential equation

$$y' = \frac{y}{x}$$

has all functions of the form $y = \gamma x$ as its solutions, where $x \in \mathbb{R} \setminus \{0\}$ and γ is an arbitrary real constant. A particular solution $y = \frac{1}{4}x$ is sketched together with the slope field of the differential equation in Figure 12.4.

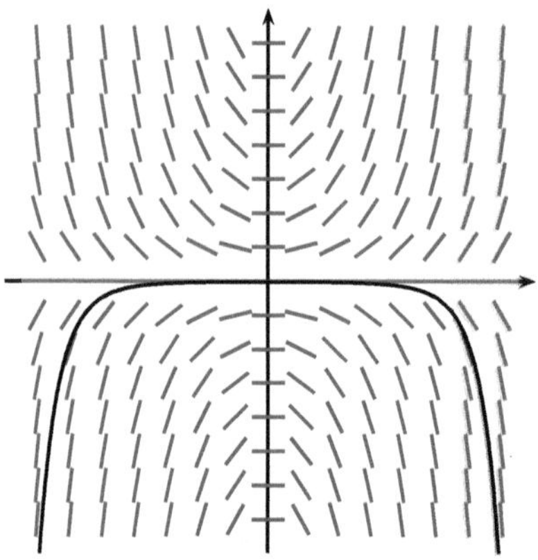

Figure 12.3: Slope field of the differential equation $y' = xy$.

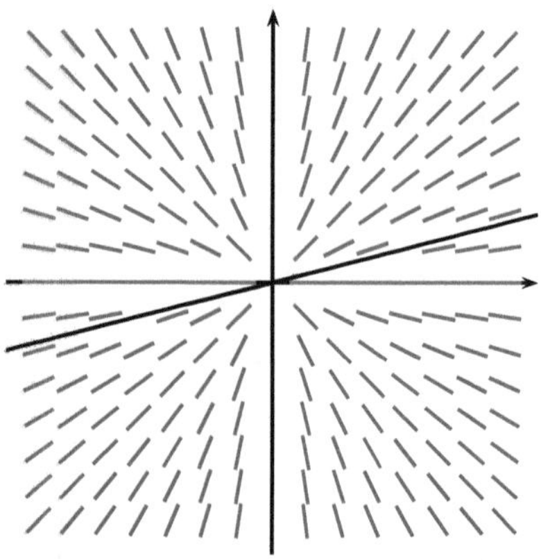

Figure 12.4: Slope field of the differential equation $y' = \frac{y}{x}$.

As can be seen from the above examples, the solution to a differential equation is generally not unique. Rather, there are infinitely many functions that satisfy the differential equation. Therefore, it is often required that the solution satisfies a so-called **initial condition**, which specifies the function value $y_0 = f(x_0)$ at some argument x_0. In this case, the solution to a first-order differential equation is, in most cases, unique.

Example 12.8. We are looking for the solution to the differential equation

$$y' = 2y$$

with initial condition $y(0) = 5$. The general solution to the differential equation is $y = \gamma e^{2x}$, $x \in \mathbb{R}$. To find the particular solution that satisfies the initial condition $y(0) = 5$, we evaluate the general solution at $x_0 = 0$:

$$y(0) = \gamma e^{2 \cdot 0} = \gamma\,.$$

Thus, to satisfy the initial condition, $\gamma = 5$ must hold. The unique solution to the differential equation with this initial condition is therefore $y = 5e^{2x}$.

In the above examples, finding a solution to the differential equation was relatively easy. But what about more complicated differential equations? Is there a general method for finding one or all solutions to a differential equation? Unfortunately, such a universal recipe does not exist. However, for certain classes of differential equations, solutions can be stated in a general form. Therefore, we will introduce some types of differential equations in the following and outline how they can be solved.

The simplest case of a differential equation occurs when the differential equation can be written in the form $y' = g(x)$, i.e., when the right-hand side does not depend on y at all. In this case, any antiderivative of g is a solution to the differential equation.

Theorem 12.1: Differential equation of the form $y' = g(x)$

Consider a differential equation of the form

$$y' = g(x)\,.$$

If G is an antiderivative of g, the general solution to this differential equation is given by

$$y = \int g(x)\,dx = G(x) + C\,, \qquad C \in \mathbb{R}\,.$$

The constant C can be determined from an initial condition.

Example 12.9. Consider the differential equation $y' = 5x^4 + 1$. Then

$$y = \int \left(5x^4 + 1\right)\,dx = x^5 + x + C\,, \qquad C \in \mathbb{R}\,,$$

is the general solution to the differential equation.

An important class of differential equations are the so-called linear differential equations.

Definition 12.2: First-order linear differential equation

A differential equation of the form

$$y' + a(x)y = b(x)$$

is called **first-order linear differential equation.** Here, $a(x)$ and $b(x)$ are given functions of x.

If $b(x) = 0$ for all x, the linear differential equation is called **homogeneous**, otherwise it is called **nonhomogeneous**. The function $b(x)$ is referred to as the **forcing function**.

Such a differential equation is called linear because the left-hand side of the equation is linear in both y and y'.

The solution to a first-order homogeneous linear differential equation can, in principle, be found much more easily compared to a that of a first-order nonhomogeneous linear differential equation.

Theorem 12.2: First-order homogeneous linear differential equation

The first-order homogeneous linear differential equation $y' + a(x)y = 0$ has the general solution

$$y = \gamma \exp\left(-\int a(x)dx\right) = \gamma e^{-A(x)}, \qquad \gamma \in \mathbb{R},$$

where $A(x)$ denotes an antiderivative of $a(x)$.

The unique solution to the initial condition $y_0 = f(x_0)$ is given by[5]

$$y = y_0 \exp\left(-\int_{x_0}^{x} a(t)dt\right) = y_0 e^{-(A(x)-A(x_0))}.$$

It is easy to verify that the given function is a solution to the differential equation by differentiating the function with respect to x and substituting it into the differential equation. It should be noted that the derivative of a definite integral with respect to its upper limit is just the integrand, as we have seen in Section 7.3. It is also immediately clear that the given function satisfies the initial condition when substituting $x = x_0$.

Remark: When calculating the indefinite integral $\int a(x)\,dx$ in Theorem 12.2, the constant of integration can be safely omitted because it is already incorporated in the multiplicative factor γ. If one uses the antiderivative $A(x) + C$ instead of $A(x)$, then one obtains

$$y = \gamma e^{-(A(x)+C)} = \gamma e^{-C} e^{-A(x)} = \tilde{\gamma} e^{-A(x)},$$

where $\tilde{\gamma} = \gamma e^{-C} \in \mathbb{R}$.

The differential equations from Examples 12.1 to 12.8 are all first-order linear differential equations. With the exception of Example 12.2, they are homogeneous linear differential equations; only the differential equation from Example 12.2 is nonhomogeneous.

We now show how to find the solution to the differential equations from Examples 12.1 and 12.3 using Theorem 12.2.

Example 12.10 (Demand function). In Example 12.1, we derived the differential equation $q' + (\delta/p)q = 0$. This is a homogeneous linear differential equation,

[5] Since the variable x is the upper limit of integration, a different notation – here t – must be chosen for the integration variable.

where q plays the role of y and p plays the role of x. The function a from Theorem 12.1 is given by $a(p) = \delta/p$. The general solution is thus obtained according to

$$q = \gamma \exp\left(-\int \frac{\delta}{p}\, dp\right) = \gamma \exp\big(-\delta \ln(p)\big) = \gamma \cdot p^{-\delta} = \frac{\gamma}{p^{\delta}}.$$

For a given initial condition $q(p_0) = q_0$, we obtain $\gamma = q_0 \cdot p_0^{\delta}$ and thus

$$q = \frac{q_0 \cdot p_0^{\delta}}{p^{\delta}} = q_0 \left(\frac{p_0}{p}\right)^{\delta}.$$

The demand function in Example 12.1 is thus a **Cobb-Douglas function**. Generally, it can be shown that a function of several variables, whose partial elasticities are all constant, is a Cobb-Douglas function.

Example 12.11 (Harrod-Domar model). In the Harrod-Domar model, we obtain the differential equation $Y'(t) = (s/v)Y(t)$ or $Y'(t) - (s/v)Y(t) = 0$. This is a homogeneous linear differential equation, where Y plays the role of y and t the role of x. The function a is given by $a(t) = -s/v$. Thus, the general solution is

$$Y(t) = \gamma \exp\left(\int \frac{s}{v}\, dt\right) = \gamma \exp\left(\frac{s}{v} t\right) = \gamma e^{\frac{s}{v} t}.$$

From an initial condition $Y(t_0) = Y_0$, the value of γ can be found to be $\gamma = Y_0\, e^{-\frac{s}{v} t_0}$. Thus,

$$Y(t) = Y_0\, e^{-\frac{s}{v} t_0} e^{\frac{s}{v} t} = Y_0\, e^{\frac{s}{v}(t - t_0)}.$$

In the Harrod-Domar model, national income exhibits exponential growth. The growth rate depends on the ratio of the savings rate to the capital coefficient: a higher savings rate leads to faster growth, while a larger capital coefficient results in slower growth.

We now turn to the solution of the nonhomogeneous linear differential equation. For any two particular solutions $y = f_1(x)$ and $y = f_2(x)$ of a nonhomogeneous linear differential equation, it holds that

$$f_1'(x) + a(x) f_1(x) = b(x),$$
$$f_2'(x) + a(x) f_2(x) = b(x).$$

Subtracting the two equations gives

$$\big(f_1'(x) - f_2'(x)\big) + a(x)\big(f_1(x) - f_2(x)\big) = b(x) - b(x) = 0.$$

hence

$$(f_1 - f_2)'(x) + a(x)(f_1 - f_2)(x) = 0.$$

This shows that the difference of two solutions to the nonhomogeneous differential equation is a solution to the homogeneous differential equation. Every solution to the nonhomogeneous differential equation is therefore obtained as the sum of a particular solution to the nonhomogeneous differential equation and a solution to the homogeneous differential equation. Conversely, however, every sum of a particular solution to the nonhomogeneous differential equation and a solution to the homogeneous differential equation is again a solution to the nonhomogeneous equation. Therefore, to find the general solution to the nonhomogeneous differential equation, one only needs to find the **general solution to the homogeneous equation** and **a particular solution to the nonhomogeneous equation.**

How can we find a particular solution to the nonhomogeneous differential equation? We start with the special case where the functions $a(x)$ and $b(x)$ are constant. Then the differential equation is

$$y' + ay = b\,.$$

If $a \neq 0$, then clearly $y(x) = b/a$ is a solution to the differential equation. This solution is called an **equilibrium solution**, because it is constant. The general solution to the homogeneous linear differential equation is then given by the previous theorem as $y = \gamma e^{-ax}$ with $\gamma \in \mathbb{R}$. This gives the following theorem about the general solution to the nonhomogeneous linear differential equation with constant coefficients.

Theorem 12.3: Nonhomogeneous linear differential equation

The nonhomogeneous linear differential equation **with constant coefficients**, $y' + ay = b$, $a \neq 0$, has the general solution

$$y = \frac{b}{a} + \gamma e^{-ax}\,, \qquad \gamma \in \mathbb{R}\,.$$

The unique solution to the initial condition $y_0 = f(x_0)$ is given by

$$y = \frac{b}{a} + \left(y_0 - \frac{b}{a}\right) e^{-a(x-x_0)}\,.$$

We now use Theorem 12.3 to solve the differential equation that arises in the Evans price adjustment model (Example 12.2).

Example 12.12 (Evans price adjustment model). The Evans price adjustment model led to the differential equation $p'(t) = a - bp(t)$ or $p'(t) + bp(t) = a$. This is a nonhomogeneous linear differential equation with constant coefficients. According to Theorem 12.3, the general solution is

$$p(t) = \frac{a}{b} + \gamma e^{-bt}\,,$$

and for the initial condition $p(t_0) = p_0$ the unique solution is

$$p(t) = \frac{a}{b} + \left(p_0 - \frac{a}{b}\right) e^{-b(t-t_0)}\,.$$

An interesting question is how the price evolves in the long run. To calculate the limit, the sign of the constant b is important. Since λ, β, and δ are positive, $b = \lambda(\beta + \delta)$ is also positive. This determines the limit of the price as t tends to infinity:

$$\begin{aligned}\lim_{t\to\infty} p(t) &= \lim_{t\to\infty}\left(\frac{a}{b} + \left(p_0 - \frac{a}{b}\right)e^{-b(t-t_0)}\right)\\ &= \frac{a}{b} + \left(p_0 - \frac{a}{b}\right)\lim_{t\to\infty} e^{-b(t-t_0)} = \frac{a}{b} + \left(p_0 - \frac{a}{b}\right)\cdot 0 = \frac{a}{b},\end{aligned}$$

that is, the price converges to the equilibrium solution. Substituting $a = \lambda(\alpha - \gamma)$ and $b = \lambda(\beta + \delta)$ for a and b, gives

$$\lim_{t\to\infty} p(t) = \frac{\lambda(\alpha-\gamma)}{\lambda(\beta+\delta)} = \frac{\alpha-\gamma}{\beta+\delta} = \overline{p}\,.$$

This is the **equilibrium price** $\overline{p}$, at which supply and demand are exactly equal. According to the Evans price adjustment model, the price thus converges to the equilibrium price in the long run.

For non-constant forcing functions, the solution to a nonhomogeneous linear differential equation can be found using the method known as **variation of constants**. The key idea is to replace the constant γ in the general general solution $y = \gamma\exp(-A(x))$ of the homogeneous equation by a function $\gamma(x)$. So, we assume that there is a solution to the nonhomogeneous equation of the form $y = \gamma(x)\exp(-A(x))$. Its derivative is given by

$$y' = \gamma'(x)e^{-A(x)} - \gamma(x)a(x)e^{-A(x)}\,.$$

If we substitute this into the nonhomogeneous differential equation, we get

$$\begin{aligned}y' + a(x)y &= \left(\gamma'(x)e^{-A(x)} - \gamma(x)a(x)e^{-A(x)}\right) + a(x)\gamma(x)e^{-A(x)}\\ &= \gamma'(x)e^{-A(x)}\,.\end{aligned}$$

Thus, it must hold that

$$y' + a(x)y = b(x) \iff \gamma'(x)e^{-A(x)} = b(x) \iff \gamma'(x) = b(x)e^{A(x)}\,.$$

Hence, $\gamma(x)$ can be obtained as the indefinite integral

$$\gamma(x) = \int b(x)e^{A(x)}\,dx\,.$$

A particular solution to the nonhomogeneous differential equation is therefore

$$y = \left(\int b(x)e^{A(x)}\,dx\right)e^{-A(x)}\,.$$

The general solution to the nonhomogeneous equation is then obtained as the sum of a particular solution to the nonhomogeneous equation and the general solution to the homogeneous equation.

Theorem 12.4: Nonhomogeneous linear differential equation

The **first-order nonhomogeneous linear differential equation**

$$y' + a(x)y = b(x)$$

has the general solution

$$y = \left(\int b(x)e^{A(x)}\,dx + \gamma\right)e^{-A(x)}, \qquad \gamma \in \mathbb{R},$$

where $A(x)$ is an antiderivative of $a(x)$.

The unique solution to the initial condition $y_0 = f(x_0)$ is

$$y = \left(\int_{x_0}^{x} b(t)e^{A(t)}\,dt + y_0e^{A(x_0)}\right)e^{-A(x)}.$$

As in Theorem 12.2, it is sufficient to find **one** antiderivative $A(x)$. The constant of integration can therefore be omitted again.

We illustrate **variation of constants** by an example.

Example 12.13 (Variation of constants). Consider the differential equation

$$y' + 2xy = 3x.$$

The general solution to the homogeneous differential equation is

$$y = \gamma e^{-x^2}.$$

The find a particular solution to the nonhomogeneous equation we use **variation of constants**. This leads to the integral

$$\gamma(x) = \int 3xe^{x^2}\,dx = \frac{3}{2}\int e^u\,du = \frac{3}{2}e^u = \frac{3}{2}e^{x^2}.$$

Here, the substitution $u = x^2$, $du = 2x\,dx$ was used. Thus, a particular solution to the nonhomogeneous differential equation is

$$y = \gamma(x)e^{-x^2} = \frac{3}{2}e^{x^2}e^{-x^2} = \frac{3}{2},$$

and the general solution to the nonhomogeneous differential equation is

$$y = \frac{3}{2} + \gamma e^{-x^2}.$$

We now turn to the so-called **exact differential equations**. A differential equation is called **exact** if it can be written in the form

$$g(x,y) + h(x,y)y' = 0$$

where

$$g(x,y)\,dx + h(x,y)\,dy$$

is the **total differential** of a differentiable function F. Therefore, exact differential equations are most often written in the form

$$g(x,y)\,dx + h(x,y)\,dy = 0\,.$$

The following theorem states the general solution to an exact differential equation.

Theorem 12.5: Exact differential equation

The general solution to an exact differential equation is implicitly given by

$$F(x,y) = C\,, \quad C \in \mathbb{R}\,.$$

In this context, two questions naturally arise.

1. How can one check whether a given differential equation is exact, i.e., whether a function F exists whose partial derivatives are $g(x,y)$ and $h(x,y)$, respectively?
2. If the differential equation is exact, how can the function F be found?

A necessary condition for the first question can be stated quite simply. If a function F exists such that $F_x = g$ and $F_y = h$, and if F is twice continuously differentiable, then because $F_{xy} = F_{yx}$, it must also hold

$$\frac{\partial g}{\partial y} = F_{yx} = F_{xy} = \frac{\partial h}{\partial x}\,.$$

That this condition is also sufficient is the content of the following theorem:

Theorem 12.6: Exact differential equation

Consider the differential equation $g(x,y) + h(x,y)y' = 0$. If the functions g and h are continuously differentiable, then the differential equation is exact if and only if

$$\frac{\partial g}{\partial y} = \frac{\partial h}{\partial x}\,.$$

The second question is answered by the following theorem, which provides an explicit representation of the function F.

Theorem 12.7: Solution of an exact differential equation

Consider the exact differential equation $g(x, y) + h(x, y)y' = 0$. A solution to the initial condition $y_0 = f(x_0)$ is implicitly given by

$$F(x, y) = 0,$$

with

$$F(x, y) = \int_{x_0}^{x} g(s, y_0)\, ds + \int_{y_0}^{y} h(x, t)\, dt$$

or, equivalently,

$$F(x, y) = \int_{x_0}^{x} g(s, y)\, ds + \int_{y_0}^{y} h(x_0, t)\, dt.$$

We illustrate the solution to an exact differential equation by an example.

Example 12.14 (Exact differential equation). We seek the solution to the differential equation

$$8xy\, dx + 4x^2\, dy = 0$$

subject to the initial condition $y_0 = f(1) = 2$. The differential equation is exact, since

$$\frac{\partial}{\partial y} 8xy = 8x = \frac{\partial}{\partial x} 4x^2 .$$

The function F is given by

$$F(x, y) = \int_1^x 16s\, ds + \int_2^y 4x^2\, dt = \left[8s^2\right]_1^x + \left[4x^2 t\right]_2^y$$
$$= (8x^2 - 8) + (4x^2 y - 8x^2) = 4x^2 y - 8 .$$

Thus, the solution to the differential equation is implicitly given by

$$4x^2 y - 8 = 0 .$$

This equation can explicitly be solved for y to obtain

$$y = \frac{2}{x^2} .$$

The solution is not defined for $x = 0$. Since a solution to a differential equation must be defined on an interval, and the initial condition is given at $x_0 = 1 > 0$, this function is a solution only for $x > 0$.

Another important class of differential equations are so-called **separable differential**

equations. These are differential equations of the form

$$y' = g(x)h(y)\,.$$

Their right-hand side can be written as the product of two functions, one of which is a function of x only, the other a function of y only. The following theorem gives the solution to a separable differential equation.

Theorem 12.8: Separable differential equation

Let g, h be continuous functions defined on open intervals A and B, respectively. Let further $h(y) \neq 0$ for all $y \in B$. The general solution to the separable differential equation $y' = g(x)h(y)$ is implicitly given by

$$\int \frac{dy}{h(y)} = \int g(x)\,dx + C\,, \qquad C \in \mathbb{R}\,.$$

The unique solution to the initial condition $f(x_0) = y_0$ is given by

$$\int_{y_0}^{y} \frac{ds}{h(s)} = \int_{x_0}^{x} g(t)\,dt\,.$$

The technique for solving a separable differential equation is straightforward to remember. First, write $\frac{dy}{dx}$ in place of y'. Then, rearrange the equation by multiplying both sides by dx and dividing by $h(y)$, so that all terms involving y appear on one side and all terms involving x appear on the other – thus the variables are "separated". In the last step put an integral sign on both sides of the equation.

$$\frac{dy}{dx} = g(x)h(y) \quad\Longleftrightarrow\quad \frac{dy}{h(y)} = g(x)dx \quad\Longleftrightarrow\quad \int \frac{dy}{h(y)} = \int g(x)dx$$

After integrating both sides, we obtain a solution in implicit form. If possible, this equation should then be solved for y explicitly.

Example 12.15 (Separable differential equation). The differential equation $y' = xy^2$ is separable with $g(x) = x$ and $h(y) = y^2$. Its solution is given by

$$\int \frac{1}{y^2}\,dy = \int x\,dx \iff -y^{-1} = \frac{x^2}{2} + C$$
$$\iff y = -\frac{1}{\frac{x^2}{2} + C} = -\frac{2}{x^2 + 2C}\,, \qquad C \in \mathbb{R}\,.$$

If, for example, the initial condition $y(0) = 1$ is given, the constant C can be found by substituting $x = 0$ and $y = 1$:

$$1 = -\frac{2}{0^2 + 2C} \quad\Longleftrightarrow\quad C = -1\,.$$

The unique solution to the initial condition $y(0) = 1$ is thus

$$y = \frac{2}{2 - x^2}.$$

This function is not defined for $x \neq \pm\sqrt{2}$. Since a solution to a differential equation must be defined on an interval, and the initial condition is given at $x = 0$, the largest interval on which the solution can be defined is the open interval $(-\sqrt{2}, \sqrt{2})$.

Example 12.16 (Logistic growth). We study the growth of a quantity y as a function of time t. A simple growth model was already considered in Example 12.5, where we studied the differential equation $y' = \lambda y$. This can be interpreted as the (absolute) change of y at any time being proportional to the value of y, or, equivalently, the relative change y'/y being constantly equal to λ. This is the model of so-called **exponential growth**. The general solution to this differential equation is $y(t) = y_0 e^{\lambda t}$, where y_0 denotes the value of y at time $t = 0$. Exponential growth of an economic quantity over long periods of time is not very realistic for positive λ, since the quantity $y(t)$ increases without bound as time progresses. Therefore, exponential growth is only observed over short periods of time. Many natural growth processes can be well described by exponential growth in the initial phase. However, a saturation effect often sets in and the relative change decreases. As an example, consider the stock of DVD players in Germany. Initially, the stock exhibited exponential growth of the stock. However, as time passed and nearly all households in Germany acquired a DVD player, the growth rate declined.

A more realistic model for growth processes with saturation is the so-called **logistic growth**. It is based on the differential equation

$$\frac{y'}{y} = \lambda(G - y) \quad \text{or} \quad y' = \lambda y(G - y).$$

This equation states that the (continuous) growth rate $\frac{y'}{y}$ is proportional, with proportionality factor λ, to the distance between y and the saturation limit G.

We now determine the solution to this differential equation subject to the initial condition $y(0) = y_0$. Separation of variables yields

$$\frac{dy}{dt} = \lambda y(G - y) \quad \Longleftrightarrow \quad \frac{dy}{y(G - y)} = \lambda dt.$$

Taking into account the initial condition $y(0) = y_0$, this leads to the equation

$$\int_{y_0}^{y} \frac{ds}{s(G - s)} = \int_0^t \lambda du.$$

The solution to the right-hand side integral is $\int_0^t \lambda du = \lambda t$. To solve the left-hand side integral, we write the integrand as

$$\frac{1}{s(G - s)} = \frac{1}{sG} + \frac{1}{G(G - s)}.$$

Assuming $0 < y_0, y < G$, we get

$$\int_{y_0}^{y} \frac{1}{sG}\, ds = \frac{1}{G} \ln\left(\frac{y}{y_0}\right)$$

and

$$\int_{y_0}^{y} \frac{1}{G(G-s)}\, ds = \frac{1}{G}\Big[-\ln(G-s)\Big]_{y_0}^{y} = \frac{1}{G} \ln\left(\frac{G-y_0}{G-y}\right).$$

This gives

$$\int_{y_0}^{y} \frac{ds}{s(G-s)} = \frac{1}{G} \ln\left(\frac{y}{y_0}\right) + \frac{1}{G} \ln\left(\frac{G-y_0}{G-y}\right) = \frac{1}{G} \ln\left(\frac{y}{y_0} \cdot \frac{G-y_0}{G-y}\right).$$

Thus, the solution to the differential equation is obtained from the equation

$$\frac{1}{G} \ln\left(\frac{y}{y_0} \cdot \frac{G-y_0}{G-y}\right) = \lambda t \quad \text{or} \quad \ln\left(\frac{y}{y_0} \cdot \frac{G-y_0}{G-y}\right) = \lambda G t\,.$$

Applying the exponential function on both sides yields

$$\frac{y}{y_0} \cdot \frac{G-y_0}{G-y} = e^{\lambda G t} \quad \text{or} \quad \frac{y}{G-y} = \frac{y_0}{G-y_0} e^{\lambda G t}.$$

Solving for y gives

$$y = G\, \frac{\frac{y_0}{G-y_0} e^{\lambda G t}}{1 + \frac{y_0}{G-y_0} e^{\lambda G t}} = G\, \frac{1}{\frac{G-y_0}{y_0} e^{-\lambda G t} + 1} = \frac{G}{1 + \left(\frac{G}{y_0} - 1\right) e^{-\lambda G t}}.$$

The function

$$\boxed{y = \frac{G}{1 + \left(\frac{G}{y_0} - 1\right) e^{-\lambda G t}}}$$

is called the **logistic function**. It describes growth processes with saturation according to the above differential equation. Here, y_0 denotes the initial stock at time $t = 0$, G the saturation limit, and λ the proportionality factor of the growth rate.

Figure 12.5 shows the slope field of the differential equation for $\lambda = 0.2$ and $G = 4$. Furthermore, the solution for the initial condition $y_0 = 1.5$ is plotted.

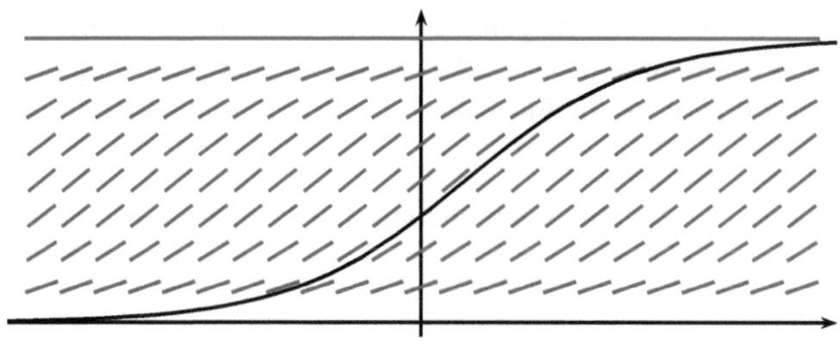

Figure 12.5: Slope field of the logistic differential equation.

First-order homogeneous linear differential equations are special cases of separable differential equations since

$$y' + a(x)y = 0 \quad \Longleftrightarrow \quad y' = -a(x)y\,.$$

By using separation of variables and simplifying, we arrive at the form of the general solution for homogeneous linear differential equations given in Theorem 12.2.

12.2 Linear differential equations of order m with constant coefficients

In this section, we study an important class of differential equations, the ***m*-th order linear differential equations with constant coefficients**.

Definition 12.3: m-th order linear differential equation

An ***m*-th order linear differential equation with constant coefficients** is a differential equation of the form

$$y^{(m)} + a_{m-1}y^{(m-1)} + \ldots + a_1 y' + a_0 y = b(x)$$

where $a_1, \ldots, a_{m-1} \in \mathbb{R}$ and $b : A \to \mathbb{R}$ is a given function.

If $b(x) = 0$ for $x \in A$, then the differential equation is called **homogeneous**, otherwise **nonhomogeneous**. The function b is called **forcing function**.

We start with the case of **homogeneous** linear differential equations and then turn to the general case of **nonhomogeneous** linear differential equations.

Structure of the solution space of the homogeneous equation

We first want to examine the structure of the solution sets of homogeneous linear differential equations in more detail. To do this, we will need some results of linear algebra, which are presented in Chapter 9.

We have already seen in the previous section on first-order linear differential equations that the solution to a homogeneous linear differential equation is not unique. But how do the different solutions relate to each other? The following two observations, which follow from the linearity of the differential equation, provide insight into this.

Observation 1: If $y = f(x)$ is a solution to the homogeneous linear differential equation, then for every real number γ, $y = \gamma f(x)$ is also a solution to the homogeneous differential equation.

If we substitute $\gamma f(x)$ for y, we get

$$\gamma f^{(m)}(x) + a_{m-1}\gamma f^{(m-1)}(x) + \ldots + a_1\gamma f'(x) + a_0\gamma f(x)$$
$$= \gamma\underbrace{\big(f^{(m)}(x) + a_{m-1}f^{(m-1)}(x) + \ldots + a_1 f'(x) + a_0 f(x)\big)}_{=0} = \gamma \cdot 0 = 0\,.$$

Observation 2: If $y = f_1(x)$ and $y = f_2(x)$ are two solutions to the homogeneous linear differential equation, then $y = f_1(x) + f_2(x)$ is also a solution to the homogeneous differential equation.

If we substitute $y = f_1(x) + f_2(x)$ into the differential equation, we get

$$\big(f_1^{(m)}(x) + f_2^{(m)}(x)\big) + a_{m-1}\big(f_1^{(m-1)}(x) + f_2^{(m-1)}(x)\big) + \ldots$$
$$\ldots + a_1\big(f_1'(x) + f_2'(x)\big) + a_0\big(f_1(x) + f_2(x)\big)$$
$$= \underbrace{\big(f_1^{(m)}(x) + a_{m-1}f_1^{(m-1)}(x) + \ldots + a_1 f_1'(x) + a_0 f_1(x)\big)}_{=0}$$
$$+ \underbrace{\big(f_2^{(m)}(x) + a_{m-1}f_2^{(m-1)}(x) + \ldots + a_1 f_2'(x) + a_0 f_2(x)\big)}_{=0} = 0\,.$$

From these two observations, the following theorem follows immediately.

Theorem 12.9: Superposition principle

If $y = f_1(x)$ and $y = f_2(x)$ are two solutions to the m-th order homogeneous linear differential equation

$$y^{(m)} + a_{m-1}y^{(m-1)} + \ldots + a_1 y' + a_0 y = 0\,,$$

then any linear combination

$$y = \gamma_1 f_1(x) + \gamma_2 f_2(x)$$

with arbitrary real numbers $\gamma_1, \gamma_2 \in \mathbb{R}$ is also a solution to the homogeneous linear differential equation.

From the superposition principle, it follows that conditions U1 and U2 for a subspace (see Section 9.2) are satisfied for the solution set of a homogeneous linear differential equation. Since the zero function $y(x) = 0$ for all x is always a solution, the solution set is non-empty and condition U0 is satisfied. Therefore, the solution set forms a subspace of the vector space $\mathcal{F}(A, \mathbb{R})$ consisting of all functions with domain A and codomain $\mathbb{R}$. Moreover, one can show that the dimension of the solution space equals the order of the differential equation. This is the content of the following theorem:

Theorem 12.10: Structure of the solution (homogeneous equation)

The general solution to the m-th order **homogeneous** linear differential equation,

$$y^{(m)} + a_{m-1}y^{(m-1)} + \ldots + a_1 y' + a_0 y = 0\,,$$

is an $\boldsymbol{m}$**-dimensional subspace** of $\mathcal{F}(A, \mathbb{R})$.

If $f_1, \ldots, f_m$ is a **basis of the general solution**, then the general solution is

$$y = \gamma_1 f_1(x) + \cdots + \gamma_m f_m(x)\,, \quad \gamma_1, \ldots, \gamma_m \in \mathbb{R}\,.$$

A basis of the solution space $f_1, \ldots, f_m$ is also referred to as **fundamental system** of the differential equation.

To specify the solution to an m-th order homogeneous linear differential equation, it is therefore sufficient to provide a fundamental system, that is, m linearly independent solutions. This leads to the question of how one can determine whether a set of functions is linearly independent. A simple and effective criterion is given by the so-called **Wronskian** (or **Wronski**[6] **determinant**), which provides a test for linear independence of differentiable functions. Let $f_1, \ldots, f_m \in \mathcal{F}(A, \mathbb{R})$ be $(m-1)$-times differentiable. Then the determinant

$$W(x) = \det \begin{bmatrix} f_1(x) & f_2(x) & \ldots & f_m(x) \\ f_1'(x) & f_2'(x) & \ldots & f_m'(x) \\ \vdots & \vdots & & \vdots \\ f_1^{(m-1)}(x) & f_2^{(m-1)}(x) & \ldots & f_m^{(m-1)}(x) \end{bmatrix}$$

is called the **Wronskian** of $f_1, \ldots, f_m$. The following theorem shows how linear independence of the functions $f_1, \ldots, f_m$ can be checked using the Wronskian.

Theorem 12.11: Wronskian

If $f_1, \ldots, f_m \in \mathcal{F}(A, \mathbb{R})$ are $(m-1)$-times differentiable, then

$$W(x) \neq 0 \text{ for } \textbf{one } x \quad \Longrightarrow \quad f_1, \ldots, f_m \text{ are linearly independent.}$$

If $f_1, \ldots, f_m \in \mathcal{F}(A, \mathbb{R})$ are **solutions** of an m-th order homogeneous linear differential equation, then

$$W(x) \neq 0 \text{ for } \textbf{all } x \quad \Longleftrightarrow \quad f_1, \ldots, f_m \text{ are linearly independent.}$$

The theorem states, in particular, that for solutions of a homogeneous linear differential equation the Wronskian is either non-zero for all x (if the solutions are linearly independent) or identically zero for all x (if the solutions are linearly dependent).

[6] Jósef Maria Hoënë-Wroński (1776–1835)

Consequently, to determine whether m solutions are linearly independent, it suffices to compute the Wronskian at a single value of x.

Solution of the homogeneous linear differential equation

The solution to a first-order homogeneous linear differential equation $y' + a_0 y = 0$ has the form $y = \gamma \exp(-a_0 x)$. Consequently, for the solution to an m-th order homogeneous linear differential equation we employ the trial solution $y = \exp(\lambda x)$. Substituting this into the differential equation yields

$$\lambda^m e^{\lambda x} + a_{m-1}\lambda^{m-1} e^{\lambda x} + \ldots + a_1 \lambda e^{\lambda x} + a_0 e^{\lambda x} = 0$$
$$\iff \quad \lambda^m + a_{m-1}\lambda^{m-1} + \ldots + a_1\lambda + a_0 = 0\,.$$

This equation is called the **characteristic equation** of the differential equation. Accordingly, the polynomial on the left-hand side is called the **characteristic polynomial**. The solution to the differential equation is therefore reduced to the problem of finding all roots of the characteristic polynomial. For each root λ_0 of the characteristic polynomial, $y(x) = \exp(\lambda_0 x)$ is a solution to the differential equation.

A polynomial of degree m has at most m different real roots. If there are m different real roots, these roots provide m different solutions of the differential equation. Using the Wronskian, it can be shown that these m solutions are linearly independent, so they form a fundamental system, that is, a basis of the solution space.

Theorem 12.12: m-th order homogeneous linear differential equation

If the characteristic polynomial

$$\lambda^m + a_{m-1}\lambda^{m-1} + \ldots + a_1\lambda + a_0$$

of the m-th order homogeneous linear differential equation has exactly m different real roots $\lambda_1, \lambda_2, \ldots, \lambda_m$, then the general solution is given by

$$y_t = \gamma_1 e^{\lambda_1 x} + \gamma_2 e^{\lambda_2 x} + \ldots + \gamma_m e^{\lambda_m x}$$

where $\gamma_1, \gamma_2, \ldots, \gamma_m \in \mathbb{R}$ are arbitrary real constants.

Example 12.17. The second-order homogeneous linear differential equation

$$y'' - 4y = 0$$

has the characteristic polynomial $\lambda^2 - 4$. Its roots are $\lambda_1 = 2$ and $\lambda_2 = -2$. The general solution is therefore

$$y = \gamma_1 e^{2x} + \gamma_2 e^{-2x}\,, \quad \gamma_1, \gamma_2 \in \mathbb{R}.$$

Solution for multiple or complex roots

If the characteristic polynomial has m different roots $\lambda_1, \ldots, \lambda_m$, then according to Theorem 12.12 the functions $\exp(\lambda_1 x), \ldots, \exp(\lambda_m x)$ form a basis of the solution space. However, a polynomial of degree m may have fewer than m distinct real roots, as some roots may be non-real or may occur with multiplicity. For example, a quadratic polynomial may have one (i.e., double) root or even no real root at all.

If we allow for roots that are complex numbers, the fundamental theorem of algebra (Theorem E.2) states that every polynomial of degree m has exactly m roots in the complex numbers, counting each root according to its multiplicity. Thus, a polynomial may have both real and non-real complex roots. Furthermore, both real and complex roots can occur with multiplicity. We will now explain how the basis of the solution space of a homogeneous linear differential equation can be determined in the general case.

Reading this section requires basic knowledge about complex numbers. A brief introduction to complex numbers and their most important properties can be found in Appendix E.

Multiple roots As already mentioned, a root of the characteristic polynomial may have multiplicity greater than one. In this case, the root yields not just a single solution to the differential equation, but as many linearly independent solutions as its multiplicity.

Theorem 12.13: Solution for multiple roots

Let $\lambda \in \mathbb{R}$ be a real root of the characteristic polynomial of a homogeneous differential equation. If λ has multiplicity k, then

$$e^{\lambda x}, xe^{\lambda x}, \ldots, x^{k-1}e^{\lambda x}$$

are linearly independent solutions to the differential equation.

Thus, a real root of multiplicity k provides k linearly independent solutions of the differential equation.

Example 12.18. The second-order homogeneous linear differential equation

$$y'' + 4y' + 4y = 0$$

has the characteristic polynomial $\lambda^2 + 4\lambda + 4$. Using the quadratic formula, we obtain the roots

$$\lambda_{1,2} = -\frac{4}{2} \pm \sqrt{\frac{16}{4} - 4} = -2 \pm \sqrt{4-4} = -2\,.$$

Hence, $\lambda_{1,2} = -2$ is a double root of the characteristic polynomial. The general solution to the differential equation is therefore

$$y = \gamma_1 e^{-2x} + \gamma_2 x e^{-2x} = (\gamma_1 + \gamma_2 x)e^{-2x}, \quad \gamma_1, \gamma_2 \in \mathbb{R}\,.$$

Complex roots When we speak of complex roots of the characteristic polynomial, we always mean non-real complex roots $\lambda = a + bi$, where the imaginary part $b \neq 0$. In principle, such roots do not differ from real roots. So, if the complex number $\lambda = a + bi$ is a root of the characteristic polynomial, then $e^{(a+bi)x}$ is a solution to the differential equation. This extends to roots with multiplicity: if $\lambda = a + bi$ is a root of multiplicity k, then the functions $e^{(a+bi)x}, xe^{(a+bi)x}, \ldots, x^{k-1}e^{(a+bi)x}$ are solutions of the differential equation.

However, it is clear that for a linear homogeneous differential equation with real coefficients and real initial conditions, all solutions $y = f(x)$ must themselves be real-valued. Therefore, it may seem somewhat unsatisfactory to represent the solution using complex numbers. In the following, we will show that this is, in fact, not necessary: the general solution can always be written in real terms, even when the characteristic equation has complex roots.

We start from a homogeneous linear differential equation

$$y^{(m)} + a_{m-1}y^{(m-1)} + \ldots + a_1y' + a_0y = 0$$

with real coefficients $a_0, \ldots, a_{m-1}$. If $\lambda = a+bi$ is a complex root of the characteristic polynomial, then the complex conjugate $\overline{\lambda} = a - bi$ is also a root of the characteristic polynomial (see Theorem E.2 in the Appendix). Thus, the functions $y = e^{(a+bi)x}$ and $y = e^{(a-bi)x}$ are solutions of the differential equation. Due to Euler's formula (E.1) we have

$$\begin{aligned}
\gamma_1 e^{(a+bi)x} + \gamma_2 e^{(a-bi)x} &= \gamma_1\left[e^{ax}\cos(bx) + ie^{ax}\sin(bx)\right] \\
&\quad + \gamma_2\left[e^{ax}\cos(bx) - ie^{ax}\sin(bx)\right] \\
&= (\gamma_1 + \gamma_2)e^{ax}\cos(bx) + i(\gamma_1 - \gamma_2)e^{ax}\sin(bx) \\
&= C_1\, e^{ax}\cos(bx) + C_2\, e^{ax}\sin(bx)\,,
\end{aligned}$$

where $C_1 = \gamma_1 + \gamma_2$ and $C_2 = i(\gamma_1 - \gamma_2)$. This shows that any linear combination of the solutions $e^{(a+bi)x}$ and $e^{(a-bi)x}$ can be written as a linear combination (with complex coefficients) of the real-valued solution functions $e^{ax}\cos(bx)$ and $e^{ax}\sin(bx)$. Conversely, any linear combination of $e^{ax}\cos(bx)$ and $e^{ax}\sin(bx)$ can be expressed as a linear combination of $e^{(a+bi)x}$ and $e^{(a-bi)x}$, by setting $\gamma_1 = \frac{1}{2}(C_1 - iC_2)$ and $\gamma_2 = \frac{1}{2}(C_1 + iC_2)$.

Therefore, the functions $e^{(a+bi)x}$ and $e^{(a+bi)x}$, as well as the real-valued functions $e^{ax}\cos(bx)$ and $e^{ax}\sin(bx)$, each are linearly independent solutions to the differential equation. We summarize this result in the following theorem.

Theorem 12.14: Solution for complex roots

If $\lambda = a + bi$ is a complex root of the characteristic polynomial, then its complex conjugate $\overline{\lambda} = a - bi$ is also a root, and the functions

$$e^{ax}\cos(bx) \qquad \text{and} \qquad e^{ax}\sin(bx)$$

are linearly independent solutions of the differential equation.

Example 12.19. The second-order homogeneous linear differential equation

$$y'' + 4y = 0$$

has the characteristic polynomial $\lambda^2 + 4$. Its roots are $\lambda_1 = 2i$ and $\lambda_2 = -2i$. The general solution to the differential equation is thus

$$y = \gamma_1 \cos(2x) + \gamma_2 \sin(2x)\,, \quad \gamma_1, \gamma_2 \in \mathbb{R}\,.$$

Theorem 12.14 extends naturally to multiple complex roots.

Theorem 12.15: Solution for multiple complex roots

If $\lambda = a + bi$ is a complex root of the characteristic polynomial of multiplicity k, then its complex conjugate $\overline{\lambda} = a - bi$ is also a root of multiplicity k, and the functions

$$e^{ax}\cos(bx)\,, xe^{ax}\cos(bx)\,, x^2e^{ax}\cos(bx)\,, \ldots, x^{k-1}e^{ax}\cos(bx)\,,$$
$$e^{ax}\sin(bx)\,, xe^{ax}\sin(bx)\,, x^2e^{ax}\sin(bx)\,, \ldots, x^{k-1}e^{ax}\sin(bx)$$

are linearly independent solutions of the differential equation.

Since a polynomial of degree m always has exactly m complex roots, counted with multiplicity, Theorems 12.13 and 12.15 guarantee that we can find m linearly independent solutions to the differential equation, and thus a fundamental system. We illustrate this with an example.

Example 12.20. Consider the fourth-order homogeneous linear differential equation

$$y^{(4)} - 3y''' + 0.5y'' + 2y' + 2y = 0\,.$$

The characteristic polynomial is

$$\lambda^4 - 3\lambda^3 + 0.5\lambda^2 + 2\lambda + 2\,.$$

By applying the root-finding methods from Section 4.10, or by making an educated guess one finds the solution $\lambda_1 = 2$. Polynomial division gives

$$(\lambda^4 - 3\lambda^3 + 0.5\lambda^2 + 2\lambda + 2) : (\lambda - 2) = \lambda^3 - \lambda^2 - 1.5\lambda - 1\,.$$

The cubic polynomial obtained here again has a root at $\lambda_2 = 2$. Dividing out the linear factor $(\lambda - 2)$ yields the quadratic polynomial

$$(\lambda^3 - \lambda^2 - 1.5\lambda - 1) : (\lambda - 2) = \lambda^2 + \lambda + 0.5\,.$$

The roots of this quadratic polynomial are found using the quadratic formula:

$$\lambda_{3,4} = -\frac{1}{2} \pm \sqrt{\frac{1}{4} - \frac{1}{2}} = -\frac{1}{2} \pm \sqrt{-\frac{1}{4}} = -\frac{1}{2} \pm i\frac{1}{2}.$$

We thus have the following roots of the characteristic polynomial,

$$\lambda_1 = 2, \quad \lambda_2 = 2, \quad \lambda_3 = -\frac{1}{2} + i\frac{1}{2}, \quad \lambda_4 = -\frac{1}{2} - i\frac{1}{2}.$$

$\lambda_{1,2} = 2$ is a double root and provides the solutions e^{2x} and xe^{2x}. The complex conjugate roots $\lambda_{3,4} = -\frac{1}{2} \pm \frac{1}{2}i$ provide the two solutions

$$e^{-\frac{1}{2}x} \cos\left(\frac{1}{2}x\right) \quad \text{and} \quad e^{-\frac{1}{2}x} \sin\left(\frac{1}{2}x\right).$$

Therefore, the general solution to the differential equation is

$$y = \gamma_1 e^{2x} + \gamma_2 x e^{2x} + \gamma_3 e^{-\frac{1}{2}x} \cos\left(\frac{1}{2}x\right) + \gamma_4 e^{-\frac{1}{2}x} \sin\left(\frac{1}{2}x\right), \quad \gamma_1, \ldots, \gamma_4 \in \mathbb{R}.$$

If we set $\gamma_1, \gamma_2 = 0$ and additionally require the initial conditions $y(0) = 1$ and $y'(0) = 0$, we obtain the solution

$$y = e^{-\frac{1}{2}x} \cos\left(\frac{1}{2}x\right) - e^{-\frac{1}{2}x} \sin\left(\frac{1}{2}x\right),$$

which is shown in Figure 12.6. The solution to the given initial conditions is a damped oscillation; for x towards infinity, the solution converges to zero.

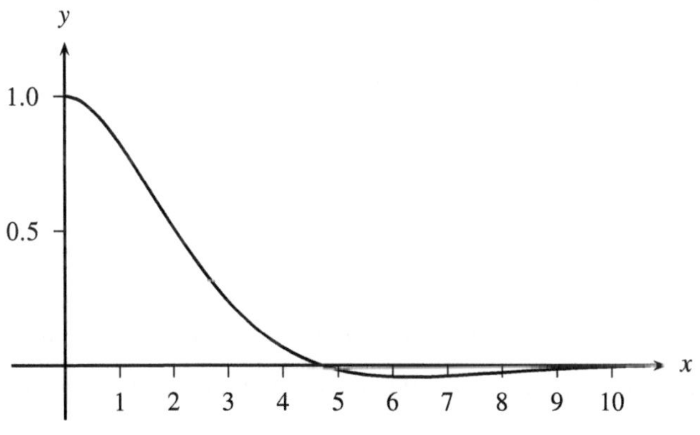

Figure 12.6: A solution to the differential equation from Example 12.20.

On the other hand, if we set $\gamma_3, \gamma_4 = 0$ and impose the initial conditions $y(0) = 1$ and $y'(0) = 0$, we get the solution

$$y_t = e^{2x} - 2xe^{2x} = e^{2x}(1 - 2x).$$

This solution does not converge; rather, it diverges to $-\infty$. Thus, depending on the initial conditions, qualitatively different behaviors can arise for solutions of this differential equation.

We conclude this section with a short discussion of the solutions of a **second-order homogeneous linear differential equation**, $y'' + a_1y' + a_0 = 0$. The characteristic polynomial is $\lambda^2 + a_1\lambda + a_0$. Its roots are

$$\lambda_{1,2} = -\frac{a_1}{2} \pm \sqrt{\frac{a_1^2}{4} - a_0}\,.$$

The following cases can occur:

Two real roots λ_1 and λ_2, if $a_0 < \frac{a_1^2}{4}$. In this case, the general solution is

$$\boxed{y = \gamma_1 e^{\lambda_1 x} + \gamma_2 e^{\lambda_2 x}\,.}$$

A double root $\lambda_1 = \lambda_2$, if $a_0 = \frac{a_1^2}{4}$. Then, the general solution is

$$\boxed{y = (\gamma_1 + \gamma_2 x)e^{\lambda_1 x}\,.}$$

Two complex roots $\lambda_{1,2} = a \pm bi$, if $a_0 > \frac{a_1^2}{4}$. Then, the general solution is

$$\boxed{y = \gamma_1 e^{ax}\cos(bx) + \gamma_2 e^{ax}\sin(bx)\,.}$$

Solution of the nonhomogeneous linear differential equation

We begin this section by studying the structure of the solution set of a nonhomogeneous linear differential equation. The following two observations continue Observations 1 and 2 from page 364.

Observation 3: If $y = f_h(x)$ is a solution to the **homogeneous** differential equation and $y = f_p(x)$ is a solution to the **nonhomogeneous** differential equation, then $y = f_h(x) + f_p(x)$ is a solution to the **nonhomogeneous** differential equation.

If we substitute $y = f_h(x) + f_p(x)$ into the differential equation, we obtain

$$\begin{aligned}
&\big(f_h^{(m)}(x) + f_p^{(m)}(x)\big) + a_{m-1}\big(f_h^{(m-1)}(x) + f_p^{(m-1)}(x)\big) + \ldots \\
&\qquad\qquad \ldots + a_1\big(f_h'(x) + f_p'(x)\big) + a_0\big(f_h(x) + f_p(x)\big) \\
&= \underbrace{\big(f_h^{(m)}(x) + a_{m-1}f_h^{(m-1)}(x) + \ldots + a_1 f_h'(x) + a_0 f_h(x)\big)}_{=0} \\
&\quad + \underbrace{\big(f_p^{(m)}(x) + a_{m-1}f_p^{(m-1)}(x) + \ldots + a_1 f_p'(x) + a_0 f_p(x)\big)}_{=b(x)} = b(x)\,.
\end{aligned}$$

Observation 4: If $y = f_1(x)$ and $y = f_2(x)$ are two solutions to the **nonhomogeneous** differential equation, then $y = f_1(x) - f_2(x)$ is a solution to the **homogeneous** differential equation.

If we substitute $y = f_1(x) - f_2(x)$ into the differential equation, we obtain

$$\begin{aligned}
&\big(f_1^{(m)}(x) - f_2^{(m)}(x)\big) + a_{m-1}\big(f_1^{(m-1)}(x) - f_2^{(m-1)}(x)\big) + \dots \\
&\qquad\qquad \dots + a_1\big(f_1'(x) - f_2'(x)\big) + a_0\big(f_1(x) - f_2(x)\big) \\
&= \underbrace{\big(f_1^{(m)}(x) + a_{m-1}f_1^{(m-1)}(x) + \dots + a_1 f_1'(x) + a_0 f_1(x)\big)}_{=b(x)} \\
&\quad - \underbrace{\big(f_2^{(m)}(x) + a_{m-1}f_2^{(m-1)}(x) + \dots + a_1 f_2'(x) + a_0 f_2(x)\big)}_{=b(x)} = 0\,.
\end{aligned}$$

From these observations, it immediately follows that the general solution to the nonhomogeneous differential equation can be written as the sum of a **particular solution** to the nonhomogeneous equation and the **general solution** to the corresponding homogeneous differential equation.

Theorem 12.16: Structure of the solution (nonhomogeneous equation)

Let $y = f_p(x)$ be a particular solution to the m-th oder **nonhomogeneous** linear differential equation,

$$y^{(m)} + a_{m-1}y^{(m-1)} + \dots + a_1 y' + a_0 y = b(x)\,,$$

and let $f_1, \dots, f_m$ be a basis of the general solution to the homogeneous equation.

Then the **general solution to the nonhomogeneous equation** is

$$y = f_p(x) + \gamma_1 f_1(x) + \dots + \gamma_m f_m(x)\,, \quad \gamma_1, \dots, \gamma_m \in \mathbb{R}\,.$$

The theorem states that to obtain the general solution to the nonhomogeneous differential equation, one only needs the general solution to the homogeneous equation and a **single particular solution** to the nonhomogeneous equation.

Solution under constant forcing

We now turn to the solution to the m-th order nonhomogeneous linear differential equation. To begin, we restrict ourselves to the case where the forcing function is constant, that is, $b(x) = b$ for all $x \in A$. In this case, the differential equation takes the form

$$y^{(m)} + a_{m-1}y^{(m-1)} + \dots + a_1 y' + a_0 y = b\,.$$

Assuming a constant particular solution $y(x) = c$ leads to the equation $a_0 c = b$. For $a_0 \neq 0$, the solution

$$y(x) = \frac{b}{a_0} \qquad \text{for all } x \in A$$

is called the **equilibrium solution**. Thus, according to Theorem 12.16, the following theorem holds.

Theorem 12.17: Solution under constant forcing

If $a_0 \neq 0$, then the nonhomogeneous linear differential equation

$$y^{(m)} + a_{m-1}y^{(m-1)} + \ldots + a_1 y' + a_0 y = b$$

has the general solution

$$y = \frac{b}{a_0} + y_h(x),$$

where $y_h(x)$ denotes the general solution to the corresponding homogeneous linear differential equation.

If $a_0 = 0$, then $\lambda = 0$ is a root of the characteristic polynomial, and $y(x) = e^{0 \cdot x} = 1$ (and thus also $y(x) = c$) is a solution to the homogeneous equation. Hence, $y(x) = c$ cannot be a solution to the nonhomogeneous equation. In this case, we say that there is **resonance**. Instead of $y(x) = c$, we then use the trial solution $y(x) = cx$ and attempt to determine c appropriately. If this also fails, we continue with the trial solutions $y(x) = cx^2$, $y(x) = cx^3$, and so on. In general, if $\lambda = 0$ is a **root of multiplicity** k, then the trial solution $y(x) = cx^k$ will succeed.

Example 12.21. Consider the differential equation $y'' + 2y' = 5$. The trial solution $y(x) = c$ leads to the contradiction $0 = 5$. Therefore, there is **resonance**. Hence, we multiply the trial solution by x and obtain $y(x) = cx$. This new trial solution leads to $2c = 5$, i.e., $c = \frac{5}{2}$. Thus, $y = \frac{5}{2}x$ is a particular solution to the nonhomogeneous equation.

Method of undetermined coefficients

While the results from Theorems 12.12 to 12.15 completely solve the problem of finding the general solution to a homogeneous linear differential equation, there is no comparable theory for the solution to a nonhomogeneous linear differential equation. For the case of constant forcing, we have already presented the solution to the nonhomogeneous differential equation in Theorem 12.17. In this section, we introduce an approach that allows to find a solution to the nonhomogeneous differential equation for a large class of forcing functions. This approach is based on the observation that, for certain forcing functions, the differential equation admits solutions of the same form as the forcing function itself. In the so-called **method of undetermined coefficients**, a trial solution is used that is of the same form as the forcing function. If the forcing function is, for example, $b(x) = e^{ax}$, we use the trial solution $y(x) = ce^{ax}$ and substitute it into the differential equation. This yields an equation from which the undetermined coefficient c can be found.

The following table gives an overview of forcing functions and trial solutions to be used:

forcing function	trial solution
e^{ax}	ce^{ax}
$\sin(bx)$ or $\cos(bx)$	$c_1 \sin(bx) + c_2 \cos(bx)$
$p_0 + p_1 x + \cdots + p_n x^n$	$c_0 + c_1 x + \cdots + c_n x^n$
$(p_0 + p_1 x + \cdots + p_n x^n)e^{ax}$	$(c_0 + c_1 x + \cdots + c_n x^n)e^{ax}$
$e^{ax} \sin(bx)$ or $e^{ax} \cos(bx)$	$e^{ax}(c_1 \sin(bx) + c_2 \cos(bx))$

At first glance, this procedure may appear to be mere trial and error. However, the method of undetermined coefficients provides a systematic way to determine a solution to the nonhomogeneous differential equation for a broad class of forcing functions, including polynomials, exponentials, sines, cosines, and linear combinations and products of these functions.

Theorem 12.18: Method of undetermined coefficients

Consider the nonhomogeneous linear differential equation

$$y^{(m)} + a_{m-1} y^{(m-1)} + \ldots + a_1 y' + a_0 y = b(x) .$$

If the forcing function has the form

$$b(x) = p_1(x) e^{ax} \sin(bx) + p_2(x) e^{ax} \cos(bx) ,$$

where $a, b \in \mathbb{R}$ and p_1, p_2 are polynomials of degree n_1 and n_2, respectively, and if $z = a + bi$ is **not a root of the characteristic polynomial**, then the nonhomogeneous differential equation has a particular solution to the form

$$y(x) = q_1(x) e^{ax} \sin(bx) + q_2(x) e^{ax} \cos(bx) ,$$

where q_1 and q_2 are polynomials of degree $n = \max\{n_1, n_2\}$.

Remark: If $z = a + bi$ is a root of the characteristic polynomial, we say that there is **resonance**. How to proceed in this case will be discussed in Theorem 12.19.

The forcing functions and trial solutions of the table on page 375 are special cases of the general forcing function in Theorem 12.18. For example, for $b = 0$ one obtains the forcing function $b(x) = p_2(x)e^{ax}$, where $p_2(x)$ is a polynomial. If, further, $a = 0$, one obtains the forcing function $b(x) = p_2(x)$, i.e., a polynomial. For polynomials p_1 and p_2 of degree zero, one obtains the forcing function $b(x) = c_1 e^{ax} \sin(bx) + c_2 e^{ax} \cos(bx)$.

We illustrate the method of undetermined coefficients with two examples.

Example 12.22. Consider the nonhomogeneous linear differential equation

$$y'' - 3y' + 2y = e^{3x}.$$

Substituting the trial solution $y(x) = ce^{3x}$ into the differential equation gives

$$9ce^{3x} - 3 \cdot 3ce^{3x} + 2ce^{3x} = e^{3x}.$$

Division by e^{3x} yields

$$9c - 3 \cdot 3c + 2c = 1 \quad \Longleftrightarrow \quad 2c = 1 \quad \Longleftrightarrow \quad c = \frac{1}{2}.$$

Thus, $y(x) = \frac{1}{2}e^{3x}$ is a particular solution to the nonhomogeneous differential equation.

Example 12.23. Consider the nonhomogeneous linear differential equation

$$y'' - 3y' + 2y = x^2 + x.$$

Since the forcing function is a second-degree polynomial, we use a general second-degree polynomial $y(x) = ax^2 + bx + c$ as a trial solution. Substituting it into the differential equation leads to

$$\begin{aligned} &2a - 3(2ax + b) + 2(ax^2 + bx + c) = x^2 + x \\ &\qquad \Longleftrightarrow \quad 2ax^2 + (2b - 6a)x + (2c - 3b + 2a) = x^2 + x. \end{aligned}$$

On both sides of the equation, there is a quadratic polynomial. Since two polynomials are equal if and only if their coefficients are equal, one obtains the following three equations[7]

$$2a = 1, \quad 2b - 6a = 1 \quad \text{and} \quad 2c - 3b + 2a = 0.$$

From these equations follows $a = 1/2$, $b = 2$ and $c = 5/2$. Thus, $y(x) = 0.5x^2 + 2x + 2.5$ is a particular solution to the nonhomogeneous differential equation.

The forcing functions considered in Theorem 12.18 may also occur as solutions to homogeneous linear differential equations. If the forcing function of a nonhomogeneous linear differential equation is itself a solution to the corresponding homogeneous equation, then the method of undetermined coefficients will fail.

Example 12.24. Consider the nonhomogeneous linear differential equation

$$y'' - 3y' + 2y = e^{2x}.$$

[7] For this reasoning, one says briefly: "Equating coefficients yields . . . ".

Substituting the trial solution $y(x) = ce^{2x}$ into the differential equation gives

$$4ce^{2x} - 3 \cdot 2ce^{2x} + 2ce^{2x} = e^{2x} \quad \Longleftrightarrow \quad 0 = e^{2x},$$

thus a contradiction. The reason for this is that $y(x) = ce^{2x}$ is a solution to the corresponding homogeneous equation $y'' - 3y' + 2y = 0$, since $z = 2$ is a root of the characteristic polynomial.

The method of undetermined coefficients fails when the forcing function is already a solution to the homogeneous differential equation – a situation referred to as **resonance**. In cases of resonance, one multiplies the trial solution by x to obtain a new trial solution and attempts to determine its coefficients. If this approach also fails (e.g., the resonance continues because the root has higher multiplicity), the trial solution is multiplied again by x, and the process is repeated. For second-order differential equations, multiplying by x or at most x^2 suffices to resolve resonance; for higher-order equations, this procedure may need to be continued up to the multiplicity of the root.

Example 12.25 (Continuation of Example 12.24). Consider again the differential equation $y'' - 3y' + 2y = e^{2x}$. We have seen in Example 12.24 that there is resonance. Therefore, we multiply the trial solution $y(x) = ce^{2x}$ by x to obtain the new trial solution $y(x) = cxe^{2x}$. Its first two derivatives are

$$y' = c(1 + 2x)e^{2x} \quad \text{and} \quad y'' = c(4 + 4x)e^{2x}.$$

Substituting them into the differential equation yields

$$\begin{aligned} &c(4 + 4x)e^{2x} - 3c(1 + 2x)e^{2x} + 2cxe^{2x} = e^{2x} \\ \Longleftrightarrow \quad &ce^{2x}(4 + 4x - 3 - 6x + 2x) = e^{2x} \quad \Longleftrightarrow \quad ce^{2x} = e^{2x}. \end{aligned}$$

Dividing by e^{2x} gives $c = 1$. Thus, $y(x) = xe^{2x}$ is a particular solution to the nonhomogeneous differential equation.

The following theorem specifies the procedure in the case of resonance.

Theorem 12.19: Method of undetermined coefficients (resonance case)

Consider the nonhomogeneous linear differential equation

$$y^{(m)} + a_{m-1}y^{(m-1)} + \ldots + a_1y' + a_0y = b(x).$$

If the forcing function has the form

$$b(x) = p_1(x)e^{ax}\sin(bx) + p_2(x)e^{ax}\cos(bx),$$

where p_1 and p_2 are polynomials of degree n_1 and n_2, respectively, and if $z = a + bi$ is a **root of the characteristic polynomial of multiplicity k**, then the nonhomogeneous differential equation has a particular solution to the form

$$y(x) = x^k \left[q_1(x)e^{ax}\sin(bx) + q_2(x)e^{ax}\cos(bx)\right] ,$$

where q_1 and q_2 are polynomials of degree $n = \max\{n_1, n_2\}$.

Remark: The forcing functions and trial solutions of the table on page 375 are special cases of the general forcing function in Theorem 12.19. Accordingly, the trial solutions given there are also to be multiplied by x^k.

If $z = a + bi$ is known to be a root of the characteristic polynomial with multiplicity k, one can immediately multiply the trial solution by x^k and thereby avoid a trial-and-error approach. If the forcing function is a polynomial, $b(x) = p(x)$, then resonance occurs when $z = 0$ is a root of the characteristic polynomial. If the forcing function is of the form $b(x) = e^{ax}$ or $b(x) = p(x)e^{ax}$, one must check whether $z = a$ is a root.

The following table shows how the functions in the table on page 375 are obtained as special cases of the general form and the value of the critical root.

forcing function	special case	critical root
$p(x)$	$a = 0, b = 0$	$z = 0$
e^{ax}, $p(x)e^{ax}$	$a \neq 0, b = 0$	$z = a$
$\sin(bx)$, $p(x)\sin(bx)$ $\cos(bx)$, $p(x)\cos(bx)$	$a = 0, b \neq 0$	$z = bi$
$e^{ax}\sin(bx)$, $p(x)e^{ax}\sin(bx)$ $e^{ax}\cos(bx)$, $p(x)e^{ax}\cos(bx)$	$a \neq 0, b \neq 0$	$z = a + bi$

Example 12.26. Consider the nonhomogeneous differential equation

$$y''' - 3y'' = x .$$

Since the right-hand side is a linear polynomial, we use a general linear polynomial $y(x) = a + bx$ as a trial solution. Since the characteristic polynomial $z^3 - 3z^2$ has a double root at $z = 0$, we multiply the trial solution by x^2 to obtain the new trial solution $y(x) = ax^2 + bx^3$. The derivatives of this trial solution are

$$y'(x) = 2ax + 3bx^2 , \quad y''(x) = 2a + 6bx , \quad \text{and} \quad y'''(x) = 6b .$$

Substituting this into the differential equation leads to

$$6b - 3(2a + 6bx) = x \quad \Longleftrightarrow \quad -18bx + (6b - 6a) = x .$$

Equating coefficients gives $-18b = 1$ and $6b - 6a = 0$, from which $b = -\frac{1}{18}$ and $a = -\frac{1}{18}$ follow. Thus,

$$y(x) = -\frac{1}{18}x^2 - \frac{1}{18}x^3$$

is a particular solution to the nonhomogeneous equation.

Remark: If the forcing function $b(x)$ is a sum of functions of suitable type, one can make a separate approach for each summand and simply add the obtained solutions. We illustrate this again with an example.

Example 12.27 (Continuation of Examples 12.23 to 12.25). Consider the nonhomogeneous differential equation

$$y'' - 3y' + 2y = e^{2x} + x^2 + x\,.$$

The differential equation $y''-3y'+2y = e^{2x}$ has a particular solution $y(x) = xe^{2x}$. Accordingly, the differential equation $y'' - 3y' + 2y = x^2 + x$ has a particular solution $y(x) = 0.5x^2 + 2x + 2.5$. Thus,

$$y(x) = xe^{2x} + 0.5x^2 + 2x + 2.5$$

is a particular solution to the differential equation $y'' - 3y' + 2y = e^{2x} + x^2 + x$.

12.3 Stability of differential equations

In many cases, the exact solution to a differential equation is not required, while the main interest rather lies in the limiting behavior of the solution for large values of x. In particular, one wants to know whether the solution of a differential equation converges to an equilibrium solution as x increases. In Example 12.20 we saw that the solution to a differential equation may converge for certain initial conditions, while for others it diverges. If **every** solution to a homogeneous linear differential equation converges to zero, the differential equation is referred to as stable. More generally, one defines:

Definition 12.4: Stability

A linear differential equation is called (asymptotically) **stable**, if any two of its solutions, $y = f_1(x)$ and $y = f_2(x)$, satisfy

$$\lim_{x\to\infty} (f_1(x) - f_2(x)) = 0\,.$$

If the differential equation has a constant equilibrium solution y^*, then the above definition is equivalent to the fact that **every** solution to the differential equation converges to the equilibrium solution y^* as x tends to infinity,

$$\lim_{x\to\infty} y(x) = y^*\,.$$

Since the difference between two solutions of a nonhomogeneous differential equation is a solution to the corresponding homogeneous differential equation, a nonhomogeneous differential equation is stable if and only if every solution $y(x)$ to the corresponding homogeneous differential equation converges to zero as x goes to infinity,

$$\lim_{x\to\infty} y(x) = 0\,.$$

Since every solution to a homogeneous linear differential equation is a linear combination of the basis functions, each basic solution must in particular have a limit of zero as x tends to infinity.

If $\lambda = a+bi$ is a root (real or complex) of the characteristic polynomial of multiplicity k, then $e^{\lambda x}, xe^{\lambda x}, \ldots, x^{k-1}e^{\lambda x}$ are linearly independent solutions. It holds:

$$\lim_{x\to\infty} x^l e^{(a+bi)x} = 0 \quad \Longleftrightarrow \quad \lim_{x\to\infty} \left|x^l e^{(a+bi)x}\right| = 0$$

For $x > 0$ it follows from $|e^{ibx}| = 1$ that

$$\left|x^l e^{(a+bi)x}\right| = \left|x^l e^{ax} e^{ibx}\right| = \left|x^l\right| \cdot \left|e^{ax}\right| \cdot \left|e^{ibx}\right| = x^l e^{ax}\,.$$

Therefore, for arbitrary $l \in \mathbb{N}_0$

$$\lim_{x\to\infty} \left|x^l e^{(a+bi)x}\right| = \lim_{x\to\infty} x^l e^{ax}\,.$$

For $a > 0$ as well as for $a = 0$ and $l > 0$ the limit is infinite, for $a = 0$ and $l = 0$ the limit is one. For $a < 0$ one obtains by repeated application of l'Hôpital's rule

$$\lim_{x\to\infty} x^l e^{ax} = \lim_{x\to\infty} \frac{x^l}{e^{-ax}} = \lim_{x\to\infty} \frac{lx^{l-1}}{(-a)e^{-ax}} = \lim_{x\to\infty} \frac{l(l-1)x^{l-2}}{(-a)^2 e^{-ax}} = \ldots$$

$$\cdots = \lim_{x\to\infty} \frac{l!}{(-a)^l e^{-ax}} = 0\,.$$

Informally speaking, the exponential function e^x tends to infinity faster than any polynomial in x.

From this it follows that a linear differential equation (with constant coefficients) is stable if and only if all roots of the characteristic polynomial have a negative real part. We formulate this result as a theorem.

Theorem 12.20: Stability of a linear differential equation

A linear differential equation of order m with constant coefficients is stable if and only if every root z of the characteristic polynomial satisfies $\mathrm{Re}(z) < 0$, that is, if z has a negative real part.

Of course, it would be convenient if one could decide directly from the coefficients of a given differential equation whether it is stable, rather than first having to determine the roots of the characteristic polynomial. For second-order differential equations, such conditions on the coefficients can be formulated relatively easily. We therefore want to take a closer look at second-order linear differential equations. The homogeneous equation has the form

$$y'' + a_1 y' + a_0 y = 0\,.$$

The characteristic polynomial is therefore $z^2 + a_1 z + a_0$ and has the roots

$$z_{1,2} = -\frac{a_1}{2} \pm \sqrt{\frac{a_1^2}{4} - a_0}\,.$$

We distinguish three cases:

Case 1 ($a_0 < a_1^2/4$): Then the characteristic polynomial has two different real roots, namely

$$z_1 = -\frac{a_1}{2} + \sqrt{\frac{a_1^2}{4} - a_0} \quad \text{and} \quad z_2 = -\frac{a_1}{2} - \sqrt{\frac{a_1^2}{4} - a_0}\,.$$

If $a_1 \leq 0$, then $z_1 > 0$. Therefore, for both roots to be negative, $a_1 > 0$ must hold. In this case, the larger root z_1 is negative if and only if

$$\begin{aligned} -\frac{a_1}{2} + \sqrt{\frac{a_1^2}{4} - a_0} < 0 \quad &\Longleftrightarrow \quad \sqrt{\frac{a_1^2}{4} - a_0} < \frac{a_1}{2} \\ &\Longleftrightarrow \quad \frac{a_1^2}{4} - a_0 < \frac{a_1^2}{4} \quad \Longleftrightarrow \quad a_0 > 0\,. \end{aligned}$$

In Case 1, both roots are negative if and only if $a_0, a_1 > 0$.

Case 2 ($a_0 = a_1^2/4$): Then the characteristic polynomial has a double real root at $z_{1,2} = -a_1/2$. This is negative if and only if $a_1 > 0$. In this case ($a_0 = a_1^2/4$) $a_0 > 0$ is also true.

Case 3 ($a_0 > a_1^2/4$): Then the characteristic polynomial has two complex roots, namely

$$z_1 = -\frac{a_1}{2} + i\sqrt{a_0 - \frac{a_1^2}{4}} \quad \text{and} \quad z_2 = -\frac{a_1}{2} - i\sqrt{a_0 - \frac{a_1^2}{4}}\,.$$

The real part is $-a_1/2$, which is negative if and only if $a_1 > 0$. In this case ($a_0 > a_1^2/4$) $a_0 > 0$ is also true.

Combining the three cases shows that the differential equation is stable if and only if a_0 and a_1 are both positive. We formulate this as a theorem.

Theorem 12.21: Stability of a second-order linear differential equation

The second-order linear differential equation,

$$y'' + a_1 y' + a_0 y = b(x)\,,$$

is stable if and only if $a_0 > 0$ and $a_1 > 0$.

For differential equations of higher order, stability can also be characterized by conditions on the coefficients of the differential equation. For this, we need the notion of a **Hurwitz[8] matrix**. For a polynomial

$$a_m z^m + a_{m-1} z^{m-1} + \ldots + a_1 z + a_0$$

the Hurwitz matrix is the following $m \times m$ matrix

$$\boldsymbol{H}_m = \begin{bmatrix} a_1 & a_0 & 0 & 0 & \ldots & 0 \\ a_3 & a_2 & a_1 & a_0 & \ldots & 0 \\ a_5 & a_4 & a_3 & a_2 & \ldots & 0 \\ a_7 & a_6 & a_5 & a_4 & \ldots & 0 \\ \vdots & \vdots & \vdots & \vdots & \ddots & \vdots \\ 0 & 0 & 0 & 0 & \ldots & a_m \end{bmatrix}.$$

The Hurwitz matrix is constructed according to the following rule:

1. The main diagonal consists of the coefficients $a_1, \ldots, a_m$.
2. In each row, the indices of the coefficients are decreasing from left to right.
3. Coefficients with index less than zero or greater than m are set to zero.

With the help of the Hurwitz matrix, we can now formulate the following criterion.

Theorem 12.22: Hurwitz criterion for polynomials

All roots of the polynomial

$$a_m z^m + a_{m-1} z^{m-1} + \ldots + a_1^z + a_0$$

have negative real part if and only if

1. $a_0, a_1, \ldots, a_m > 0$ and
2. the principal minors of the Hurwitz matrix $\Delta_1, \Delta_2, \ldots, \Delta_m > 0$ are positive.

We now apply the Hurwitz criterion to the problem of deciding whether a linear differential equation of order m is stable or not. As usual, we assume that the coefficient of the highest derivative is equal to one, so the differential equation takes the form

$$y^{(m)} + a_{m-1} y^{(m-1)} + \ldots + a_1 y' + a_0 y = b(x)\,.$$

[8] Adolf Hurwitz (1859–1919)

In that case $\Delta_m = \Delta_{m-1}$, as can be easily seen by expanding the determinant along the last row. Hence, we can skip the check of the highest principal minor. The Hurwitz criterion for the stability of a linear differential equation is then as follows.

Theorem 12.23: Hurwitz criterion for linear differential equations

The linear differential equation

$$y^{(m)} + a_{m-1}y^{(m-1)} + \ldots + a_1 y' + a_0 y = b(x)$$

is stable if and only if

1. $a_0, a_1, \ldots, a_{m-1} > 0$ and
2. the first $m - 1$ principal minors of the Hurwitz matrix $\Delta_1, \ldots, \Delta_{m-1} > 0$ are positive.

For a second-order linear differential equation, $y'' + a_1 y_1' + a_0 y = b(x)$, the Hurwitz matrix is

$$\boldsymbol{H}_2 = \begin{bmatrix} a_1 & 0 \\ 0 & 1 \end{bmatrix}.$$

Since $\Delta_1 = \Delta_2 = a_1$, the Hurwitz criterion implies that the differential equation is stable if and only if $a_1, a_0 > 0$, the same condition as in Theorem 12.21.

Theorem 12.23 also provides a simple necessary condition for the stability of the differential equation, namely, that all coefficients are positive. Thus, if one of the coefficients is non-positive, then the differential equation is not stable.

Example 12.28 (Hurwitz criterion). Consider the differential equation $y''' + 0.8y' + 0.3y = 0$. Since the coefficient of the second derivative y'' is zero, $a_2 \leq 0$. Hence, the differential equation is **not stable**.

Example 12.29 (Hurwitz criterion). Consider the differential equation $y''' + 0.5y'' + 0.8y' + 0.3y = 0$. All coefficients are positive, and the Hurwitz matrix is

$$\boldsymbol{H}_3 = \begin{bmatrix} 0.8 & 0.3 & 0 \\ 1 & 0.5 & 0.8 \\ 0 & 0 & 1 \end{bmatrix}.$$

Since $\Delta_1 = 0.8 > 0$ and $\Delta_2 = 0.8 \cdot 0.5 - 1 \cdot 0.3 = 0.1 > 0$, the differential equation is **stable**.

Example 12.30 (Hurwitz criterion). Consider the differential equation $y''' + 0.3y'' + 0.8y' + 0.5y = 0$. All coefficients are positive, and the Hurwitz matrix is

$$\boldsymbol{H}_3 = \begin{bmatrix} 0.8 & 0.5 & 0 \\ 1 & 0.3 & 0.8 \\ 0 & 0 & 1 \end{bmatrix}.$$

Since $\Delta_2 = 0.8 \cdot 0.3 - 1 \cdot 0.5 = -0.26 \leq 0$, the differential equation is **not stable**.

12.4 Systems of first-order differential equations

Definition 12.5: System of first-order linear differential equations

A system of **first-order linear differential equations with constant coefficients** is an equation of the form

$$\boldsymbol{y}' = \boldsymbol{A}\boldsymbol{y} + \boldsymbol{b}(x)\,.$$

Here, $\boldsymbol{A}$ is a real $n \times n$ matrix, $\boldsymbol{b} : A \to \mathbb{R}^n, x \mapsto \boldsymbol{b}(x) \in \mathbb{R}^n$, is a given function, the so-called **forcing function**, and $A \subset \mathbb{R}$ is an interval.

If $\boldsymbol{b}(x) = \boldsymbol{0}$ for all $x \in A$, the system is called **homogeneous**, otherwise **nonhomogeneous**.

The **general solution** of a system of differential equations consists in determining all vector-valued functions $\boldsymbol{y} = \boldsymbol{f}(x)$, $x \in A$, for which the system of differential equations is satisfied.

There is a close relationship between first-order systems of differential equations and m-th order linear differential equations, since every m-th order linear differential equation can be represented as a system of first-order differential equations. Consider the m-th order linear differential equation

$$y^{(m)} + a_{m-1}y^{(m-1)} + \ldots + a_1 y' + a_0 y = b(x)\,.$$

Now, define

$$z_1(x) = y(x),\; z_2(x) = y'(x), \ldots,\; z_m(x) = y^{(m-1)}(x)\,.$$

This gives

$$\begin{aligned} z_m'(x) = y^{(m)}(x) &= -a_0 y(x) - a_1 y'(x) - \cdots - a_{m-1}y^{(m-1)}(x) + b(x) \\ &= -a_0 z_1(x) - a_1 z_2(x) - \cdots - a_{m-1} z_m(x) + b(x)\,. \end{aligned}$$

Furthermore, we get the equations $z_i'(x) = z_{i+1}(x)$, $i = 1, \ldots, m-1$. This gives the system

$$\begin{aligned} z_1'(x) &= z_2(x) \\ z_2'(x) &= z_3(x) \\ &\;\;\vdots \\ z_{m-1}'(x) &= z_m(x) \\ z_m'(x) &= -a_0 z_1(x) - a_1 z_2(x) - \cdots - a_{m-1} z_m(x) + b(x)\,. \end{aligned}$$

or in matrix notation

$$\begin{bmatrix} z_1'(x) \\ z_2'(x) \\ \vdots \\ z_{m-2}'(x) \\ z_{m-1}'(x) \\ z_m'(x) \end{bmatrix} = \begin{bmatrix} 0 & 1 & 0 & \dots & 0 & 0 \\ 0 & 0 & 1 & \dots & 0 & 0 \\ \vdots & & & \ddots & & \vdots \\ 0 & 0 & 0 & \dots & 1 & 0 \\ 0 & 0 & 0 & \dots & 0 & 1 \\ -a_0 & -a_1 & -a_2 & \dots & -a_{m-2} & -a_{m-1} \end{bmatrix} \begin{bmatrix} z_1(x) \\ z_2(x) \\ z_3(x) \\ \vdots \\ z_{m-1}(x) \\ z_m(x) \end{bmatrix} + \begin{bmatrix} 0 \\ 0 \\ 0 \\ \vdots \\ 0 \\ b(x) \end{bmatrix}$$

or briefly $\boldsymbol{z}'(x) = \boldsymbol{A}\boldsymbol{z}(x) + \boldsymbol{b}(x)$.

Theorem 12.24: Structure of the solution space

Consider a system of first-order linear differential equations with constant coefficients,

$$\boldsymbol{y}'(x) = \boldsymbol{A}\boldsymbol{y}(x) + \boldsymbol{b}(x)\,.$$

The general solution to the **homogeneous** system is an **n-dimensional subspace** of $\mathcal{F}(A, \mathbb{R}^n)$.

If $\boldsymbol{f}_1, \dots, \boldsymbol{f}_n$ is a **basis of the general solution**, then the general solution is

$$\boldsymbol{y} = \gamma_1 \boldsymbol{f}_1(x) + \dots + \gamma_n \boldsymbol{f}_n(x)\,, \quad \gamma_1, \dots, \gamma_n \in \mathbb{R}\,.$$

A basis $\boldsymbol{f}_1, \dots, \boldsymbol{f}_n$ of the general solution to the homogeneous system is also referred to as a **fundamental system**.

Let $y = \boldsymbol{f}_p(x)$ be a solution to the **nonhomogeneous** system and let $\boldsymbol{f}_1, \dots, \boldsymbol{f}_n$ be a fundamental system of the homogeneous system. Then the **general solution to the nonhomogeneous system** is

$$\boldsymbol{y} = \boldsymbol{f}_p(x) + \gamma_1 \boldsymbol{f}_1(x) + \dots + \gamma_n \boldsymbol{f}_n(x)\,, \quad \gamma_1, \dots, \gamma_n \in \mathbb{R}\,.$$

Remark: Combining the basis functions $\boldsymbol{f}_1(x), \dots, \boldsymbol{f}_n(x)$ into a matrix $\boldsymbol{F}(x)$ and the constants $\gamma_1, \dots, \gamma_n$ into a column vector $\boldsymbol{\gamma}$, the general solution to the homogeneous system can be written as

$$\boldsymbol{y}(x) = \boldsymbol{F}(x)\boldsymbol{\gamma}\,.$$

Here, $\boldsymbol{F}(x)$ is a **matrix-valued function** of one variable, whose derivative $\boldsymbol{F}'(x)$ is to be understood component-wise. $\boldsymbol{F}(x)$ is called **fundamental matrix**. Since each basis function $\boldsymbol{f}_i(x)$ is a solution to the homogeneous system, it holds $\boldsymbol{f}_i'(x) = \boldsymbol{A}\boldsymbol{f}_i(x)$, $i = 1, \dots, n$. Then, the matrix equation $\boldsymbol{F}'(x) = \boldsymbol{A}\boldsymbol{F}(x)$ also holds.

We first consider the solution to the homogeneous system. In the study of linear differential equations with constant coefficients, exponential functions often appear as solutions to homogeneous linear differential equations. Therefore, we use the trial solution $\boldsymbol{y}(x) = \boldsymbol{a}e^{\lambda x}$ with $\boldsymbol{a} \in \mathbb{R}^n$ and $\lambda \in \mathbb{R}$. For the derivative, we get

$$\boldsymbol{y}'(x) = \lambda \boldsymbol{a} e^{\lambda x}\,.$$

Substituting into the differential equation $\boldsymbol{y}'(x) = \boldsymbol{A}\boldsymbol{y}(x)$ then yields

$$\lambda \boldsymbol{a} e^{\lambda x} = \boldsymbol{A}\boldsymbol{a} e^{\lambda x},$$

and, after division by $e^{\lambda x}$,

$$\lambda \boldsymbol{a} = \boldsymbol{A}\boldsymbol{a}.$$

Thus, $\boldsymbol{y}(x) = \boldsymbol{a} e^{\lambda x}$ is a solution to the differential equation if and only if λ is an eigenvalue of $\boldsymbol{A}$ and $\boldsymbol{a}$ is an eigenvector of $\boldsymbol{A}$ corresponding to the eigenvalue λ.

For the general solution to the system, we need to find a fundamental system, i.e., n linearly independent solutions.

If we set $x = 0$, we see that n solutions $\boldsymbol{a}_1 e^{\lambda_1 x}, \ldots, \boldsymbol{a}_n e^{\lambda_n x}$ are linearly independent if the vectors $\boldsymbol{a}_1, \ldots, \boldsymbol{a}_n$ are linearly independent. Since eigenvectors corresponding to different eigenvalues are linearly independent (Theorem 10.15), the above trial solution yields the general solution, provided $\boldsymbol{A}$ has n linearly independent (possibly complex) eigenvectors, in other words, if $\boldsymbol{A}$ is complex diagonalizable (see Definition 10.5).

As with simple linear differential equations, we would like to express the solution to systems of differential equations entirely in terms of real basis functions. If $\lambda = a + bi$ is a complex eigenvalue, then the complex conjugate $\overline{\lambda} = a - bi$ is also a complex eigenvalue. It also holds that the corresponding eigenvectors are complex conjugates of each other. So if $\boldsymbol{u} + i\boldsymbol{v}$ is an eigenvector corresponding to λ, then $\boldsymbol{u} - i\boldsymbol{v}$ is an eigenvector corresponding to $\overline{\lambda}$. A pair of complex conjugate eigenvalues thus provides the two basis functions

$$\begin{aligned}(\boldsymbol{u} + i\boldsymbol{v})e^{(a+bi)x} &= (\boldsymbol{u} + i\boldsymbol{v})e^{ax}\left(\cos(bx) + i\sin(bx)\right)\\ &= e^{ax}\left[(\cos(bx)\boldsymbol{u} - \sin(bx)\boldsymbol{v}) + i\left(\sin(bx)\boldsymbol{u} + \cos(bx)\boldsymbol{v}\right)\right]\end{aligned}$$

and

$$\begin{aligned}(\boldsymbol{u} - i\boldsymbol{v})e^{(a-bi)x} &= (\boldsymbol{u} - i\boldsymbol{v})e^{ax}\left(\cos(bx) - i\sin(bx)\right)\\ &= e^{ax}\left[(\cos(bx)\boldsymbol{u} - \sin(bx)\boldsymbol{v}) - i\left(\sin(bx)\boldsymbol{u} + \cos(bx)\boldsymbol{v}\right)\right].\end{aligned}$$

Thus, we have

$$\begin{aligned}&\gamma_1(\boldsymbol{u} + i\boldsymbol{v})e^{(a+bi)x} + \gamma_2(\boldsymbol{u} - i\boldsymbol{v})e^{(a-bi)x}\\ &\qquad = C_1 e^{ax}\left(\cos(bx)\boldsymbol{u} - \sin(bx)\boldsymbol{v}\right) + C_2 e^{ax}\left(\sin(bx)\boldsymbol{u} + \cos(bx)\boldsymbol{v}\right)\end{aligned}$$

with $C_1 = \gamma_1 + \gamma_2$ and $C_2 = i(\gamma_1 - \gamma_2)$. Hence, instead of the two complex basis functions, we can also use the two real basis functions

$$e^{ax}\left(\cos(bx)\boldsymbol{u} - \sin(bx)\boldsymbol{v}\right) \quad \text{and} \quad e^{ax}\left(\sin(bx)\boldsymbol{u} + \cos(bx)\boldsymbol{v}\right)$$

to represent the general solution.

Theorem 12.25: Solution to the homogeneous system I

Consider the homogeneous system of linear differential equations $\boldsymbol{y}' = \boldsymbol{A}\boldsymbol{y}$. Let $\boldsymbol{A}$ be complex diagonalizable and possess the (not necessarily different) eigenvalues

$$\lambda_1, \ldots, \lambda_m, (\lambda_{m+1}, \overline{\lambda}_{m+1}), \ldots, (\lambda_{m+r}, \overline{\lambda}_{m+r}),$$

where λ_j, $j = 1, \ldots, m$, are real eigenvalues and $\lambda_{m+k}, \overline{\lambda}_{m+k} = a_{m+k} \pm b_{m+k}i$, $k = 1, \ldots, r$, are pairs of complex conjugate eigenvalues, $m + 2r = n$. Furthermore, let $\boldsymbol{u}_1, \ldots, \boldsymbol{u}_m$ and $\boldsymbol{u}_{m+1} \pm i\boldsymbol{v}_{m+1}, \ldots, \boldsymbol{u}_{m+r} \pm i\boldsymbol{v}_{m+r}$ be corresponding linearly independent eigenvectors. Then the functions

$$\begin{aligned}
&\boldsymbol{u}_j e^{\lambda_j x}, && j = 1, \ldots, m,\\
&e^{a_{m+k}x}\left(\cos(b_{m+k}x)\boldsymbol{u}_{m+k} - \sin(b_{m+k}x)\boldsymbol{v}_{m+k}\right), && k = 1 \ldots, r,\\
&e^{a_{m+k}x}\left(\sin(b_{m+k}x)\boldsymbol{u}_{m+k} + \cos(b_{m+k}x)\boldsymbol{v}_{m+k}\right), && k = 1 \ldots, r,
\end{aligned}$$

form a basis of the general solution to the homogeneous system.

This so-called **eigenvalue-eigenvector method** is illustrated by two examples.

Example 12.31. Consider the homogeneous system of linear differential equations $\boldsymbol{y}' = \boldsymbol{A}\boldsymbol{y}$ with

$$\boldsymbol{A} = \begin{bmatrix} 2 & 4 \\ 1 & -1 \end{bmatrix}$$

subject to the initial condition $\boldsymbol{y}(0) = [2, 1]^T$.

The characteristic polynomial of $\boldsymbol{A}$ is $\lambda^2 - \lambda - 6$. Using the quadratic formula, we obtain the two eigenvalues $\lambda_1 = 3$ and $\lambda_2 = -2$. To find the eigenspace for $\lambda_1 = 3$, we solve the linear equation $(\boldsymbol{A} - 3\boldsymbol{I})\boldsymbol{x} = \boldsymbol{0}$. Because

$$\boldsymbol{A} - 3\boldsymbol{I} = \begin{bmatrix} -1 & 4 \\ 1 & -4 \end{bmatrix},$$

$\boldsymbol{u}_1 = [4, 1]^T$ is an eigenvector for $\lambda_1 = 3$. Similarly, for $\lambda_2 = -2$, we find that

$$\boldsymbol{A} + 2\boldsymbol{I} = \begin{bmatrix} 4 & 4 \\ 1 & 1 \end{bmatrix}.$$

Hence, $\boldsymbol{u}_2 = [1, -1]^T$ is an eigenvector for $\lambda_2 = -2$. The general solution is therefore

$$\boldsymbol{y} = \gamma_1 e^{3x} \begin{bmatrix} 4 \\ 1 \end{bmatrix} + \gamma_2 e^{-2x} \begin{bmatrix} 1 \\ -1 \end{bmatrix}, \quad \gamma_1, \gamma_2 \in \mathbb{R}.$$

Thus,

$$\boldsymbol{F}(x) = \begin{bmatrix} 4e^{3x} & e^{-2x} \\ e^{3x} & -e^{-2x} \end{bmatrix}$$

is a fundamental matrix of the system.

Substituting the general solution into the initial condition yields the linear equation

$$\boldsymbol{y}(0) = \boldsymbol{F}(0)\boldsymbol{\gamma} = \begin{bmatrix} 4 & 1 \\ 1 & -1 \end{bmatrix} \begin{bmatrix} \gamma_1 \\ \gamma_2 \end{bmatrix} = \begin{bmatrix} 2 \\ 1 \end{bmatrix}.$$

Solving using Gauss-Jordan elimination yields the solution $\gamma_1 = 3/5$ and $\gamma_2 = -2/5$ and thus the solution

$$\boldsymbol{y} = e^{3x} \begin{bmatrix} \frac{12}{5} \\ \frac{3}{5} \end{bmatrix} + e^{-2x} \begin{bmatrix} -\frac{2}{5} \\ \frac{2}{5} \end{bmatrix} = \begin{bmatrix} \frac{12}{5}e^{3x} - \frac{2}{5}e^{-2x} \\ \frac{3}{5}e^{3x} + \frac{2}{5}e^{-2x} \end{bmatrix}.$$

Example 12.32. Consider the homogeneous system of linear differential equations $\boldsymbol{y}' = \boldsymbol{A}\boldsymbol{y}$ with

$$\boldsymbol{A} = \begin{bmatrix} 4 & -1 \\ 5 & 2 \end{bmatrix},$$

subject to the initial condition $\boldsymbol{y}(0) = [1, -1]^T$.

The characteristic polynomial of $\boldsymbol{A}$ is $\lambda^2 - 6\lambda + 13$. Using the quadratic formula, we obtain the two complex roots $\lambda_{1,2} = 3 \pm 2i$. Hence, there are two complex conjugate eigenvalues. To determine the eigenspace for $\lambda_1 = 3 + 2i$, we solve the linear equation $(\boldsymbol{A} - (3 + 2i)\boldsymbol{I})\boldsymbol{x} = \boldsymbol{0}$ using Gauss-Jordan elimination[9]. Since

$$\boldsymbol{A} - (3 + 2i)\boldsymbol{I} = \begin{bmatrix} 1 - 2i & -1 \\ 5 & -1 - 2i \end{bmatrix},$$

$\boldsymbol{u}_1 + i\boldsymbol{v}_1 = [1, 1]^T + i[0, -2]^T$ is an eigenvector for $\lambda_1 = 3 + 2i$. Correspondingly, $\boldsymbol{u}_1 - i\boldsymbol{v}_1 = [1, 1]^T - i[0, -2]^T$ is an eigenvector for $\lambda_2 = 3 - 2i$.

The general solution is therefore

$$\begin{aligned} \boldsymbol{y} = {} & \gamma_1 e^{3x} \left(\cos(2x) \begin{bmatrix} 1 \\ 1 \end{bmatrix} - \sin(2x) \begin{bmatrix} 0 \\ -2 \end{bmatrix} \right) \\ & + \gamma_2 e^{3x} \left(\sin(2x) \begin{bmatrix} 1 \\ 1 \end{bmatrix} + \cos(2x) \begin{bmatrix} 0 \\ -2 \end{bmatrix} \right), \quad \gamma_1, \gamma_2 \in \mathbb{R}. \end{aligned}$$

Since $e^0 = 1$, $\cos(0) = 1$ and $\sin(0) = 0$, substituting the general solution into the initial condition yields the linear equation

$$\boldsymbol{y}(0) = \gamma_1 \begin{bmatrix} 1 \\ 1 \end{bmatrix} + \gamma_2 \begin{bmatrix} 0 \\ -2 \end{bmatrix} = \begin{bmatrix} 1 & 0 \\ 1 & -2 \end{bmatrix} \begin{bmatrix} \gamma_1 \\ \gamma_2 \end{bmatrix} = \begin{bmatrix} 1 \\ -1 \end{bmatrix}.$$

Gauss-Jordan elimination yields $\gamma_1 = \gamma_2 = 1$, and thus the solution

$$\boldsymbol{y} = e^{3x} \left(\cos(2x) \begin{bmatrix} 1 \\ 1 \end{bmatrix} - \sin(2x) \begin{bmatrix} 0 \\ -2 \end{bmatrix} \right) + e^{3x} \left(\sin(2x) \begin{bmatrix} 1 \\ 1 \end{bmatrix} + \cos(2x) \begin{bmatrix} 0 \\ -2 \end{bmatrix} \right)$$

[9] The solution of homogeneous linear equations for coefficient matrices with complex entries is also carried out with Gauss-Jordan elimination.

$$= e^{3x}\left(\cos(2x)\begin{bmatrix}1\\-1\end{bmatrix} + \sin(2x)\begin{bmatrix}1\\3\end{bmatrix}\right) = \begin{bmatrix}\cos(2x)+\sin(2x)\\3\sin(2x)-\cos(2x)\end{bmatrix} e^{3x}.$$

A homogeneous system of first-order linear differential equations,

$$\boldsymbol{y}' = \boldsymbol{A}\boldsymbol{y}$$

is called **decoupled**, if $\boldsymbol{A} = \text{diag}(d_1, \ldots, d_n)$ is a diagonal matrix. In this case, the system consists of n independent differential equations, which can be solved separately:

$$\begin{aligned}
y_1' &= d_1 y_1 &\implies\quad y_1 &= \gamma_1 e^{d_1 x},\\
y_2' &= d_2 y_2 &\implies\quad y_2 &= \gamma_2 e^{d_2 x},\\
&&\vdots\\
y_n' &= d_n y_n &\implies\quad y_n &= \gamma_n e^{d_n x}.
\end{aligned}$$

If $\boldsymbol{A}$ is not a diagonal matrix, the system is called **coupled**.

If the matrix $\boldsymbol{A}$ is complex diagonalizable, it is possible to decouple a coupled system of differential equations, i.e., to transform it into a decoupled and thus easier to solve system. If $\boldsymbol{A} = \boldsymbol{S}\boldsymbol{D}\boldsymbol{S}^{-1}$ with an invertible matrix $\boldsymbol{S}$ and a diagonal matrix $\boldsymbol{D}$, then

$$\boldsymbol{y}' = \boldsymbol{A}\boldsymbol{y} = \boldsymbol{S}\boldsymbol{D}\boldsymbol{S}^{-1}\boldsymbol{y} \quad\iff\quad \boldsymbol{S}^{-1}\boldsymbol{y}' = \boldsymbol{D}\boldsymbol{S}^{-1}\boldsymbol{y}.$$

Substituting $\boldsymbol{z} = \boldsymbol{S}^{-1}\boldsymbol{y}$, and hence $\boldsymbol{z}' = \boldsymbol{S}^{-1}\boldsymbol{y}'$, we obtain a **decoupled** system of differential equations for the transformed variables $\boldsymbol{z}$,

$$\boldsymbol{z}' = \boldsymbol{D}\boldsymbol{z},$$

which can be solved directly. The solution to the original system $\boldsymbol{y}' = \boldsymbol{A}\boldsymbol{y}$ is then obtained by back transformation $\boldsymbol{y} = \boldsymbol{S}\boldsymbol{z}$.

The solution to a system of differential equations is more difficult if the matrix $\boldsymbol{A}$ is not complex diagonalizable. Then, in addition to basis functions of the form $\boldsymbol{u}_i e^{\lambda_i x}$, basis functions of the form

$$\boldsymbol{p}_i(x)e^{\lambda_i x}$$

occur. Here, $\boldsymbol{p}_i(x)$ is a **vector-valued polynomial** in x, i.e., a function of the form

$$\boldsymbol{p}(x) = \sum_{j=0}^{k} \boldsymbol{a}_j x^j = \boldsymbol{a}_0 + \boldsymbol{a}_1 x + \cdots + \boldsymbol{a}_k x^k,$$

where $\boldsymbol{a}_0, \ldots, \boldsymbol{a}_k$ are vectors, $\boldsymbol{a}_k \neq \boldsymbol{0}$. The degree k of the polynomial $\boldsymbol{p}_i$ is strictly less than the algebraic multiplicity of the corresponding eigenvalue λ_i. For eigenvalues λ, for which the geometric multiplicity is smaller than the algebraic multiplicity, we therefore use the trial solution

$$\boldsymbol{y}(x) = \boldsymbol{p}(x)e^{\lambda x}$$

with a vector-valued polynomial $p(x)$ of degree one less than the multiplicity of the eigenvalue and substitute this into the system of differential equations to determine the unknown coefficients of the polynomial $p(x)$. We illustrate the **method of undetermined coefficients** again with two examples.

Example 12.33. Consider the homogeneous system of linear differential equations $y' = Ay$ with

$$A = \begin{bmatrix} 5 & -3 \\ 3 & -1 \end{bmatrix}.$$

The characteristic polynomial of A is $\lambda^2 - 4\lambda + 4 = (\lambda - 2)^2$. Thus, A has a double eigenvalue $\lambda_{1,2} = 2$. Therefore, we use the following trial solution with a linear polynomial, i.e.,

$$y = (a + bx)e^{2x}.$$

The derivative is

$$y' = be^{2x} + 2(a + bx)e^{2x} = \left[(2a + b) + 2bx\right]e^{2x}.$$

Substituting into the system gives

$$\left[(2a + b) + 2bx\right]e^{2x} = A(a + bx)e^{2x}.$$

Dividing by e^{2x} and expanding the parentheses on the right-hand side gives

$$(2a + b) + 2bx = Aa + Abx.$$

Equating coefficients yields the two equations

$$2a + b = Aa \quad \text{and} \quad 2b = Ab.$$

The second equation implies that b is an eigenvector to the eigenvalue $\lambda = 2$. The calculation of the eigenspace yields $b = \gamma_1[1, 1]^T$, $\gamma_1 \in \mathbb{R}$. To find a, we solve the linear equation $Aa - 2a = b$ or $(A - 2I)a = \gamma_1[1, 1]^T$ and obtain

$$a = \gamma_1 \begin{bmatrix} \frac{1}{3} \\ 0 \end{bmatrix} + \gamma_2 \begin{bmatrix} 1 \\ 1 \end{bmatrix} \quad \text{with } \gamma_1, \gamma_2 \in \mathbb{R}.$$

The general solution to the system is thus

$$y = \left(\gamma_1 \begin{bmatrix} \frac{1}{3} \\ 0 \end{bmatrix} + \gamma_2 \begin{bmatrix} 1 \\ 1 \end{bmatrix} + \gamma_1 \begin{bmatrix} 1 \\ 1 \end{bmatrix} x\right) e^{2x}$$

$$= \gamma_1 \left(\begin{bmatrix} \frac{1}{3} \\ 0 \end{bmatrix} + \begin{bmatrix} 1 \\ 1 \end{bmatrix} x\right) e^{2x} + \gamma_2 \begin{bmatrix} 1 \\ 1 \end{bmatrix} e^{2x}, \quad \text{where } \gamma_1, \gamma_2 \in \mathbb{R}.$$

Example 12.34. Consider the homogeneous system of linear differential equations $y' = Ay$ with

$$A = \begin{bmatrix} 3 & 2 & -1 \\ -1 & 0 & 2 \\ 1 & 2 & -1 \end{bmatrix}.$$

A has the characteristic polynomial $p_A(\lambda) = -\lambda^3 + 2\lambda^2 + 4\lambda - 8$. Calculating the eigenvalues yields the simple eigenvalue $\lambda_1 = -2$ and the double eigenvalue $\lambda_{2,3} = 2$.

For the simple eigenvalue $\lambda_1 = -2$, we obtain the eigenspace $V(-2) = \{\gamma_1[2,-3,4]^T \mid \gamma_1 \in \mathbb{R}\}$. Thus,

$$y = \gamma_1 \begin{bmatrix} 2 \\ -3 \\ 4 \end{bmatrix} e^{-2x}$$

is a solution to the homogeneous system for every $\gamma_1 \in \mathbb{R}$.

For the double eigenvalue $\lambda_{2,3} = 2$, we use the following trial solution with a linear polynomial,

$$y = (a + bx)e^{2x}.$$

The derivative is

$$y' = be^{2x} + 2(a + bx)e^{2x} = \big[(2a + b) + 2bx\big]e^{2x}.$$

Substituting into the system gives

$$\big[(2a + b) + 2bx\big]e^{2x} = A(a + bx)e^{2x}.$$

Dividing by e^{2x} and expanding the parentheses on the right-hand side gives

$$(2a + b) + 2bx = Aa + Abx.$$

Equating coefficients gives the following two equations

$$2a + b = Aa \quad \text{and} \quad 2b = Ab.$$

From the second equation it follows that b is an eigenvector for the eigenvalue $\lambda = 2$. Calculating the eigenspace gives $b = \gamma_2[2,-1,0]^T$. To determine a, we solve the linear equation $Aa - 2a = b$ or $(A - 2I)a = \gamma_2[2,-1,0]^T$ and obtain

$$a = \gamma_2 \begin{bmatrix} 3 \\ 0 \\ 1 \end{bmatrix} + \gamma_3 \begin{bmatrix} 2 \\ -1 \\ 0 \end{bmatrix} \quad \text{with } \gamma_2, \gamma_3 \in \mathbb{R}.$$

The general solution to the system is thus

$$y = \gamma_1 \begin{bmatrix} 2 \\ -3 \\ 4 \end{bmatrix} e^{-2x} + \left(\gamma_2 \begin{bmatrix} 3 \\ 0 \\ 1 \end{bmatrix} + \gamma_3 \begin{bmatrix} 2 \\ -1 \\ 0 \end{bmatrix} + \gamma_2 \begin{bmatrix} 2 \\ -1 \\ 0 \end{bmatrix} x \right) e^{2x}$$

$$= \gamma_1 \begin{bmatrix} 2 \\ -3 \\ 4 \end{bmatrix} e^{-2x} + \gamma_2 \left(\begin{bmatrix} 3 \\ 0 \\ 1 \end{bmatrix} + \begin{bmatrix} 2 \\ -1 \\ 0 \end{bmatrix} x \right) e^{2x} + \gamma_3 \begin{bmatrix} 2 \\ -1 \\ 0 \end{bmatrix} e^{2x},$$

where $\gamma_1, \gamma_2, \gamma_3 \in \mathbb{R}$.

Another method for solving the homogeneous equation, which also works for non-diagonalizable coefficient matrices, is based on the so-called **matrix exponential**. The first-order homogeneous linear differential equation $y' = ay$ has the general solution $y(x) = \gamma e^{ax}$. Formally, one could expect, that the general solution to the system $\boldsymbol{y}' = \boldsymbol{A}\boldsymbol{y}$ is given by $\boldsymbol{y}(x) = e^{\boldsymbol{A}x}\boldsymbol{\gamma}$. Of course, one has to specify what is meant by $e^{\boldsymbol{A}x}$, that is, how a matrix exponential function can be defined. As is well known (see Definition 3.7), the exponential function $e^x = \exp(x)$ is defined by an infinite series,

$$\exp(x) = \sum_{k=0}^{\infty} \frac{x^k}{k!}.$$

It is now natural to replace x by a matrix $\boldsymbol{A}$ in this definition. Indeed, it can be shown that the corresponding series converges for every (complex) matrix, thereby defining a matrix-valued function.

Theorem 12.26: Matrix exponential

For every (complex) $n \times n$ matrix $\boldsymbol{A}$, the infinite series

$$e^{\boldsymbol{A}} = \exp(\boldsymbol{A}) = \sum_{k=0}^{\infty} \frac{\boldsymbol{A}^k}{k!}.$$

converges. The function defined in this way on the set of (complex) $n \times n$ matrices is called the **matrix exponential function** or short **matrix exponential**.

Many properties of the matrix exponential function coincide with those of the exponential function. We list the most important properties below.

Theorem 12.27: Properties of the matrix exponential

The following properties hold:

(i) $\exp(\boldsymbol{0}) = \boldsymbol{I}$,

(ii) $\boldsymbol{B}\exp(\boldsymbol{A}x) = \exp(\boldsymbol{A}x)\boldsymbol{B}$, if $\boldsymbol{AB} = \boldsymbol{BA}$,

(iii) $\exp(\boldsymbol{A}x)\exp(\boldsymbol{B}x) = \exp\big((\boldsymbol{A}+\boldsymbol{B})x\big)$, if $\boldsymbol{AB} = \boldsymbol{BA}$,

(iv) $\exp(\boldsymbol{A}x)\exp(\boldsymbol{A}y) = \exp\big(\boldsymbol{A}(x+y)\big)$,

(v) $\big(\exp(\boldsymbol{A}x)\big)^{-1} = \exp(-\boldsymbol{A}x)$,

(vi) $\frac{d}{dx}\exp(\boldsymbol{A}x) = \boldsymbol{A}\exp(\boldsymbol{A}x)$.

Of course, the last property is important in our context, as it shows, that $\boldsymbol{y}(x) = \exp(\boldsymbol{A}x)\boldsymbol{\gamma}$ for any $\gamma \in \mathbb{R}^n$ is indeed the general solution to the homogeneous system of differential equations $\boldsymbol{y}' = \boldsymbol{A}\boldsymbol{y}$.

Theorem 12.28: Solution to the homogeneous system II

Consider the homogeneous system of linear differential equations $\boldsymbol{y}' = \boldsymbol{A}\boldsymbol{y}$. The general solution to the system is given by

$$\boldsymbol{y}(x) = \exp(\boldsymbol{A}x)\boldsymbol{\gamma}\,, \quad \boldsymbol{\gamma} \in \mathbb{R}^n\,.$$

The unique solution to the initial condition $\boldsymbol{y}(x_0) = \boldsymbol{y}_0$ is given by

$$\boldsymbol{y}(x) = \exp(\boldsymbol{A}(x - x_0))\boldsymbol{y}_0\,.$$

However, we still lack a way to actually calculate the matrix exponential. In the following, we describe the so-called **Cayley-Hamilton method** for calculating the matrix exponential. In Theorem 10.27, we have seen that every power $\boldsymbol{A}^k$ of an $n \times n$ matrix $\boldsymbol{A}$ can be represented as a polynomial of at most $(n-1)$-th degree in $\boldsymbol{A}$,

$$\boldsymbol{A}^k = p_k(\boldsymbol{A}) = a_{k,0}\boldsymbol{I} + a_{k,1}\boldsymbol{A} + a_{k,2}\boldsymbol{A}^2 + \cdots + a_{k,n-1}\boldsymbol{A}^{n-1}\,.$$

Substituting this into the definition of the matrix exponential yields[10]

$$\begin{aligned}\exp(\boldsymbol{A}x) &= \sum_{k=0}^{\infty}\frac{x^k}{k!}\boldsymbol{A}^k = \sum_{k=0}^{\infty}\frac{x^k}{k!}p_k(\boldsymbol{A}) = \sum_{k=0}^{\infty}\frac{x^k}{k!}\sum_{l=0}^{n-1}a_{k,l}\boldsymbol{A}^l = \sum_{l=0}^{n-1}\left(\sum_{k=0}^{\infty}\frac{x^k}{k!}a_{k,l}\right)\boldsymbol{A}^l \\ &= \sum_{l=0}^{n-1}\alpha_l(x)\boldsymbol{A}^l\,,\end{aligned}$$

where for simplicity $\alpha_l(x) = \sum_{k=0}^{\infty}\frac{x^k}{k!}a_{k,l}$ was set. The matrix exponential can thus be calculated as a matrix polynomial of degree $n-1$. It remains the question of how to find the coefficients $\alpha_l(x)$. Since the eigenvalues are roots of the characteristic polynomial, the powers of the eigenvalues have the same representation as a polynomial of degree $n-1$ in λ as the matrix powers, i.e., it holds

$$\lambda^k = r_k(\lambda) = a_{k,0}\boldsymbol{I} + a_{k,1}\lambda + a_{k,2}\lambda^2 + \cdots + a_{k,n-1}\lambda^{n-1}$$

for each eigenvalue λ. Then it also holds

$$\exp(\lambda x) = \sum_{l=0}^{n-1}\alpha_l(x)\lambda^l\,.$$

[10] In the infinite series appearing in this equation the order of summation can be changed, which is not self-evident. However, it follows here from the so-called absolute convergence of the exponential series.

If $\boldsymbol{A}$ has n different eigenvalues $\lambda_1, \ldots, \lambda_n$, this gives n equations, from which the coefficients $\alpha_l(x)$ can be uniquely determined. If multiple eigenvalues occur, this approach does not provide enough equations to uniquely determine the coefficients. However, it can be shown that an eigenvalue λ_i of multiplicity m additionally satisfies the equations

$$\frac{d^j}{d\lambda^j} \exp(\lambda x) = \frac{d^j}{d\lambda^j} \sum_{k=0}^{n-1} \alpha_l(x)\lambda^l , \quad j = 1, \ldots, m-1 .$$

A double eigenvalue thus additionally provides the equation

$$xe^{\lambda x} = \sum_{k=1}^{n-1} l\alpha_l(x)\lambda^{l-1} ,$$

and a triple eigenvalue further adds the equation

$$x^2 e^{\lambda x} = \sum_{k=2}^{n-1} l(l-1)\alpha_l(x)\lambda^{l-2} .$$

An eigenvalue of multiplicity m thus provides m equations, so that the coefficients $\alpha_0(x), \ldots, \alpha_{n-1}(x)$ can be uniquely determined.

We illustrate the Cayley-Hamilton method with several examples:

Example 12.35 (cf. Example 12.31). Consider the homogeneous system of linear differential equations $\boldsymbol{y}' = \boldsymbol{A}\boldsymbol{y}$ with

$$\boldsymbol{A} = \begin{bmatrix} 2 & 4 \\ 1 & -1 \end{bmatrix} ,$$

subject to the initial condition $\boldsymbol{y}(0) = [2, 1]^T$.

The characteristic polynomial of $\boldsymbol{A}$ is $\lambda^2 - \lambda - 6$. Therefore $\boldsymbol{A}$ has the simple eigenvalues $\lambda_1 = 3$ and $\lambda_2 = -2$. This yields the two equations

$$\begin{aligned} e^{3x} &= \alpha_0(x) + \alpha_1(x) \cdot 3 , \\ e^{-2x} &= \alpha_0(x) + \alpha_1(x) \cdot (-2) . \end{aligned}$$

Subtracting the two equations gives

$$5\alpha_1(x) = e^{3x} - e^{-2x} \quad \text{or} \quad \alpha_1(x) = \frac{1}{5}\left(e^{3x} - e^{-2x}\right) .$$

From this it follows that

$$\alpha_0(x) = e^{3x} - \frac{3}{5}\left(e^{3x} - e^{-2x}\right) = \frac{2}{5}e^{3x} + \frac{3}{5}e^{-2x} .$$

Thus, we have

$$\exp(\boldsymbol{A}x) = \left(\frac{2}{5}e^{3x} + \frac{3}{5}e^{-2x}\right)\boldsymbol{I} + \frac{1}{5}\left(e^{3x} - e^{-2x}\right)\boldsymbol{A}$$

$$= \begin{bmatrix} \frac{4}{5}e^{3x} + \frac{1}{5}e^{-2x} & \frac{4}{5}e^{3x} - \frac{4}{5}e^{-2x} \\ \frac{1}{5}e^{3x} - \frac{1}{5}e^{-2x} & \frac{1}{5}e^{3x} + \frac{4}{5}e^{-2x} \end{bmatrix}.$$

The unique solution to the initial condition $y(0) = [2, 1]^T$ is given by

$$y = \exp(Ax)\begin{bmatrix} 2 \\ 1 \end{bmatrix} = \begin{bmatrix} \frac{12}{5}e^{3x} - \frac{2}{5}e^{-2x} \\ \frac{3}{5}e^{3x} + \frac{2}{5}e^{-2x} \end{bmatrix}.$$

Example 12.36 (cf. Example 12.33). Consider the homogeneous system of linear differential equations $y' = Ay$ with

$$A = \begin{bmatrix} 5 & -3 \\ 3 & -1 \end{bmatrix}$$

subject to the initial condition $y(0) = [1, 3]^T$.

The characteristic polynomial of A is $\lambda^2 - 4\lambda + 4 = (\lambda - 2)^2$. Thus, A has a double eigenvalue $\lambda_{1,2} = 2$. Using the Cayley-Hamilton method, we obtain the equations

$$\begin{aligned} e^{2x} &= \alpha_0(x) + \alpha_1(x) \cdot 2\,, \\ xe^{2x} &= \alpha_1(x)\,. \end{aligned}$$

The second equation directly yields $\alpha_1(x)$. Substituting this into the first equation gives

$$\alpha_0(x) = e^{2x} - 2xe^{2x} = (1 - 2x)e^{2x}\,.$$

Thus, we have

$$\begin{aligned} \exp(Ax) &= (1 - 2x)e^{2x}I + xe^{2x}A \\ &= \begin{bmatrix} 1 + 3x & -3x \\ 3x & 1 - 3x \end{bmatrix} e^{2x}\,. \end{aligned}$$

The unique solution to the initial condition $y(0) = [1, 3]^T$ is given by

$$y = \exp(Ax)\begin{bmatrix} 1 \\ 3 \end{bmatrix} = \begin{bmatrix} 1 - 6x \\ 3 - 6x \end{bmatrix} e^{2x}\,.$$

Example 12.37 (cf. Example 12.32). Consider the homogeneous system of linear differential equations $y' = Ay$ with

$$A = \begin{bmatrix} 4 & -1 \\ 5 & 2 \end{bmatrix}$$

subject to the initial condition $y(0) = [1, -1]^T$.

The characteristic polynomial of A is $\lambda^2 - 6\lambda + 13$ with roots $\lambda_{1,2} = 3 \pm 2i$. Thus, there are two complex conjugate eigenvalues. From the Cayley-Hamilton method we obtain the equations

$$\begin{aligned} e^{(3+2i)x} &= \alpha_0(x) + \alpha_1(x)(3+2i)\,, \\ e^{(3-2i)x} &= \alpha_0(x) + \alpha_1(x)(3-2i)\,. \end{aligned}$$

Using Euler's formula and taking advantage of the symmetry relations $\cos(-x) = \cos(x)$ and $\sin(-x) = -\sin(x)$, results in

$$\begin{aligned} e^{3x}\cos(2x) + ie^{3x}\sin(2x) &= (\alpha_0(x) + 3\alpha_1(x)) + i\cdot 2\alpha_1(x)\,, \\ e^{3x}\cos(2x) - ie^{3x}\sin(2x) &= (\alpha_0(x) + 3\alpha_1(x)) - i\cdot 2\alpha_1(x)\,. \end{aligned}$$

Equating the real and imaginary parts yields the two equations

$$e^{3x}\cos(2x) = \alpha_0(x) + 3\alpha_1(x) \quad \text{and} \quad e^{3x}\sin(2x) = 2\alpha_1(x)\,.$$

Hence, $\alpha_1(x) = \frac{1}{2}e^{3x}\sin(2x)$ and

$$\alpha_0(x) = e^{3x}\cos(2x) - \frac{3}{2}e^{3x}\sin(2x) = e^{3x}\left(\cos(2x) - \frac{3}{2}\sin(2x)\right).$$

Thus, we have

$$\begin{aligned} \exp(\boldsymbol{A}x) &= e^{3x}\left(\cos(2x) - \frac{3}{2}\sin(2x)\right)\boldsymbol{I} + \frac{1}{2}e^{3x}\sin(2x)\boldsymbol{A} \\ &= \begin{bmatrix} \cos(2x) + \frac{1}{2}\sin(2x) & -\frac{1}{2}\sin(2x) \\ \frac{5}{2}\sin(2x) & \cos(2x) - \frac{1}{2}\sin(2x) \end{bmatrix} e^{3x}\,. \end{aligned}$$

The unique solution to the initial condition $\boldsymbol{y}(0) = [1,-1]^T$ is given by

$$\boldsymbol{y} = \exp(\boldsymbol{A}x)\begin{bmatrix} 1 \\ -1 \end{bmatrix} = \begin{bmatrix} \cos(2x) + \sin(2x) \\ 3\sin(2x) - \cos(2x) \end{bmatrix} e^{3x}\,.$$

Example 12.38 (cf. Examples 12.34 and 10.15). Consider the homogeneous system of linear differential equations $\boldsymbol{y}' = \boldsymbol{A}\boldsymbol{y}$ with

$$\boldsymbol{A} = \begin{bmatrix} 3 & 2 & -1 \\ -1 & 0 & 2 \\ 1 & 2 & -1 \end{bmatrix}.$$

$\boldsymbol{A}$ has the characteristic polynomial $p_{\boldsymbol{A}}(\lambda) = -\lambda^3 + 2\lambda^2 + 4\lambda - 8$. Thus, $\lambda_1 = -2$ is a simple and $\lambda_{2,3} = 2$ is a double eigenvalue. The Cayley-Hamilton method provides the system of equations

$$\begin{aligned} e^{-2x} &= \alpha_0(x) + \alpha_1(x)\cdot(-2) + \alpha_2(x)\cdot(-2)^2\,, \\ e^{2x} &= \alpha_0(x) + \alpha_1(x)\cdot 2 + \alpha_2(x)\cdot 2^2\,, \\ xe^{2x} &= \alpha_1(x) + 2\alpha_2(x)\cdot 2\,. \end{aligned}$$

Solving the system of equations using Gauss-Jordan elimination yields

$$\alpha_0(x) = \frac{3}{4}e^{2x} + \frac{1}{4}e^{-2x} - xe^{2x},$$
$$\alpha_1(x) = \frac{1}{4}\left(e^{2x} - e^{-2x}\right),$$
$$\alpha_2(x) = \frac{1}{4}xe^{2x} - \frac{1}{16}\left(e^{2x} - e^{-2x}\right).$$

The general solution is therefore given by

$$\boldsymbol{y} = \left[\left(\frac{3}{4}e^{2x} + \frac{1}{4}e^{-2x} - xe^{2x}\right)\boldsymbol{I} + \left(\frac{1}{4}\left(e^{2x} - e^{-2x}\right)\right)\boldsymbol{A} + \left(\frac{1}{4}xe^{2x} - \frac{1}{16}\left(e^{2x} - e^{-2x}\right)\right)\boldsymbol{A}^2\right]\gamma, \quad \gamma \in \mathbb{R}^n.$$

With the hyperbolic functions $\cosh(x) = \frac{1}{2}(e^x + e^{-x})$ and $\sinh(x) = \frac{1}{2}(e^x - e^{-x})$ the solution can be written slightly simpler as

$$\boldsymbol{y} = \left[\left(\cosh(2x) + \frac{1}{2}\sinh(2x) - xe^{2x}\right)\boldsymbol{I} + \frac{1}{2}\sinh(2x)\boldsymbol{A} + \left(\frac{1}{4}xe^{2x} - \frac{1}{8}\sinh(2x)\right)\boldsymbol{A}^2\right]\gamma, \quad \gamma \in \mathbb{R}^n.$$

We have seen that every linear differential equation of higher order can also be written as a system of first-order differential equations. In the so-called **elimination method**, one proceeds in the opposite direction and tries to transform a system of first-order differential equations into a linear differential equation of higher order. This method is often easy to apply to small systems, especially two-dimensional systems. We illustrate it with an example.

Example 12.39 (cf. Examples 12.33 and 12.36). Consider the system of linear differential equations

$$\boldsymbol{y}' = \begin{bmatrix} 5 & -3 \\ 3 & -1 \end{bmatrix}\boldsymbol{y}, \quad \text{that is} \quad \begin{aligned} y_1' &= 5y_1 - 3y_2, \\ y_2' &= 3y_1 - y_2, \end{aligned}$$

subject to the initial condition $\boldsymbol{y}(0) = [1, 3]^T$. Differentiating the first equation and plugging-in the second equation gives

$$y_1'' = 5y_1' - 3y_2' = 5y_1' - 3(3y_1 - y_2) = 5y_1' - 9y_1 + 3y_2.$$

Solving the first equation for $3y_2$ yields $3y_2 = 5y_1 - y_1'$. Substituting this for $3y_2$ in the above equation, yields

$$y_1'' = 5y_1' - 9y_1 + (5y_1 - y_1') = 4y_1' - 4y_1 \quad \Longleftrightarrow \quad y_1'' - 4y_1' + 4y_1 = 0.$$

This is a second order differential equation for y_1. Its characteristic polynomial is $\lambda^2 - 4\lambda + 4 = (\lambda - 2)^2$ with the double root $\lambda_{1,2} = 2$. Thus,

$$y_1 = \gamma_1 e^{2x} + \gamma_2 x e^{2x} = (\gamma_1 + \gamma_2 x)e^{2x} .$$

The solution for y_2 is obtained by solving the first differential equation for y_2 and inserting the solution obtained for y_1:

$$\begin{aligned} y_2 = \frac{5}{3}y_1 - \frac{1}{3}y_1' &= \frac{5}{3}(\gamma_1 + \gamma_2 x)e^{2x} - \frac{1}{3}((2\gamma_1 + \gamma_2) + 2\gamma_2 x)e^{2x} \\ &= \left(\frac{5}{3}\gamma_1 - \frac{2}{3}\gamma_1 - \frac{1}{3}\gamma_2\right)e^{2x} + \left(\frac{5}{3}\gamma_2 - \frac{2}{3}\gamma_2\right)xe^{2x} \\ &= \left((\gamma_1 - \frac{1}{3}\gamma_2) + \gamma_2 x\right)e^{2x} . \end{aligned}$$

The unique solution to the given initial conditions is obtained by substituting the above solutions into the initial conditions $y_1(0) = 1$ and $y_2(0) = 3$. Then $y_1(0) = \gamma_1 = 1$ and thus

$$y_2(0) = \gamma_1 - \frac{1}{3}\gamma_2 = 1 - \frac{1}{3}\gamma_2 \overset{!}{=} 3 ,$$

so $\gamma_2 = -6$. The unique solution to the given initial conditions is therefore

$$\begin{aligned} y_1 &= (1 - 6x)e^{2x} , \\ y_2 &= (3 - 6x)e^{2x} . \end{aligned}$$

Solution to the nonhomogeneous system

We present three methods for solving nonhomogeneous systems of first-order linear differential equations. The simplest one is the so-called **decoupling method**, which applies only to diagonalizable coefficient matrices. The **method of undetermined coefficients** is also relatively straightforward to use, but it requires the forcing function to have a specific form. Finally, **variation of constants** can always be applied, though it involves the greatest computational effort.

Decoupling method

If the coefficient matrix $\boldsymbol{A}$ is diagonalizable, i.e., $\boldsymbol{A} = \boldsymbol{S}\boldsymbol{D}\boldsymbol{S}^{-1}$, then the system $\boldsymbol{y}' = \boldsymbol{A}\boldsymbol{y} + \boldsymbol{b}(x)$ can be written as

$$\boldsymbol{y}' = \boldsymbol{S}\boldsymbol{D}\boldsymbol{S}^{-1}\boldsymbol{y} + \boldsymbol{b}(x) .$$

Multiplying from the left by $\boldsymbol{S}^{-1}$ gives

$$\boldsymbol{S}^{-1}\boldsymbol{y}' = \boldsymbol{D}\boldsymbol{S}^{-1}\boldsymbol{y} + \boldsymbol{S}^{-1}\boldsymbol{b}(x) .$$

By substituting $\boldsymbol{z} = \boldsymbol{S}^{-1}\boldsymbol{y}$ and $\boldsymbol{c}(x) = \boldsymbol{S}^{-1}\boldsymbol{b}(x)$, we obtain for $\boldsymbol{z}$ the **decoupled** system

$$\boldsymbol{z}' = \boldsymbol{D}\boldsymbol{z} + \boldsymbol{c}(x),$$

which consists of n first-order nonhomogeneous linear differential equations, $z_i' = d_i z_i + c_i(x)$, $i = 1, \ldots, n$. These differential equations can be solved by variation of constants (Theorem 12.4) or by the method of undetermined coefficients. The solution to the original system $\boldsymbol{y}' = \boldsymbol{A}\boldsymbol{y} + \boldsymbol{b}(x)$ is obtained by back transformation $\boldsymbol{y} = \boldsymbol{S}\boldsymbol{z}$.

Example 12.40 (cf. Examples 12.31 and 12.35). Consider the nonhomogeneous system of linear differential equations $\boldsymbol{y}' = \boldsymbol{A}\boldsymbol{y} + \boldsymbol{b}(x)$ with

$$\boldsymbol{A} = \begin{bmatrix} 2 & 4 \\ 1 & -1 \end{bmatrix} \quad \text{and} \quad \boldsymbol{b}(x) = \begin{bmatrix} 1 \\ 2 \end{bmatrix} e^{3x}.$$

The matrix $\boldsymbol{A}$ has the eigenvalues $\lambda_1 = 3$ and $\lambda_2 = -2$ with associated eigenvectors $\boldsymbol{v}_1 = [4, 1]^T$ and $\boldsymbol{v}_2 = [1, -1]^T$. The matrices $\boldsymbol{S}$ and $\boldsymbol{D}$ are therefore

$$\boldsymbol{S} = \begin{bmatrix} 4 & 1 \\ 1 & -1 \end{bmatrix} \quad \text{and} \quad \boldsymbol{D} = \begin{bmatrix} 3 & 0 \\ 0 & -2 \end{bmatrix}.$$

The inverse of $\boldsymbol{S}$ is

$$\boldsymbol{S}^{-1} = \frac{1}{5}\begin{bmatrix} 1 & 1 \\ 1 & -4 \end{bmatrix}.$$

Then,

$$\boldsymbol{c}(x) = \boldsymbol{S}^{-1}\boldsymbol{b}(x) = \frac{1}{5}\begin{bmatrix} 1 & 1 \\ 1 & -4 \end{bmatrix}\begin{bmatrix} 1 \\ 2 \end{bmatrix} e^{3x} = \begin{bmatrix} 0.6 \\ -1.4 \end{bmatrix} e^{3x}.$$

For the transformed variables $\boldsymbol{z} = \boldsymbol{S}^{-1}\boldsymbol{y}$ we therefore obtain the decoupled system

$$\begin{bmatrix} z_1' \\ z_2' \end{bmatrix} = \begin{bmatrix} 3 & 0 \\ 0 & -2 \end{bmatrix}\begin{bmatrix} z_1 \\ z_2 \end{bmatrix} + \begin{bmatrix} 0.6 \\ -1.4 \end{bmatrix} e^{3x},$$

that is, the two first-order linear differential equations

$$z_1' = 3z_1 + 0.6e^{3x} \quad \text{and} \quad z_2' = -2z_2 - 1.4e^{3x}.$$

For the first equation, we apply the method of undetermined coefficients with the trial solution $z_1 = axe^{3x}$ (due to resonance), yielding the particular solution $z_1 = 0.6xe^{3x}$. Similarly, for the second equation, we use the trial solution $z_2 = ae^{3x}$, yielding the particular solution $z_2 = -0.28e^{3x}$.

A solution to the original system is obtained by the back transformation

$$\boldsymbol{y} = \boldsymbol{S}\boldsymbol{z} = \begin{bmatrix} 4 & 1 \\ 1 & -1 \end{bmatrix}\begin{bmatrix} 0.6x \\ -0.28 \end{bmatrix} e^{3x} = \begin{bmatrix} 2.4x - 0.28 \\ 0.6x + 0.28 \end{bmatrix} e^{3x}.$$

Method of undetermined coefficients

The method of undetermined coefficients can also be used for systems of differential equations. In the following, a **vector-valued polynomial** (cf. p. 389) is a function of the form $\boldsymbol{p}(x) = \sum_{i=0}^{k} \boldsymbol{a}_i x^i$, where $\boldsymbol{a}_0, \ldots, \boldsymbol{a}_k$ are vectors, that is, a vector-valued function whose components are polynomials

Theorem 12.29: Method of undetermined coefficients for systems

Consider the system of first-order linear differential equations

$$\boldsymbol{y}' = \boldsymbol{A}\boldsymbol{y} + \boldsymbol{b}(x).$$

If the forcing function has the form

$$\boldsymbol{b}(x) = \boldsymbol{p}_1(x)e^{ax}\sin(bx) + \boldsymbol{p}_2(x)e^{ax}\cos(bx),$$

where $\boldsymbol{p}_1$ and $\boldsymbol{p}_2$ are vector-valued polynomials of degree n_1 and n_2, respectively, and if $z = a + bi$ is **not an eigenvalue** of $\boldsymbol{A}$, then the nonhomogeneous system has a particular solution of the form

$$\boldsymbol{y}(x) = \boldsymbol{q}_1(x)e^{ax}\sin(bx) + \boldsymbol{q}_2(x)e^{ax}\cos(bx),$$

where $\boldsymbol{q}_1$ and $\boldsymbol{q}_2$ are vector-valued polynomials of degree $n = \max\{n_1, n_2\}$.

If $z = a + bi$ is an **eigenvalue of multiplicity** k of $\boldsymbol{A}$, then the nonhomogeneous system has a particular solution of the form

$$\boldsymbol{y}(x) = \boldsymbol{q}_1(x)e^{ax}\sin(bx) + \boldsymbol{q}_2(x)e^{ax}\cos(bx),$$

where $\boldsymbol{q}_1$ and $\boldsymbol{q}_2$ are vector-valued polynomials of degree $n = \max\{n_1, n_2\} + k$.

We illustrate the method with two examples.

Example 12.41 (cf. Examples 12.31, 12.35 and 12.40). Consider the nonhomogeneous system of linear differential equations $\boldsymbol{y}' = \boldsymbol{A}\boldsymbol{y} + \boldsymbol{b}(x)$ with

$$\boldsymbol{A} = \begin{bmatrix} 2 & 4 \\ 1 & -1 \end{bmatrix} \quad \text{and} \quad \boldsymbol{b}(x) = \begin{bmatrix} 1 \\ 2 \end{bmatrix} e^{2x}.$$

Since the forcing function has the form $\boldsymbol{b}(x) = \boldsymbol{u}e^{2x}$ and since $\lambda = 2$ is not an eigenvalue of $\boldsymbol{A}$, we use the trial solution $\boldsymbol{y} = \boldsymbol{a}e^{2x}$. The derivative of $\boldsymbol{y}$ is given by $\boldsymbol{y}' = 2\boldsymbol{a}e^{2x}$. Substituting into the system yields

$$2\boldsymbol{a}e^{2x} = \boldsymbol{A}\boldsymbol{a}e^{2x} + \boldsymbol{u}e^{2x}.$$

Dividing by e^{2x} gives

$$2\boldsymbol{a} = \boldsymbol{A}\boldsymbol{a} + \boldsymbol{u} \iff (\boldsymbol{A} - 2\boldsymbol{I})\boldsymbol{a} = -\boldsymbol{u} \iff \begin{bmatrix} 0 & 4 \\ 1 & -3 \end{bmatrix} \begin{bmatrix} c_1 \\ c_2 \end{bmatrix} = \begin{bmatrix} -1 \\ -2 \end{bmatrix}.$$

Solving the linear equation using Gauss-Jordan elimination finally yields $\boldsymbol{a} = -[2.75, 0.25]^T$. A particular solution to the nonhomogeneous system is thus

$$\boldsymbol{y} = -\begin{bmatrix} 2.75 \\ 0.25 \end{bmatrix} e^{2x} .$$

Example 12.42 (cf. Examples 12.31, 12.35, 12.40 and 12.41). Consider the nonhomogeneous system of linear differential equations $\boldsymbol{y}' = \boldsymbol{A}\boldsymbol{y} + \boldsymbol{b}(x)$ with

$$\boldsymbol{A} = \begin{bmatrix} 2 & 4 \\ 1 & -1 \end{bmatrix} \quad \text{and} \quad \boldsymbol{b}(x) = \begin{bmatrix} 1 \\ 2 \end{bmatrix} e^{3x} .$$

Since the forcing function has the form $\boldsymbol{b}(x) = \boldsymbol{u}e^{3x}$ and since $\lambda = 3$ is a simple eigenvalue of $\boldsymbol{A}$, we use the trial solution $\boldsymbol{y} = (\boldsymbol{a} + \boldsymbol{b}x)e^{3x}$. The derivative of $\boldsymbol{y}$ is given by $\boldsymbol{y}' = \big((3\boldsymbol{a} + \boldsymbol{b}) + 3\boldsymbol{b}x\big)e^{3x}$. Substituting into the system yields

$$\big((3\boldsymbol{a} + \boldsymbol{b}) + 3\boldsymbol{b}x\big)e^{3x} = \boldsymbol{A}(\boldsymbol{a} + \boldsymbol{b}x)e^{3x} + \boldsymbol{u}e^{3x} .$$

Dividing by e^{3x} gives

$$(3\boldsymbol{a} + \boldsymbol{b}) + 3\boldsymbol{b}x = \boldsymbol{A}(\boldsymbol{a} + \boldsymbol{b}x) + \boldsymbol{u} \iff (3\boldsymbol{a} + \boldsymbol{b}) + 3\boldsymbol{b}x = (\boldsymbol{A}\boldsymbol{a} + \boldsymbol{u}) + \boldsymbol{A}\boldsymbol{b}x .$$

Equating coefficients yields the two equations

$$3\boldsymbol{a} + \boldsymbol{b} = \boldsymbol{A}\boldsymbol{a} + \boldsymbol{u} \qquad \text{and} \qquad 3\boldsymbol{b} = \boldsymbol{A}\boldsymbol{b} ,$$

which are equivalent to

$$(\boldsymbol{A} - 3\boldsymbol{I})\boldsymbol{a} = \boldsymbol{b} - \boldsymbol{u} \qquad \text{and} \qquad (\boldsymbol{A} - 3\boldsymbol{I})\boldsymbol{b} = \boldsymbol{0} .$$

From the second equation follows that $\boldsymbol{b}$ is an eigenvector for the eigenvalue $\lambda = 3$. Therefore, $\boldsymbol{b} = [4c, c]^T$ where $c \in \mathbb{R}$ (cf. Example 12.31).

Inserting this into the first equation $(\boldsymbol{A} - 3\boldsymbol{I})\boldsymbol{a} = \boldsymbol{b} - \boldsymbol{u}$ and applying one step of Gauss-Jordan elimination gives

$$\left[\begin{array}{cc|c} -1 & 4 & 4c - 1 \\ 1 & -4 & c - 2 \end{array}\right] \longrightarrow \left[\begin{array}{cc|c} 1 & -4 & c - 2 \\ 0 & 0 & 5c - 3 \end{array}\right] .$$

This equation has a solution only if $5c - 3 = 0$, i.e., if $c = 3/5$, in which case the vectors $\boldsymbol{a}$ and $\boldsymbol{b}$ are given by

$$\boldsymbol{a} = \begin{bmatrix} -1.4 \\ 0 \end{bmatrix} + d \begin{bmatrix} 4 \\ 1 \end{bmatrix}, \; d \in \mathbb{R}, \quad \text{and} \quad \boldsymbol{b} = \begin{bmatrix} 2.4 \\ 0.6 \end{bmatrix} .$$

Since d can be chosen freely, we set $d = 0$ and get $\boldsymbol{a} = [-1.4, 0]^T$. Therefore, a particular solution to the nonhomogeneous system is given by

$$\boldsymbol{y} = \left(\begin{bmatrix} -1.4 \\ 0 \end{bmatrix} + \begin{bmatrix} 2.4 \\ 0.6 \end{bmatrix} x \right) e^{3x} = \begin{bmatrix} 2.4x - 1.4 \\ 0.6x \end{bmatrix} e^{3x} .$$

The preceding example nicely illustrates the following two observations.

Remarks:

1. For systems, the polynomials $q_1(x)$ and $q_2(x)$ must always contain **all lower-order powers**, even in the resonance case. If $a + bi$ is a root of multiplicity k of the characteristic polynomial, a trial solution of the form
$$\boldsymbol{y}(x) = x^k \left[\boldsymbol{q}_1(x)e^{ax}\sin(bx) + \boldsymbol{q}_2(x)e^{ax}\cos(bx)\right] ,$$
with polynomials $\boldsymbol{q}_1(x)$, $\boldsymbol{q}_2(x)$ of degree $n = \max\{n_1, n_2\}$ will, in general, not yield a solution. This is because the trial solution multiplied by x^k does not contain powers with exponents less than k.
2. Under resonance, the resulting system of equations usually has no unique solution. In the above example, we could choose d freely. For simplicity, constants that can be chosen freely are often set to zero.

Variation of constants

While the method of undetermined coefficients can only be used for certain forcing functions, the following method, **variation of constants**, can be used universally. Here, in the general solution to the homogeneous system, $\boldsymbol{y}(x) = \boldsymbol{F}(x)\boldsymbol{\gamma}$, the constant vector $\boldsymbol{\gamma}$ is replaced by a function $\boldsymbol{\gamma}(x)$, i.e., one uses the trial solution $\boldsymbol{y}(x) = \boldsymbol{F}(x)\boldsymbol{\gamma}(x)$. Its derivative is given by
$$\boldsymbol{y}'(x) = \boldsymbol{F}'(x)\boldsymbol{\gamma}(x) + \boldsymbol{F}(x)\boldsymbol{\gamma}'(x) .$$
Substituting this into the nonhomogeneous system $\boldsymbol{y}'(x) = \boldsymbol{A}\boldsymbol{y}(x) + \boldsymbol{b}(x)$ gives
$$\boldsymbol{F}'(x)\boldsymbol{\gamma}(x) + \boldsymbol{F}(x)\boldsymbol{\gamma}'(x) = \boldsymbol{A}\boldsymbol{F}(x)\boldsymbol{\gamma}(x) + \boldsymbol{b}(x) .$$
Because of $\boldsymbol{F}'(x) = \boldsymbol{A}\boldsymbol{F}(x)$ (see page 385) one obtains
$$\boldsymbol{A}\boldsymbol{F}(x)\boldsymbol{\gamma}(x) + \boldsymbol{F}(x)\boldsymbol{\gamma}'(x) = \boldsymbol{A}\boldsymbol{F}(x)\boldsymbol{\gamma}(x) + \boldsymbol{b}(x)$$
and thus
$$\boldsymbol{F}(x)\boldsymbol{\gamma}'(x) = \boldsymbol{b}(x) .$$
Since the fundamental matrix $\boldsymbol{F}(x)$ is invertible for every $x \in \mathbb{R}$, one finally obtains
$$\boldsymbol{\gamma}'(x) = \boldsymbol{F}(x)^{-1}\boldsymbol{b}(x)$$
or
$$\boldsymbol{\gamma}(x) = \int \boldsymbol{F}(x)^{-1}\boldsymbol{b}(x)\,dx ,$$
where the integral of a vector-valued function is to be understood component-wise. A particular solution to the nonhomogeneous system is therefore obtained by
$$\boldsymbol{y}(x) = \boldsymbol{F}(x)\int \boldsymbol{F}(x)^{-1}\boldsymbol{b}(x)\,dx .$$
Again, we illustrate the method by an example.

Example 12.43 (cf. Examples 12.31, 12.35, 12.40, 12.41 and 12.42). Consider the nonhomogeneous system of linear differential equations $\boldsymbol{y}' = \boldsymbol{A}\boldsymbol{y} + \boldsymbol{b}(x)$ with

$$\boldsymbol{A} = \begin{bmatrix} 2 & 4 \\ 1 & -1 \end{bmatrix} \quad \text{and} \quad \boldsymbol{b}(x) = \begin{bmatrix} 1 \\ 2 \end{bmatrix} e^{2x}.$$

According to Example 12.31, a fundamental matrix is given by

$$\boldsymbol{F}(x) = \begin{bmatrix} 4e^{3x} & e^{-2x} \\ e^{3x} & -e^{-2x} \end{bmatrix}.$$

Its inverse is

$$\boldsymbol{F}(x)^{-1} = -\frac{1}{5}e^{-x} \begin{bmatrix} -e^{-2x} & -e^{-2x} \\ -e^{3x} & 4e^{3x} \end{bmatrix} = \frac{1}{5} \begin{bmatrix} e^{-3x} & e^{-3x} \\ e^{2x} & -4e^{2x} \end{bmatrix}.$$

For $\boldsymbol{F}(x)^{-1}\boldsymbol{b}(x)$ we get

$$\boldsymbol{F}(x)^{-1}\boldsymbol{b}(x) = \frac{1}{5} \begin{bmatrix} e^{-3x} & e^{-3x} \\ e^{2x} & -4e^{2x} \end{bmatrix} \begin{bmatrix} 1 \\ 2 \end{bmatrix} e^{2x} = \frac{1}{5} \begin{bmatrix} 3e^{-x} \\ -7e^{4x} \end{bmatrix}.$$

Integration yields

$$\int \boldsymbol{F}(x)^{-1}\boldsymbol{b}(x)\,dx = \frac{1}{5} \begin{bmatrix} -3e^{-x} \\ -\frac{7}{4}e^{4x} \end{bmatrix}.$$

Hence, a particular solution is given by

$$\boldsymbol{y}(x) = \boldsymbol{F}(x) \int \boldsymbol{F}(x)^{-1}\boldsymbol{b}(x)\,dx = \begin{bmatrix} 4e^{3x} & e^{-2x} \\ e^{3x} & -e^{-2x} \end{bmatrix} \frac{1}{5} \begin{bmatrix} -3e^{-x} \\ -\frac{7}{4}e^{4x} \end{bmatrix}$$

$$= \frac{1}{5} \begin{bmatrix} -12e^{2x} - \frac{7}{4}e^{2x} \\ -3e^{2x} + \frac{7}{4}e^{2x} \end{bmatrix} = -\begin{bmatrix} 2.75 \\ 0.25 \end{bmatrix} e^{2x}.$$

12.5 Stability of systems of differential equations

We have already dealt with the stability of simple differential equations in Section 12.3. Also in systems, the limit behavior of the solution for large values of x, or its dependence on given initial conditions, is often of primary interest.

Definition 12.6: Stability

A system of first-order linear differential equations is called (asymptotically) **stable** if any two of its solutions, $\boldsymbol{y} = \boldsymbol{f}_1(x)$ and $\boldsymbol{y} = \boldsymbol{f}_2(x)$, satisfy

$$\lim_{x\to\infty} (\boldsymbol{f}_1(x) - \boldsymbol{f}_2(x)) = \boldsymbol{0}.$$

If the system of differential equations has a constant equilibrium solution $\boldsymbol{y}^*$, the above definition is equivalent to the fact that **every** solution to the system converges to the equilibrium solution $\boldsymbol{y}^*$ as x tends to infinity[11],

$$\lim_{x\to\infty} \boldsymbol{y}(x) = \boldsymbol{y}^* .$$

Since the difference between two solutions of a nonhomogeneous system is a solution to the corresponding homogeneous system, a nonhomogeneous system is stable if and only if every solution $\boldsymbol{y}(x)$ of the corresponding homogeneous system converges to zero as x goes to infinity,

$$\lim_{x\to\infty} \boldsymbol{y}(x) = \boldsymbol{0} .$$

From Theorem 12.25 it immediately follows that in the case of a complex diagonalizable coefficient matrix $\boldsymbol{A}$ all solutions of the homogeneous system converge to zero if and only if all eigenvalues have a negative real part. In fact, it can be shown that this result also holds for systems with a non-diagonalizable coefficient matrix $\boldsymbol{A}$. We formulate this result as a theorem.

Theorem 12.30: Stability of a system of linear differential equations

A system of first-order linear differential equations with constant coefficients, $\boldsymbol{y}' = \boldsymbol{A}\boldsymbol{y} + \boldsymbol{b}(x)$, is stable if and only if every eigenvalue λ of $\boldsymbol{A}$ satisfies $\mathrm{Re}(\lambda) < 0$, that is, if λ has a negative real part.

It would of course be desirable to have simple criteria that allow the decision whether a system of differential equations is stable or not to be made, without actually having to compute the eigenvalues. Such a criterion is already provided by the Hurwitz criterion, Theorem 12.22. It is sufficient to calculate the characteristic polynomial of the matrix $\boldsymbol{A}$. Applying the Hurwitz criterion, one can then easily decide whether all roots of the characteristic polynomial, i.e., all eigenvalues, have a negative real part.

The following theorem gives a criterion for stability that is simple to apply, but only necessary.

Theorem 12.31: Stability of a system, necessary condition

If the system $\boldsymbol{y}' = \boldsymbol{A}\boldsymbol{y} + \boldsymbol{b}(x)$ is stable, then the sum of the diagonal elements of $\boldsymbol{A}$ is negative,

$$\sum_{i=1}^{n} a_{ii} < 0 .$$

Example 12.44. Consider the system of linear differential equations $\boldsymbol{y}' = \boldsymbol{A}\boldsymbol{y}$

[11] For the definition of convergence of vectors see Section 3.2.

with

$$A = \begin{bmatrix} -1 & -0.5 & -1 \\ 1 & 2.5 & -1 \\ -0.5 & 0.75 & -1.5 \end{bmatrix}.$$

Since the sum of the diagonal elements is zero and therefore not negative, the system is not stable. In fact, $\boldsymbol{A}$ has the eigenvalues $\lambda_1 = 2$, $\lambda_2 = 0$ and $\lambda_3 = -2$.

For a square matrix $\boldsymbol{A}$, the matrix

$$A_s = \frac{1}{2}\left(A + A^T\right)$$

is called the **symmetric part** of $\boldsymbol{A}$. The following theorem provides a sufficient criterion for the stability of a system of differential equations, which is based on the definiteness of $\boldsymbol{A}_s$.

Theorem 12.32: Stability of a system, sufficient condition

Consider the system of first-order linear differential equations $y' = Ay + b(x)$. If the symmetric part $\boldsymbol{A}_s$ of the coefficient matrix $\boldsymbol{A}$ is negative definite, then the system $y' = Ay + b(x)$ is stable.

Whether the symmetric part $\boldsymbol{A}_s$ is negative definite can easily be checked using Sylvester's criterion, Theorem 10.12.

Example 12.45. Consider the system of linear differential equations $y' = Ay$ with

$$A = \begin{bmatrix} -1 & -0.5 & -1 \\ 1 & -1.5 & -1 \\ -0.5 & 0.75 & -1.5 \end{bmatrix}.$$

The sum of the diagonal elements is -4 and therefore negative. The necessary criterion is thus satisfied. The symmetric part of the matrix $\boldsymbol{A}$ is

$$A_s = \begin{bmatrix} -1 & 0.25 & -0.75 \\ 0.25 & -1.5 & -0.125 \\ -0.75 & -0.125 & -1.5 \end{bmatrix}.$$

For the principal minors, we obtain $\Delta_1 = -1 < 0$, $\Delta_2 = 1.4375 > 0$ and $\Delta_3 = -1.25 < 0$. Thus, $\boldsymbol{A}_s$ is negative definite and the system is stable. In fact, it can be shown that $\boldsymbol{A}$ has the eigenvalues $\lambda_1 = -2$ and $\lambda_{2,3} = -1 \pm i$.

Important terms for review

After reading this chapter, the following terms should be familiar:

- Differential equation
- Order of a differential equation

- Initial condition
- Homogeneous and nonhomogeneous linear differential equation
- Exact differential equation
- Separable differential equation
- Characteristic polynomial, characteristic equation
- Solution of linear differential equations with constant coefficients
- Method of undetermined coefficients
- Stability
- Systems of first-order differential equations
- Matrix exponential
- Decoupling of systems of differential equations
- Method of undetermined coefficients for systems of differential equations

Self-test

You can check your understanding of this chapter using the following true-or-false questions. For each statement, indicate whether it is true (T) or false (F). Short solutions to these questions can be found in Appendix H.

I. Are the following statements about differential equations true or false?

	T	F
In a differential equation of order m, all derivatives up to order m appear.	☐	☐
The function $y = f(x) = x^4$ is a solution to the differential equation $y = \frac{1}{4}xy'$.	☐	☐
The sum of the solution to a nonhomogeneous linear differential equation and a solution to the corresponding homogeneous linear differential equation is a solution to the nonhomogeneous linear differential equation.	☐	☐
A solution to a differential equation of the form $y' = g(x)h(y)$ can be obtained by equating the integral $\int g(x)\,dx$ with $\int h(y)\,dy$.	☐	☐
When solving a separable differential equation, after the first step – integration – one obtains a solution that is defined implicitly by an equation.	☐	☐

Exercises

Exercise 1
The elasticity of a demand function $q = f(p)$ is given by

$$\varepsilon_q(p) = -0.5 - 2p\,.$$

Furthermore, it is known that at a price of $p_0 = 2$, the demand is $q_0 = 100$. Find the demand function $q = f(p)$. What is the demand at a price of $p = 2.5$?

Exercise 2
The elasticity of a demand function $q = f(p)$ is a linear function, i.e., $\varepsilon_q(p) = a - bp$ where $a, b \in \mathbb{R}$ are real constants. Find the demand function that satisfies the initial condition $q_0 = f(p_0)$. What is the demand function in the case $a = -2$, $b = \ln(2)$, $p_0 = 1$ and $q_0 = 5$?

Exercise 3
The costs C incurred in the production of q units of a certain good are described by a function $C = f(q)$. It is known that the marginal costs equal half the average costs, i.e.,

$$\frac{dC}{dq} = \frac{1}{2}\frac{C}{q}\,.$$

Furthermore, it is known that the production of 9 units of the good incurs a cost of 6 monetary units. Find the cost function $C = f(q)$.

Exercise 4
Consider the first-order linear differential equation

$$y' - \frac{x}{1+x^2}y = 2x\,,$$

subject to the initial condition $y(0) = 1$.

a) Find the general solution to the homogeneous differential equation.

b) Find a particular solution to the nonhomogeneous equation using variation of constants.

c) State the general solution to the nonhomogeneous differential equation.

d) Find the solution of the differential equation that satisfies the initial condition.

Exercise 5
Find the general solution to the differential equation $y' = 2xy + x^3$.

Exercise 6
Consider the differential equation $9y' = x^2 e^{-y}$. Find the general solution as well as the unique solution given the initial condition $y(3) = 0$.

Exercise 7
Find the solution to the differential equation

$$x^3 + y^3 + 3xy^2y' = 0,$$

given the initial condition $y(1) = 2$.

Exercise 8
For each of the following homogeneous differential equations, find the solution for the given initial conditions. Also, check whether the differential equation is stable.

a) $y'' - 1.2y' + 0.32y = 0$, where $y(0) = 0$ and $y'(0) = 0.8$.

b) $y'' + 4y = 0$, where $y(0) = 4$ and $y'(0) = -4$.

c) $y'' - 0.4y' + 0.04y = 0$, where $y(0) = 4$ and $y(2) = 0$.

d) $y'' - 4y' + 5y = 0$, where $y(0) = 2$ and $y(\pi/2) = -e^{\pi}$.

Exercise 9
For each of the following nonhomogeneous differential equations, find the general solution:

a) $y'' - 1.2y' + 0.32y = 8 + 3e^{2x}$,

b) $y'' + 4y = x^2$,

c) $y'' - 0.4y' + 0.04y = 6e^{0.2x}$,

d) $y'' - 4y' + 5y = 10x^2$.

Exercise 10
Check whether the following two differential equations are stable:

a) $y^{(4)} + 2y''' - 2y'' + 4y' + 5y = 0$,

b) $y^{(4)} + 2y''' + 4y'' + 4y' + 2y = 0$.

Exercise 11
Consider the following homogeneous system of linear differential equations

$$\mathbf{y}' = \begin{bmatrix} 1 & 9 \\ -1 & -5 \end{bmatrix} \mathbf{y}.$$

Find the general solution using the method of undetermined coefficients.

Chapter 13

Difference equations

In the previous chapter, we considered the Evans price adjustment model and the Harrod-Domar model as continuous-time models, i.e., the relevant economic variables were observed continuously in time. In this chapter, we discuss so-called **discrete-time models**, where the relevant economic variables are only considered at discrete points or intervals in time. Thus, the time parameter t takes discrete values, typically natural numbers or integers. In the following, we assume that t varies over a finite or infinite set T of consecutive integers, usually starting at $t = 1$.

We begin this chapter with some examples that underline the relevance of difference equations in economics.

Example 13.1 (Harrod-Domar model, discrete time). The Harrod-Domar model describes the growth of an economy over time. We use the following notations: For $t \in T$ let

K_t denote capital stock in period t,
Y_t national income (social product) in period t,
S_t savings in period t,
I_t net investment in period t.

As with the continuous-time Harrod-Domar model (see Example 12.3), the question of interest is how the economic variables under consideration evolve over time. The model is based on the following assumptions:

- By definition, net investment is equal to the change in the capital stock:

$$I_t = K_t - K_{t-1} .$$

K. Mosler et al., *Mathematical Methods for Business and Economics*, Mathematics Study Resources 20, https://doi.org/10.1007/978-3-662-73831-3_13

- The national income is proportional to the level of the capital stock:

$$Y_t = \frac{1}{v} K_t \,.$$

  The constant v is referred to as the **capital coefficient**, and its reciprocal $1/v$ as **capital productivity**.

- Savings are proportional to the national income:

$$S_t = sY_t \,.$$

  The constant s is referred to as the **savings rate**.

- Savings are equal to net investment:

$$I_t = S_t \,.$$

From the first two equations it follows that

$$I_t = v(Y_t - Y_{t-1}) \,.$$

This is referred to as the **capacity effect** of investment. From the third and fourth equation above, we obtain the so-called **income effect** of investment,

$$I_t = sY_t \,.$$

Equating the last two equations and solving for Y_t finally yields

$$\boxed{Y_t = \frac{v}{v - s} Y_{t-1} \,.}$$

Thus, the Harrod-Domar model leads to an equation relating the values of an economic variable, national income, at different points in time. We are interested in sequences $(Y_t), t \in T$ that satisfy the above equation. Such a sequence can also be represented as a function $Y : T \to \mathbb{R}$.

Example 13.2 (Cobweb model). We again consider the cobweb model, which was already discussed in Sections 3.1 and 3.10. For $t \in T$ let

d_t denote the demand for a good in period t,
s_t the supply of this good in period t,
p_t the price of this good in period t.

These quantities are related among one another by the following assumptions:

- The demand in a period depends linearly on the current price:

$$d_t = a - bp_t \,.$$

  We assume $b > 0$, ensuring that demand decreases as price increases.

- The supply reacts with a delay to price changes. More precisely, we assume that the supply in a period depends linearly on the price in the previous period:

$$s_t = c + dp_{t-1} .$$

Further assume $d > 0$, ensuring that supply increases as price increases.

- In each period, the price is determined such that demand equals supply:

$$d_t = s_t .$$

Substituting the first two equations into the last equation yields

$$a - bp_t = c + dp_{t-1},$$

i.e.,

$$\boxed{p_t + \frac{d}{b}p_{t-1} = \frac{a-c}{b} .}$$

Again, an equation has been derived that links the values of an economic variable, here a price, at different points in time, and a function $p : T \to \mathbb{R}$ is sought that satisfies this equation.

Example 13.3 (Multiplier-accelerator model by Samuelson[1]). The multiplier-accelerator model by Samuelson, like the Harrod-Domar model, is a macroeconomic model that describes the growth of an economy. For $t \in T$, let

Y_t	denote the national income in period t,
C_t	the consumption in period t,
I_t	the investment in period t,
G_t	the government spending in period t.

The following assumptions are made:

- By definition, national income is the sum of consumption, investment, and government spending:

$$Y_t = C_t + I_t + G_t .$$

- Consumption is proportional to the national income of the previous period:

$$C_t = cY_{t-1} .$$

The constant c is called **consumption rate**. We assume $0 < c < 1$.

- Investments are proportional to the change in consumption:

$$I_t = k(C_t - C_{t-1}) .$$

[1] Paul Anthony Samuelson (1915–2009); Nobel Prize in Economics 1970.

The constant k is referred to as **accelerator**. We assume $k > 0$, ensuring that an increase in consumption leads to positive investments.

- Government spending is constant, i.e., $G_t = G$ for all $t \in T$.

From the above equations, we get

$$\begin{aligned} Y_t &= cY_{t-1} + k(C_t - C_{t-1}) + G \\ &= cY_{t-1} + ck(Y_{t-1} - Y_{t-2}) + G \\ &= c(1+k)Y_{t-1} - ckY_{t-2} + G , \end{aligned}$$

and thus

$$\boxed{Y_t - c(1+k)Y_{t-1} + ckY_{t-2} = G .}$$

As in the two previous models, an equation is obtained that relates the values of an economic variable, here the national income Y_t, at different points in time. We are looking for functions $Y : T \to \mathbb{R}$ that satisfy this equation.

In each of the previous three examples, an equation arieses that relates the values of an economic variable at different discrete points in time. Such equations are called **difference equations**. We will focus exclusively on a special – but central in applications – class of difference equations, the so-called **linear difference equations with constant coefficients**.

Definition 13.1: m-th order linear difference equation

An **m-th order linear difference equation with constant coefficients** is an equation of the form

$$y_t + a_1 y_{t-1} + \ldots + a_m y_{t-m} = b_t ,$$

where $a_1, \ldots, a_m \in \mathbb{R}$ and $b : T \to \mathbb{R}, t \mapsto b_t$, is a given function.

If $b_t = 0$ for all $t \in T$, then the difference equation is called **homogeneous**, otherwise **nonhomogeneous**. The function b_t is called **forcing function**.

A function $y_t = f(t)$ that satisfies the difference equation is called a **particular solution** to the difference equation. The set of all solutions to a difference equation is referred to as **general solution**.

In the following, we provide simple examples of difference equations and their solutions. In these examples, the solution can be conveniently verified by substituting the solution into the difference equation.

Example 13.4. The difference equation $y_t - y_{t-1} = 5$ has the function $y_t = 5t$, $t \in \mathbb{Z}$, as a particular solution. However, also $y_t = 5t + 2$, or more generally $y_t = 5t + \gamma$ with an arbitrary constant $\gamma \in \mathbb{R}$ is a solution.

Example 13.5. The difference equation $y_t = 5y_{t-1}$ has the function $y_t = 5^t$, $t \in \mathbb{Z}$, as a particular solution. As in the previous example, the solution is not unique here either. In general, every function $y_t = \gamma 5^t$, $\gamma \in \mathbb{R}$, is a solution to the difference equation.

Example 13.6. The difference equation $y_t + 4y_{t-1} = 5$ has the functions $y_t = 1 + \gamma(-4)^t$, $t \in \mathbb{Z}$, as general solution. Here, γ is again an arbitrary real constant.

As with differential equations, the solution to a difference equation is generally not unique. Rather, there are infinitely many functions that satisfy a difference equation. However, by specifying one or more initial conditions, depending on the order of the difference equation, a unique solution is obtained. We illustrate this with an example.

Example 13.7. The general solution to the second-order homogeneous linear difference equation

$$y_t = 3y_{t-1} - 2y_{t-2}$$

is given by

$$y_t = \gamma_1 + \gamma_2 2^t\,, \qquad \gamma_1, \gamma_2 \in \mathbb{R}\,.$$

If we additionally impose the initial conditions $y_0 = 5$ and $y_1 = 8$, the unique solution is $y_t = 2 + 3 \cdot 2^t$. This is found by solving the system

$$\gamma_1 + \gamma_2 2^0 = 5 \quad \text{and} \quad \gamma_1 + \gamma_2 2^1 = 8\,.$$

Specifying only a single initial condition is not sufficient to determine a unique solution. For example, if only $y_0 = 2$ is given, then all functions of the form $y_t = \gamma + (2-\gamma)2^t$ satisfy the difference equation and the specified initial condition.

In the previous example, we needed to specify two initial conditions for consecutive values of the solution to a second-order difference equation to obtain a unique solution. The following theorem establishes that this is generally the case.

Theorem 13.1: Specification of initial conditions

If m **consecutive** values (e.g. $y_0, y_1, \ldots, y_{m-1}$) are specified for an m-th order linear difference equation, then there exists a unique solution y_t satisfying these initial conditions.

13.1 First-order linear difference equations

We begin with the solution of first-order linear difference equations and then address the solution to general linear difference equations.

A first-order homogeneous linear difference equation has the form $y_t + a_1 y_{t-1} = 0$ or equivalently $y_t = (-a_1)y_{t-1}$. Starting from the initial condition $y_0 = \gamma$, one obtains $y_1 = \gamma(-a_1)$, $y_2 = \gamma(-a_1)^2$, $y_3 = \gamma(-a_1)^3$, and so forth. Therefore, the general solution to the homogeneous equation is given by the following theorem.

Theorem 13.2: First-order homogeneous difference equation

The first-order **homogeneous** linear difference equation $y_t + a_1 y_{t-1} = 0$ has the general solution

$$y_t = \gamma(-a_1)^t\,, \quad \gamma \in \mathbb{R}\,.$$

The unique solution for a given initial condition y_0 at time $t = 0$ is

$$y_t = y_0(-a_1)^t\,.$$

Example 13.8 (Harrod-Domar model, discrete time). The Harrod-Domar model (Example 13.1) leads to the difference equation

$$Y_t = \frac{v}{v-s} Y_{t-1} \quad \text{or} \quad Y_t - \frac{v}{v-s} Y_{t-1} = 0\,.$$

This is a first-order homogeneous linear difference equation. Its solution is

$$Y_t = \left(\frac{v}{v-s}\right)^t Y_0\,.$$

In the economically relevant case $0 < s < v$, the ratio $v/(v-s)$ is greater than one. Hence, national income increases each period by a fixed percentage, that is, the economy exhibits exponential growth. Even in the case of discrete time, the Harrod–Domar model therefore implies an exponentially expanding economy.

We now turn to the solution of the nonhomogeneous linear difference equation. For the moment, we restrict attention to the case of a constant forcing function, that is, $b_t = b$ for all $t \in T$. The difference equation then takes the form

$$y_t + a_1 y_{t-1} = b\,.$$

If $a_1 \neq -1$, the so-called **equilibrium solution,**

$$y_t = \frac{b}{1+a_1} \qquad \text{for all } t,$$

is a particular solution to the nonhomogeneous difference equation, as can be readily verified by substitution into the equation. When $a_1 = -1$, the equation simplifies to $y_t - y_{t-1} = b$, in which case $y_t = bt$ is a particular solution to the nonhomogeneous difference equation.

In the context of linear differential equations, we have seen that the general solution to a nonhomogeneous linear differential equation can be expressed as the sum of a particular solution to the nonhomogeneous equation and the general solution of the corresponding homogeneous equation. The same principle applies to linear difference equations. If $y_t^{(1)}$ and $y_t^{(2)}$ are two different solutions to a nonhomogeneous linear difference equation, then

$$y_t^{(1)} + a_1 y_{t-1}^{(1)} = b\,,$$

$$y_t^{(2)} + a_1 y_{t-1}^{(2)} = b\,.$$

Subtracting the equations yields

$$(y_t^{(1)} - y_t^{(2)}) + a_1(y_{t-1}^{(1)} - y_{t-1}^{(2)}) = b - b = 0\,.$$

The difference between the two solutions is therefore a solution to the associated homogeneous difference equation. Every solution to a nonhomogeneous equation is thus obtained as the sum of a particular solution of the nonhomogeneous equation and a solution of the homogeneous difference equation. We summarize these results on the solution of a first-order nonhomogeneous linear difference equation in a theorem:

Theorem 13.3: First-order nonhomogeneous linear difference equation

The first-order **nonhomogeneous** linear difference equation $y_t + a_1 y_{t-1} = b$ has the general solution

$$y_t = \begin{cases} \dfrac{b}{1+a_1} + \gamma(-a_1)^t\,, & \text{if } a_1 \neq -1\,,\\ bt + \gamma\,, & \text{if } a_1 = -1\,,\end{cases}$$

where $\gamma \in \mathbb{R}$ is a real constant.

The unique solution for a given initial condition y_0 at time $t = 0$ is

$$y_t = \begin{cases} \dfrac{b}{1+a_1} + \left(y_0 - \dfrac{b}{1+a_1}\right)(-a_1)^t\,, & \text{if } a_1 \neq -1\,,\\ bt + y_0\,, & \text{if } a_1 = -1\,.\end{cases}$$

Example 13.9 (Cobweb model). The cobweb model leads to the nonhomogeneous linear difference equation

$$p_t + \frac{d}{b}p_{t-1} = \frac{a-c}{b}\,.$$

Since $b, d > 0$, we have $d/b \neq -1$. Therefore, there exists an equilibrium solution

$$p_t = \frac{\frac{a-c}{b}}{1+\frac{d}{b}} = \frac{a-c}{b+d} = p^*\,.$$

p^* is the so-called **equilibrium price**. If the price in some period equals p^*, then the price in all subsequent periods is also p^*. Thus, the system is in equilibrium.

Hence, the general solution to the cobweb model is

$$p_t = \frac{a-c}{b+d} + \gamma\left(-\frac{d}{b}\right)^t = p^* + \gamma\left(-\frac{d}{b}\right)^t\,.$$

The expression $\gamma(-d/b)^t$ describes the deviation of the price in period t from the equilibrium price. Since $-d/b < 0$, the current price oscillates above and below the equilibrium price.

What is particularly interesting is the way the price evolves over time. Therefore, we study the limit behavior of the price as t approaches infinity. To this end, we distinguish three cases:

Case 1 ($d < b$)**:** In this case, $|d/b| < 1$ and it holds that $\lim_{t\to\infty}(-d/b)^t = 0$. This gives

$$\lim_{t\to\infty} p_t = \frac{a-c}{b+d} = p^* .$$

So in this case, the price converges to the equilibrium price. There is a **damped oscillation** around the equilibrium price.

Case 2 ($d = b$)**:** Then $p_t = p^* + \gamma(-1)^t$. The price thus oscillates by γ units above and below the equilibrium price, resulting in a **constant oscillation** around the equilibrium price.

Case 3 ($d > b$)**:** Here, $|d/b| > 1$. This means that the price moves further away from the equilibrium price with each period. There is an **exploding oscillation** around the equilibrium price.

The three possible price developments are illustrated in Figure 13.1.

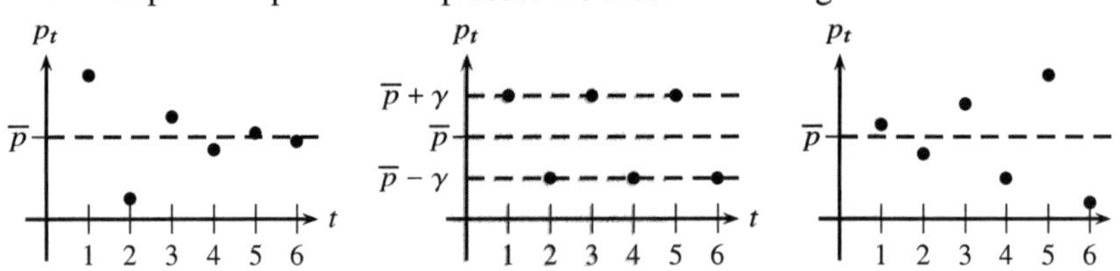

Figure 13.1: Price development in the cobweb model for $d < b$, $d = b$, $d > b$.

We will study the long-term behavior of the solutions of higher order difference equations below in Section 13.3.

13.2 Linear difference equations of order *m*

In this section, we solve linear difference equations of arbitrary order. We start with the case of **homogeneous** linear difference equations and then consider the general case of **nonhomogeneous** linear difference equations.

Structure of the solution space of the homogeneous equation

We first examine the structure of the solution set of a homogeneous linear difference equation in detail. For this, we need some results of linear algebra from Chapter 9.

The solution to a homogeneous linear difference equation

$$y_t + a_1 y_{t-1} + \ldots + a_m y_{t-m} = 0\,, \quad t \in T\,,$$

is not unique. But how do the different solutions relate to each other? The following two observations, which follow from the linearity of the difference equation (see also the corresponding observations for differential equations on page 364), provide insight into this.

Observation 1: If $y_t = f(t)$ is a solution to the homogeneous linear difference equation, then for every real number γ, $y_t = \gamma f(t)$ is also a solution to the homogeneous difference equation.

If we substitute $\gamma f(t)$ for y_t, we get

$$\gamma f(t) + a_1 \gamma f(t-1) + \ldots + a_m \gamma f(t-m) \\ = \gamma \underbrace{\big(f(t) + a_1 f(t-1) + \ldots + a_m f(t-m)\big)}_{=0} = \gamma \cdot 0 = 0 .$$

Observation 2: If $y_t = f_1(t)$ and $y_t = f_2(t)$ are two solutions to the homogeneous linear difference equation, then $y_t = f_1(t) + f_2(t)$ is also a solution to the homogeneous difference equation.

If we substitute $f_1(t) + f_2(t)$ for y_t, we get

$$\big(f_1(t) + f_2(t)\big) + a_1\big(f_1(t-1) + f_2(t-1)\big) + \ldots + a_m\big(f_1(t-m) + f_2(t-m)\big) \\ = \underbrace{\big(f_1(t) + a_1 f_1(t-1) + \ldots + a_m f_1(t-m)\big)}_{=0} \\ + \underbrace{\big(f_2(t) + a_1 f_2(t-1) + \ldots + a_m f_2(t-m)\big)}_{=0} = 0 .$$

From these two observations, the following theorem follows directly.

Theorem 13.4: Superposition principle

If $y_t = f_1(t)$ and $y_t = f_2(t)$ are two solutions to the m-th order homogeneous linear difference equation

$$y_t + a_1 y_{t-1} + \ldots + a_m y_{t-m} = 0 ,$$

then any linear combination

$$y_t = \gamma_1 f_1(t) + \gamma_2 f_2(t)$$

with arbitrary real numbers $\gamma_1, \gamma_2 \in \mathbb{R}$ is also a solution to the homogeneous linear difference equation.

From the superposition principle, it follows that conditions U1 and U2 for a subspace (see Section 9.2) are satisfied for the solution set of a homogeneous linear difference equation. Since the zero function $y_t = 0$ for all t is always a solution, the solution set is non-empty and condition U0 is satisfied. Therefore, the solution set forms a subspace of the vector space $\mathcal{F}(T, \mathbb{R})$ consisting of all functions with domain T and codomain $\mathbb{R}$. Moreover, one can show that the dimension of the solution space equals the order of the difference equation. This is summarized in the following theorem:

Theorem 13.5: Structure of the solution (homogeneous equation)

The general solution to the m-th order **homogeneous** linear difference equation,

$$y_t + a_1 y_{t-1} + a_2 y_{t-2} + \cdots + a_m y_{t-m} = 0,$$

is an **m-dimensional subspace** of $\mathcal{F}(T, \mathbb{R})$.

If $f_1, \ldots, f_m$ is a **basis of the general solution**, then the general solution is

$$y_t = \gamma_1 f_1(t) + \cdots + \gamma_m f_m(t), \quad \gamma_1, \ldots, \gamma_m \in \mathbb{R}.$$

A basis of the solution space $f_1, \ldots, f_m$ is also referred to as **fundamental system** of the difference equation.

To specify the solution to an m-th order homogeneous linear difference equation, it is therefore sufficient to provide a fundamental system, that is, m linearly independent solutions. This leads to the question of how one can determine whether a set of functions is linearly independent. A simple and effective criterion is given by the so-called **Casoratian** (or **Casorati**[2] **determinant**), which provides a test for linear independence of functions. Let $f_1, \ldots, f_m \in \mathcal{F}(T, \mathbb{R})$. Then the determinant

$$C(t) := \det \begin{bmatrix} f_1(t) & f_2(t) & \ldots & f_m(t) \\ f_1(t+1) & f_2(t+1) & \ldots & f_m(t+1) \\ \vdots & \vdots & & \vdots \\ f_1(t+m-1) & f_2(t+m-1) & \ldots & f_m(t+m-1) \end{bmatrix}$$

is called the **Casoratian** of $f_1, \ldots, f_m$. The following theorem shows how linear independence of the functions $f_1, \ldots, f_m$ can be checked using the Casoratian.

Theorem 13.6: Casoratian

If $f_1, f_2, \ldots, f_m \in \mathcal{F}(T, \mathbb{R})$, then

$$C(t) \neq 0 \text{ for } \textbf{one } t \quad \Longrightarrow \quad f_1, \ldots, f_m \text{ are linearly independent.}$$

If $f_1, f_2, \ldots, f_m \in \mathcal{F}(T, \mathbb{R})$ are **solutions** to an m-th order homogeneous linear difference equation, then

$$C(t) \neq 0 \text{ for } \textbf{all } t \quad \Longleftrightarrow \quad f_1, \ldots, f_m \text{ are linearly independent.}$$

[2]Felice Casorati (1835–1890)

The theorem states, in particular, that for solutions to a homogeneous linear difference equation the Casoratian is either non-zero for all t (if the solutions are linearly independent) or identically zero for all t (if the solutions are linearly dependent). Consequently, to determine whether m solutions are linearly independent, it suffices to compute the Casoratian for a single value of t.

Solution to the homogeneous linear difference equation

The solution to a first-order homogeneous linear difference equation $y_t + a_1 y_{t-1} = 0$ has the form $y_t = \gamma(-a_1)^t$. Consequently, for the solution to an m-th order homogeneous linear difference equation we employ the trial solution $y_t = z^t$. Substituting this into the difference equation yields

$$z^t + a_1 z^{t-1} + \ldots + a_{m-1} z^{t-m+1} + a_m z^{t-m} = 0\,.$$

For $z = 0$, the equation is clearly satisfied. However, this yields only the trivial solution $y_t = 0$, which is a solution to every homogeneous linear difference equation. For $z \neq 0$, division by z^{t-m} leads to the equation

$$z^m + a_1 z^{m-1} + \ldots + a_{m-1} z + a_m = 0\,.$$

This equation is called the **characteristic equation** of the difference equation. Accordingly, the polynomial on the left-hand side is called the **characteristic polynomial**. The solution to the difference equation is therefore reduced to the problem of finding all roots of the characteristic polynomial. For each root z_0 of the characteristic polynomial, $y_t = z_0^t$ is a solution to the difference equation.

A polynomial of degree m has at most m different real roots. If there are m different real roots, these roots provide m different solutions to the difference equation. Using the Casoratian, it can be shown that these m solutions are linearly independent, so they form a fundamental system, that is, a basis of the solution space.

Theorem 13.7: m-th order homogeneous linear difference equation

If the characteristic polynomial

$$z^m + a_1 z^{m-1} + \ldots + a_{m-1} z + a_m$$

of an m-th order homogeneous linear difference equation has exactly m different real roots $z_1, z_2, \ldots, z_m$, then the general solution is given by

$$y_t = \gamma_1 z_1^t + \gamma_2 z_2^t + \ldots + \gamma_m z_m^t\,,$$

where $\gamma_1, \gamma_2, \ldots, \gamma_m \in \mathbb{R}$ are arbitrary real constants.

Example 13.10 (Fibonacci[3] sequence). In the so-called **Fibonacci sequence**, each term is the sum of the two preceding terms, the first two elements of the sequence being zero and one:

$$0,\ 1,\ 1,\ 2,\ 3,\ 5,\ 8,\ 12,\ 21,\ 34,\ \ldots$$

This sequence was considered by Leonardo Fibonacci as a model of a rabbit population. Formally, the defining relation is given by

$$y_n = y_{n-1} + y_{n-2}\,, \quad n \geq 2; \qquad y_0 = 0;\ y_1 = 1\,.$$

Clearly, the Fibonacci sequence is a solution to a second-order homogeneous linear difference equation. Since $y_n - y_{n-1} - y_{n-2} = 0$, the characteristic polynomial is given by $z^2 - z - 1$. Its roots can be found using the quadratic formula:

$$z_{1,2} = \frac{1}{2} \pm \sqrt{\frac{1}{4} + 1} = \frac{1 \pm \sqrt{5}}{2}\,.$$

The general solution to the difference equation is thus

$$y_n = \gamma_1 \left(\frac{1+\sqrt{5}}{2}\right)^n + \gamma_2 \left(\frac{1-\sqrt{5}}{2}\right)^n .$$

Because of the initial conditions $y_0 = 0$ and $y_1 = 1$, it must hold

$$\gamma_1 \left(\frac{1+\sqrt{5}}{2}\right)^0 + \gamma_2 \left(\frac{1-\sqrt{5}}{2}\right)^0 \stackrel{!}{=} 0\,,$$
$$\gamma_1 \left(\frac{1+\sqrt{5}}{2}\right)^1 + \gamma_2 \left(\frac{1-\sqrt{5}}{2}\right)^1 \stackrel{!}{=} 1\,.$$

From the first equation, we get $\gamma_1 + \gamma_2 = 0$, i.e., $\gamma_2 = -\gamma_1$. Substituting into the second equation yields

$$\gamma_1 \frac{1+\sqrt{5}}{2} - \gamma_1 \frac{1-\sqrt{5}}{2} = \frac{1}{2}\gamma_1 (1 + \sqrt{5} - 1 + \sqrt{5}) = \gamma_1 \sqrt{5} \stackrel{!}{=} 1$$

and therefore $\gamma_1 = 1/\sqrt{5}$. The solution to the difference equation thus yields the following explicit formula for the Fibonacci numbers:

$$y_n = \frac{1}{\sqrt{5}} \left[\left(\frac{1+\sqrt{5}}{2}\right)^n - \left(\frac{1-\sqrt{5}}{2}\right)^n \right].$$

The formula can further be simplified by taking into account that all Fibonacci numbers must be integers. Since

$$\left| \frac{1}{\sqrt{5}} \left(\frac{1-\sqrt{5}}{2}\right)^n \right| < 0.5 \quad \text{for all } n \in \mathbb{N}_0,$$

[3] Leonardo Fibonacci, actually Leonardo da Pisa (ca. 1170 – ca. 1240)

the n-th Fibonacci number is obtained if the second term in the square brackets is omitted and the resulting value is rounded to the nearest integer.

Solution for multiple or complex roots

If the characteristic polynomial has m different roots $z_1, \ldots, z_m$, then according to Theorem 13.7 the functions $z_1^t, z_2^t, \ldots, z_m^t$ form a basis of the solution space. However, a polynomial of degree m may have fewer than m distinct real roots, as some roots may be non-real or occur with multiplicity.

If we extend the number system from the real numbers to the complex numbers, the fundamental theorem of algebra (Theorem E.2) states that every polynomial of degree m has exactly m roots in the complex numbers, counting each root according to its multiplicity. Thus, a polynomial may have both real and non-real complex roots. Furthermore, both real and complex roots can occur with multiplicity. We will now explain how the basis of the solution space of a homogeneous linear difference equation can be determined in the general case.

Reading this section requires basic knowledge about complex numbers. A brief introduction to complex numbers and their most important properties can be found in Appendix E.

Multiple roots As already mentioned, a root of the characteristic polynomial may have multiplicity greater than one. In this case, the root yields not just a single solution to the difference equation, but as many linearly independent solutions as its multiplicity.

Theorem 13.8: Solution for multiple roots

Let $z \in \mathbb{R}$ be a real root of the characteristic polynomial of multiplicity k. Then,

$$z^t,\ tz^t,\ t^2z^t, \ldots, t^{k-1}z^t$$

are linearly independent solutions to the difference equation.

Thus, a real root of multiplicity k provides k linearly independent solutions to the difference equation.

Example 13.11. The second-order homogeneous linear difference equation

$$y_t - 4y_{t-1} + 4y_{t-2} = 0$$

has the characteristic polynomial $z^2 - 4z + 4$. Using the quadratic formula, we obtain the roots

$$z_{1,2} = \frac{4}{2} \pm \sqrt{\frac{16}{4} - 4} = 2 \pm \sqrt{4-4} = 2\,.$$

Hence, $z_{1,2} = 2$ is a double root of the characteristic polynomial. The general solution to the difference equation is therefore

$$y_t = \gamma_1 2^t + \gamma_2 t 2^t = (\gamma_1 + \gamma_2 t)2^t\,, \quad \gamma_1, \gamma_2 \in \mathbb{R}\,.$$

Complex roots When we speak of complex roots of the characteristic polynomial, we always mean non-real complex roots $z = a + bi$, with $b \neq 0$. Regarding solutions, complex roots have the same meaning as real roots. So, if the complex number $z = a + bi$ is a root of the characteristic polynomial, then $(a + bi)^t$ is a solution to the difference equation. This extends to roots with multiplicity: if $z = a + bi$ is a root of multiplicity k, then the functions $(a + bi)^t, t(a + bi)^t, \ldots, t^{k-1}(a + bi)^t$ are solutions to the difference equation.

However, it is clear that for a linear homogeneous difference equation with real coefficients and real initial conditions, all solutions y_t must themselves be real-valued. Therefore, it may seem somewhat unsatisfactory to represent the solution using complex numbers. In the following, we will show that this is, in fact, not necessary: the general solution can always be written in real terms, even when the characteristic equation has complex roots.

We start from a homogeneous linear difference equation

$$y_t + a_1 y_{t-1} + \ldots + a_m y_{t-m} = 0$$

with real coefficients $a_1, \ldots, a_m$. If $z = a + bi$ is a complex root of the characteristic polynomial, then the complex conjugate $\overline{z} = a - bi$ is also a root of the characteristic polynomial (see Theorem E.2 in the appendix). Thus, the functions $y_t = z^t$ and $y_t = \overline{z}^t$ are solutions to the difference equation. In polar form, we have $z = re^{i\varphi}$ and $\overline{z} = re^{-i\varphi}$. Since every linear combination of two solutions is again a solution to the homogeneous difference equation,

$$\gamma_1 z^t + \gamma_2 \overline{z}^t = \gamma_1 (re^{i\varphi})^t + \gamma_2 (re^{-i\varphi})^t = \gamma_1 r^t e^{i\varphi t} + \gamma_2 r^t e^{-i\varphi t}$$

is also a solution to the difference equation. Due to Euler's formula (E.1) we have

$$\begin{aligned}
\gamma_1 z^t + \gamma_2 \overline{z}^t &= \gamma_1 r^t \left[\cos(\varphi t) + i\sin(\varphi t)\right] + \gamma_2 r^t \left[\cos(-\varphi t) + i\sin(-\varphi t)\right] \\
&= r^t \left[\gamma_1 \cos(\varphi t) + i\gamma_1 \sin(\varphi t) + \gamma_2 \cos(\varphi t) - i\gamma_2 \sin(\varphi t)\right] \\
&= r^t \left[(\gamma_1 + \gamma_2)\cos(\varphi t) + i(\gamma_1 - \gamma_2)\sin(\varphi t)\right] \\
&= (\gamma_1 + \gamma_2) r^t \cos(\varphi t) + i(\gamma_1 - \gamma_2) r^t \sin(\varphi t) \\
&= C_1\, r^t \cos(\varphi t) + C_2\, r^t \sin(\varphi t)\,,
\end{aligned}$$

where $C_1 = \gamma_1 + \gamma_2$ and $C_2 = i(\gamma_1 - \gamma_2)$. This shows that any linear combination of the solutions z^t and $\overline{z}^t$ can be written as a linear combination (with complex coefficients) of the real-valued solution functions $r^t \cos(\varphi t)$ and $r^t \sin(\varphi t)$. Conversely, any linear combination of $r^t \cos(\varphi t)$ and $r^t \sin(\varphi t)$ can be expressed as a linear combination of z^t and $\overline{z}^t$ by setting $\gamma_1 = \frac{1}{2}(C_1 - iC_2)$ and $\gamma_2 = \frac{1}{2}(C_1 + iC_2)$.

Therefore, the functions z^t and $\overline{z}^t$, as well as the real-valued functions $r^t \cos(\varphi t)$ and $r^t \sin(\varphi t)$, each are linearly independent solutions to the difference equation. We summarize this result in the following theorem.

Theorem 13.9: Solution for complex roots

If z is a complex root of the characteristic polynomial, then its complex conjugate $\bar{z}$ is also a root of the characteristic polynomial. If, in polar form, $z = re^{i\varphi}$, then

$$y_t = r^t \cos(\varphi t) \quad \text{and} \quad y_t = r^t \sin(\varphi t)$$

are linearly independent solutions to the difference equation.

We illustrate Theorem 13.9 by the following example:

Example 13.12. The homogeneous linear difference equation

$$y_t + 2y_{t-1} + 5y_{t-2} = 0$$

has the characteristic polynomial $z^2 + 2z + 5 = 0$. Using the quadratic formula, we obtain the roots

$$z_{1,2} = -1 \pm \sqrt{(-1)^2 - 5} = -1 \pm \sqrt{-4} = -1 \pm 2i\,.$$

To obtain a representation of the general solution in terms of real basis functions, the polar form of $z_1 = -1 + 2i$ must be computed. First, calculate the absolute value

$$r = |z_1| = \sqrt{(-1)^2 + 2^2} = \sqrt{5}\,.$$

The argument φ of a complex number $z = x + iy$ is obtained from

$$\varphi = \arg(z) = \begin{cases} \arccos\left(\frac{x}{r}\right), & \text{if } y \geq 0\,, \\ -\arccos\left(\frac{x}{r}\right), & \text{if } y < 0\,. \end{cases}$$

In general, we choose φ as the argument of the complex root with positive imaginary part, which lies in $[0, \pi)$. It can then simply be computed as

$$\varphi = \arccos\left(\frac{x}{r}\right).$$

Hence, the argument of $z_1 = -1 + 2i$ is

$$\varphi = \arg(z_1) = \arccos\left(\frac{-1}{\sqrt{5}}\right) = 2.0334\,.$$

The two roots $-1 \pm 2i$ yield the two linearly independent solutions

$$\left(\sqrt{5}\right)^t \cos(2.0344\,t) \quad \text{and} \quad \left(\sqrt{5}\right)^t \sin(2.0344\,t)\,.$$

Thus, the general solution to the difference equation is

$$y_t = \gamma_1 \left(\sqrt{5}\right)^t \cos(2.0344\,t) + \gamma_2 \left(\sqrt{5}\right)^t \sin(2.0344\,t)\,, \quad \gamma_1, \gamma_2 \in \mathbb{R}\,.$$

Theorem 13.9 extends naturally to multiple complex roots:

Theorem 13.10: Solution for multiple complex roots

If z is a complex root of the characteristic polynomial of multiplicity k, then its complex conjugate $\overline{z}$ is also a root of multiplicity k. If, in polar form, $z = re^{i\varphi}$, then

$$r^t \cos(\varphi t)\,,\ tr^t \cos(\varphi t)\,,\ t^2 r^t \cos(\varphi t)\,,\dots, t^{k-1} r^t \cos(\varphi t)\,,$$
$$r^t \sin(\varphi t)\,,\ tr^t \sin(\varphi t)\,,\ t^2 r^t \sin(\varphi t)\,,\dots, t^{k-1} r^t \sin(\varphi t)$$

are linearly independent solutions to the difference equation.

Since a polynomial of degree m always has exactly m complex roots, counted with their multiplicity, Theorems 13.8 and 13.10 guarantee that we can find m linearly independent solutions to the difference equation, and thus a fundamental system. We illustrate this with an example.

Example 13.13. Consider the fourth-order homogeneous linear difference equation

$$y_t - 5y_{t-1} + 8.5y_{t-2} - 6y_{t-3} + 2y_{t-4} = 0\,.$$

The characteristic polynomial is

$$z^4 - 5z^3 + 8.5z^2 - 6z + 2\,.$$

By applying the root-finding methods from Section 4.10, or by making an educated guess one finds the solution $z_1 = 2$. Polynomial division gives

$$(z^4 - 5z^3 + 8.5z^2 - 6z + 2) : (z - 2) = z^3 - 3z^2 + 2.5z - 1\,.$$

The cubic polynomial obtained here again has a root at $z_2 = 2$. Dividing out the linear factor $(z - 2)$ yields the quadratic polynomial

$$(z^3 - 3z^2 + 2.5z - 1) : (z - 2) = z^2 - z + 0.5\,.$$

The roots of this polynomial are found with the quadratic formula:

$$z_{3,4} = \frac{1}{2} \pm \sqrt{\frac{1}{4} - \frac{1}{2}} = \frac{1}{2} \pm \sqrt{-\frac{1}{4}} = \frac{1}{2} \pm i\frac{1}{2}$$

We thus have the following roots of the characteristic polynomial,

$$z_1 = 2\,,\quad z_2 = 2\,,\quad z_3 = \frac{1}{2} + i\frac{1}{2}\,,\quad z_4 = \frac{1}{2} - i\frac{1}{2}\,.$$

$z_{1,2} = 2$ is a double root and provides the solutions 2^t and $t2^t$. For the pair of complex conjugate roots z_3 and z_4, we determine the polar form. We first calculate the absolute value

$$r = \sqrt{\left(\frac{1}{2}\right)^2 + \left(\frac{1}{2}\right)^2} = \sqrt{\frac{1}{2}} = \frac{1}{\sqrt{2}}\,.$$

Then, the argument of $\frac{1}{2} + i\frac{1}{2}$ is

$$\varphi = \arccos\left(\frac{\frac{1}{2}}{\frac{1}{\sqrt{2}}}\right) = \arccos\left(\frac{1}{\sqrt{2}}\right) = \frac{\pi}{4}.$$

The complex conjugate roots $z_{3,4} = \frac{1}{2} \pm i\frac{1}{2}$ thus provide the two solutions

$$\frac{1}{\sqrt{2}^t}\cos\left(\frac{\pi}{4}t\right) \quad \text{and} \quad \frac{1}{\sqrt{2}^t}\sin\left(\frac{\pi}{4}t\right).$$

Therefore, the general solution to the difference equation is

$$y_t = \gamma_1 2^t + \gamma_2 t 2^t + \gamma_3 \frac{1}{\sqrt{2}^t}\cos\left(\frac{\pi}{4}t\right) + \gamma_4 \frac{1}{\sqrt{2}^t}\sin\left(\frac{\pi}{4}t\right), \quad \gamma_1, \ldots, \gamma_4 \in \mathbb{R}.$$

If we set $\gamma_1, \gamma_2 = 0$ and additionally require the initial conditions $y_0 = y_1 = 1$, the solution

$$y_t = \frac{1}{\sqrt{2}^t}\cos\left(\frac{\pi}{4}t\right) + \frac{1}{\sqrt{2}^t}\sin\left(\frac{\pi}{4}t\right)$$

is obtained, which is shown in Figure 13.2. The solution to the given initial conditions is a damped oscillation, for t towards infinity, it converges to zero.

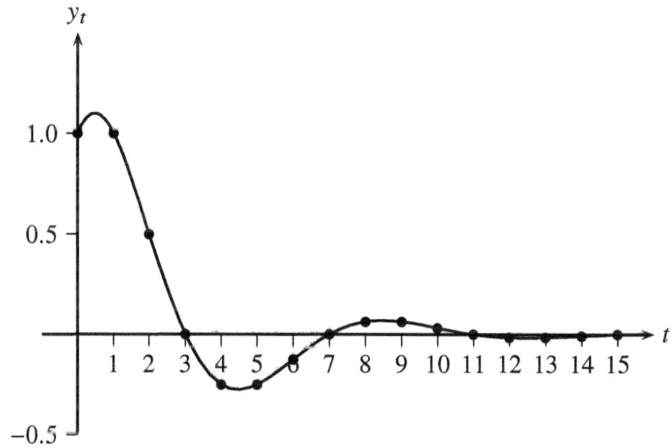

Figure 13.2: A solution to the difference equation from Example 13.13.

On the other hand, if we set $\gamma_3, \gamma_4 = 0$ and impose the initial conditions $y_0 = 1$ and $y_1 = 0$, the solution is

$$y_t = 2^t - t2^t = (1 - t)2^t.$$

This solution does not converge; it rather diverges to $-\infty$. Thus, depending on the initial conditions, qualitatively different solutions arise.

We conclude this section with a short discussion of the solutions of a **second-**

order homogeneous linear difference equation, $y_t + a_1 y_{t-1} + a_2 y_{t-2} = 0$. The characteristic polynomial is $z^2 + a_1 z + a_2 = 0$. Its roots are

$$z_{1,2} = -\frac{a_1}{2} \pm \sqrt{\frac{a_1^2}{4} - a_2}\,.$$

The following cases can occur:

Two real roots z_1 and z_2, if $\frac{a_1^2}{4} > a_2$. In this case, the general solution is

$$\boxed{y_t = \gamma_1 z_1^t + \gamma_2 z_2^t\,.}$$

A double root $z_1 = z_2$, if $\frac{a_1^2}{4} = a_2$. Then, the general solution is

$$\boxed{y_t = (\gamma_1 + \gamma_2 t) z_1^t\,.}$$

Two complex roots $z_{1,2}$, if $\frac{a_1^2}{4} < a_2$. With $z_{1,2} = \alpha \pm \beta i$, we compute

$$r = |z_{1,2}| = \sqrt{\alpha^2 + \beta^2} \quad \text{and} \quad \varphi = \arccos\left(\frac{\alpha}{r}\right),$$

and obtain the general solution

$$\boxed{y_t = \gamma_1 r^t \cos(\varphi t) + \gamma_2 r^t \sin(\varphi t)\,.}$$

Solution to the nonhomogeneous linear difference equation

We begin this section by studying the structure of the solution set of a nonhomogeneous linear difference equation. The following two observations complement Observations 1 and 2 from page 417. Compare also the corresponding explanations for differential equations on page 372.

Observation 3: If $y_t = f_h(t)$ is a solution to the **homogeneous** difference equation and $y_t = f_p(t)$ is a solution to the **nonhomogeneous** difference equation, then $y_t = f_h(t) + f_p(t)$ is a solution to the **nonhomogeneous** difference equation.

If we substitute $y_t = f_h(t) + f_p(t)$ into the difference equation, we obtain

$$\begin{aligned} &\big(f_h(t) + f_p\big) + a_1\big(f_h(t-1) + f_p(t-1)\big) + \ldots + a_m\big(f_h(t-m) + f_p(t-m)\big) \\ &\quad = \underbrace{\big(f_h(t) + a_1 f_h(t-1) + \ldots + a_m f_h(t-m)\big)}_{=0} \end{aligned}$$

$$+ \underbrace{\big(f_p(t) + a_1 f_p(t-1) + \ldots + a_m f_p(t-m)\big)}_{= b_t} = b_t \,.$$

Observation 4: If $y_t = f_1(t)$ and $y_t = f_2(t)$ are two solutions to the **nonhomogeneous** difference equation, then $y_t = f_1(t) - f_2(t)$ is a solution to the **homogeneous** difference equation.

If we substitute $y_t = f_1(t) - f_2(t)$ into the difference equation, we obtain

$$\begin{aligned}
&\big(f_1(t) - f_2(t)\big) + a_1\big(f_1(t-1) - f_2(t-1)\big) + \ldots + a_m\big(f_1(t-m) - f_2(t-m)\big) \\
&\quad = \underbrace{\big(f_1(t) + a_1 f_1(t-1) + \ldots + a_m f_1(t-m)\big)}_{= b_t} \\
&\qquad - \underbrace{\big(f_2(t) + a_1 f_2(t-1) + \ldots + a_m f_2(t-m)\big)}_{= b_t} = 0 \,.
\end{aligned}$$

From these observations, it immediately follows that the general solution to the nonhomogeneous difference equation can be written as the sum of **a particular solution** to the nonhomogeneous equation and the **general solution** to the corresponding homogeneous difference equation.

Theorem 13.11: Structure of the solution (nonhomogeneous equation)

Let $y_t = f_p(t)$ be a particular solution to the m-th order **nonhomogeneous** linear difference equation,

$$y_t + a_1 y_{t-1} + a_2 y_{t-2} + \cdots + a_m y_{t-m} = b_t \,,$$

and $f_1, \ldots, f_m$ be a basis of the general solution to the homogeneous equation.

Then the **general solution to the nonhomogeneous equation** is

$$y_t = f_p(t) + \gamma_1 f_1(t) + \cdots + \gamma_m f_m(t) \,, \quad \gamma_1, \ldots, \gamma_m \in \mathbb{R} \,.$$

The theorem states that to obtain the general solution to the nonhomogeneous difference equation, one only needs the general solution to the homogeneous equation and a **single particular solution** to the nonhomogeneous equation.

Solution under constant forcing

We now turn to the solution of the m-th order nonhomogeneous linear difference equation. To begin, we restrict ourselves to the case where the forcing function is constant, that is, $b_t = b$ for all $t \in T$. In this case, the difference equation takes the form

$$y_t + a_1 y_{t-1} + \ldots + a_m y_{t-m} = b \,.$$

Assuming a constant particular solution $y_t = c$ leads to the equation

$$c + a_1 c + \ldots + a_m c = c(1 + a_1 + \ldots + a_m) = b\,.$$

If $1 + a_1 + \ldots + a_m \neq 0$, the solution

$$y_t = \frac{b}{1 + a_1 + \ldots + a_m} = y^* \qquad \text{for all } t$$

is called the **equilibrium solution**. Thus, according to Theorem 13.11, the following theorem holds.

Theorem 13.12: Solution under constant forcing

If $1 + a_1 + \ldots + a_m \neq 0$, then the nonhomogeneous linear difference equation

$$y_t + a_1 y_{t-1} + \ldots + a_m y_{t-m} = b$$

has the general solution

$$y_t = y_t^h + \frac{b}{1 + a_1 + \ldots + a_m} = y_t^h + y^*\,,$$

where y_t^h denotes the general solution to the corresponding homogeneous linear difference equation.

If $1 + a_1 + \ldots + a_m = 0$, then $z = 1$ is a root of the characteristic polynomial, and $y_t = 1$ (and thus also $y_t = c$) is a solution to the homogeneous equation. Hence, $y_t = c$ cannot be a solution to the nonhomogeneous equation. In this case, we say that there is **resonance**. Instead of $y_t = c$, we then use the trial solution $y_t = ct$ and attempt to determine c appropriately. If this also fails, we continue with the trial solutions $y_t = ct^2$, $y_t = ct^3$, and so on. In general, if $z = 1$ is a **root of multiplicity** k, then the trial solution $y_t = ct^k$ will succeed.

Example 13.14. Consider the difference equation $y_t - 3y_{t-1} + 2y_{t-2} = 5$. The trial solution $y_t = c$ leads to $c - 3c + 2c = 5$ and thus to the contradiction $0 = 5$. Therefore, there is **resonance**. Hence, we multiply the trial solution by t and obtain $y_t = ct$. This new trial solution leads to

$$ct - 3c(t-1) + 2c(t-2) = 5 \quad \Longleftrightarrow \quad -c = 5 \quad \Longleftrightarrow \quad c = -5\,.$$

Thus, $y_t = -5t$ is a particular solution to the nonhomogeneous equation.

Method of undetermined coefficients

While the results of Theorems 13.7 to 13.10 completely solve the problem of finding the general solution to a homogeneous linear difference equation, there is no comparable theory for the solution to a nonhomogeneous linear difference equation. For the case of constant forcing, we have already presented the solution to the nonhomogeneous difference equation in Theorem 13.12. In this section, we introduce an approach that allows to find a solution to the nonhomogeneous difference equation for a large class of forcing functions. This approach is based on the observation that, for certain forcing functions, the difference equation admits solutions of the same form as the forcing function itself. In the so-called **method of undetermined coefficients**, a trial solution is used that is of the same form as the forcing function. If the forcing function is, for example, $b_t = u^t$, we use the trial solution $y_t = cu^t$ and substitute it into the difference equation. This yields an equation from which the undetermined coefficient c can be found.

The following table gives an overview of forcing functions and trial solutions to be used:

forcing function	trial solution
u^t	cu^t
$\sin(\varphi t)$ or $\cos(\varphi t)$	$c_1 \sin(\varphi t) + c_2 \cos(\varphi t)$
$p_0 + p_1 t + \cdots + p_n t^n$	$c_0 + c_1 t + \cdots + c_n t^n$
$(p_0 + p_1 t + \cdots + p_n t^n)u^t$	$(c_0 + c_1 t + \cdots + c_n t^n)u^t$
$u^t \sin(\varphi t)$ or $u^t \cos(\varphi t)$	$u^t(c_1 \sin(\varphi t) + c_2 \cos(\varphi t))$

At first glance, this procedure may appear to be mere trial and error. However, the method of undetermined coefficients provides a systematic way to determine a solution to the nonhomogeneous difference equation for a broad class of forcing functions, including polynomials, exponentials, sines, cosines, and linear combinations and products of these functions.

Theorem 13.13: Method of undetermined coefficients

Consider the nonhomogeneous linear difference equation

$$y_t + a_1 y_{t-1} + \ldots + a_m y_{t-m} = b_t \,.$$

If the forcing function has the form

$$b_t = p_1(t)u^t \sin(\varphi t) + p_2(t)u^t \cos(\varphi t) \,,$$

where $u, \varphi \in \mathbb{R}$ and p_1, p_2 are polynomials of degree n_1 and n_2, respectively, and if $z = ue^{i\varphi}$ is **not a root of the characteristic polynomial**, then the nonhomogeneous difference equation has a particular solution of the form

$$y_t = q_1(t)u^t \sin(\varphi t) + q_2(t)u^t \cos(\varphi t) \,,$$

where q_1 and q_2 are polynomials of degree $n = \max\{n_1, n_2\}$.

Remark: If $ue^{i\varphi}$ is a root of the characteristic polynomial, we say that there is

resonance. How to proceed in this case will be discussed in Theorem 13.14.

The forcing functions and trial solutions of the table on page 429 are special cases of the general forcing function in Theorem 13.13. For example, setting $\varphi = 0$ yields the forcing function $b_t = p_2(t)u^t$. If additionally $u = 1$, the polynomial $b_t = p_2(t)$ is obtained. With polynomials p_1 and p_2 of degree zero, one gets the forcing function $b_t = c_1 u^t \sin(\varphi t) + c_2 u^t \cos(\varphi t)$.

We illustrate the method of undetermined coefficients with an example.

Example 13.15. Consider the nonhomogeneous linear difference equation

$$y_t - 3y_{t-1} + 2y_{t-2} = 3^t \,.$$

Substituting the trial solution $y_t = c3^t$ into the difference equation gives

$$c3^t - 3c3^{t-1} + 2c3^{t-2} = 3^t \,.$$

Division by 3^{t-2} yields

$$9c - 9c + 2c = 9 \quad \Longleftrightarrow \quad 2c = 9 \quad \Longleftrightarrow \quad c = \frac{9}{2} \,.$$

Thus, $y_t = 4.5 \cdot 3^t$ is a particular solution to the nonhomogeneous difference equation.

The forcing functions considered in Theorem 13.13 may also occur as solutions to homogeneous linear difference equations. If the the forcing function of a nonhomogeneous linear difference equation is itself a solution to the corresponding homogeneous equation, then the method of undetermined coefficients will fail.

Example 13.16. Consider the nonhomogeneous linear difference equation

$$y_t - 3y_{t-1} + 2y_{t-2} = 2^t \,.$$

Substituting the trial solution $y_t = c2^t$ into the difference equation gives

$$c2^t - 3c2^{t-1} + 2c2^{t-2} = 2^t \quad \Longleftrightarrow \quad 2^{t-2}(4c - 6c + 2c) = 2^t \quad \Longleftrightarrow \quad 0 = 2^t \,,$$

thus a contradiction. The reason for this is that $y_t = c2^t$ is a solution to the corresponding homogeneous equation $y_t - 3y_{t-1} + 2y_{t-2} = 0$, since $z = 2$ is a root of the characteristic polynomial.

The method of undetermined coefficients fails when the forcing function is already a solution to the homogeneous difference equation – a situation referred to as **resonance**. In cases of resonance, one multiplies the trial solution by t to obtain a new trial solution and attempts to determine its coefficients. If this approach also fails (e.g., the resonance continues because the root has higher multiplicity), the trial solution is multiplied again by t, and the process is repeated. For second-order difference equations, multiplying by t or at most t^2 suffices to resolve resonance; for higher-order equations, this procedure may need to be continued up to the multiplicity of the root.

Example 13.17 (Continuation of Example 13.16). Consider again the difference equation $y_t - 3y_{t-1} + 2y_{t-2} = 2^t$. We have seen that in Example 13.16 there is resonance. Therefore, we multiply the trial solution $y_t = c2^t$ by t to obtain the new trial solution $y_t = ct2^t$. Substituting this into the difference equation yields

$$ct2^t - 3c(t-1)2^{t-1} + 2c(t-2)2^{t-2} = 2^t .$$

Division by 2^{t-2} then gives

$$4ct - 6c(t-1) + 2c(t-2) = 4 \iff 2c = 4 \iff c = 2 .$$

Thus, $y_t = 2t2^t = t2^{t+1}$ is a particular solution to the nonhomogeneous difference equation.

The following theorem specifies the procedure in the case of resonance.

Theorem 13.14: Method of undetermined coefficients (resonance case)

Consider the nonhomogeneous linear difference equation

$$y_t + a_1 y_{t-1} + \ldots + a_m y_{t-m} = b_t .$$

If the forcing function has the form

$$b_t = p_1(t)u^t \sin(\varphi t) + p_2(t)u^t \cos(\varphi t) ,$$

where $u, \varphi \in \mathbb{R}$ and p_1, p_2 are polynomials of degree n_1 and n_2, respectively, and if $z = ue^{i\varphi}$ is a **root of the characteristic polynomial of multiplicity** k, then the nonhomogeneous difference equation has a particular solution of the form

$$y_t = t^k \left[q_1(t)u^t \sin(\varphi t) + q_2(t)u^t \cos(\varphi t) \right] ,$$

where q_1 and q_2 are polynomials of degree $n = \max\{n_1, n_2\}$.

Remark: The forcing functions and trial solutions of the table on page 429 are special cases of the general forcing function in Theorem 13.14. Accordingly, the trial solutions given there are also to be multiplied by t^k.

Example 13.18. Consider the nonhomogeneous linear difference equation

$$y_t - 3y_{t-1} + 2y_{t-2} = t .$$

Since the right-hand side is a first-degree polynomial, we use a general first-degree polynomial $y_t = a + bt$ as a trial solution. Substituting it into the difference equation leads to

$$(a + bt) - 3(a + b(t-1)) + 2(a + b(t-2)) = t \iff -b = t$$

and thus a contradiction, since the equation must hold for all t. This shows that there is **resonance**. Therefore, we multiply the trial solution by t to obtain the new trial solution $y_t = at + bt^2$. Substituting this into the difference equation gives:

$$(at + bt^2) - 3(a(t-1) + b(t-1)^2) + 2(a(t-2) + b(t-2)^2) = t$$
$$\iff \quad -2bt - a + 5b = t$$

On both sides of the equation, there is a linear polynomial. Since two polynomials are equal if and only if their coefficients are equal, we obtain the following two equations

$$-2b = 1 \quad \text{and} \quad -a + 5b = 0\,.$$

From these equations follows $b = -1/2$ and $a = -5/2$. Thus, $y_t = -2.5t - 0.5t^2$ is a particular solution to the nonhomogeneous difference equation.

If $z = ue^{i\varphi}$ is known to be a root of the characteristic polynomial with multiplicity k, one can immediately multiply the trial solution by t^k and thereby avoid a trial-and-error approach. If the forcing function is a polynomial, $b_t = p(t)$, then resonance occurs when $z = 1$ is a root of the characteristic polynomial. If the forcing function is of the form $b_t = u^t$ or $b_t = p(t)u^t$, one must check whether $z = u$ is a root. If the forcing function contains trigonometric functions $\sin(\varphi t)$ or $\cos(\varphi t)$, the complex root $ue^{i\varphi}$ must be examined.

The following table shows how the functions in the table on page 429 are obtained as special cases of the general form and the value of the critical root.

Disturbance function	obtained for	critical root
$p(t)$	$u = 1, \varphi = 0$	$z = 1$
$u^t, p(t)u^t$	$u \neq 1, \varphi = 0$	$z = u$
$\sin(\varphi t), p(t)\sin(\varphi t)$ $\cos(\varphi t), p(t)\cos(\varphi t)$	$u = 1, \varphi \neq 0$	$z = e^{i\varphi}$
$u^t\sin(\varphi t), p(t)u^t\sin(\varphi t)$ $u^t\cos(\varphi t), p(t)u^t\cos(\varphi t)$	$u \neq 1, \varphi \neq 0$	$z = ue^{i\varphi}$

Remark: If the forcing function b_t is a sum of functions of suitable type, one can make a separate approach for each summand and simply add the obtained solutions. We illustrate this again with an example.

Example 13.19 (Continuation of Examples 13.15 and 13.18). Consider the nonhomogeneous difference equation

$$y_t - 3y_{t-1} + 2y_{t-2} = 3^t + t\,.$$

The difference equation $y_t - 3y_{t-1} + 2y_{t-2} = 3^t$ has as a particular solution $y_{t,1} = 4.5 \cdot 3^t$ (see Example 13.15).

Accordingly, the difference equation $y_t - 3y_{t-1} + 2y_{t-2} = t$ has a particular solution $y_{t,2} = -2.5t - 0.5t^2$ (see Example 13.18).

Thus,

$$y_t = y_{t,1} + y_{t,2} = 4.5 \cdot 3^t - 2.5t - 0.5t^2$$

is a particular solution to the difference equation $y_t - 3y_{t-1} + 2y_{t-2} = 3^t + t$.

13.3 Stability of difference equations

In many cases, the exact solution to a difference equation is not required. Rather, the main interest lies in the limiting behavior of the solution when t becomes large. In particular, one wants to know whether the solution to a difference equation converges to an equilibrium solution as t increases. Example 13.13 has illustrated that the solution of a difference equation may converge for certain initial conditions while for others it diverges. If *every* solution to a homogeneous linear difference equation converges to zero, the difference equation is referred to as stable. More generally, one defines:

Definition 13.2: Stability

A linear difference equation is called (asymptotically) **stable** if any two of its solutions, $y_t = f_1(t)$ and $y_t = f_2(t)$, satisfy

$$\lim_{t\to\infty} \left(y_t^{(1)} - y_t^{(2)} \right) = 0 .$$

If the difference equation has a constant equilibrium solution y^*, then the above definition is equivalent to the fact that *every* solution to the difference equation converges to the equilibrium solution y^* as t tends to infinity,

$$\lim_{t\to\infty} y_t = y^* .$$

Since the difference between two solutions to a nonhomogeneous difference equation is a solution to the corresponding homogeneous difference equation, a nonhomogeneous difference equation is stable if and only if every solution y_t to the corresponding homogeneous difference equation vanishes as t goes to infinity,

$$\lim_{t\to\infty} y_t = 0 .$$

Since every solution to a homogeneous linear difference equation is a linear combination of basis functions, each basic solution must in particular have limit zero as t tends to infinity.

If z is a root (real or complex) of the characteristic polynomial of multiplicity k, then $z^t, tz^t, \dots, t^{k-1}z^t$ are linearly independent solutions. It holds:

$$\lim_{t\to\infty} z^t = 0 \quad \Longleftrightarrow \quad \lim_{t\to\infty} |z^t| = \lim_{t\to\infty} |z|^t = 0 \quad \Longleftrightarrow \quad |z| < 1 .$$

Further, if $|z| < 1$, then also

$$\lim_{t\to\infty} t^l z^t = 0 \qquad \text{for all } l \in \mathbb{N},$$

as can be shown by repeated application of l'Hôpital's rule.

From this, it follows that a linear difference equation is stable if and only if all roots of the characteristic polynomial have absolute value less than one. We formulate this result as a theorem.

Theorem 13.15: Stability of a linear difference equation

A linear difference equation of order m is stable if and only if every root z (real or complex) of the characteristic polynomial satisfies

$$|z| < 1 .$$

This stability condition is commonly stated as: *All roots of the characteristic polynomial lie inside the unit circle.*

Of course, it would be convenient to decide directly from the coefficients of a given difference equation whether it is stable or not, rather than first determining the roots of the characteristic polynomial. For second-order difference equations, such conditions on the coefficients can be formulated relatively easily. We therefore want to take a closer look at second-order linear difference equations (see the parallel discussion of second-order linear differential equations on page 381).

The homogeneous equation has the form $y_t + a_1 y_{t-1} + a_2 y_{t-2} = 0$. The characteristic polynomial is therefore $z^2 + a_1 z + a_2$ and has the roots

$$z_{1,2} = -\frac{a_1}{2} \pm \sqrt{\frac{a_1^2}{4} - a_2} .$$

We now distinguish two cases:

Case 1 ($a_2 \le a_1^2/4$): Then the characteristic polynomial has two real roots, counting with multiplicity, namely

$$z_1 = -\frac{a_1}{2} + \sqrt{\frac{a_1^2}{4} - a_2} \quad \text{and} \quad z_2 = -\frac{a_1}{2} - \sqrt{\frac{a_1^2}{4} - a_2} .$$

The difference equation is stable if and only if $|z_1| < 1$ and $|z_2| < 1$, that is, if

$$-1 < -\frac{a_1}{2} + \sqrt{\frac{a_1^2}{4} - a_2} < 1 \quad \text{and} \quad -1 < -\frac{a_1}{2} - \sqrt{\frac{a_1^2}{4} - a_2} < 1$$

holds. Multiplying the inequalities by two and adding a_1 on both sides gives

$$-2 + a_1 < \sqrt{a_1^2 - 4a_2} < 2 + a_1 \quad \text{and} \quad -2 + a_1 < -\sqrt{a_1^2 - 4a_2} < 2 + a_1 .$$

Since $z_1 + z_2 = -a_1$, it follows from $|z_1|, |z_2| < 1$ that $|a_1| < 2$. This clearly implies $-2+a_1 < 0$ and $2+a_1 > 0$ so that the first and fourth inequalities hold automatically. The remaining two inequalities are equivalent to

$$\sqrt{a_1^2 - 4a_2} < 2 + a_1 \quad \text{and} \quad \sqrt{a_1^2 - 4a_2} < 2 - a_1 \,.$$

By squaring the first of them we get

$$a_1^2 - 4a_2 < (2 + a_1)^2 = 4 + 4a_1 + a_1^2 \quad \Longleftrightarrow \quad 1 + a_1 + a_2 > 0\,.$$

Accordingly, squaring the second inequality yields

$$a_1^2 - 4a_2 < (2 - a_1)^2 = 4 - 4a_1 + a_1^2 \quad \Longleftrightarrow \quad 1 - a_1 + a_2 > 0\,.$$

Further, if $|a_1| < 2$, then $a_2 \leq a_1^2/4 < 1$. Therefore, in Case 1, stability of the difference equation implies that $1 + a_1 + a_2 > 0$, $1 - a_1 + a_2 > 0$ and $a_2 < 1$ hold. Conversely, it is straightforward to verify that these three conditions are also sufficient in Case 1.

In Figure 13.3, the areas where the characteristic polynomial has real roots are marked in light gray and white. In the white area, the difference equation is also stable.

Case 2 ($a_2 > a_1^2/4$): Then the characteristic polynomial has two complex roots, namely

$$z_1 = -\frac{a_1}{2} + i\sqrt{a_2 - \frac{a_1^2}{4}} = re^{i\varphi} \quad \text{and} \quad z_2 = -\frac{a_1}{2} - i\sqrt{a_2 - \frac{a_1^2}{4}} = re^{-i\varphi}\,.$$

The absolute value of the roots is

$$|z_{1,2}| = r = \sqrt{\left(-\frac{a_1}{2}\right)^2 + \left(a_2 - \frac{a_1^2}{4}\right)} = \sqrt{\frac{a_1^2}{4} + a_2 - \frac{a_1^2}{4}} = \sqrt{a_2}\,.$$

In Case 2, The difference equation is therefore stable if and only if $a_2 < 1$ holds.

In Figure 13.3, the areas where the characteristic polynomial has complex roots are marked in medium and dark gray. In the dark gray area, the difference equation is also stable.

We formulate the above result about the stability of a second-order linear difference equation as a theorem:

Theorem 13.16: Stability of a second-order difference equation

A second-order linear difference equation,

$$y_t + a_1 y_{t-1} + a_2 y_{t-2} = b_t\,,$$

is stable if and only if the following conditions are met:

$$1 + a_1 + a_2 > 0, \quad 1 - a_1 + a_2 > 0, \quad 1 - a_2 > 0.$$

In Figure 13.3, the set of coefficient pairs (a_1, a_2) that yield a stable difference equation forms a triangle in the a_1-a_2 plane. This region is marked in white and dark gray.

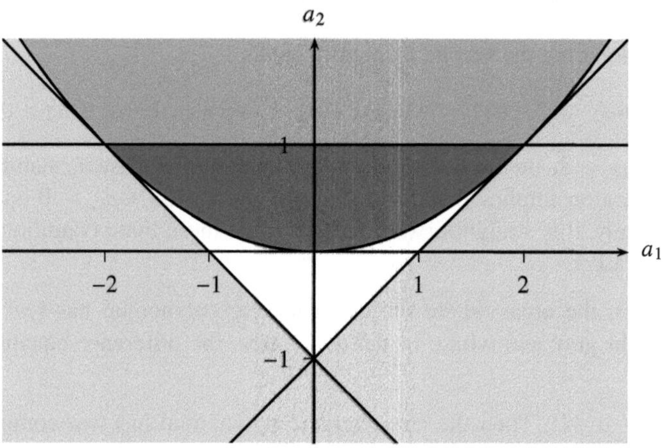

Figure 13.3: Stability and type of solution in the second-order linear difference equation.

For higher order difference equations, stability criteria like that in Theorem 13.16 cannot be formulated as simply. However, there are necessary and sufficient criteria for the stability of a linear difference equation:

Theorem 13.17: Stability criteria

Consider the linear difference equation

$$y_t + a_1 y_{t-1} + \cdots + a_m y_{t-m} = b_t .$$

Then the following statements hold:

Sufficient stability criterion (Sato[4]): If $a_1, a_2, \ldots, a_n$ are all positive and

$$1 > a_1 > a_2 > \cdots > a_{n-1} > a_n ,$$

then the difference equation is stable.

Sufficient stability criterion (Smithies[5]): If

$$1 - |a_1| - |a_2| - \cdots - |a_{n-1}| - |a_n| > 0 ,$$

[4] Ryuzo Sato (1931–)
[5] Arthur Smithies (1907–1981)

then the difference equation is stable.

Necessary stability criterion (Smithies): If the difference equation is stable, then

$$1 + a_1 + a_2 + \cdots + a_{n-1} + a_n > 0 .$$

At the end of this section, we will revisit Samuelson's multiplier-accelerator model. In particular, we will examine the conditions under which the model is stable. We will also determine when the characteristic polynomial has complex roots, leading to oscillatory behavior in the solution.

Example 13.20 (Multiplier-accelerator model). The multiplier-accelerator model leads to the second-order nonhomogeneous difference equation,

$$Y_t - c(1+k)Y_{t-1} + ckY_{t-2} = G ,$$

where $0 < c < 1$ is the consumption rate and $k > 0$ is the accelerator.

We first examine the conditions for stability that are formulated in Theorem 13.16. With $a_1 = -c(1+k)$ and $a_2 = ck$ we get

$$1 + a_1 + a_2 = 1 - c(1+k) + ck = 1 - c > 0, \quad \text{since } 0 < c < 1 .$$

Also, for the second condition

$$1 - a_1 + a_2 = 1 + c(1+k) + ck = 1 + c + 2ck > 0, \quad \text{since } 0 < c < 1,\ k > 0 .$$

The third condition is satisfied if and only if

$$1 - a_2 = 1 - ck > 0 \iff ck < 1 \iff c < \frac{1}{k} .$$

The first two conditions are always satisfied due to the assumptions on the parameters c and k, the third condition if and only if $c < 1/k$. The multiplier-accelerator model is therefore stable if and only if the consumption rate is less than the reciprocal of the accelerator.

We now consider under which conditions the characteristic polynomial has real resp. complex roots. According to the above considerations, the difference equation has two real solutions if $a_2 < a_1^2/4$. For the multiplier-accelerator model this means

$$ck < \frac{c^2(1+k)^2}{4} \iff k < \frac{c(1+k)^2}{4} \iff c > \frac{4k}{(1+k)^2} .$$

In Figure 13.4, the region in the k-c plane where exponential functions form the solution (two real roots) is marked white and light gray. In the medium and dark gray regions, however, the solutions are sine or cosine oscillations (two complex roots). The white and dark gray regions indicate the parameter values (k, c) for which the model is stable.

If, for example, we assume that $c = 0.75$ and $k = 1.2$, the difference equation is stable because $c < 1/k = 0.8333$. Furthermore, the characteristic polynomial has two complex solutions since $c < 4k/(1+k)^2 = 0.9917$. In this case, the solution to the multiplier-accelerator model is described by a damped oscillation.

Specifically, for the parameter values $c = 0.75$, $k = 1.2$ and $G = 25$, we obtain the difference equation

$$Y_t - 1.65Y_{t-1} + 0.9Y_{t-2} = 25\,.$$

The characteristic polynomial is

$$z^2 - 1.65z + 0.9 = 0\,.$$

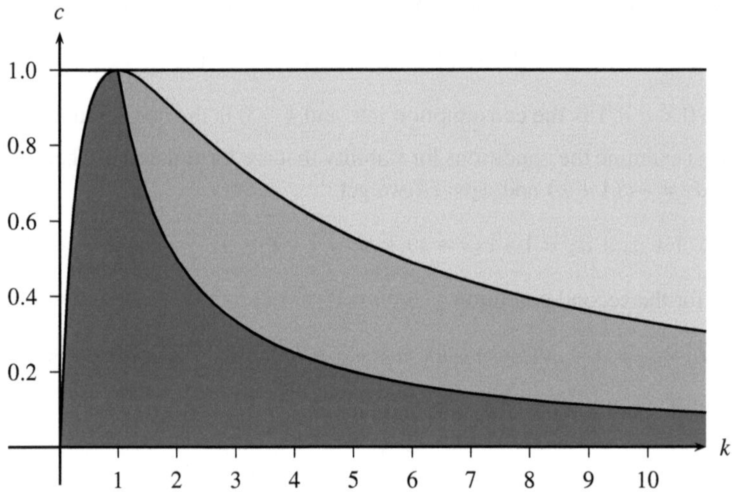

Figure 13.4: Stability and type of solution in the multiplier-accelerator model.

Using the quadratic formula, we obtain the two complex roots

$$z_{1,2} = 0.825 \pm \sqrt{0.825^2 - 0.9} = 0.825 \pm \sqrt{-0.219375} = 0.825 \pm 0.4684i\,.$$

To find the polar form, we first calculate the absolute value

$$r = |z_{1,2}| = \sqrt{0.825^2 + 0.4684^2} = \sqrt{0.9}\,.$$

Then, the argument φ is obtained from

$$\varphi = \arccos\left(\frac{0.825}{\sqrt{0.9}}\right) = 0.5164\,.$$

The general solution to the homogeneous difference equation is therefore

$$Y_t = \gamma_1\left(\sqrt{0.9}\right)^t \cos(0.5164t) + \gamma_2\left(\sqrt{0.9}\right)^t \sin(0.5164t) \qquad \gamma_1, \gamma_2 \in \mathbb{R}\,.$$

For the nonhomogeneous difference equation, the equilibrium solution is

$$Y^* = \frac{25}{1 - 1.65 + 0.9} = 100\,.$$

The general solution to the nonhomogeneous difference equation is therefore

$$Y_t = 100 + \left(\sqrt{0.9}\right)^t \left(\gamma_1 \cos(0.5164t) + \gamma_2 \sin(0.5164t)\right), \quad \gamma_1, \gamma_2 \in \mathbb{R}\,.$$

For given initial conditions $Y_0 = 102$ and $Y_1 = 101.1815$, we obtain the solution

$$Y_t = 100 + \left(\sqrt{0.9}\right)^t \left(2\cos(0.5164t) - \sin(0.5164t)\right),$$

which is shown in Figure 13.5. This displays a damped oscillation around the equilibrium solution $y^* = 100$.

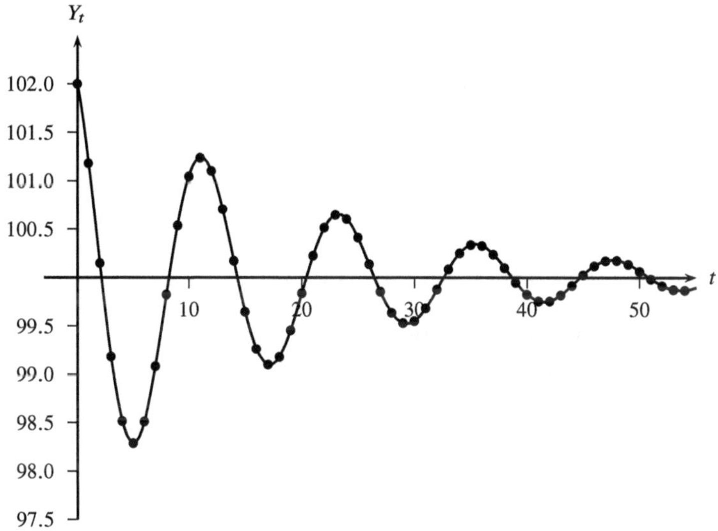

Figure 13.5: A solution to the multiplier-accelerator model.

13.4 Systems of first-order difference equations

To start this section with an application, let us discuss a simple population model:

Example 13.21 (Leslie[6] population model). Consider a population of individuals with maximum age n years. For $t \in T$, let

$y_{i,t}$	be the number of individuals in their i-th year of life at the end of year t,
f_i	be the birth rate in the i-th year of life,
d_i	be the mortality rate in the i-th year of life.

To avoid distinguishing between males and females, we simply understand the

[6]Patrick H. Leslie (1900–1974)

birth rate f_i as the average number of births per individual (male or female) in their i-th year of life. Correspondingly, the mortality rate d_i is the proportion of i-year old individuals who die during a year.

The following applies to the population dynamics: The total number of individuals in their first year of life at the end of year t results from the number of births in the last year, i.e.,

$$y_{1,t} = f_1 y_{1,t-1} + f_2 y_{2,t-1} + \cdots + f_n y_{n,t-1} .$$

Furthermore, for $i \geq 2$, the number of individuals in their i-th year of life at the end of year t results from the number of individuals in their $(i-1)$-th year of life at the end of year $t-1$ who did not die in the last year, i.e.,

$$y_{i,t} = (1 - d_{i-1})\, y_{i-1,t-1} , \quad i \geq 2 .$$

In summary, the development of the population is given by

$$\begin{aligned} y_{1,t} &= f_1 y_{1,t-1} + f_2 y_{2,t-1} + \cdots + f_n y_{n,t-1} \\ y_{2,t} &= (1 - d_1)\, y_{1,t-1} \\ y_{3,t} &= (1 - d_2)\, y_{2,t-1} \\ \vdots \quad & \quad \vdots \\ y_{n,t} &= (1 - d_{n-1}) y_{n-1,t-1} . \end{aligned}$$

This is a so-called **system of difference equations**. If we combine the quantities $y_{1,t}, \ldots, y_{n,t}$ into a vector $\boldsymbol{y}_t$, the system can be written as

$$\boldsymbol{y}_t = \boldsymbol{A} \boldsymbol{y}_{t-1} ,$$

where the matrix $\boldsymbol{A}$ is given by

$$\boldsymbol{A} = \begin{bmatrix} f_1 & f_2 & \cdots & f_{n-1} & f_n \\ 1 - d_1 & 0 & \ldots & 0 & 0 \\ 0 & 1 - d_2 & & 0 & 0 \\ \vdots & \vdots & & \vdots & \vdots \\ 0 & 0 & \ldots & 1 - d_{n-1} & 0 \end{bmatrix} .$$

Definition 13.3: System of first-order linear difference equations

A system of **first-order linear difference equations with constant coefficients** is an equation of the form

$$\boldsymbol{y}_t = \boldsymbol{A} \boldsymbol{y}_{t-1} + \boldsymbol{b}_t .$$

Here, A is a real $n \times n$ matrix and $b : T \to \mathbb{R}^n$, $t \mapsto b_t$, is a given function, the so-called **forcing function**.

If $b_t = 0$ for all $t \in T$, the system is called **homogeneous**, otherwise **nonhomogeneous**.

The **general solution** to a system of difference equations consists in determining all vector-valued functions $y_t = f(t)$, $t \in T$, for which the system of difference equations is satisfied.

Remark: The equation $y_t = Ay_{t-1} + b_t$ can be rewritten as $y_t - Ay_{t-1} = b_t$. Thus, the forcing function b_t is commonly called the right-hand side of the system, even when Ay_{t-1} also appears there.

Interestingly, every m-th order linear difference equation can be represented as a system of first-order difference equations. If an m-th order linear difference equation

$$y_t + a_1 y_{t-1} + \ldots + a_m y_{t-m} = b_t$$

is given, we define

$$z_{1,t} = y_t\,, z_{2,t} = y_{t-1}\,, \ldots, z_{m,t} = y_{t-(m-1)}\,.$$

Thus, we have

$$\begin{aligned} z_{1,t} = y_t &= -a_1 y_{t-1} - a_2 y_{t-2} - \cdots - a_m y_{t-m} + b_t \\ &= -a_1 z_{1,t-1} - a_2 z_{2,t-1} - \cdots - a_m z_{m,t-1} + b_t\,. \end{aligned}$$

Furthermore, we get the equations $z_{i,t} = z_{i-1,t-1}$, $i = 2, \ldots, m$. This gives the system

$$\begin{aligned} z_{1,t} &= -a_1 z_{1,t-1} - a_2 z_{2,t-1} - \cdots - a_{m-1} z_{m-1,t-1} - a_m z_{m,t-1} + b_t \\ z_{2,t} &= z_{1,t-1} \\ \vdots\ & \quad \vdots \\ z_{m,t} &= z_{m-1,t-1} \end{aligned}$$

or in matrix notation

$$\begin{bmatrix} z_{1,t} \\ z_{2,t} \\ z_{3,t} \\ \vdots \\ z_{m,t} \end{bmatrix} = \begin{bmatrix} -a_1 & -a_2 & \ldots & -a_{m-1} & -a_m \\ 1 & 0 & \ldots & 0 & 0 \\ 0 & 1 & \ldots & 0 & 0 \\ \vdots & & \ddots & & \vdots \\ 0 & 0 & \ldots & 1 & 0 \end{bmatrix} \begin{bmatrix} z_{1,t-1} \\ z_{2,t-1} \\ \vdots \\ z_{m-1,t-1} \\ z_{m,t-1} \end{bmatrix} + \begin{bmatrix} b_t \\ 0 \\ 0 \\ \vdots \\ 0 \end{bmatrix},$$

briefly $z_t = Az_{t-1} + b_t$.

Theorem 13.18: Structure of the solution space

Consider a system of first-order linear difference equations with constant coefficients,

$$\boldsymbol{y}_t = \boldsymbol{A}\boldsymbol{y}_{t-1} + \boldsymbol{b}_t \,.$$

The general solution to the **homogeneous** system is an $\boldsymbol{n}$**-dimensional subspace** of $\mathcal{F}(T, \mathbb{R}^n)$.

If $\boldsymbol{f}_1, \ldots, \boldsymbol{f}_n$ is a **basis of the general solution**, then the general solution is

$$\boldsymbol{y}_t = \gamma_1 \boldsymbol{f}_1(t) + \cdots + \gamma_n \boldsymbol{f}_n(t) \,, \quad \gamma_1, \ldots, \gamma_n \in \mathbb{R} \,.$$

A basis $\boldsymbol{f}_1, \ldots, \boldsymbol{f}_n$ of the general solution to the homogeneous system is also referred to as a **fundamental system**.

Let $\boldsymbol{y}_t = \boldsymbol{f}_p(t)$ be a solution to the **nonhomogeneous** system and let $\boldsymbol{f}_1, \ldots, \boldsymbol{f}_n$ be a fundamental system to the homogeneous system. Then the **general solution to the nonhomogeneous system** is

$$\boldsymbol{y}_t = \boldsymbol{f}_p(t) + \gamma_1 \boldsymbol{f}_1(t) + \cdots + \gamma_n \boldsymbol{f}_n(t) \,, \quad \gamma_1, \ldots, \gamma_n \in \mathbb{R} \,.$$

Remark: Combining the basis functions $\boldsymbol{f}_1(t), \ldots, \boldsymbol{f}_n(t)$ into a matrix $\boldsymbol{F}_t$ and the constants $\gamma_1, \ldots, \gamma_n$ into a column vector $\boldsymbol{\gamma}$, the general solution to the homogeneous system can be written as

$$\boldsymbol{y}_t = \boldsymbol{F}_t \boldsymbol{\gamma} \,.$$

The matrix $\boldsymbol{F}_t$ is called **fundamental matrix**. Since each basis function $\boldsymbol{f}_i(t)$ is a solution of the homogeneous system, it holds $\boldsymbol{f}_i(t) = \boldsymbol{A}\boldsymbol{f}_i(t-1), i = 1, \ldots, n$. Then, the matrix equation $\boldsymbol{F}_t = \boldsymbol{A}\boldsymbol{F}_{t-1}$ also holds.

We first consider the solution of the homogeneous system. In the study of linear difference equations, exponential functions often appear as solutions to homogeneous linear difference equations. Therefore, we use the trial solution $\boldsymbol{y}_t = \boldsymbol{a}\lambda^t$ with $\boldsymbol{a} \in \mathbb{R}^n$ and $\lambda \in \mathbb{R}$. Substituting into the difference equation $\boldsymbol{y}_t = \boldsymbol{A}\boldsymbol{y}_{t-1}$ then yields

$$\boldsymbol{a}\lambda^t = \boldsymbol{A}\boldsymbol{a}\lambda^{t-1}$$

and, after division by λ^{t-1},

$$\lambda \boldsymbol{a} = \boldsymbol{A}\boldsymbol{a} \,.$$

Thus, $\boldsymbol{y}_t = \boldsymbol{a}\lambda^t$ is a solution to the difference equation if and only if λ is an eigenvalue of $\boldsymbol{A}$ and $\boldsymbol{a}$ is an eigenvector of $\boldsymbol{A}$ corresponding to the eigenvalue λ.

For the general solution of the system, we need to find a fundamental system, i.e., n linearly independent solutions.

Setting $t = 0$, we see that n solutions $\boldsymbol{a}_1\lambda_1^t, \ldots, \boldsymbol{a}_n\lambda_n^t$ are linearly independent if the vectors $\boldsymbol{a}_1, \ldots, \boldsymbol{a}_n$ are linearly independent. Since eigenvectors corresponding to different eigenvalues are linearly independent (Theorem 10.15), the above trial solution yields the general solution, provided $\boldsymbol{A}$ has n linearly independent (possibly complex) eigenvectors, in other words, if $\boldsymbol{A}$ is complex diagonalizable (see Definition 10.5).

As with simple linear difference equations, we would like to express the solution of systems of difference equations entirely in terms of real basis functions. If $\lambda = re^{i\varphi}$ is a complex eigenvalue, then the complex conjugate $\overline{\lambda} = re^{-i\varphi}$ is also a complex eigenvalue. It also holds that the corresponding eigenvectors are complex conjugates of each other. So if $\boldsymbol{u} + i\boldsymbol{v}$ is an eigenvector corresponding to λ, then $\boldsymbol{u} - i\boldsymbol{v}$ is an eigenvector corresponding to $\overline{\lambda}$. A pair of complex conjugate eigenvalues thus provides the two basis functions

$$\begin{aligned}(\boldsymbol{u} + i\boldsymbol{v})r^t e^{i\varphi t} &= (\boldsymbol{u} + i\boldsymbol{v})r^t\,(\cos(\varphi t) + i\sin(\varphi t))\\ &= r^t\,[(\cos(\varphi t)\boldsymbol{u} - \sin(\varphi t)\boldsymbol{v}) + i\,(\sin(\varphi t)\boldsymbol{u} + \cos(\varphi t)\boldsymbol{v})]\end{aligned}$$

and

$$\begin{aligned}(\boldsymbol{u} - i\boldsymbol{v})r^t e^{-i\varphi t} &= (\boldsymbol{u} - i\boldsymbol{v})r^t\,(\cos(-\varphi t) + i\sin(-\varphi t))\\ &= (\boldsymbol{u} - i\boldsymbol{v})r^t\,(\cos(\varphi t) - i\sin(\varphi t))\\ &= r^t\,[(\cos(\varphi t)\boldsymbol{u} - \sin(\varphi t)\boldsymbol{v}) - i\,(\sin(\varphi t)\boldsymbol{u} + \cos(\varphi t)\boldsymbol{v})]\ .\end{aligned}$$

Thus, we have

$$\begin{aligned}&\gamma_1(\boldsymbol{u} + i\boldsymbol{v})r^t e^{i\varphi t} + \gamma_2(\boldsymbol{u} - i\boldsymbol{v})r^t e^{-i\varphi t}\\ &\qquad = C_1 r^t\,(\cos(\varphi t)\boldsymbol{u} - \sin(\varphi t)\boldsymbol{v}) + C_2 r^t\,(\sin(\varphi t)\boldsymbol{u} + \cos(\varphi t)\boldsymbol{v})\end{aligned}$$

with $C_1 = \gamma_1 + \gamma_2$ and $C_2 = i(\gamma_1 - \gamma_2)$. Hence, to represent the general solution, instead of the two complex basis functions, we use the two real basis functions

$$r^t\,(\cos(\varphi t)\boldsymbol{u} - \sin(\varphi t)\boldsymbol{v}) \quad \text{and} \quad r^t\,(\sin(\varphi t)\boldsymbol{u} + \cos(\varphi t)\boldsymbol{v})\ .$$

Theorem 13.19: Solution to the homogeneous system

Consider the homogeneous system of linear difference equations $\boldsymbol{y}_t = \boldsymbol{A}\boldsymbol{y}_{t-1}$. Let $\boldsymbol{A}$ be complex diagonalizable and possess the (not necessarily different) eigenvalues

$$\lambda_1, \dots, \lambda_m, (\lambda_{m+1}, \overline{\lambda}_{m+1}), \dots, (\lambda_{m+r}, \overline{\lambda}_{m+r}),$$

where λ_j, $j = 1, \dots, m$, are real eigenvalues and $\lambda_{m+k}, \overline{\lambda}_{m+k} = a_{m+k} \pm b_{m+k}i$, $k = 1, \dots, r$, are pairs of complex conjugate eigenvalues, $m + 2r = n$. Furthermore, let $\boldsymbol{u}_1, \dots, \boldsymbol{u}_m$ and $\boldsymbol{u}_{m+1} \pm i\boldsymbol{v}_{m+1}, \dots, \boldsymbol{u}_{m+r} \pm i\boldsymbol{v}_{m+r}$ be corresponding linearly independent eigenvectors. Then the functions

$$\begin{aligned}&\boldsymbol{u}_j\lambda_j^t\,, && j = 1, \dots, m,\\ &r_{m+k}^t\,(\cos(\varphi_{m+k}t)\boldsymbol{u}_{m+k} - \sin(\varphi_{m+k}t)\boldsymbol{v}_{m+k})\,, && k = 1\,\dots, r,\\ &r_{m+k}^t\,(\sin(\varphi_{m+k}t)\boldsymbol{u}_{m+k} + \cos(\varphi_{m+k}t)\boldsymbol{v}_{m+k})\,, && k = 1\,\dots, r,\end{aligned}$$

form a basis of the general solution of the homogeneous system.

This so-called **eigenvalue-eigenvector method** is illustrated by two examples.

Example 13.22. Consider the homogeneous system of linear difference equations $\boldsymbol{y}_t = \boldsymbol{A}\boldsymbol{y}_{t-1}$ with

$$\boldsymbol{A} = \begin{bmatrix} 2 & 4 \\ 1 & -1 \end{bmatrix},$$

subject to the initial condition $\boldsymbol{y}_0 = [2, 1]^T$.

The characteristic polynomial of $\boldsymbol{A}$ is $\lambda^2 - \lambda - 6$. Using the quadratic formula, we obtain the two eigenvalues $\lambda_1 = 3$ and $\lambda_2 = -2$. To find the eigenspace for $\lambda_1 = 3$, we solve the linear equation $(\boldsymbol{A} - 3\boldsymbol{I})\boldsymbol{x} = \boldsymbol{0}$. Because

$$(\boldsymbol{A} - 3\boldsymbol{I}) = \begin{bmatrix} -1 & 4 \\ 1 & -4 \end{bmatrix},$$

$\boldsymbol{u}_1 = [4, 1]^T$ is an eigenvector for $\lambda_1 = 3$. Similarly, for $\lambda_2 = -2$, we find that

$$(\boldsymbol{A} + 2\boldsymbol{I}) = \begin{bmatrix} 4 & 4 \\ 1 & 1 \end{bmatrix}.$$

Hence, $\boldsymbol{u}_2 = [1, -1]^T$ is an eigenvector for $\lambda_2 = -2$. The general solution is therefore

$$\boldsymbol{y}_t = \gamma_1 3^t \begin{bmatrix} 4 \\ 1 \end{bmatrix} + \gamma_2 (-2)^t \begin{bmatrix} 1 \\ -1 \end{bmatrix}, \quad \gamma_1, \gamma_2 \in \mathbb{R}.$$

Thus,

$$\boldsymbol{F}_t = \begin{bmatrix} 4 \cdot 3^t & (-2)^t \\ 3^t & -(-2)^t \end{bmatrix}$$

is a fundamental matrix of the system.

Substituting the complete solution into the initial condition yields the linear equation

$$\boldsymbol{y}_0 = \gamma_1 \begin{bmatrix} 4 \\ 1 \end{bmatrix} + \gamma_2 \begin{bmatrix} 1 \\ -1 \end{bmatrix} = \begin{bmatrix} 4 & 1 \\ 1 & -1 \end{bmatrix} \begin{bmatrix} \gamma_1 \\ \gamma_2 \end{bmatrix} = \begin{bmatrix} 2 \\ 1 \end{bmatrix}.$$

By Gauss-Jordan elimination the solution $\gamma_1 = 3/5$ and $\gamma_2 = -2/5$ is obtained, and thus the solution

$$\boldsymbol{y}_t = 3^t \begin{bmatrix} \frac{12}{5} \\ \frac{3}{5} \end{bmatrix} + (-2)^t \begin{bmatrix} -\frac{2}{5} \\ \frac{2}{5} \end{bmatrix} = \begin{bmatrix} \frac{12}{5} \cdot 3^t - \frac{2}{5} \cdot (-2)^t \\ \frac{3}{5} \cdot 3^t + \frac{2}{5} \cdot (-2)^t \end{bmatrix}.$$

Example 13.23. Consider the homogeneous system of linear difference equations $\boldsymbol{y}_t = \boldsymbol{A}\boldsymbol{y}_{t-1}$ with

$$\boldsymbol{A} = \begin{bmatrix} 4 & -1 \\ 5 & 2 \end{bmatrix},$$

subject to the initial condition $\boldsymbol{y}_0 = [1, -1]^T$.

The characteristic polynomial of $\boldsymbol{A}$ is $\lambda^2 - 6\lambda + 13$. Using the quadratic formula, we obtain the two complex roots $\lambda_{1,2} = 3 \pm 2i$. In polar form, this is $\lambda_{1,2} = \sqrt{13}\, e^{\pm 0.5880i}$. Hence, there are two complex conjugate eigenvalues. To determine the eigenspace for $\lambda_1 = 3 + 2i$, we solve the linear equation $(\boldsymbol{A} - (3 + 2i)\boldsymbol{I})\boldsymbol{x} = \boldsymbol{0}$. Since

$$(\boldsymbol{A} - (3 + 2i)\boldsymbol{I}) = \begin{bmatrix} 1 - 2i & -1 \\ 5 & -1 - 2i \end{bmatrix},$$

$\boldsymbol{u}_1 + i\boldsymbol{v}_1 = [1, 1]^T + i[0, -2]^T$ is an eigenvector for $\lambda_1 = 3 + 2i$. Correspondingly, $\boldsymbol{u}_1 - i\boldsymbol{v}_1 = [1, 1]^T - i[0, -2]^T$ is an eigenvector for $\lambda_2 = 3 - 2i$.

The general solution is thus

$$\begin{aligned} \boldsymbol{y}_t = {} & \gamma_1(\sqrt{13})^t \left(\cos(0.5880\,t) \begin{bmatrix} 1 \\ 1 \end{bmatrix} - \sin(0.5880\,t) \begin{bmatrix} 0 \\ -2 \end{bmatrix}\right) \\ & + \gamma_2(\sqrt{13})^t \left(\sin(0.5880\,t) \begin{bmatrix} 1 \\ 1 \end{bmatrix} + \cos(0.5880\,t) \begin{bmatrix} 0 \\ -2 \end{bmatrix}\right), \qquad \gamma_1, \gamma_2 \in \mathbb{R}. \end{aligned}$$

Since $(\sqrt{13})^0 = 1$, $\cos(0) = 1$ and $\sin(0) = 0$, substituting the general solution into the initial condition results in the linear equation

$$\boldsymbol{y}_0 = \gamma_1 \begin{bmatrix} 1 \\ 1 \end{bmatrix} + \gamma_2 \begin{bmatrix} 0 \\ -2 \end{bmatrix} = \begin{bmatrix} 1 & 0 \\ 1 & -2 \end{bmatrix} \begin{bmatrix} \gamma_1 \\ \gamma_2 \end{bmatrix} = \begin{bmatrix} 1 \\ -1 \end{bmatrix}.$$

Gauss-Jordan elimination yields $\gamma_1 = \gamma_2 = 1$ and thus the solution

$$\begin{aligned} \boldsymbol{y}_t & = (\sqrt{13})^t \left(\cos(0.5880\,t) \begin{bmatrix} 1 \\ 1 \end{bmatrix} - \sin(0.5880\,t) \begin{bmatrix} 0 \\ -2 \end{bmatrix}\right) \\ & \quad + (\sqrt{13})^t \left(\sin(0.5880\,t) \begin{bmatrix} 1 \\ 1 \end{bmatrix} + \cos(0.5880\,t) \begin{bmatrix} 0 \\ -2 \end{bmatrix}\right) \\ & = (\sqrt{13})^t \left(\cos(0.5880\,t) \begin{bmatrix} 1 \\ -1 \end{bmatrix} + \sin(0.5880\,t) \begin{bmatrix} 1 \\ 3 \end{bmatrix}\right) \\ & = \begin{bmatrix} (\sqrt{13})^t (\cos(0.5880\,t) + \sin(0.5880\,t)) \\ -(\sqrt{13})^t (\cos(0.5880\,t) - 3\sin(0.5880\,t)) \end{bmatrix}. \end{aligned}$$

Solving a system of difference equations becomes more intricate if the matrix $\boldsymbol{A}$ is not complex diagonalizable. Then, in addition to basis functions of the form $\boldsymbol{u}_i\lambda_i^t$, basis functions of the form

$$\boldsymbol{p}_i(t)\lambda_i^t$$

occur. Here, $\boldsymbol{p}_i(t)$ is a **vector-valued polynomial** in t, i.e., a function of the form

$$\boldsymbol{p}(t) = \sum_{j-0}^{k} \boldsymbol{a}_j t^j = \boldsymbol{a}_0 + \boldsymbol{a}_1 t + \cdots + \boldsymbol{a}_k t^k,$$

where $\boldsymbol{a}_0, \ldots, \boldsymbol{a}_k$ are vectors, $\boldsymbol{a}_k \neq \boldsymbol{0}$. The degree k of the polynomial $\boldsymbol{p}_i$ is strictly less than the algebraic multiplicity of the corresponding eigenvalue λ_i. For eigenvalues λ whose geometric multiplicity is smaller than its algebraic multiplicity we therefore use the trial solution

$$\boldsymbol{y}_t = \boldsymbol{p}(t)\lambda^t$$

with a vector-valued polynomial $\boldsymbol{p}(t)$ of degree one less than the multiplicity of the eigenvalue. We substitute the trial solution into the system of difference equations to determine the unknown coefficients of the polynomial $\boldsymbol{p}(t)$. The **method of undetermined coefficients** is illustrated with two further examples.

Example 13.24. Consider the homogeneous system of linear difference equations $\boldsymbol{y}_t = \boldsymbol{A}\boldsymbol{y}_{t-1}$ with

$$\boldsymbol{A} = \begin{bmatrix} 5 & -3 \\ 3 & -1 \end{bmatrix}.$$

The characteristic polynomial of $\boldsymbol{A}$ is $\lambda^2 - 4\lambda + 4 = (\lambda - 2)^2$. Thus, $\boldsymbol{A}$ has a double eigenvalue $\lambda_{1,2} = 2$. Therefore, we use a trial solution with a linear polynomial,

$$\boldsymbol{y}_t = (\boldsymbol{a} + \boldsymbol{b}t)2^t$$

and substitute it into the system:

$$(\boldsymbol{a} + \boldsymbol{b}t)2^t = \boldsymbol{A}\big(\boldsymbol{a} + \boldsymbol{b}(t-1)\big)2^{t-1}.$$

Dividing by 2^{t-1} and expanding the parentheses on the right-hand side gives

$$2\boldsymbol{a} + 2\boldsymbol{b}t = \boldsymbol{A}\big(\boldsymbol{a} + \boldsymbol{b}(t-1)\big) = \boldsymbol{A}(\boldsymbol{a} - \boldsymbol{b}) + \boldsymbol{A}\boldsymbol{b}t.$$

Equating coefficients yields the two equations

$$2\boldsymbol{a} = \boldsymbol{A}(\boldsymbol{a} - \boldsymbol{b}) \quad \text{and} \quad \boldsymbol{A}\boldsymbol{b} = 2\boldsymbol{b}.$$

The second equation implies that $\boldsymbol{b}$ is an eigenvector to the eigenvalue $\lambda = 2$. The calculation of the eigenspace yields $\boldsymbol{b} = \gamma_1[1,1]^T$. To find $\boldsymbol{a}$, we solve the linear equation $\boldsymbol{A}\boldsymbol{a} - 2\boldsymbol{a} = \boldsymbol{A}\boldsymbol{b}$ or $(\boldsymbol{A} - 2\boldsymbol{I})\boldsymbol{a} = \gamma_1[2,2]^T$ and obtain

$$\boldsymbol{a} = \gamma_1 \begin{bmatrix} \frac{2}{3} \\ 0 \end{bmatrix} + \gamma_2 \begin{bmatrix} 1 \\ 1 \end{bmatrix} \quad \text{with } \gamma_1, \gamma_2 \in \mathbb{R}.$$

The general solution to the system is thus

$$\begin{aligned}
\boldsymbol{y}_t &= \left(\gamma_1 \begin{bmatrix} \frac{2}{3} \\ 0 \end{bmatrix} + \gamma_2 \begin{bmatrix} 1 \\ 1 \end{bmatrix} + \gamma_1 \begin{bmatrix} 1 \\ 1 \end{bmatrix} t\right) 2^t \\
&= \gamma_1 \left(\begin{bmatrix} \frac{2}{3} \\ 0 \end{bmatrix} + \begin{bmatrix} 1 \\ 1 \end{bmatrix} t\right) 2^t + \gamma_2 \begin{bmatrix} 1 \\ 1 \end{bmatrix} 2^t, \quad \text{where } \gamma_1, \gamma_2 \in \mathbb{R}.
\end{aligned}$$

Example 13.25. Consider the homogeneous system of linear difference equations $\boldsymbol{y}_t = \boldsymbol{A}\boldsymbol{y}_{t-1}$ with

$$\boldsymbol{A} = \begin{bmatrix} 3 & 2 & -1 \\ -1 & 0 & 2 \\ 1 & 2 & -1 \end{bmatrix}.$$

$\boldsymbol{A}$ has the characteristic polynomial $p_{\boldsymbol{A}}(\lambda) = -\lambda^3 + 2\lambda^2 + 4\lambda - 8$. Calculating the eigenvalues yields the simple eigenvalue $\lambda_1 = -2$ and the double eigenvalue $\lambda_{2,3} = 2$.

For the simple eigenvalue $\lambda_1 = -2$, we obtain the eigenspace $V(-2) = \{\gamma_1[2, -3, 4]^T \mid \gamma_1 \in \mathbb{R}\}$. Thus,

$$\boldsymbol{y}_t = \gamma_1 \begin{bmatrix} 2 \\ -3 \\ 4 \end{bmatrix} (-2)^t$$

is a solution to the homogeneous system for every $\gamma_1 \in \mathbb{R}$.

For the double eigenvalue $\lambda_{2,3} = 2$, we use the following trial solution with a linear polynomial $\boldsymbol{y}_t = (\boldsymbol{a}+\boldsymbol{b}t)2^t$. As in the previous example, after some simplifications and by equating coefficients, we obtain the two equations $2\boldsymbol{a} = \boldsymbol{A}(\boldsymbol{a}-\boldsymbol{b})$ and $2\boldsymbol{b} = \boldsymbol{A}\boldsymbol{b}$. From the second equation, it follows that $\boldsymbol{b}$ is an eigenvector for the eigenvalue $\lambda = 2$. Calculating the eigenspace gives $\boldsymbol{b} = \gamma_2[2, -1, 0]^T$. To determine $\boldsymbol{a}$, we solve the linear equation $\boldsymbol{A}\boldsymbol{a} - 2\boldsymbol{a} = \boldsymbol{A}\boldsymbol{b}$ or $(\boldsymbol{A} - 2\boldsymbol{I})\boldsymbol{a} = \gamma_2[4, -2, 0]^T$ and obtain

$$\boldsymbol{a} = \gamma_2 \begin{bmatrix} 6 \\ 0 \\ 2 \end{bmatrix} + \gamma_3 \begin{bmatrix} 2 \\ -1 \\ 0 \end{bmatrix} \quad \text{with } \gamma_2, \gamma_3 \in \mathbb{R}.$$

The general solution of the system is thus

$$\begin{aligned} \boldsymbol{y}_t &= \gamma_1 \begin{bmatrix} 2 \\ -3 \\ 4 \end{bmatrix} (-2)^t + \left(\gamma_2 \begin{bmatrix} 6 \\ 0 \\ 2 \end{bmatrix} + \gamma_3 \begin{bmatrix} 2 \\ -1 \\ 0 \end{bmatrix} + \gamma_2 \begin{bmatrix} 2 \\ -1 \\ 0 \end{bmatrix} t \right) 2^t \\ &= \gamma_1 \begin{bmatrix} 2 \\ -3 \\ 4 \end{bmatrix} (-2)^t + \gamma_2 \left(\begin{bmatrix} 6 \\ 0 \\ 2 \end{bmatrix} + \begin{bmatrix} 2 \\ -1 \\ 0 \end{bmatrix} t \right) 2^t + \gamma_3 \begin{bmatrix} 2 \\ -1 \\ 0 \end{bmatrix} 2^t, \end{aligned}$$

where $\gamma_1, \gamma_2, \gamma_3 \in \mathbb{R}$.

Another method for solving the homogeneous equation, which also works for non-diagonalizable coefficient matrices, is the **solution using matrix powers**. By substitution, it is easily verified that the general solution of a first-order system is $\boldsymbol{y}_t = \boldsymbol{A}^t\boldsymbol{y}_0$, where the initial condition $\boldsymbol{y}_0$ can be chosen arbitrarily[7]. We formulate this result as a theorem.

[7]Here, $\boldsymbol{A}^t$ stands for the product $\boldsymbol{A} \cdot \boldsymbol{A} \cdot \ldots \cdot \boldsymbol{A}$ with t identical factors $\boldsymbol{A}$, and *not* for the transpose $\boldsymbol{A}^T$ of $\boldsymbol{A}$.

Theorem 13.20: Solution to the homogeneous system (matrix powers)

Consider the homogeneous system of linear difference equations $\boldsymbol{y}_t = \boldsymbol{A}\boldsymbol{y}_{t-1}$. The general solution of the system is

$$\boldsymbol{y}_t = \boldsymbol{A}^t\boldsymbol{\gamma}, \quad \boldsymbol{\gamma} \in \mathbb{R}^n .$$

The unique solution subject to the initial condition $\boldsymbol{y}_0$ is

$$\boldsymbol{y}_t = \boldsymbol{A}^t\boldsymbol{y}_0 .$$

The challenge here consists in computing high matrix powers $\boldsymbol{A}^t$, as their direct calculation is computationally intensive. For a diagonal matrix $\boldsymbol{D} = \operatorname{diag}(d_1, \ldots, d_n)$, however, those matrix powers are straightforwardly computed:

$$\boldsymbol{D}^t = \operatorname{diag}(d_1^t, \ldots, d_n^t) .$$

A is diagonalizable, if $\boldsymbol{A} = \boldsymbol{S}\boldsymbol{D}\boldsymbol{S}^{-1}$ for some invertible matrix $\boldsymbol{S}$ and a diagonal matrix $\boldsymbol{D}$. Thus,

$$\boldsymbol{A}^t = \boldsymbol{S}\boldsymbol{D}^t\boldsymbol{S}^{-1} ,$$

which follows easily by mathematical induction.

Theorem 13.21: Solution using matrix powers (A diagonalizable)

Consider the homogeneous system of linear difference equations $\boldsymbol{y}_t = \boldsymbol{A}\boldsymbol{y}_{t-1}$. If A is diagonalizable, $\boldsymbol{A} = \boldsymbol{S}\boldsymbol{D}\boldsymbol{S}^{-1}$, then the general solution of the system is given by

$$\boldsymbol{y}_t = \boldsymbol{S}\boldsymbol{D}^t\boldsymbol{S}^{-1}\boldsymbol{y}_0 ,$$

where $\boldsymbol{y}_0$ can be chosen arbitrarily.

Example 13.26 (cf. Example 13.22). Consider the homogeneous system of linear difference equations $\boldsymbol{y}_t = \boldsymbol{A}\boldsymbol{y}_{t-1}$ with

$$\boldsymbol{A} = \begin{bmatrix} 2 & 4 \\ 1 & -1 \end{bmatrix} ,$$

subject to the initial condition $\boldsymbol{y}_0 = [2, 1]^T$.

As calculated in Example 13.22, the matrix $\boldsymbol{A}$ has eigenvalues $\lambda_1 = 3$ and $\lambda_2 = -2$, with corresponding eigenvectors $\boldsymbol{u}_1 = [4, 1]^T$ and $\boldsymbol{u}_2 = [1, -1]^T$. The transformation matrix $\boldsymbol{S}$ and its inverse are thus

$$\boldsymbol{S} = \begin{bmatrix} 4 & 1 \\ 1 & -1 \end{bmatrix} \quad \text{and} \quad \boldsymbol{S}^{-1} = \frac{1}{5}\begin{bmatrix} 1 & 1 \\ 1 & -4 \end{bmatrix} .$$

Further, $\boldsymbol{D} = \operatorname{diag}(3, -2)$. The solution satisfying the initial condition is then given by

$$\begin{aligned}
y_t &= SD^t S^{-1} y_0 \\
&= \frac{1}{5}\begin{bmatrix} 4 & 1 \\ 1 & -1 \end{bmatrix}\begin{bmatrix} 3^t & 0 \\ 0 & (-2)^t \end{bmatrix}\begin{bmatrix} 1 & 1 \\ 1 & -4 \end{bmatrix}\begin{bmatrix} 2 \\ 1 \end{bmatrix} \\
&= \frac{1}{5}\begin{bmatrix} 4 & 1 \\ 1 & -1 \end{bmatrix}\begin{bmatrix} 3^t & 0 \\ 0 & (-2)^t \end{bmatrix}\begin{bmatrix} 3 \\ -2 \end{bmatrix} \\
&= \frac{1}{5}\begin{bmatrix} 4 & 1 \\ 1 & -1 \end{bmatrix}\begin{bmatrix} 3 \cdot 3^t \\ (-2)(-2)^t \end{bmatrix} \\
&= \frac{1}{5}\begin{bmatrix} 12 \cdot 3^t - 2(-2)^t \\ 3 \cdot 3^t + 2(-2)^t \end{bmatrix} \\
&= \begin{bmatrix} \frac{12}{5} \cdot 3^t - \frac{2}{5} \cdot (-2)^t \\ \frac{3}{5} \cdot 3^t + \frac{2}{5} \cdot (-2)^t \end{bmatrix}.
\end{aligned}$$

A homogeneous system of first-order linear difference equations,

$$y_t = Ay_{t-1}$$

is called **decoupled**, if $A = \text{diag}(d_1, \ldots, d_n)$ is a diagonal matrix. In this case, the system consists of n independent difference equations, which can be solved separately:

$$\begin{aligned}
y_{1,t} &= a_{11} y_{1,t-1} &\implies&& y_{1,t} &= \gamma_1 a_{11}^t, \\
y_{2,t} &= a_{22} y_{2,t-1} &\implies&& y_{2,t} &= \gamma_2 a_{22}^t, \\
& &\vdots&& & \\
y_{n,t} &= a_{nn} y_{n,t-1} &\implies&& y_{n,t} &= \gamma_n a_{nn}^t.
\end{aligned}$$

If A is not a diagonal matrix, the system is called **coupled**.

The approach just introduced for solving a system of homogeneous linear difference equations with diagonalizable coefficient matrix A has a natural interpretation here: it allows transforming a coupled system of difference equations into a decoupled system. If A is diagonalzable, $A = SDS^{-1}$, then

$$y_t = Ay_{t-1} = SDS^{-1} y_{t-1}$$

or

$$S^{-1} y_t = DS^{-1} y_{t-1}.$$

Substituting $z_t = S^{-1} y_t$, we obtain a **decoupled** system of difference equations in the **transformed variables** z_t,

$$z_t = Dz_{t-1},$$

which can be solved directly. The solution of the original system $y_t = Ay_{t-1}$ is then obtained by back transformation $y_t = Sz_t$.

Example 13.27 (cf. Examples 13.22 and 13.26). Consider the homogeneous system of linear difference equations $\boldsymbol{y}_t = \boldsymbol{A}\boldsymbol{y}_{t-1}$ with

$$\boldsymbol{A} = \begin{bmatrix} 2 & 4 \\ 1 & -1 \end{bmatrix}.$$

As calculated in Example 13.22, the matrix $\boldsymbol{A}$ has eigenvalues $\lambda_1 = 3$ and $\lambda_2 = -2$, with corresponding eigenvectors $\boldsymbol{u}_1 = [4, 1]^T$ and $\boldsymbol{u}_2 = [1, -1]^T$. The transformation matrix $\boldsymbol{S}$ is thus

$$\boldsymbol{S} = \begin{bmatrix} 4 & 1 \\ 1 & -1 \end{bmatrix}.$$

Further, $\boldsymbol{D} = \operatorname{diag}(3, -2)$. The general solution of the decoupled system is therefore $\boldsymbol{z}_t = [\gamma_1 3^t, \gamma_2(-2)^t]^T$. Thus, the solution of the original system is

$$\boldsymbol{y}_t = \boldsymbol{S}\boldsymbol{z}_t = \begin{bmatrix} 4 & 1 \\ 1 & -1 \end{bmatrix} \begin{bmatrix} \gamma_1 3^t \\ \gamma_2(-2)^t \end{bmatrix} = \gamma_1 \begin{bmatrix} 4 \\ 1 \end{bmatrix} 3^t + \gamma_2 \begin{bmatrix} 1 \\ -1 \end{bmatrix} (-2)^t .$$

The calculation of matrix powers is more demanding when $\boldsymbol{A}$ is not diagonalizable. We now describe the so-called **Cayley-Hamilton method** for computing them. As shown in Theorem 10.27, every matrix polynomial $p(\boldsymbol{A})$ (in particular, every matrix power) of an $n \times n$ matrix $\boldsymbol{A}$ can be represented as a linear combination of the matrix powers $\boldsymbol{I}, \boldsymbol{A}, \boldsymbol{A}^2, \ldots, \boldsymbol{A}^{n-1}$,

$$p(\boldsymbol{A}) = a_0\boldsymbol{I} + a_1\boldsymbol{A} + a_2\boldsymbol{A}^2 + \cdots + a_{n-1}\boldsymbol{A}^{n-1} = r(\boldsymbol{A}) .$$

Here, $r(\boldsymbol{A})$ is the remainder polynomial obtained from dividing $p(\boldsymbol{A})$ by the characteristic polynomial $p_{\boldsymbol{A}}(\boldsymbol{A})$. Following Theorem 10.27, we have also seen how to calculate the coefficients $a_0, a_1, \ldots, a_{n-1}$ of the polynomial $r(\boldsymbol{A})$. We briefly recap the procedure: For each (real or complex) eigenvalue of multiplicity m, one obtains the m equations

$$p(\lambda) = r(\lambda), \quad p'(\lambda) = r'(\lambda), \quad \ldots, \quad p^{(m-1)}(\lambda) = r^{(m-1)}(\lambda) .$$

In total, this yields n equations from which the coefficients $a_0, a_1, \ldots, a_{n-1}$ can be uniquely determined.

We use the Cayley-Hamilton method in the following examples to represent the general matrix power $\boldsymbol{A}^t$ as a linear combination of $\boldsymbol{I}, \boldsymbol{A}, \boldsymbol{A}^2, \ldots, \boldsymbol{A}^{n-1}$. In these examples, $p(\boldsymbol{A}) = \boldsymbol{A}^t$ will always hold. The coefficients of the polynomial $r(\boldsymbol{A})$ are then precisely the coefficients (depending on t) of this linear combination.

Example 13.28 (cf. Example 13.24). Consider the homogeneous system of linear difference equations $\boldsymbol{y}_t = \boldsymbol{A}\boldsymbol{y}_{t-1}$ with

$$\boldsymbol{A} = \begin{bmatrix} 5 & -3 \\ 3 & -1 \end{bmatrix},$$

subject to the initial condition $\boldsymbol{y}_0 = [1, 3]^T$.

The characteristic polynomial of $\boldsymbol{A}$ is $\lambda^2 - 4\lambda + 4 = (\lambda - 2)^2$. Thus, $\boldsymbol{A}$ has a double eigenvalue $\lambda_{1,2} = 2$. Using the Cayley-Hamilton method, we obtains the equations

$$\begin{aligned} 2^t &= a_0(t) + a_1(t) \cdot 2\,, \\ t2^{t-1} &= a_1(t)\,. \end{aligned}$$

The second equation directly yields $a_1(t)$. Substituting this into the first equation, gives

$$a_0(t) = 2^t - 2 \cdot t2^{t-1} = (1-t)2^t\,.$$

Thus, we have

$$\boldsymbol{A}^t = (1-t)2^t \boldsymbol{I} + t2^{t-1}\boldsymbol{A} = \begin{bmatrix} 2^t + \frac{3}{2}t2^t & -\frac{3}{2}t2^t \\ \frac{3}{2}t2^t & 2^t - \frac{3}{2}t2^t \end{bmatrix}.$$

The unique solution satisfying the initial condition $\boldsymbol{y}_0 = [1, 3]^T$ is given by

$$\boldsymbol{y}_t = \boldsymbol{A}^t \begin{bmatrix} 1 \\ 3 \end{bmatrix} = \begin{bmatrix} 2^t - 3 \cdot t2^t \\ 3 \cdot 2^t - 3 \cdot t2^t \end{bmatrix} = \begin{bmatrix} 1 - 3t \\ 3 - 3t \end{bmatrix} 2^t\,.$$

Example 13.29 (cf. Example 13.23). Consider the homogeneous system of linear difference equations $\boldsymbol{y}_t = \boldsymbol{A}\boldsymbol{y}_{t-1}$ with

$$\boldsymbol{A} = \begin{bmatrix} 4 & -1 \\ 5 & 2 \end{bmatrix},$$

subject to the initial condition $\boldsymbol{y}_0 = [1, -1]^T$.

The characteristic polynomial of $\boldsymbol{A}$ is $\lambda^2 - 6\lambda + 13$ with roots $\lambda_{1,2} = 3 \pm 2i$. In polar form, one obtains $\lambda_{1,2} = (\sqrt{13})^t e^{\pm 0.5880i}$. Thus, there are two complex conjugate eigenvalues. Using the Cayley-Hamilton method, we obtain the equations

$$\begin{aligned} (\sqrt{13})^t e^{0.5880it} &= a_0(t) + a_1(t)(3 + 2i)\,, \\ (\sqrt{13})^t e^{-0.5880it} &= a_0(t) + a_1(t)(3 - 2i)\,. \end{aligned}$$

The two equations are clearly equivalent, as they transform into each other by complex conjugation. We therefore only consider the first equation. Using Euler's formula, we obtain

$$(\sqrt{13})^t \big(\cos(0.5880\,t) + i \sin(0.5880\,t)\big) = \big(a_0(t) + 3a_1(t)\big) + i \cdot 2a_1(t)\,.$$

Equating the real and imaginary parts yields the two equations

$$(\sqrt{13})^t \cos(0.5880\,t) = a_0(t) + 3a_1(t) \quad \text{and} \quad (\sqrt{13})^t \sin(0.5880\,t) = 2a_1(t)\,.$$

Hence, $a_1(t) = \frac{1}{2}(\sqrt{13})^t \sin(0.5880\,t)$ and

$$a_0(t) = (\sqrt{13})^t \cos(0.5880\,t) - \frac{3}{2}(\sqrt{13})^t \sin(0.5880\,t)\,.$$

Thus, we have

$$A^t = (\sqrt{13})^t \left(\left(\cos(0.5880\,t) - \frac{3}{2}\sin(0.5880\,t)\right)I + \frac{1}{2}\sin(0.5880\,t)A\right)$$

$$= \begin{bmatrix} \cos(0.5880\,t) + \frac{1}{2}\sin(0.5880\,t) & -\frac{1}{2}\sin(0.5880\,t) \\ \frac{5}{2}\sin(0.5880\,t) & \cos(0.5880\,t) - \frac{1}{2}\sin(0.5880\,t) \end{bmatrix} (\sqrt{13})^t .$$

The unique solution satisfying the initial condition $y_0 = [1, -1]^T$ is given by

$$y_t = A^t \begin{bmatrix} 1 \\ -1 \end{bmatrix} = \begin{bmatrix} \cos(0.5880\,t) + \sin(0.5880\,t) \\ -\cos(0.5880\,t) + 3\sin(0.5880\,t) \end{bmatrix} (\sqrt{13})^t .$$

Example 13.30 (cf. Examples 13.25 and 10.15). Consider the homogeneous system of linear difference equations $y_t = Ay_{t-1}$ with

$$A = \begin{bmatrix} 3 & 2 & -1 \\ -1 & 0 & 2 \\ 1 & 2 & -1 \end{bmatrix} .$$

We have already seen in Example 10.15 that the coefficients are given by

$$\begin{aligned} a_0(t) &= 3 \cdot 2^{t-2} + (-2)^{t-2} - t \cdot 2^{t-1} , \\ a_1(t) &= 2^{t-2} - (-2)^{t-2} , \\ a_2(t) &= -2^{t-4} + (-2)^{t-4} + t \cdot 2^{t-3} . \end{aligned}$$

The general solution is therefore given by

$$\begin{aligned} y_t = \Big[\left(3 \cdot 2^{t-2} + (-2)^{t-2} - t \cdot 2^{t-1}\right) I + \left(2^{t-2} - (-2)^{t-2}\right) A \\ + \left(-2^{t-4} + (-2)^{t-4} + t \cdot 2^{t-3}\right) A^2\Big]\gamma, \quad \gamma \in \mathbb{R}^n . \end{aligned}$$

We have seen that every linear difference equation of higher order can also be written as a system of first-order difference equations. In the so-called **elimination method**, one proceeds in the opposite direction and tries to transform a system of first-order difference equations into a linear difference equation of higher order. This method is often easily applied to small systems, especially two-dimensional systems. We illustrate it with an example.

Example 13.31 (cf. Example 13.24 and 13.28). Consider the homogeneous system of linear difference equations

$$y_t = \begin{bmatrix} 5 & -3 \\ 3 & -1 \end{bmatrix} y_{t-1} , \quad \text{that is} \quad \begin{aligned} y_{1,t} &= 5y_{1,t-1} - 3y_{2,t-1} \\ y_{2,t} &= 3y_{1,t-1} - y_{2,t-1} , \end{aligned}$$

subject to the initial condition $y_0 = [1, 3]^T$. Solving the first equation for $y_{2,t-1}$

gives

$$y_{2,t-1} = -\frac{1}{3}y_{1,t} + \frac{5}{3}y_{1,t-1}\,.$$

Substituting into the second equation yields

$$y_{2,t} = 3y_{1,t-1} - \left(-\frac{1}{3}y_{1,t} + \frac{5}{3}y_{1,t-1}\right) = \frac{1}{3}y_{1,t} + \frac{4}{3}y_{1,t-1}\,.$$

Substituting this back into the first equation (with a one-step lag) yields

$$y_{1,t} = 5y_{1,t-1} - 3\left(\frac{1}{3}y_{1,t-1} + \frac{4}{3}y_{1,t-2}\right) = 4y_{1,t-1} - 4y_{1,t-2}$$

or

$$y_{1,t} - 4y_{1,t-1} + 4y_{1,t-2} = 0\,,$$

that is, a second-order linear difference equation for $y_{1,t}$. Its characteristic polynomial is $z^2 - 4z + 4 = (z-2)^2$ having a double root $z_{1,2} = 2$. Thus,

$$y_{1,t} = \gamma_1 2^t + \gamma_2\, t\, 2^t = (\gamma_1 + \gamma_2\, t)\, 2^t\,.$$

By substituting this into $y_{2,t} = \frac{1}{3}y_{1,t} + \frac{4}{3}y_{1,t-1}$, the solution for $y_{2,t}$ is obtained,

$$y_{2,t} = \frac{1}{3}(\gamma_1 + \gamma_2 t)\, 2^t + \frac{4}{3}(\gamma_1 + \gamma_2(t-1))\, 2^{t-1} = \left(\gamma_1 + \gamma_2\left(t - \frac{2}{3}\right)\right) 2^t\,.$$

Inserting the initial conditions $y_{1,0} = 1$ and $y_{2,0} = 3$ yields $y_{1,0} = \gamma_1 = 1$ and

$$y_{2,0} = \gamma_1 - \frac{2}{3}\gamma_2 = 1 - \frac{2}{3}\gamma_2 = 3\,,$$

hence $\gamma_2 = -3$. The unique solution satisfying the initial conditions is therefore

$$y_{1,t} = (1 - 3t)2^t\,,$$
$$y_{2,t} = (3 - 3t)2^t\,.$$

Solution to the nonhomogeneous system

We present two methods for solving nonhomogeneous systems of first-order linear difference equations. The simplest approach is the **decoupling method**, which, however, applies only to diagonalizable coefficient matrices. The **method of undetermined coefficients** is also straightforward but requires the forcing function to have a specific form.

Decoupling method If the coefficient matrix $\boldsymbol{A}$ is diagonalizable, $\boldsymbol{A} = \boldsymbol{SDS}^{-1}$, the system of difference equations $\boldsymbol{y}_t = \boldsymbol{A}\boldsymbol{y}_{t-1} + \boldsymbol{b}_t$ can be written as

$$\boldsymbol{y}_t = \boldsymbol{SDS}^{-1}\boldsymbol{y}_{t-1} + \boldsymbol{b}_t\,.$$

Multiplication by S^{-1} then yields

$$S^{-1}y_t = DS^{-1}y_{t-1} + S^{-1}b_t \,.$$

Substituting $z_t = S^{-1}y_t$ and $c_t = S^{-1}b_t$, yields the **decoupled** system

$$z_t = Dz_{t-1} + c_t$$

for the transformed variables z_t. The decoupled system consists of n first-order nonhomogeneous linear difference equations, $z_{i,t} = d_i z_{i,t-1} + c_{i,t}$, $i = 1, \dots, n$. These difference equations can be solved, for example, by using the method of undetermined coefficients. The solution of the original system $y_t = Ay_{t-1} + b_t$ is then obtained by the back transformation $y_t = Sz_t$.

Example 13.32 (cf. Example 13.22, 13.26 and 13.27). Consider the nonhomogeneous system of linear difference equations $y_t = Ay_{t-1} + b_t$ with

$$A = \begin{bmatrix} 2 & 4 \\ 1 & -1 \end{bmatrix} \quad \text{and} \quad b_t = \begin{bmatrix} 1 \\ 2 \end{bmatrix} 3^t \,.$$

The matrix A has the eigenvalues $\lambda_1 = 3$ and $\lambda_2 = -2$ with associated eigenvectors $v_1 = [4, 1]^T$ and $v_2 = [1, -1]^T$. The matrices S and D are therefore

$$S = \begin{bmatrix} 4 & 1 \\ 1 & -1 \end{bmatrix} \quad \text{and} \quad D = \begin{bmatrix} 3 & 0 \\ 0 & -2 \end{bmatrix} .$$

The inverse of S is

$$S^{-1} = \frac{1}{5} \begin{bmatrix} 1 & 1 \\ 1 & -4 \end{bmatrix} .$$

Then,

$$c_t = S^{-1}b_t = \frac{1}{5} \begin{bmatrix} 1 & 1 \\ 1 & -4 \end{bmatrix} \begin{bmatrix} 1 \\ 2 \end{bmatrix} 3^t = \begin{bmatrix} 0.6 \\ -1.4 \end{bmatrix} 3^t \,.$$

For the transformed variables $z_t = S^{-1}y_t$ we therefore obtain the decoupled system

$$z_t = \begin{bmatrix} 3 & 0 \\ 0 & -2 \end{bmatrix} z_{t-1} + \begin{bmatrix} 0.6 \\ -1.4 \end{bmatrix} 3^t ,$$

that is, the two linear first-order difference equations

$$z_{1,t} = 3z_{1,t-1} + 0.6 \cdot 3^t \qquad \text{and} \qquad z_{2,t} = -2z_{2,t-1} - 1.4 \cdot 3^t \,.$$

For the first equation, we apply the method of undetermined coefficients with the trial solution $z_{1,t} = ct\,3^t$ (due to resonance), yielding the particular solution $z_{1,t} = 0.6t\,3^t$. Similarly, for the second equation, we use the trial solution $z_{2,t} = c3^t$, yielding the particular solution $z_{2,t} = -0.84 \cdot 3^t$.

A solution of the original system is then obtained by the back transformation

$$y_t = Sz_t = \begin{bmatrix} 4 & 1 \\ 1 & -1 \end{bmatrix} \begin{bmatrix} 0.6t \\ -0.84 \end{bmatrix} 3^t = \begin{bmatrix} 2.4t - 0.84 \\ 0.6t + 0.84 \end{bmatrix} 3^t \,.$$

Method of undetermined coefficients The method of undetermined coefficients can also be used for systems of difference equations. In the following, a **vector-valued polynomial** (cf. p. 389) is a function of the form $\boldsymbol{p}(t) = \sum_{i=0}^{k} \boldsymbol{a}_i t^i$, where $\boldsymbol{a}_0, \ldots, \boldsymbol{a}_k$ are vectors, that is, a vector-valued function whose components are polynomials.

Theorem 13.22: Method of undetermined coefficients for systems

Consider the system of first-order linear difference equations

$$\boldsymbol{y}_t = \boldsymbol{A}\boldsymbol{y}_{t-1} + \boldsymbol{b}_t \,.$$

If the forcing function has the form

$$\boldsymbol{b}_t = \boldsymbol{p}_1(t)u^t \sin(\varphi t) + \boldsymbol{p}_2(t)u^t \cos(\varphi t)\,,$$

where $\boldsymbol{p}_1$ and $\boldsymbol{p}_2$ are vector-valued polynomials of degree n_1 and n_2, respectively, and if $z = ue^{i\varphi}$ is **not an eigenvalue** of $\boldsymbol{A}$, then the nonhomogeneous system has a particular solution of the form

$$\boldsymbol{y}_t = \boldsymbol{q}_1(t)u^t \sin(\varphi t) + \boldsymbol{q}_2(t)u^t \cos(\varphi t)\,,$$

where $\boldsymbol{q}_1$ and $\boldsymbol{q}_2$ are vector-valued polynomials of degree $n = \max\{n_1, n_2\}$.

If $z = ue^{i\varphi}$ is an **eigenvalue of multiplicity** k of $\boldsymbol{A}$, then the nonhomogeneous system has a particular solution of the form

$$\boldsymbol{y}_t = \boldsymbol{q}_1(t)u^t \sin(\varphi t) + \boldsymbol{q}_2(t)u^t \cos(\varphi t)\,,$$

where $\boldsymbol{q}_1$ and $\boldsymbol{q}_2$ are vector-valued polynomials of degree $n = \max\{n_1, n_2\} + k$.

We illustrate the method with two examples.

Example 13.33 (cf. Example 13.22, 13.26, 13.27 and 13.32). Consider the nonhomogeneous system of linear difference equations $\boldsymbol{y}_t = \boldsymbol{A}\boldsymbol{y}_{t-1} + \boldsymbol{b}_t$ with

$$\boldsymbol{A} = \begin{bmatrix} 2 & 4 \\ 1 & -1 \end{bmatrix} \quad \text{and} \quad \boldsymbol{b}_t = \begin{bmatrix} 1 \\ 2 \end{bmatrix} 2^t \,.$$

Since the forcing function has the form $\boldsymbol{b}_t = \boldsymbol{u}\,2^t$ and since $\lambda = 2$ is not an eigenvalue of $\boldsymbol{A}$, we use the trial solution $\boldsymbol{y}_t = \boldsymbol{a}\,2^t$. Substituting into the system yields

$$\boldsymbol{a}\,2^t = \boldsymbol{A}\boldsymbol{a}\,2^{t-1} + \boldsymbol{u}\,2^t \,.$$

Dividing by 2^{t-1} gives

$$2\boldsymbol{a} = \boldsymbol{A}\boldsymbol{a} + 2\boldsymbol{u} \iff (\boldsymbol{A} - 2\boldsymbol{I})\boldsymbol{a} = -2\boldsymbol{u} \iff \begin{bmatrix} 0 & 4 \\ 1 & -3 \end{bmatrix}\begin{bmatrix} c_1 \\ c_2 \end{bmatrix} = \begin{bmatrix} -2 \\ -4 \end{bmatrix}.$$

Solving the linear equation using Gauss-Jordan elimination finally yields $\boldsymbol{a} =$

$-[5.5, 0.5]^T$. A particular solution of the nonhomogeneous system is thus

$$y_t = -\begin{bmatrix} 5.5 \\ 0.5 \end{bmatrix} 2^t .$$

Example 13.34 (cf. Examples 13.22, 13.26, 13.27, 13.32 and 13.33). Consider the nonhomogeneous system of linear difference equations $\boldsymbol{y}_t = \boldsymbol{A}\boldsymbol{y}_{t-1} + \boldsymbol{b}_t$ with

$$\boldsymbol{A} = \begin{bmatrix} 2 & 4 \\ 1 & -1 \end{bmatrix} \quad \text{and} \quad \boldsymbol{b}_t = \begin{bmatrix} 1 \\ 2 \end{bmatrix} 3^t .$$

Since the forcing function has the form $\boldsymbol{b}_t = \boldsymbol{u}\,3^t$ and since $\lambda = 3$ is a simple eigenvalue of $\boldsymbol{A}$, we use the trial solution $\boldsymbol{y}_t = (\boldsymbol{a} + \boldsymbol{b}\,t)3^t$. Substituting into the system yields

$$(\boldsymbol{a} + \boldsymbol{b}\,t)3^t = \boldsymbol{A}\big(\boldsymbol{a} + \boldsymbol{b}(t-1)\big)3^{t-1} + \boldsymbol{u}\,3^t .$$

Dividing by 3^{t-1} and expanding parentheses gives

$$3\boldsymbol{a} + 3\boldsymbol{b}\,t = \boldsymbol{A}\boldsymbol{a} + \boldsymbol{A}\boldsymbol{b}\,t - \boldsymbol{A}\boldsymbol{b} + 3\boldsymbol{u}$$
$$\iff \quad 3\boldsymbol{a} + 3\boldsymbol{b}\,t = (\boldsymbol{A}\boldsymbol{a} - \boldsymbol{A}\boldsymbol{b} + 3\boldsymbol{u}) - \boldsymbol{A}\boldsymbol{b}\,t .$$

Equating coefficients yields the two equations

$$3\boldsymbol{a} = \boldsymbol{A}\boldsymbol{a} - \boldsymbol{A}\boldsymbol{b} + 3\boldsymbol{u} \qquad \text{and} \qquad 3\boldsymbol{b} = \boldsymbol{A}\boldsymbol{b},$$

which are equivalent to

$$(\boldsymbol{A} - 3\boldsymbol{I})\boldsymbol{a} = 3\boldsymbol{b} - 3\boldsymbol{u} \qquad \text{and} \qquad (\boldsymbol{A} - 3\boldsymbol{I})\boldsymbol{b} = \boldsymbol{0} .$$

From the second equation follows that $\boldsymbol{b}$ is an eigenvector for the eigenvalue $\lambda = 3$. Therefore, $\boldsymbol{b} = [4c, c]^T$, $c \in \mathbb{R}$ (cf. Example 13.22).

Inserting this in the first equation $(\boldsymbol{A} - 3\boldsymbol{I})\boldsymbol{a} = 3(\boldsymbol{b} - \boldsymbol{u})$ and performing one step of Gauss-Jordan elimination gives

$$\left[\begin{array}{cc|c} -1 & 4 & 12c - 3 \\ 1 & -4 & 3c - 6 \end{array}\right] \longrightarrow \left[\begin{array}{cc|c} 1 & -4 & 3c - 6 \\ 0 & 0 & 15c - 9 \end{array}\right] .$$

This equation has a solution only if $15c - 9 = 0$, i.e., if $c = 3/5$, in which case the vectors $\boldsymbol{a}$ and $\boldsymbol{b}$ are given by

$$\boldsymbol{a} = \begin{bmatrix} -4.2 \\ 0 \end{bmatrix} + d \begin{bmatrix} 4 \\ 1 \end{bmatrix}, \; d \in \mathbb{R}, \quad \text{and} \quad \boldsymbol{b} = \begin{bmatrix} 2.4 \\ 0.6 \end{bmatrix} .$$

Since d can be chosen freely, we set $d = 0$ and get $\boldsymbol{a} = [-4.2, 0]^T$. Therefore, a particular solution to the nonhomogeneous system is given by

$$\boldsymbol{y} = \left(\begin{bmatrix} -4.2 \\ 0 \end{bmatrix} + \begin{bmatrix} 2.4 \\ 0.6 \end{bmatrix} t \right) 3^t = \begin{bmatrix} -4.2 + 2.4t \\ 0.6t \end{bmatrix} 3^t .$$

The preceding example nicely illustrates the following two observations.

Remarks:

1. For systems, the polynomials $q_1(t)$ and $q_2(t)$ must always contain **all lower-order powers**, even in the resonance case. If $ue^{i\varphi}$ is a root of multiplicity k of the characteristic polynomial, a trial solution of the form
$$y_t = t^k \left[q_1(t)u^t \sin(\varphi t) + q_2(t)u^t \cos(\varphi t)\right]$$
with polynomials $q_1(t)$, $q_2(t)$ of degree $n = \max\{n_1, n_2\}$ will, in general, not yield a solution. This is because the trial solution multiplied by t^k does not contain powers with exponents less than k.
2. Under resonance, the resulting system of equations usually has no unique solution. In the above example, we have d freely chosen. For simplicity, constants that can be chosen freely are often set to zero.

13.5 Stability of systems of difference equations

We have already dealt with the stability of simple difference equations in Section 13.3. Also in systems it is often of primary interest to find out the limit behavior of the solution for large values of t and, additionally, its dependence on initial conditions.

Definition 13.4: Stability

A system of first-order linear difference equations is called (asymptotocally) **stable** if any two of its solutions, $\boldsymbol{y}_t = \boldsymbol{f}_1(t)$ and $\boldsymbol{y}_t = \boldsymbol{f}_2(t)$, satisfy
$$\lim_{t\to\infty} (\boldsymbol{f}_1(t) - \boldsymbol{f}_2(t)) = \boldsymbol{0}\,.$$

If the nonhomogeneous system of difference equations has a constant equilibrium solution $\boldsymbol{y}^*$, the above definition is equivalent to the fact that every solution of the system converges to the equilibrium solution $\boldsymbol{y}^*$ as t tends to infinity[8],
$$\lim_{t\to\infty} \boldsymbol{y}_t = \boldsymbol{y}^*\,.$$

Since the difference between two solutions to a nonhomogeneous system is a solution to the corresponding homogeneous system, a nonhomogeneous system is stable if and only if every solution $\boldsymbol{y}_t$ of the corresponding homogeneous system converges to zero as t goes to infinity,
$$\lim_{t\to\infty} \boldsymbol{y}_t = \boldsymbol{0}\,.$$

From Theorem 13.19 it immediately follows that in the case of a complex diagonalizable coefficient matrix $\boldsymbol{A}$ all solutions of the homogeneous system converge to

[8] For the definition of convergence of vectors see Section 3.2.

zero if and only if all eigenvalues have absolute value less than one. In fact, it can be shown that this result also holds for systems with a non-diagonalizable coefficient matrix $\boldsymbol{A}$. We formulate this result as a theorem:

Theorem 13.23: Stability of a system of linear difference equations

A system of first-order linear difference equations with constant coefficients, $\boldsymbol{y}_t = \boldsymbol{A}\boldsymbol{y}_{t-1} + \boldsymbol{b}_t$, is stable if and only if every eigenvalue λ of $\boldsymbol{A}$ satisfies $|\lambda| < 1$.

It is of course desirable to have simple criteria for deciding whether a system of difference equations is stable, without actually having to compute the eigenvalues.

In the following, we provide two simple criteria that are either necessary or sufficient for the stability of a system of first-order linear difference equations.

Theorem 13.24: Stability of a system, necessary condition

If the system $\boldsymbol{y}_t = \boldsymbol{A}\boldsymbol{y}_{t-1} + \boldsymbol{b}_t$ is stable, then the absolute value of the sum of the diagonal elements of $\boldsymbol{A}$ is less than n,

$$\left|\sum_{i=1}^{n} a_{ii}\right| < n\,.$$

Theorem 13.25: Stability of a system, sufficient condition

Consider the system of first-order linear difference equations, $\boldsymbol{y}_t = \boldsymbol{A}\boldsymbol{y}_{t-1} + \boldsymbol{b}_t$. If

$$\sum_{j=1}^{n} |a_{ij}| < 1 \quad \text{for all } i = 1, \ldots, n$$

or

$$\sum_{i=1}^{n} |a_{ij}| < 1 \quad \text{for all } j = 1, \ldots, n,$$

then the system is stable.

We illustrate these criteria with two examples.

Example 13.35. Consider the homogeneous system of linear difference equations $\boldsymbol{y}_t = \boldsymbol{A}\boldsymbol{y}_{t-1}$ where

$$\boldsymbol{A} = \begin{bmatrix} 1 & -0.5 & -1 \\ 1 & 2.5 & -1 \\ -0.5 & 0.75 & -0.5 \end{bmatrix}.$$

Since the absolute value of the sum of the diagonal elements is three and thus equal to n, the necessary condition is violated. Hence, the system is not stable.

Example 13.36. Consider the homogeneous system of linear difference equations $\boldsymbol{y}_t = \boldsymbol{A}\boldsymbol{y}_{t-1}$ where

$$\boldsymbol{A} = \begin{bmatrix} 0.4 & -0.2 & 0.3 \\ 0.6 & 0.2 & -0.1 \\ 0.2 & 0.5 & -0.1 \end{bmatrix}.$$

Since the sum of the absolute values of the elements in each row is less than one (namely 0.9, 0.9 and 0.8), the sufficient criterion is satisfied, and the system is therefore stable.

Important terms for review

After reading this chapter, the following terms should be familiar:

- Difference equation
- Order of a difference equation
- Initial condition
- Homogeneous and nonhomogeneous linear difference equation
- Characteristic polynomial, characteristic equation
- Solution of linear difference equations with constant coefficients
- Method of undetermined coefficients
- Stability
- Systems of first-order difference equations
- Decoupling of systems of difference equations
- Method of undetermined coefficients for systems of difference equations

Self-test

You can check your understanding of this chapter using the following true-or-false questions. For each statement, indicate whether it is true (T) or false (F). Short solutions to these questions can be found in Appendix H.

I. Are the following statements about differential equations true or false?

	T	F
Difference equations are used in dynamic economic models in continuous time.	☐	☐
Periodic compound interest is described by a first-order homogeneous linear difference equation.	☐	☐
If a difference equation is solvable, a stable equilibrium is always reached in the long run, regardless of the initial condition.	☐	☐

To solve a homogeneous linear difference equation of order m, one uses a characteristic polynomial of order $m - 1$. □ □

For a first-order difference equation, one initial condition suffices to obtain a unique solution of the difference equation. □ □

Exercises

Exercise 1
Solve the following difference equations:

a) $y_t = 5y_{t-1}$, where $y_2 = 100$,

b) $y_t = 2y_{t-1}$, where $y_0 = 0$,

c) $y_t - 3 = -4y_{t-1}$, where $y_0 = \frac{8}{5}$,

d) $y_t = 4 - 3y_{t-1}$, where $y_3 = -26$.

Exercise 2
Consider the Cobweb model with demand function $d_t = 100 - 2p_t$ and supply function $s_t = 40 + p_{t-1}$. At time $t = 0$, the price is $p_0 = 50$.

a) Find the equilibrium price.

b) Find the solution to the difference equation subject to the given initial condition.

c) From which period onward is the difference between the price and the equilibrium price less than 0.01?

Exercise 3
For each of the following difference equations, find the general solution and the solution corresponding to the given initial conditions.

a) $y_t - 4y_{t-2} = 0$, where $y_0 = 0$ and $y_1 = 1$,

b) $y_t + 4y_{t-2} = 0$, where $y_0 = 1$ and $y_1 = -2$,

c) $y_t - 0.4y_{t-1} + 0.04y_{t-2} = 0$, where $y_0 = 1$ and $y_1 = 2$,

d) $y_t + y_{t-1} - 6y_{t-2} = 0$, where $y_0 = 0$ and $y_1 = 2$,

e) $y_t - 4y_{t-1} + 5y_{t-2} = 0$, where $y_0 = y_1 = 2$,

f) $y_t - 2y_{t-1} - y_{t-2} + 2y_{t-3} = 0$, where $y_0 = 0$, $y_1 = 0$ and $y_2 = 1$,

g) $y_t - 3y_{t-1} + 2.25y_{t-2} - 0.5y_{t-3} = 0$, where $y_0 = -3$, $y_1 = 1$ and $y_2 = 4$.

Exercise 4
Consider the following second-order linear difference equation

$$y_t - 1.2y_{t-1} + 0.32y_{t-2} = 3\,,$$

subject to the initial conditions $y_0 = 37$ and $y_1 = 33$.

a) Find the solution to the difference equation.

b) Compute the limit of y_t as t tends to infinity.

c) What is the limit of y_t as t tends to infinity, if the initial conditions are changed to $y_0 = 0$ and $y_1 = 10$?

d) Is the difference equation stable?

Exercise 5
Consider the following second-order linear difference equation

$$y_t - 3.2y_{t-1} + 2.56y_{t-2} = 1.8\,.$$

a) Find the general solution to the corresponding **homogeneous** difference equation.

b) Find the equilibrium solution to the **nonhomogeneous** difference equation.

c) Find the general solution to the **nonhomogeneous** difference equation.

d) Find the solution to the **nonhomogeneous** difference equation subject to the initial conditions $y_0 = 13$ and $y_1 = 21$.

e) Is the solution in part d) convergent? If yes, what is its limit?

Exercise 6
Consider the multiplier-accelerator model by Samuelson:

$$Y_t - c(1 + k)Y_{t-1} + ckY_{t-2} = G\,.$$

a) Consider the special case $c = 0.8$ and $k = 3$. Assume further that the national income in the first two periods is $Y_0 = 12G$ resp. $Y_1 = 15G$. Investigate the development of national income by solving the corresponding difference equation.

b) Is the difference equation stable?

Exercise 7
Find a particular solution to each of the following nonhomogeneous difference equations. Also state the general solution to each difference equation.

a) $y_t - 1.2y_{t-1} + 0.32y_{t-2} = 3 + 3 \cdot 2^t$ (cf. Exercise 4),

b) $y_t - 4y_{t-2} = t\,2^t$ (cf. Exercise 3 a),

c) $y_t - 0.4y_{t-1} + 0.04y_{t-2} = 0.2^t$ (cf. Exercise 3 c),

d) $y_t + y_{t-1} - 6y_{t-2} = 2^t$ (cf. Exercise 3 d),

e) $y_t - 4y_{t-1} + 5y_{t-2} = 10 + 2t^2$ (cf. Exercise 3 e).

Exercise 8
Consider the difference equations from Exercise 3. Which of these difference equations are stable?

Exercise 9
Check, without computing the roots of the characteristic polynomial, whether the following difference equations are stable:

a) $y_t + 0.4y_{t-1} - 0.7y_{t-2} - 0.7y_{t-3} - 0.2y_{t-4} = 0,$

b) $y_t - 0.2y_{t-1} + 0.5y_{t-2} - 0.1y_{t-3} + 0.1y_{t-4} = 0,$

c) $y_t + 0.9y_{t-1} + 0.6y_{t-2} + 0.3y_{t-3} + 0.2y_{t-4} = 0.$

Exercise 10
Consider the following system of difference equations:

$$\begin{aligned} x_t &= -x_{t-1} + 4y_{t-1}\,, \\ y_t &= x_{t-1} + 2y_{t-1}\,. \end{aligned}$$

Find the general solution to the system and the unique solution corresponding to the initial conditions $x_0 = 2$ and $y_0 = 1$.

Exercise 11
Consider the following system of difference equations $\boldsymbol{y}_t = \boldsymbol{A}\boldsymbol{y}_{t-1}$, where

$$\boldsymbol{A} = \begin{bmatrix} 3 & -2 & -2 \\ -4 & 7 & 8 \\ 4 & -5 & -6 \end{bmatrix}.$$

a) Find the general solution using the eigenvalue-eigenvector method.

b) Find the solution corresponding to the initial condition $\boldsymbol{y}_0 = [1, 1, 1]^T$ using matrix powers.

Exercise 12
Consider the system $\boldsymbol{y}_t = \boldsymbol{A}\boldsymbol{y}_{t-1} + \boldsymbol{b}_t$ of difference equations, where

$$\boldsymbol{A} = \begin{bmatrix} 3 & -2 & -2 \\ -4 & 7 & 8 \\ 4 & -5 & -6 \end{bmatrix} \quad \text{and} \quad \boldsymbol{b}_t = \begin{bmatrix} 2 + 4t \\ 1 - 2t \\ 4 \end{bmatrix}.$$

Find a particular solution using the method of undetermined coefficients.

Appendix A

Greek letters

Name	Small	Capital
Alpha	α	A
Beta	β	B
Gamma	γ	Γ
Delta	δ	Δ
Epsilon	ε	E
Zeta	ζ	Z
Eta	η	H
Theta	θ	Θ
Iota	ι	I
Kappa	κ	K
Lambda	λ	Λ
Mu	μ	M
Nu	ν	N
Xi	ξ	Ξ
Omicron	o	O
Pi	π	Π
Rho	ρ	P
Sigma	σ	Σ
Tau	τ	T
Upsilon	υ	Υ
Phi	φ	Φ
Chi	χ	X
Psi	ψ	Ψ
Omega	ω	Ω

K. Mosler et al., *Mathematical Methods for Business and Economics*,
Mathematics Study Resources 20, https://doi.org/10.1007/978-3-662-73831-3_14

Appendix B

Sets

A **set** is a collection of distinct objects, called **elements**. The elements can be numbers, but also objects of other types. Essential for the concept of a set is that for every possible object, it can be determined whether it is an element of the set or not.

The standard convention is to denote sets with uppercase letters such as A, B, C, $A_1, A_2, A_3, \dots$, etc., while their elements are represented by lowercase letters like $x, y, z, x_1, x_2, x_3, \dots$, and so on. The symbolic notation

$$x \in A$$

means that x is an element of the set A. This is also briefly referred to as "x in A".

A set is completely characterized by its elements. There are two ways to denote them mathematically. Either one lists the elements explicitly or one describes the elements by a condition. In the enumeration, the elements are enclosed in braces. The enumeration can be complete or incomplete, for example

$$\begin{aligned}
A_1 &= \{\text{red, yellow, blue}\}\,,\\
A_2 &= \{2, 4, 6, 8, 10\}\,,\\
A_3 &= \{2, 4, 6, 8, \dots, 80\}\,,\\
\mathbb{N}_0 &= \{0, 1, 2, 3, \dots\}\,.
\end{aligned}$$

Alternatively, these sets are denoted by one or more logical conditions that are imposed on a variable:

$$\begin{aligned}
A_1 &= \{C : C \text{ is a primary color}\}\,,\\
A_2 &= \{x : x \text{ is an even number}, 2 \le x \le 10\}\,,
\end{aligned}$$

K. Mosler et al., *Mathematical Methods for Business and Economics*, Mathematics Study Resources 20, https://doi.org/10.1007/978-3-662-73831-3_15

$$A_3 = \{x : x \text{ is an even number}, 2 \leq x \leq 80\},$$
$$\mathbb{N}_0 = \{n : n \text{ is a natural number or } n = 0\}.$$

Here too, the elements of the set are enclosed by braces. The colon in the above sets is read as "such that". To the left of the colon is a variable that represents the elements of the set; to the right of the colon are logical conditions on the elements that determine whether an element belongs to the set or not. The set consists of all elements for which the condition right of the colon is true. Instead of the colon, one often uses a vertical bar, e.g., $A_1 = \{F \mid F \text{ is a primary color}\}$.

The number of (*viz.* different) elements of a set A is called **cardinality** of A, and denoted with the symbol $|A|$. In the previous examples, we have

$$|A_1| = 3, \quad |A_2| = 5, \quad |A_3| = 40.$$

Depending on the number of elements, we speak of a **finite set** or an **infinite set**.

Important sets of numbers are the set of **natural numbers**,

$$\mathbb{N} = \{1, 2, 3, 4, \ldots\},$$

the set of **integers**,

$$\mathbb{Z} = \{\ldots, -3, -2, -1, 0, 1, 2, 3, \ldots\},$$

and the set of **rational numbers**,

$$\mathbb{Q} = \left\{\frac{p}{q} : p \in \mathbb{Z}, q \in \mathbb{N}\right\}.$$

The set $\mathbb{Q}$ of rational numbers consists of all fractions, where the numerator is an integer and the denominator a natural number. A rational number can always be written as a finite or periodic decimal fraction. For example, $\frac{1}{8} = 0.125$ is a finite decimal fraction and $\frac{1}{7} = 0.\overline{142857}$ is a periodic decimal fraction. In contrast, the **irrational numbers** are all those numbers whose decimal expansion is infinite and non-periodic. These include, for example, $\sqrt{2} = 1.41421356237309504880\ldots$, Euler's number $e = 2.71828182845904523536\ldots$, and $\pi = 3.14159265358979323846\ldots$. The rational numbers together with the irrational numbers form the set of **real numbers**, $\mathbb{R}$. Geometrically, they are depicted as **real line**.

The sets $\mathbb{N}$, $\mathbb{Z}$, $\mathbb{Q}$, and $\mathbb{R}$ are all infinite, but they differ fundamentally in size. The sets $\mathbb{N}$, $\mathbb{Z}$ and $\mathbb{Q}$ are **countably infinite**. Their elements can be systematically listed in a sequence. For example, $\mathbb{Z}$ can be enumerated as:

$$\mathbb{Z} = \{0, 1, -1, 2, -2, 3, -3, \ldots\},$$

demonstrating that every integer eventually appears in the sequence.

In contrast, $\mathbb{R}$ is **uncountable**. Its elements cannot be exhaustively listed in any sequence, representing a fundamentally "larger" type of infinity.

In the following, we outline some terms and rules that apply to sets. Let A and B be sets. If every element of B is contained in A, then B is called a **subset** of A (and A a **superset** of B), and one writes

$$B \subset A\,.$$

For example, $\{2,4,6\} \subset \{1,2,3,4,5,6\}$, but also $\{2,4,6\} \subset \{2,4,6\}$. It holds

$$\mathbb{N} \subset \mathbb{Z} \subset \mathbb{Q} \subset \mathbb{R}\,.$$

Equality of sets: Two sets A and B are called equal (A = B), if $A \subset B$ and $B \subset A$ holds, i.e., every element of A is also an element of B and vice versa.

The set of elements that are contained in either A or B or in both A and B is called **union** of A and B and is denoted by[1]

$$A \cup B := \{x \,:\, x \in A \vee x \in B\}\,.$$

The set of elements that are contained in both A and B is called **intersection** of A and B and is denoted by

$$A \cap B := \{x \,:\, x \in A \wedge x \in B\}\,.$$

If A and B have no common element, they are called **disjoint**; their intersection is then called **empty set** and denoted by $\emptyset$. The empty set is the unique set that contains no elements. Obviously, the empty set is a subset of any set A, $\emptyset \subset A$.

Example B.1. Consider the sets $A = \{1,2,3,4,5\}$, $B = \{3,4,5,6,7\}$ and $C = \{6,7,8\}$, so $A \cap B = \{3,4,5\}$, $A \cup B = \{1,2,3,4,5,6,7\}$. The sets A and C are disjoint, $A \cap C = \emptyset$.

The following rules apply for calculating with sets:

Theorem B.1: Rules of set operations

Let A, B, C be sets. It holds:

(i) $A \subset B$ and $B \subset C \implies A \subset C$,

(ii) $A = B$ and $B = C \implies A = C$,

(iii) $A \cup B = B \cup A$,

(iv) $A \cap B = B \cap A$,

(v) $A \cap (B \cup C) = (A \cap B) \cup (A \cap C)$,

(vi) $A \cup (B \cap C) = (A \cup B) \cap (A \cup C)$.

[1]The symbols $\wedge$ and $\vee$ are **sentential connectives**. $\wedge$ means "and", and $\vee$ means "or".

The set of elements that are contained in a set B, but not in another set A, is called **set difference** of A and B and is denoted by

$$B \setminus A = \{x \in B \, : \, x \notin A\} \, .$$

This is referred to as "B without A". If in particular $A \subset B$, then we write

$$\bar{A} = B \setminus A$$

and call $\bar{A}$ the **complement**[2] of A with respect to B.

Theorem B.2: Rules for set difference

Let A, B be sets. Then the following holds:

(i) $A \setminus A = \emptyset\,, \quad A \setminus \emptyset = A\,,$

(ii) $\overline{\overline{A}} = A\,,$

(iii) $\overline{A \cap B} = \overline{A} \cup \overline{B}\,,$

(iv) $\overline{A \cup B} = \overline{A} \cap \overline{B}\,.$

The rules (iii) and (iv) are called **De Morgan's**[3] **laws**.

The **power set** $\mathcal{P}(A)$ of a set A is the set of all subsets of A. Its elements are thus sets of elements of A. The empty set $\emptyset$ and the set A itself are always elements of the power set of A.

Example B.2. The power set of the set $A = \{1, 2, 3\}$ is

$$\mathcal{P}(A) = \{\emptyset, \{1\}, \{2\}, \{3\}, \{1,2\}, \{1,3\}, \{2,3\}, \{1,2,3\}\} \, .$$

It has eight elements.

If A is a finite set, the cardinality of the power set can be easily calculated:

Theorem B.3: Cardinality of a power set

If A is a finite set, then $|\mathcal{P}(A)| = 2^{|A|}$.

The **Cartesian**[4] **product** of two sets A and B is defined by

$$A \times B = \{(a, b) \, : \, a \in A \text{ and } b \in B\}$$

It is the set of all pairs of elements, the first of which is an element of A and the second an element of B.

[2] Note: The complement of a set always refers to a specific superset.
[3] Augustus De Morgan (1806–1871)
[4] René Descartes (1596–1650)

Example B.3.

a) Let $A = \{1, 3, 5\}$ and $B = \{2, 4\}$. Then

$$A \times B = \{(1, 2), (1, 4), (3, 2), (3, 4), (5, 2), (5, 4)\}.$$

b) $\mathbb{R} \times \mathbb{R} = \mathbb{R}^2$ is the set of all points (x_1, x_2) with $x_1, x_2 \in \mathbb{R}$.

Theorem B.4: Rules for the Cartesian product

For arbitrary sets A, B, C the following holds:

(i) A, B finite $\implies$ $|A \times B| = |A| \cdot |B|$,

(ii) $A \times \emptyset = \emptyset, \quad \emptyset \times A = \emptyset,$
$(A \times B) \cap (C \times B) = (A \cap C) \times B,$
$(B \times A) \cap (B \times C) = B \times (A \cap C).$

Analogous statements apply for unions "$\cup$".

Besides the sets $\mathbb{N}, \mathbb{Z}, \mathbb{Q}$ and $\mathbb{R}$, important **number sets** are the subsets of $\mathbb{R}$ known as intervals.

Let $a, b \in \mathbb{R} \cup \{\infty, -\infty\}$. The set of all real numbers between a and b is called an **interval**. If a and b are in $\mathbb{R}$, depending on whether these boundary points belong to the interval or not, we speak of a **closed**, a **half-open** or **open interval**. An interval is bounded if both of its endpoints are real numbers; otherwise, it is called a **ray** (if only one endpoint is in $\mathbb{R}$) or the entire real line $\mathbb{R}$. We distinguish the following types of intervals:

$$[a, b] = \{x \in \mathbb{R} : a \le x \le b\} \quad \text{(closed interval)},$$

$$(a, b) = \{x \in \mathbb{R} : a < x < b\} \quad \text{(open interval)},$$

$$[a, b) = \{x \in \mathbb{R} : a \le x < b\} \quad \text{(right half-open interval)},$$

$$(a, b] = \{x \in \mathbb{R} : a < x \le b\} \quad \text{(left half-open interval)},$$

$$[a, \infty) = \{x \in \mathbb{R} : a \le x\} \quad \text{(ray)},$$

$$(a, \infty) = \{x \in \mathbb{R} : a < x\} \quad \text{(ray)},$$

$$(-\infty, b] = \{x \in \mathbb{R} : x \le b\} \quad \text{(ray)},$$

$$(-\infty, b) = \{x \in \mathbb{R} : x < b\} \quad \text{(ray)},$$

$$(-\infty, \infty) = \mathbb{R} \quad \text{(real line)},$$

$$[0, \infty) = \mathbb{R}_+ \quad \text{(ray)}.$$

Appendix C

Sums and products

Formulas involving sums can often be written more clearly and transformed more easily by using a shorthand notation for sums. In this notation, the terms being summed are described using an index that runs through certain values. The sum is symbolized by a capital Greek Sigma. For example, to add the first twelve square numbers, we write

$$\sum_{i=1}^{12} i^2 = 1^2 + 2^2 + \ldots + 12^2 .$$

Definition C.1: Summation sign

Using the summation sign, a finite sum can be written in the form

$$\sum_{i=m}^{n} a_i = a_m + a_{m+1} + \ldots + a_n .$$

Here, $a_m, a_{m+1}, \ldots, a_n$ are real numbers, m and n are integers, $m \leq n$. The index i is called **summation index**, m and n are called lower and upper **summation limit**, respectively.

Example C.1.

$$\sum_{i=-2}^{3} a_i = a_{-2} + a_{-1} + a_0 + a_1 + a_2 + a_3 .$$

The name of the summation index does not matter. Thus, the following two sums are

K. Mosler et al., *Mathematical Methods for Business and Economics*, Mathematics Study Resources 20, https://doi.org/10.1007/978-3-662-73831-3_16

equal,

$$\sum_{j=-2}^{3} a_j = \sum_{i=-2}^{3} a_i .$$

Sometimes, the indices with respect to which the summation is performed are not adjacent. For example, to sum over all even indices between −4 and 6, we index the individual terms with $2i$ and let i run from −2 to 3,

$$\sum_{i=-2}^{3} a_{2i} = a_{-4} + a_{-2} + a_0 + a_2 + a_4 + a_6 .$$

Instead of a lower and an upper summation limit, one can also specify an index set. Thus, with the **index set** $I_1 = \{-2, -1, 0, 1, 2, 3\}$, we write

$$\sum_{i \in I_1} a_i = a_{-2} + a_{-1} + a_0 + a_1 + a_2 + a_3 ,$$

and with the index set $I_2 = \{-4, -2, 0, 2, 4, 6\}$,

$$\sum_{i \in I_2} a_i = a_{-4} + a_{-2} + a_0 + a_2 + a_4 + a_6 .$$

If the index set I is empty or the lower summation limit is greater than the upper summation limit, we obtain a so-called **empty sum**,

$$\sum_{i=m}^{n} a_i = \sum_{i \in \emptyset} a_i = 0 , \quad \text{if } m > n .$$

The empty sum has the value zero. If you add an empty sum to a sum, the first sum does not change. The following *rules* apply for calculating with finite sums.[1]

Theorem C.1: Rules for summation

Let $a_1, \dots, a_n$ and $b_1, \dots, b_n \in \mathbb{R}$. Then, the following statements hold.

a) For all $\alpha, \beta \in \mathbb{R}$ holds

$$\sum_{i=1}^{n} (\alpha a_i + \beta b_i) = \sum_{i=1}^{n} \alpha a_i + \sum_{i=1}^{n} \beta b_i = \alpha \sum_{i=1}^{n} a_i + \beta \sum_{i=1}^{n} b_i .$$

b) If all summands are equal, $a_1 = a_2 = \ldots = a_n = a$, then

$$\sum_{i=1}^{n} a_i = \sum_{i=1}^{n} a = na .$$

[1]For simplicity, we choose $\{1, 2, \dots, n\}$ as index set.

c) For every integer m, $1 \leq m < n$, a sum can be split as follows:

$$\sum_{i=1}^{n} a_i = \sum_{i=1}^{m} a_i + \sum_{i=m+1}^{n} a_i .$$

d) For every integer m, the index shift formula holds:

$$\sum_{i=1}^{n} a_i = \sum_{i=m+1}^{m+n} a_{i-m} .$$

Next, consider a two-dimensional scheme of real numbers:

$$\begin{matrix} a_{11} & a_{12} & \cdots & a_{1m} \\ a_{21} & a_{22} & \cdots & a_{2m} \\ \vdots & \vdots & \ddots & \vdots \\ a_{n1} & a_{n2} & \cdots & a_{nm} \end{matrix}$$

The sum over all these numbers can be written as a **double sum**:

$$\begin{aligned} \sum_{i=1}^{n} \sum_{j=1}^{m} a_{ij} &= a_{11} + a_{12} + \ldots + a_{1m} \\ &+ a_{21} + a_{22} + \ldots + a_{2m} \\ &\vdots \\ &+ a_{n1} + a_{n2} + \ldots + a_{nm} \end{aligned}$$

In this process, the inner summation sign (with summation index j) is considered first. That is, one forms the sum of the elements in the first row, then in the second row, and so on. The outer summation sign (with summation index i) then ensures that these row sums are added together.

Since the addition of real numbers is commutative, it does not matter whether the rows or columns are added first. The summation signs of finite sums can be interchanged. Note that without further assumptions, this does not hold for infinite sums! Double sums can be written as well using index sets:

$$\sum_{i=1}^{n} \sum_{j=1}^{m} a_{ij} = \sum_{j=1}^{m} \sum_{i=1}^{n} a_{ij} = \sum_{i \in I} \sum_{j \in J} a_{ij} = \sum_{j \in J} \sum_{i \in I} a_{ij} ,$$

with $I = \{1, \ldots, n\}$ and $J = \{1, \ldots, m\}$. However, one has to be careful when the range of the second index depends on the first index. Consider the sum

$$\begin{matrix} \sum_{i=1}^{n} \sum_{j=i}^{n} a_{ij} = & a_{11} & + & \ldots & + & \ldots & + & \ldots & + & a_{1n} \\ & & + & a_{22} & + & \ldots & + & \ldots & + & a_{2n} \\ & & & & + & a_{33} & + & \ldots & + & a_{3n} \\ & & & & & & & & & \vdots \\ & & & & & & & & + & a_{nn} . \end{matrix}$$

This sum can be calculated by first computing the row sums and then the sum of the row sums. Alternatively, one can first calculate the column sums and then the sum of the column sums. In doing this, the summation signs are interchanged as follows:

$$\sum_{i=1}^{n}\sum_{j=i}^{n} a_{ij} = \sum_{j=1}^{n}\sum_{i=1}^{j} a_{ij}$$

Similar to finite sums, finite products can be written in a shorthand notation. For this, one uses Π, the capital Greek Pi.

Definition C.2: Product sign

The product of n real numbers $a_1, a_2, \ldots, a_n$ is written as follows:

$$\prod_{i=1}^{n} a_i = a_1 \cdot a_2 \cdot a_3 \cdot \ldots \cdot a_n .$$

With the index set $I = \{1, \ldots, n\}$ one also writes

$$\prod_{i \in I} a_i = a_1 \cdot a_2 \cdot a_3 \cdot \ldots \cdot a_n .$$

If the index set I is empty or if $m > n$, one obtains an empty product,

$$\prod_{i=m}^{n} a_i = \prod_{i \in \emptyset} a_i = 1 , \quad \text{if } m > n .$$

The **empty product** is equal to 1, since the multiplication of a product with the empty product should not change the product.

For calculating with finite products, analogous *rules* apply as for calculating with finite sums:[2]

Theorem C.2: Rules for products

Let $a_1, \ldots, a_n$ and $b_1, \ldots, b_n \in \mathbb{R}$. Then, the following statements hold.

a) For $\alpha, \beta \in \mathbb{R}$ holds:

$$\prod_{i=1}^{n} \alpha a_i \beta b_i = \alpha^n \beta^n \prod_{i=1}^{n} a_i \prod_{i=1}^{n} b_i .$$

[2] Here too, for simplicity, the index set $\{1, 2, \ldots, n\}$ is used.

b) If all factors of a product are equal, $a_1 = a_2 = \ldots = a_n = a$, then

$$\prod_{i=1}^{n} a_i = \prod_{i=1}^{n} a = a^n .$$

c) For every integer m, $1 \leq m < n$, the product can be split as follows:

$$\prod_{i=1}^{n} a_i = \prod_{i=1}^{m} a_i \cdot \prod_{i=m+1}^{n} a_i .$$

d) For every integer m the index shift formula holds:

$$\prod_{i=1}^{n} a_i = \prod_{i=m+1}^{m+n} a_{i-m} .$$

In the following, we provide the values of certain special sums and products.

a) For the sum $1 + 2 + \ldots + n$ it holds

$$\sum_{i=1}^{n} i = \frac{n(n+1)}{2} .$$

b) For the sum $1^2 + 2^2 + \ldots + n^2$ it holds

$$\sum_{i=1}^{n} i^2 = \frac{n(n+1)(2n+1)}{6} .$$

c) For the sum $1^3 + 2^3 + \ldots + n^3$ it holds

$$\sum_{i=1}^{n} i^3 = \frac{n^2(n+1)^2}{4} .$$

d) Let $a_1, b \in \mathbb{R}$, $a_i = a_1 + (i-1)b$ for $i = 2, \ldots, n$. Then $a_1, a_2, \ldots, a_n$ is called **finite arithmetic sequence of first order**. It holds

$$\sum_{i=1}^{n} a_i = \frac{n}{2}(2a_1 + (n-1)b) .$$

e) Let $a_1, q \in \mathbb{R}$, $a_i = a_1 q^{i-1}$ for $i = 2, \ldots, n$. Then $a_1, a_2, \ldots, a_n$ is called **finite geometric sequence**. For $q \neq 1$ it holds

$$\sum_{i=1}^{n} a_i = a_1 \frac{q^n - 1}{q - 1} .$$

In numerous formulas the product $\prod_{i=1}^{n} i = 1 \cdot 2 \cdot \ldots \cdot n$ is used. This is abbreviated as

$$n! = \prod_{i=1}^{n} i\,,$$

which is read as "*n* **factorial**".

Appendix D

Combinatorics

Combinatorics deals with counting the number of possible arrangements and selections of objects. Problems of this kind arise, for example, in the calculations of Laplace[1] probabilities.

The different ways in which n distinct objects can be arranged are known as **permutations**. The number of possible permutations is

$$\boxed{n!}$$

as shown in the following. There are n possibilities to select the first of the n objects. For each of these n selections, there are $n-1$ possibilities (since one of the n objects was already selected) to select the second object, so there are $n \cdot (n-1)$ possibilities to select the first two objects. For each of these possibilities, there are $n-2$ possibilities to select the third object and so on. So in total there are

$$n \cdot (n-1) \cdot (n-2) \cdot \ldots \cdot 2 \cdot 1 = n!$$

possibilities.

Example D.1. A museum would like to display three different paintings (A, B and C) in a row on a wall. There are

$$3! = 6 \text{ possibilities,}$$

namely ABC, ACB, BAC, BCA, CAB and CBA.

[1] Pierre-Simon Laplace (1749–1827)

K. Mosler et al., *Mathematical Methods for Business and Economics*, Mathematics Study Resources 20, https://doi.org/10.1007/978-3-662-73831-3_17

If you do not select all objects, but only k objects ($k < n$), there are

$$\boxed{\frac{n!}{(n-k)!} = n \cdot (n-1) \cdot \ldots \cdot (n-k+1)}$$

possibilities to choose the objects and arrange them. When selecting the k-th object, $k-1$ objects have already been selected, so there are only $n-(k-1) = n-k+1$ objects left to choose from. These selections are called **variations without replacement**.

Example D.2. A museum wants to display 2 out of 4 pictures (A, B, C and D) side by side on a wall. There are

$$\frac{4!}{(4-2)!} = 4 \cdot 3 = 12 \text{ possibilities,}$$

viz. AB, AC, AD, BA, BC, BD, CA, CB, CD, DA, DB and DC.

Next, we consider the case where it only matters which k out of n objects are selected, but not the order in which they are arranged. If we first count the number of ways to select and arrange k objects (as above), each group of k selected objects will appear $k!$ times – once for each possible arrangement. Since the arrangement does not matter, we must divide by $k!$ to avoid counting the same selection multiple times. So in total there are only

$$\boxed{\binom{n}{k} = \frac{n!}{(n-k)! \cdot k!}}$$

possibilities. These selections are called **combinations without replacement**. $\binom{n}{k}$ is called **binomial coefficient**.

Example D.3. A museum wants to display 2 out of 4 pictures (A, B, C and D) of an artist. There are

$$\binom{4}{2} = \frac{4!}{(4-2)! \cdot 2!} = 6 \text{ possibilities,}$$

(*viz.* AB, AC, AD, BC, BD and CD. Now, BA for example, is not an additional possibility, since the order in which the pictures are shown (hung on the wall) is no longer of interest. The only thing that matters is which pictures are actually on display.

For the binomial coefficient, it holds

$$\binom{n}{k} = \binom{n}{n-k},$$

because

$$\binom{n}{n-k} = \frac{n!}{(n-(n-k))!(n-k)!} = \frac{n!}{k!(n-k)!} = \binom{n}{k}.$$

Special values of the binomial coefficients are

$$\binom{n}{0} = \binom{n}{n} = 1\,, \quad \binom{n}{1} = \binom{n}{n-1} = n\,, \quad \binom{n}{2} = \binom{n}{n-2} = \frac{n(n-1)}{2}\,.$$

So far, we have only considered cases where selected objects could not be chosen again; this is referred to as **drawing without replacement**. Now, we consider cases where objects can be selected multiple times, which is known as **drawing with replacement**.

If you draw k times with replacement from a set of n objects, there are n choices for each draw. Thus, when the order of selection matters, there are

$$\boxed{n^k}$$

possible outcomes. These are called **variations with replacement**.

Example D.4. There are $10^4 = 10\,000$ possibilities for the four-digit ($k = 4$) PIN of a mobile phone, where each digit is one of the $n = 10$ digits: 0000, 0001, 0002, ... 9998, 9999.

If k objects are selected from n given objects with replacement, and the order of selection does not matter, this is called a **combination with replacement**. The number of such combinations is given by

$$\boxed{\binom{n+k-1}{k} = \binom{n+k-1}{n-1}\,.}$$

The easiest way to explain this is by using an example:

Example D.5. For a party, a student wants to buy a crate with $k = 12$ bottles of soft drinks. In the store, there are $n = 3$ types, namely a cola drink (C), orange soda (O) and lemon soda (L) to choose from. How many possibilities are there to fill the crate with bottles of the three types of soft drinks?

The order in which the student arranges the bottles in the crate is not relevant. As she can choose each type of drink multiple times, this is a selection with replacement. If she now puts all the bottles sorted at the checkout and puts a divider between each type, then $k = 12$ bottles and $n - 1 = 3 - 1 = 2$ dividers are on the conveyor belt:

$$\underbrace{C \cdots C \,|\, O \cdots O \,|\, L \cdots L}_{n+k-1=14 \text{ objects}}$$

Depending on how many bottles of each soft drink are selected, the dividers are at different of the $n + k - 1 = 14$ positions (bottles and dividers). For example, if the dividers are at the seventh and eleventh positions, then there are six cola bottles (positions 1 to 6), three orange soda bottles (positions 8 to 10) and three lemon soda bottles (positions 12 to 14). Thus, the number of possible combinations with replacement is uniquely determined by selecting the $n-1$ positions for the product separators from the $n+k-1$ positions. Since this selection is done without replacement ($n-1$ different positions) and without considering the order, there are

$$\binom{n+k-1}{n-1} = \binom{14}{2} = 91$$

possibilities.

We now summarize the formulae for the different ways of selecting k out of n distinguishable objects in a table:

	Order relevant?	
	yes	no
without replacement	$\frac{n!}{(n-k)!}$	$\binom{n}{k}$
with replacement	n^k	$\binom{n+k-1}{k}$

Appendix E

Complex numbers

In the real numbers, the square root of a negative number is undefined, since the square of any real number is nonnegative. To enable square roots of negative numbers, the real numbers have to be extended. For this purpose, the **imaginary unit**, i is introduced, whose square equals -1 by definition. The imaginary unit allows for the square root not only of -1, but of any negative real number. For example,

$$\sqrt{-9} = \sqrt{9 \cdot (-1)} = \sqrt{9 \cdot i^2} = \pm 3i\,.$$

The product of a real number and the imaginary unit i is called an **imaginary number**.

Now, consider the quadratic equation $z^2 - 6z + 10 = 0$. The quadratic formula yields

$$z_{1,2} = 3 \pm \sqrt{3^2 - 10} = 3 \pm \sqrt{-1} = 3 \pm i\,.$$

The two solutions are sums of a real number and an imaginary number. Such a sum is mentioned as a **complex number**. The set of all complex numbers is denoted by $\mathbb{C}$,

$$\mathbb{C} = \{\, a + bi \,:\, a, b \in \mathbb{R} \,\}\,.$$

At first glance, this construction seems extremely strange, giving the impression that complex numbers are not 'real' and lack any practical relevance. At second glance, we see that the step from real to complex numbers is just a further step in augmenting the numbers we are calculating with.

- In elementary school, we start with the natural numbers $\mathbb{N}$ (without naming them as such).

K. Mosler et al., *Mathematical Methods for Business and Economics*, Mathematics Study Resources 20, https://doi.org/10.1007/978-3-662-73831-3_18

- Soon we notice that certain operations, such as $4-10$, cannot be carried out within the natural numbers. Therefore, negative numbers are introduced. Together with zero and the natural numbers, they form the set of integers $\mathbb{Z}$.
- But even with integers, not all operations are possible. For example, we cannot express the quotient $4 \div 7$ as an integer. To overcome this limitation, we introduce fractions, which together with the integers form the set of rational numbers $\mathbb{Q}$.
- In the rational numbers, all four basic operations (except division by zero) are possible. However, some numbers, such as $\sqrt{2}$, cannot be represented as a fraction. To address this, we introduce irrational numbers. Combined with the rational numbers, they form the set of real numbers $\mathbb{R}$.

Of a complex number $z = a + bi$, a is called the **real part** and b the **imaginary part**. One writes $\text{Re}(a+ib) = a$ and $\text{Im}(a+bi) = b$. Real numbers form a subset of complex numbers, since every real number a can be written as $a = a + 0\,i$. Complex numbers with zero real part are called **imaginary numbers**. Geometrically, complex numbers can be visualized in the **complex plane** (or **Gauss plane**); see Figure E.1. Each complex number $z = a + bi$ corresponds to the point having abscissa a and ordinate b. The abscissa is referred to as the real axis (where all real numbers lie) and the ordinate as the imaginary axis (where all imaginary numbers lie).

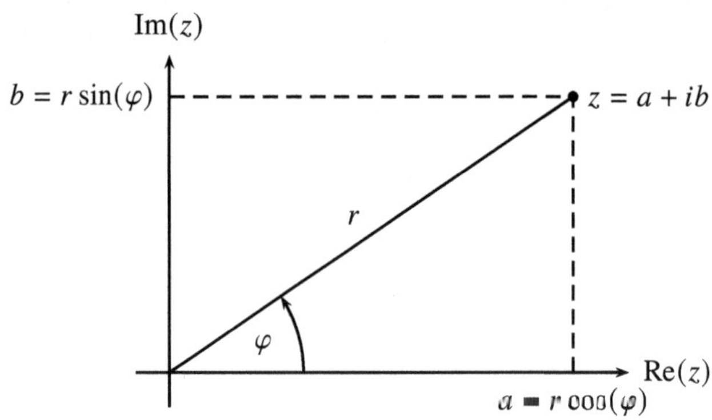

Figure E.1: The complex plane.

Denoting by $r = \sqrt{a^2+b^2}$ the Euclidean distance from $a+bi$ to the origin and by φ the angle between the positive real axis and the line connecting the origin to $a+ib$, we have

$$\text{Re}(a+ib) = a = r\cos(\varphi) \quad \text{and} \quad \text{Im}(a+ib) = b = r\sin(\varphi)\,.$$

The complex number $z = a + ib$ can therefore also be represented in **polar form**,

$$z = r\cos(\varphi) + ir\sin(\varphi)\,.$$

Here, $r = |z| = \sqrt{a^2+b^2}$ is the **absolute value** of z and $\varphi = \arg(z)$ is the **argument** (or **angle** or **phase**) of z. Since $\cos(\varphi) = \cos(\varphi + 2\pi)$ and $\sin(\varphi) = \sin(\varphi + 2\pi)$, the argument of a complex number is not unique. By convention, the argument φ is always taken in the interval $(-\pi, \pi]$.[1] The argument of $z = a + bi$ is calculated as

$$\varphi = \arg(z) = \begin{cases} \arccos\left(\frac{a}{|z|}\right), & \text{if } b \geq 0, \\ -\arccos\left(\frac{a}{|z|}\right), & \text{if } b < 0. \end{cases}$$

Basic arithmetic operations

With complex numbers the basic mathematical operations can be performed. No special rules are required besides that $i^2 = -1$. In particular it holds:

$$\begin{aligned}
(a_1 + b_1 i) + (a_2 + b_2 i) &= (a_1 + a_2) + (b_1 + b_2)i\,, \\
(a_1 + b_1 i) - (a_2 + b_2 i) &= (a_1 - a_2) + (b_1 - b_2)i\,, \\
(a_1 + b_1 i) \cdot (a_2 + b_2 i) &= a_1 a_2 + a_1 b_2 i + b_1 a_2 i + b_1 b_2 i^2 \\
&= (a_1 a_2 - b_1 b_2) + (a_1 b_2 + b_1 a_2)i\,, \\
\frac{a_1 + b_1 i)}{a_2 + b_2 i} &= \frac{a_1 + b_1 i}{a_2 + b_2 i} \cdot \frac{a_2 - b_2 i}{a_2 - b_2 i} \\
&= \frac{a_1 a_2 - a_1 b_2 i + b_1 a_2 i - b_1 b_2 i^2}{a_2 a_2 - a_2 b_2 i + b_2 a_2 i - b_2 b_2 i^2} \\
&= \frac{(a_1 a_2 + b_1 b_2) + (b_1 a_2 - a_1 b_2)i}{a_2^2 + b_2^2} \\
&= \frac{a_1 a_2 + b_1 b_2}{a_2^2 + b_2^2} + \frac{b_1 a_2 - a_1 b_2}{a_2^2 + b_2^2}\, i\,,
\end{aligned}$$

where division is defined only if $a_2 + b_2 i \neq 0$.

Thus, addition and subtraction of complex numbers correspond to the usual vector addition and subtraction in the complex plane. Moreover, the multiplication of a complex number by a real number is well defined. It is easily seen that these operations satisfy the rules that define a vector space (see Definition 9.5 in Chapter 9. By this, $\mathbb{C}$ forms a vector space, which geometrically as well as algebraically resembles the Euclidean $\mathbb{R}^2$. A geometric interpretation of the multiplication and division of complex numbers will be given in the following section.

Complex exponential function

The domain of the exponential function can be extended from the real numbers to the complex numbers as follows:

$$\exp(z) = e^z = e^{a+ib} = e^a \cos(b) + i e^a \sin(b)\,.$$

This definition is consistent with that of the real case: For complex numbers with $\mathrm{Im}(z) = b = 0$, we get $e^z = e^a$, since $\cos(0) = 1$ and $\sin(0) = 0$.

[1] Of course, other intervals like $[0, 2\pi)$ would also be possible. However, $\varphi \in (-\pi, \pi]$ is the standard convention.

With the complex exponential, the polar form of a complex number can be written more concisely as

$$z = r\cos(\varphi) + ir\sin(\varphi) = re^{i\varphi}.$$

The equation

$$e^{i\varphi} = \cos(\varphi) + i\sin(\varphi) \tag{E.1}$$

is called **Euler's formula**.

For complex numbers $z_1 = r_1 e^{i\varphi_1}$ and $z_2 = r_2 e^{i\varphi_2}$, the product of z_1 and z_2 is

$$z_1 z_2 = r_1 e^{i\varphi_1} \cdot r_2 e^{i\varphi_2} = r_1 r_2 e^{i(\varphi_1 + \varphi_2)}.$$

Similarly, the quotient of two complex numbers is

$$\frac{z_1}{z_2} = \frac{r_1 e^{i\varphi_1}}{r_2 e^{i\varphi_2}} = \frac{r_1}{r_2} e^{i(\varphi_1 - \varphi_2)}.$$

Thus, the product of two complex numbers is obtained by multiplying their absolute values and adding their arguments. Similarly, the quotient is obtained by dividing their absolute values and subtracting their arguments.

Integer powers of complex numbers can also be easily calculated using the polar form. For an integer $n \in \mathbb{Z}$, the n-th power of $z = re^{i\varphi}$ is given by

$$z^n = \left(re^{i\varphi}\right)^n = r^n e^{in\varphi}.$$

Complex conjugate

For a given $z = a + ib$, the **complex conjugate** is defined by $\overline{z} = a - ib$. In the complex plane, $\overline{z}$ is obtained from z by reflection across the real axis.

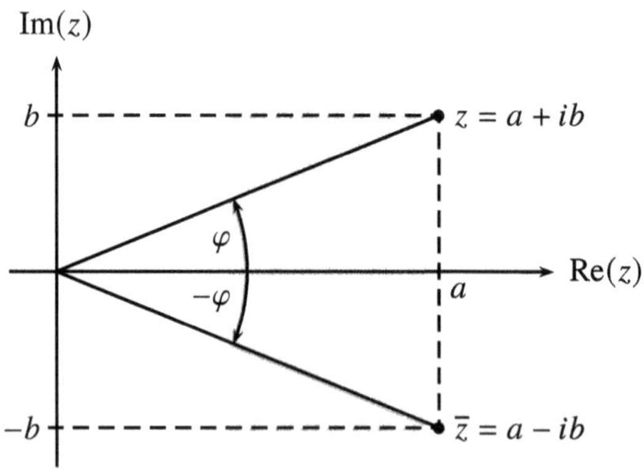

Written in polar form, the complex conjugate of $z = re^{i\varphi}$ is $\overline{z} = re^{-i\varphi}$. It is easy to verify that

$$z + \overline{z} = 2a, \qquad z - \overline{z} = 2ib, \qquad z\overline{z} = a^2 + b^2.$$

Therefore,

$$\operatorname{Re}(z) = \frac{1}{2}(z + \bar{z}), \qquad \operatorname{Im}(z) = \frac{1}{2i}(z - \bar{z}), \qquad |z| = \sqrt{z\bar{z}}.$$

For real numbers, $z = \bar{z}$ and thus $|z| = \sqrt{z^2}$, which coincides with the usual absolute value.

Fundamental theorem of algebra

Consider the polynomial $p(z) = z^3 - 4z^2 + 5z - 2$. As is easy to verify, $z = 1$ is a root of p. Polynomial division of $p(z)$ by the linear factor $(z - 1)$ yields the quadratic polynomial

$$\frac{z^3 - 4z^2 + 5z - 2}{z - 1} = z^2 - 3z + 2.$$

The quadratic formula gives the roots of the quadratic polynomial:

$$z_{1,2} = \frac{3}{2} \pm \sqrt{\frac{9}{4} - 2} = \frac{3}{2} \pm \frac{1}{2}. \quad \text{i.e., } z_1 = 1,\ z_2 = 2.$$

Thus, $z^2 - 3z + 2 = (z - 1)(z - 2)$, and the original polynomial is factorized as

$$z^3 - 4z^2 + 5z - 2 = (z - 1)(z - 1)(z - 2) = (z - 1)^2(z - 2).$$

Since the polynomial contains the linear factor $(z - 1)$ twice, we call $z = 1$ a **double root**. In contrast, $z = 2$ is a **simple root**, as the factor $(z - 2)$ appears only once. More generally, a root z_0 of a polynomial $p_n(x)$ of degree n is called a **root of multiplicity *k*** if the linear factor $(z - z_0)$ appears exactly k times in $p_n(z)$. That is,

$$p_n(z) = (z - z_0)^k q_{n-k}(z),$$

where $q_{n-k}(z)$ is a polynomial of degree $n - k$ with no root at $z = z_0$. This condition is equivalent to $p_n(z)$ being divisible by $(z - z_0)^k$ without remainder, but not by $(z - z_0)^{k+1}$.

A polynomial of degree n, written as $z^n + a_{n-1}z^{n-1} + \cdots + a_1 z + a_0$ with real coefficients $a_0, a_1, \ldots, a_{n-1}$ has *at most* n real roots, counted with multiplicity. However, polynomials of degree n may have fewer than n real roots. For instance $z^2 + 1$ and likewise $z^2 - 6z + 10$ lack real roots at all. In the complex numbers, though, $z^2 + 1$ has roots $z_{1,2} = \pm i$, and $z^2 - 6z + 10$ has roots $z_{1,2} = 3 \pm i$. This is no coincidence; it is guaranteed by the **fundamental theorem of algebra**, one of the most important results in mathematics.

Theorem E.1: Fundamental theorem of algebra (complex version)

A polynomial of degree n, $p_n(z) = z^n + a_{n-1}z^{n-1} + \ldots + a_1 z + a_0$ with complex coefficients $a_0, a_1, \ldots, a_{n-1} \in \mathbb{C}$ has exactly n roots in the complex numbers $\mathbb{C}$, counting roots with multiplicity. Thus, there exist n (not necessarily distinct) complex numbers $z_1, z_2, \ldots, z_n \in \mathbb{C}$ such that for all $z \in \mathbb{C}$,

$$z^n + a_{n-1}z^{n-1} + \ldots + a_1 z + a_0 = (z - z_1)(z - z_2) \cdot \ldots \cdot (z - z_n).$$

If we restrict ourselves to polynomials with **real coefficients**, the following result holds.

Theorem E.2: Fundamental theorem of algebra (real version)

The polynomial of degree n, $p_n(z) = z^n + a_{n-1}z^{n-1} + \ldots + a_1 z + a_0$ with real coefficients $a_0, a_1, \ldots, a_{n-1} \in \mathbb{R}$ has exactly n roots in the complex numbers $\mathbb{C}$, counting roots with multiplicity.

Moreover, if $z = a + ib$ with $b \neq 0$ is a complex root of multiplicity k, then the complex conjugate $\overline{z} = a - ib$ is also a root of multiplicity k.

This theorem can be summarized in less formal terms as follows:

Mnemonic E.1

A real polynomial of degree n has exactly n roots in the complex numbers. Non-real complex roots always occur in conjugate pairs.

Often one is interested in whether a polynomial with integer coefficients has rational or even integer roots. The following theorem, the rational root theorem, shows how such roots, if they exist, can easily be found.

Theorem E.3: Rational root theorem

Consider the polynomial equation

$$a_n x^n + a_{n-1}x^{n-1} + \cdots + a_1 x + a_0 = 0$$

where $a_0, \ldots, a_n \in \mathbb{Z}$, $a_n, a_0 \neq 0$. Let $x = p/q$ be a rational root, written in lowest terms so that p and q are relatively prime. Then,

- p is a divisor of the constant term a_0, and
- q is a divisor of the leading coefficient a_n.

If $a_n = 1$, then every rational root is an integer, $x = p$, and p is a divisor of the constant term a_0.

Appendix F

Propositional logic

A **statement** is a sentence that, in principle, can be classified as true or false.

Example F.1.

a) A : "$3 \cdot 2 = 6$"

b) B : "$3 \cdot 2 = 7$"

c) C : "It will rain tomorrow."

d) D : "The abolition of tariffs in foreign trade increases the prosperity of the involved countries."

Here, statement A is clearly true, while statement B is clearly false. Whether C is true or false will only be decided on the following day. Opinions may differ about statement D, and can be the subject of scientific discussion, for instance by specifying conditions under which D does or does not hold. In what follows, the focus will be on statements about mathematical facts and on how such statements can be justified.

The **negation** $\neg A$ (read "not A") of a statement A is a statement that is true if and only if A is false. Its **truth value** ("true" or "false") can be represented in a **truth table** as follows:

A	$\neg A$
true	false
false	true

Example F.2. If the equality sign in the previous example is replaced by an inequality sign, the negations of the statements A and B are obtained as follows:

a) $\neg A$: "$3 \cdot 2 \neq 6$"

K. Mosler et al., *Mathematical Methods for Business and Economics*, Mathematics Study Resources 20, https://doi.org/10.1007/978-3-662-73831-3_19

b) $\neg B$: "$3 \cdot 2 \neq 7$"

Obviously, $\neg A$ is false and $\neg B$ is true.

Two statements A and B can be combined using **and** or **or** to form a new statement. The statement "A and B" is called a **conjunction** and is denoted by

$$\boxed{A \wedge B\,.}$$

The statement $A \wedge B$ is true if both A and B are true, otherwise it is false. This is summarized in the following truth table:

A	B	$A \wedge B$
true	true	true
true	false	false
false	true	false
false	false	false

Example F.3. Let A_x: "x is an even number", B_x: "x is a prime number". The statement $A_x \wedge B_x$ is

- true when $x = 2$, since 2 is even and 2 is prime;
- false when $x = 4$, where only the first statement is true, and the second is false,
- false when $x = 7$, where only the second statement is true, and the first is false,
- false when $x = 9$, where both component statements are false.

In this example, A_x and B_x are statements that depend on a variable x. Their truth values – and thus that of their conjunction $A_x \wedge B_x$ – depends on the value assigned to x. It is easy to see that $A_x \wedge B_x$ is true if and only if $x = 2$, since 2 is the only even prime number.

If two statements are combined with **or**, it forms a **disjunction**, written as

$$\boxed{A \vee B\,.}$$

$A \vee B$ is true if A or B or both are true, otherwise it is false:

A	B	$A \vee B$
true	true	true
true	false	true
false	true	true
false	false	false

Example F.4. Let again A_x: "x is an even number" and B_x: "x is a prime number". The statement $A_x \vee B_x$ is

- true when $x = 2$, since 2 is even and 2 is prime;
- true when $x = 4$, since the first component statement is true;

- true when $x = 7$, since the second component statement is true,
- false when $x = 9$, where neither the first nor the second component statement is true.

Note that the disjunction $A \vee B$ denotes the **inclusive or**: it is also true if both component statements are true. This differs from the **exclusive or** common in everyday language – "either A or B" – which can be expressed as

$$\boxed{(A \vee B) \wedge (\neg(A \wedge B)),}$$

or equivalently by

$$\boxed{(A \wedge \neg B) \vee (\neg A \wedge B).}$$

These are two examples of **compound statements**; they consist of repeated combinations of the **elementary statements** A and B.

Another way to connect two statements is the **implication** "if A, then B",

$$\boxed{A \implies B\,.}$$

It is also called **subjunction**. $A \implies B$ is false if A is true and B is false, that is, when a false statement B follows from a true statement A; otherwise, $A \implies B$ is true. Here, A is called the **premise** and B the **conclusion**.

A	B	$A \implies B$
true	true	true
true	false	false
false	true	true
false	false	true

Note that, if A is false, B can be true or false; nevertheless, the implication remains true. From a false premise, anything (true or false) can be inferred!

Example F.5. Let A_x: "$x \in \mathbb{N}$" and B_x: "$x \in \mathbb{Z}$". As is well known, every natural number is an integer. Therefore, if A_x is true, then B_x is also true. If x is not a natural number (i.e., if A_x is false), then x can be an integer (e.g., $x = -3$) or not (e.g., $x = 3/2$). In any case, $A_x \implies B_x$ is true.

Numerous statements in mathematics are implications of the form $A \implies B$. In this case A is called a **sufficient condition** for B. Conversely, B is called a **necessary condition** for A. Note the difference: If A holds, then B holds as well – the occurrence of A **suffices** for the occurrence of B. Conversely, if B does not hold, A cannot hold either; thus, the occurrence of B is **necessary** for the occurrence of A.

If both $A \implies B$ and the reverse implication $B \implies A$ hold, then the statements A

and B are said to be **equivalent**, written

$$\boxed{A \iff B\,.}$$

The statement $A \iff B$ is true if and only if the truth values of A and B agree:

A	B	$A \iff B$
true	true	true
true	false	false
false	true	false
false	false	true

Instead of "A is equivalent to B", one also says "A if and only if B".

Example F.6. If $2x = 6$ is true, then $x = 3$ is also true, so $2x = 6 \implies x = 3$. On the other hand: If $x = 3$ is true, then $2x = 6$ is also true, so $x = 3 \implies 2x = 6$. Hence, $2x = 6$ is true if and only if $x = 3$ is true,

$$2x = 6 \iff x = 3\,.$$

A composite statement that is always true is called a **tautology**. A composite statement that is always false is called a **contradiction**. Whether a given composite statement is a tautology can be determined using a truth table by substituting all possible values "true" and "false" for the elementary statements it contains. A contradiction is verified in the same way.

Example F.7. The statement $A \vee \neg A$ is a tautology, while the statement $A \wedge \neg A$ is a contradiction, as shown by the following truth table:

A	$\neg A$	$A \vee \neg A$	$A \wedge \neg A$
true	false	true	false
false	true	true	false

Example F.8. The statement $\neg(A \vee B) \iff (\neg A \wedge \neg B)$ is a tautology:

A	B	$A \vee B$	$\neg(A \vee B)$	$\neg A \wedge \neg B$	$\neg(A \vee B) \iff (\neg A \wedge \neg B)$
true	true	true	false	false	true
true	false	true	false	false	true
false	true	true	false	false	true
false	false	false	true	true	true

The tautology shown in the previous example is one of the two De Morgan's laws for transforming compound statements. **De Morgan's laws** are

$$\boxed{\neg A \wedge \neg B \iff \neg(A \vee B), \qquad \neg A \vee \neg B \iff \neg(A \wedge B).}$$

Appendix G

Mathematical proofs

To prove a mathematical statement of the form $A \implies B$, three logically distinct approaches are available: the direct proof, proof by contraposition, and proof by contradiction.

Direct proof: A direct proof consists of a chain of implications leading successively from A to B.

Example G.1. One wants to show that the formula

$$\sum_{i=1}^{n} i = \frac{n(n+1)}{2}$$

holds for all natural numbers $n \in \mathbb{N}$. This statement has the form $A_x \implies B_x$ with A_x: "$x = n \in \mathbb{N}$" and B_x: "$\sum_{i=1}^{n} i = (n(n+1))/2$ with $x = n$".

Assume that the statement A_x holds, i.e., $x = n$ is a natural number. The summands $1, 2, 3, \ldots, n$ can be arranged once in ascending and once in descending order:

ascending	descending	row sum
1	n	$n+1$
2	$n-1$	$n+1$
$\vdots$	$\vdots$	$\vdots$
$n-1$	2	$n+1$
n	1	$n+1$

K. Mosler et al., *Mathematical Methods for Business and Economics*, Mathematics Study Resources 20, https://doi.org/10.1007/978-3-662-73831-3_20

Clearly, each row sum is equal to $n + 1$. Adding the n rows, one gets

$$2\sum_{i=1}^{n} i = n(n+1)\,.$$

Dividing both sides of the equation by 2 gives the statement B_x. This proves the formula above.

Proof by contraposition: The implication $A \implies B$ is equivalent to

$$\boxed{\neg B \implies \neg A\,,}$$

which is easily verified by the following truth table.

A	B	$A \implies B$	$\neg B$	$\neg A$	$\neg B \implies \neg A$
true	true	**true**	false	false	**true**
true	false	**false**	true	false	**false**
false	true	**true**	false	true	**true**
false	false	**true**	true	true	**true**

Example G.2. Consider the necessary condition for a local extremum in the interior of the domain of a differentiable function f. It states that when searching for local extrema in the interior of the domain, attention can be restricted to inner points a where $f'(a) = 0$. I.e., the condition B: "$f'(a) = 0$" is necessary for A: "having an extremum at a". The implication takes the form $A \implies B$:

$$f \text{ has a local extremum at } a \implies f'(a) = 0\,.$$

Its contraposition is

$$f'(a) \neq 0 \implies f \text{ has no local extremum at } a.$$

The **proof by contraposition** consists of proving the logically equivalent statement $\neg B \implies \neg A$ instead of $A \implies B$. To do so, assume $\neg B$ holds and constructs a chain of implications leading from $\neg B$ to $\neg A$.

Example G.3. We want to show: If n is a natural number and n^2 is even, then n is also even. Here, A_x: "$x \in \mathbb{N}$ and x^2 is even", B_x: "$x \in \mathbb{N}$ and x is even". Instead of $A_x \implies B_x$ we prove $\neg B_x \implies \neg A_x$, that is: If x is odd, then x^2 is also odd.

Assume $\neg B_x$ holds, i.e., let $x \in \mathbb{N}$ be odd. Then x has the form $x = 2m-1$ for some natural number m. It follows that $x^2 = (2m-1)^2 = 4m^2-4m+1 = 2(2m^2-2m)+1$, which is clearly odd. Thus, $\neg A_x$ holds, as required.

Proof by contradiction To prove $A \implies B$, one assumes that A holds but B does not. The proof by contradiction proceeds by deriving a contradiction from the assumption $A \wedge \neg B$. Such a contradiction consists of any statement C together with its negation $\neg C$.

Example G.4. We want to show:

$$x = 2 \implies \sqrt{x} \text{ is irrational.}$$

Assume A_x: "$x = 2$" and $\neg B_x$: "$\sqrt{x}$ is a rational." Then $\sqrt{2} = p/q$, where $p, q \in \mathbb{N}$ are coprime (i.e., the only common divisor of p and q is 1). Squaring yields

$$2 = \frac{p^2}{q^2} \iff 2q^2 = p^2 .$$

Thus, p^2 (and hence p) is even. Consequently, p^2 is divisible by 4. From $2q^2 = p^2$ it follows that q^2 (and hence q) is even. Therefore, both p and q are divisible by 2, contradicting that they are coprime. Hence, $\sqrt{2}$ is irrational. The contradictory statement is C: "p and q are coprime".

An **equivalence** $A \iff B$ between two statements is proven by successively showing both $A \implies B$ and $A \impliedby B$ (i.e., $B \implies A$). Often, however, a direct proof of the equivalence can also be conducted by constructing a chain of equivalences leading from A to B (and thus also in reverse).

Example G.5. We want to show that for every natural number n, the following holds: n is even $\iff n^2$ is even.

We have already shown the direction $A \impliedby B$ in Example G.3. The direction $A \implies B$ is proven as follows: Let n be an even natural number, then $m = n/2$ is also a natural number. It follows that $n^2 = (2m)^2 = 4m^2 = 2 \cdot 2m^2$, so n^2 is even. This proves the claimed equivalence.

The **equivalence of more than two statements**, such as $A \iff B \iff C$, can often be proven by **closed chain inference**. This consists of proving a sequence of implications that form a closed chain, for instance $A \implies B \implies C \implies A$. If these three implications hold, then clearly A, B, and C are pairwise equivalent.

Example G.6. Consider a set of N people. We are interested in two properties. Let M_1 be the subset of people who have property 1 (e.g., "male"), and M_2 the subset with property 2 (e.g., "red-haired"). The two properties are **independent** if $\frac{1}{N}|M_1| \cdot \frac{1}{N}|M_2| = \frac{1}{N}|M_1 \cap M_2|$, i.e., $|M_1| \cdot |M_2| = N|M_1 \cap M_2|$, where $|\cdot|$ denotes the number of elements in a set. Intuitively: If fifty out of a hundred people are male and ten have red hair, then the properties "red-haired" and "male" occur independently if five people are red-haired men.

We want to show that the following statements are equivalent:

a) A: $|M_1| \cdot |M_2| = N|M_1 \cap M_2|$
b) B: $|\overline{M_1}| \cdot |M_2| = N|\overline{M_1} \cap M_2|$
c) C: $|\overline{M_1}| \cdot |\overline{M_2}| = N|\overline{M_1} \cap \overline{M_2}|$
d) D: $|M_1| \cdot |\overline{M_2}| = N|M_1 \cap \overline{M_2}|$

This means: If one of the following property pairs is independent, then all others are too – male and red-haired, female and red-haired, female and not red-haired, male and not red-haired.

To prove the equivalences, start with $A \implies B$: Let $|M_1| \cdot |M_2| = N|M_1 \cap M_2|$. Then, $|\overline{M_1}| = N - |M_1|$ and

$$\begin{aligned}|\overline{M_1}| \cdot |M_2| = N|M_2| - |M_1| \cdot |M_2| &= N|M_2| - N|M_1 \cap M_2| \\ &= N|M_2 \setminus M_1| = N|\overline{M_1} \cap M_2|\,;\end{aligned}$$

thus, A implies B.

To show $B \implies C$, use the proven implication $A \implies B$ by swapping M_1 and M_2, then replacing M_1 with $\overline{M_1}$. This is allowed since $A \implies B$ holds for arbitrary subsets. Thus, statement B implies C.

Similarly, $C \implies D$ follows by replacing M_1 and M_2 in A and B with their complements $\overline{M_1}$ and $\overline{M_2}$, since $\overline{\overline{M_1}} = M_1$.

Finally, $D \implies A$ by replacing M_1 and M_2 in B and C with their complements $\overline{M_1}$ and $\overline{M_2}$. This completes the closed chain; A, B, C, and D are equivalent.

Proof by mathematical induction: Many mathematical statements have the form

$$x \in \mathbb{N} \quad \implies \quad A_x\,.$$

Such a statement A_x, intended to hold for all natural numbers $x = n$, is usually proven by mathematical induction, which consists of two parts. In the **base case**, the statement is shown to hold for $n = 1$, i.e., A_1 is true. In the **induction step**, assuming the statement holds for some $n = k$ (the **induction hypothesis**), it is proven to hold for $n = k + 1$, i.e., $A_k \implies A_{k+1}$ for any $k \in \mathbb{N}$.

This method is best understood through the ladder analogy: The base case shows one can reach the first rung ($n = 1$). The induction step proves one can always climb from rung k to the next rung $k + 1$. Thus, every rung is reachable, so the statement holds for all $n \in \mathbb{N}$.

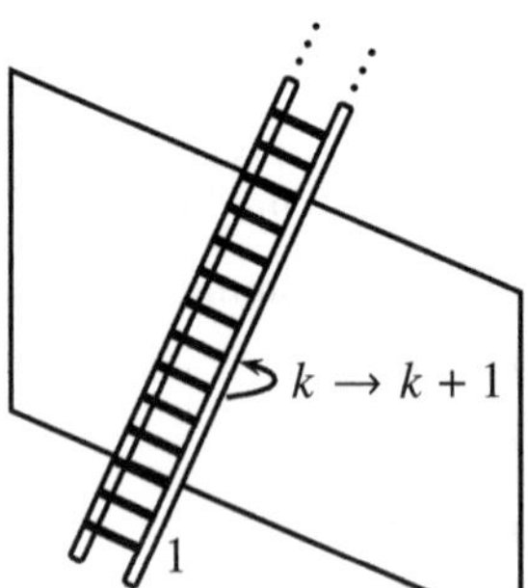

Example G.7. We want to show that the formula

$$A_n : \sum_{i=1}^{n} i = \frac{n(n+1)}{2}$$

holds for all $n \in \mathbb{N}$.

Base case ($n = 1$): For $n = 1$ the left-hand side sum is $\sum_{i=1}^{1} i = 1$, while the right-hand side is $1 \cdot (1+1)/2 = 1$. Thus, A_1 holds.

Induction step ($k \to k+1$)): Assume that A_k holds. Then,

$$\begin{aligned}\sum_{i=1}^{k+1} i &= \sum_{i=1}^{k} i + (k+1)\\ &= \frac{k(k+1)}{2} + (k+1) \quad \text{(since } A_k \text{ holds)}\\ &= \frac{k(k+1)}{2} + \frac{2(k+1)}{2}\\ &= \frac{(k+2)(k+1)}{2}.\end{aligned}$$

Hence, A_{k+1} holds. This completes the proof by mathematical induction.

Appendix H

Short solutions to the self-tests

Solutions for Chapter 1

I. T, F, F, F.

II. F, F, T, T, F, T.

III. T, F, T.

IV. T, F, F.

Solutions for Chapter 2

I. T, F, T.

II. F, T, T, T.

III. T, F.

IV. F, F, T.

Solutions for Chapter 3

I. T, T, F, T, F, F.

II. F, F, T, F.

III. T, F, T.

K. Mosler et al., *Mathematical Methods for Business and Economics*, Mathematics Study Resources 20, https://doi.org/10.1007/978-3-662-73831-3_21

Solutions for Chapter 4

I. F, F, F.

II. T, T, F.

III. F, T, F.

IV. T, F, T, F, F.

Solutions for Chapter 5

I. T, T, F.

II. T, F, T, T.

III. T, F, F, T, T, F.

IV. T, F, T.

V. T, F, T.

Solutions for Chapter 6

I. F, T, F, T.

II. T, F, T.

III. T, F, F, T, F, T, F, F.

IV. T, F.

Solutions for Chapter 7

I. T, T, F, F.

II. T, T, F, F.

III. F, T, T, T.

IV. F, T, T, F.

Solutions for Chapter 8

I. T, F, T, T.

II. F, T, T, F.

Solutions for Chapter 9

I. F, T, F, F.

II. T, F, T.

III. T, T, T.

Solutions for Chapter 10

I. T, F, F, T, T.

II. T, T, F, F.

III. F, T, T.

Solutions for Chapter 11

I. F, T, T, F, T.

II. T, T, F.

III. F, T, T, F.

Solutions for Chapter 12

I. F, T, T, F, T.

Solutions for Chapter 13

I. F, T, F, F, T.

Appendix I

Short solutions to the exercises

Solutions for Chapter 1

Solution to Exercise 1

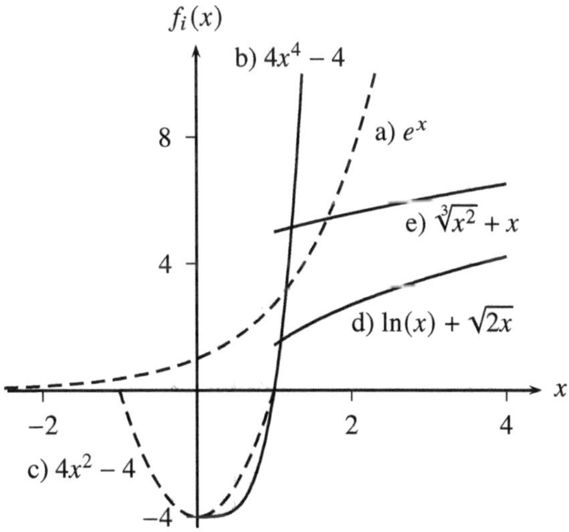

K. Mosler et al., *Mathematical Methods for Business and Economics*, Mathematics Study Resources 20, https://doi.org/10.1007/978-3-662-73831-3_22

part	image	mon. increasing	convex	bounded
a)	$\mathbb{R}_+ \setminus 0$	yes	yes	no
b)	$[-4, \infty)$	yes	yes	no
c)	$[-4, 0]$	no	yes	yes
d)	$[\sqrt{2}, \infty)$	yes	no	no
e)	$[2, 2(\sqrt[3]{2} + 2)]$	yes	no	yes

Solution to Exercise 2

a) $f_1^{-1}(y) = \sqrt[3]{4y + 8}, \quad y \in \mathbb{R}$

b) $f_2^{-1}(y) = e^{\frac{1}{2}y} - 1, \quad y \in \mathbb{R}_+$

c) Since f_3 is not strictly increasing and some function values occur more than once, it is not invertible.

Solution to Exercise 3

The isoquant for a given output level c consists of all points $(x, y) \in \mathbb{R}_+^2$ for which

$$\text{a) } y = \frac{c}{2x^2}, \qquad \text{b) } y = 2c - 4x \qquad \text{or} \qquad \text{c) } \min\{2x, 4y\} = c$$

holds. For $c = 1, 2, 3$, one obtains in parts a), b) and c), respectively:

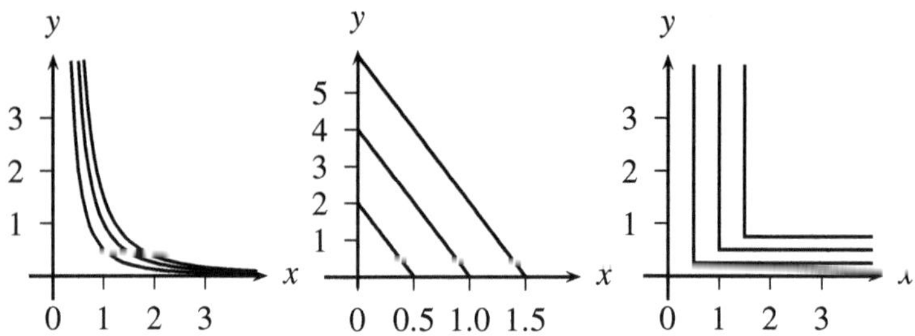

Solution to Exercise 4

The production function $f(x_1, x_2, x_3)$ [in hl] for the inputs x_1 (Pilsner malt), x_2 (wheat malt) and x_3 (hops) is

$$f(x_1, x_2, x_3) = \min\left\{\frac{1}{18}x_1, \frac{1}{4}x_2, 5x_3\right\}.$$

To produce 5 [hl] Kölsch requires 90 [kg] Pilsner malt, 20 [kg] wheat malt, and 1 [kg] hops here. The possible factor input combinations for producing 5 [hl] Kölsch are shown in the following graph:

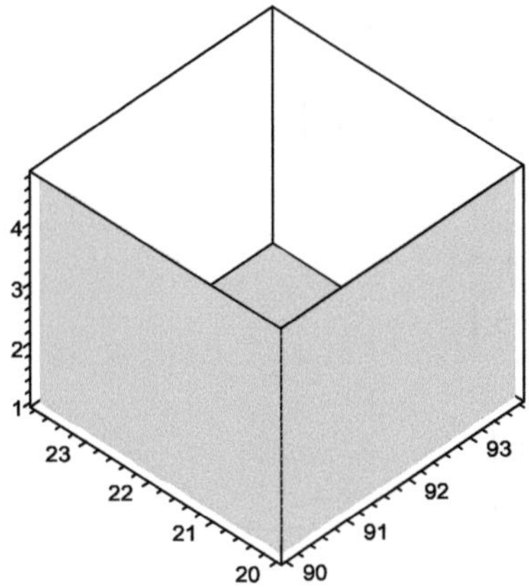

Solutions for Chapter 2

Solution to Exercise 1

a) $[-1, 15, 15, -3]^T$

b) 0

c) $\sqrt{30} \approx 5.4772$

Solution to Exercise 2

a) $[\frac{1}{2}, 2, \frac{3}{2}, 3, \frac{5}{2}]^T$

b) not defined

c) $[0, -6, -12, 18]^T$

d) not defined

e) -7

Solution to Exercise 3

a) $\boldsymbol{AB} = \begin{bmatrix} 34 & -5 & 20 \\ 33 & -3 & 19 \\ 32 & -1 & 18 \end{bmatrix}$

b) $\boldsymbol{BA} = \begin{bmatrix} 12 & 37 \\ 12 & 37 \end{bmatrix}$

c) $\boldsymbol{ABC} = \begin{bmatrix} 104 & -113 & 132 \\ 96 & -107 & 124 \\ 88 & -101 & 116 \end{bmatrix}$

d) $3CAB = \begin{bmatrix} 192 & -6 & 108 \\ -297 & 27 & -171 \\ 204 & -30 & 120 \end{bmatrix}$

e) $5de = -65$

f) $e^T A = \begin{bmatrix} -6, & 13 \end{bmatrix}$

g) $dC = \begin{bmatrix} 1, & 2, & -5 \end{bmatrix}$

h) eC is not defined.

i) $A + B^T = \begin{bmatrix} 5 & 11 \\ 3 & 4 \\ 5 & 7 \end{bmatrix}$

j) $AB + C - ed = \begin{bmatrix} 45 & -7 & 18 \\ 29 & -2 & 18 \\ 29 & -3 & 22 \end{bmatrix}$

k) $C(d^T + e) = \begin{bmatrix} -1 \\ -3 \\ 9 \end{bmatrix}$

Solution to Exercise 4

a) $(AB)^T(B^{-1}A^{-1})^T = I$

b) $(A(A^{-1} + B^{-1})B)(B + A)^{-1} = I$

c) $(I^{-1})^T A - (I^T)^{-1} A + 0 \cdot I = 0$

Solution to Exercise 5

a) $A^{-1} + B^{-1} = \begin{bmatrix} -3 & 1 \\ \frac{5}{2} & -\frac{1}{2} \end{bmatrix}$

b) $(A^{-1}B)^{-1} = \begin{bmatrix} -1 & -1 \\ 1 & \frac{3}{2} \end{bmatrix}$

c) $((A^{-1}B)^T)^{-1} = \begin{bmatrix} -1 & 1 \\ -1 & \frac{3}{2} \end{bmatrix}$

Solution to Exercise 6

a) $\|x\| + \|y\| \geq \|x + y\| = \sqrt{27}, \quad \|x\| \cdot \|z\| \geq \|x^T z\| = 4,$
$\|x\| + \|y\| = \sqrt{14} + \sqrt{13}, \quad \|x\| \cdot \|z\| = \sqrt{14} \cdot \sqrt{3}.$

b) $x^T y = 0$, so x and y are orthogonal.

Solution to Exercise 7

a) yes b) yes

Solutions for Chapter 3

Solution to Exercise 1

a) $a_n = \dfrac{1}{e^n} + 1 \longrightarrow 1$

b) $a_n = \dfrac{4 - \frac{2}{n^{\frac{8}{3}}}}{1 + \frac{5}{n}} \longrightarrow 4$

c) $a_n = -n^{\frac{1}{2}} \longrightarrow -\infty$

d) $a_n = \dfrac{1}{n!}\dfrac{1}{n+1}\dfrac{1}{2^{n-10000}} \longrightarrow 0$

e) Since $\sin(n) \in [-1, 1]$ holds for all $n \in \mathbb{N}$ and $\frac{1}{n^2}$ is a null sequence, $a_n \longrightarrow 0$.

Solution to Exercise 2

a) The monthly interest rate is $\frac{3}{12}$ % = 0.25 %. There are $60 \cdot 12$ compounding periods. The final capital is thus 4225.25 €.

b) For advance interest, the capital at the end of period i is $\widetilde{K}_i = \widetilde{K}_{i-1} + p\widetilde{K}_i$. The final capital is therefore 4244.31 €.

c) The effective interest rate for a) and b) is 3.0416 and 3.0493.

Solution to Exercise 3

Quarterly fees amount to 40 €. Thus, 800 € per quarter covers both interest and repayment. Loan repayment requires 94 quarters, or 23.5 years. Including the two-year waiting period, the loan is fully repaid after 25.5 years.

Solution to Exercise 4

It holds that $d^{-1}(q_n) = 100 - q_n$. Then $p_n = 100 - \frac{p_{n-1}^2}{100}$. Now check whether if this function is a contracting self-map: For arbitrary prices $x, y \in [0, 100]$ we have

$$\left|100 - \frac{x^2}{100} - \left(100 + \frac{y^2}{100}\right)\right| = \frac{1}{100}\left|y^2 - x^2\right| = \frac{1}{100}|y + x||y - x| \geq |y - x|,$$

if $|x+y| \geq 100$. Since $x, y \in [0, 100]$ can be chosen arbitrarily, the above function is not contracting. The following graph shows that the sequence of prices is divergent, i.e., the prices do not converge to an equilibrium price.

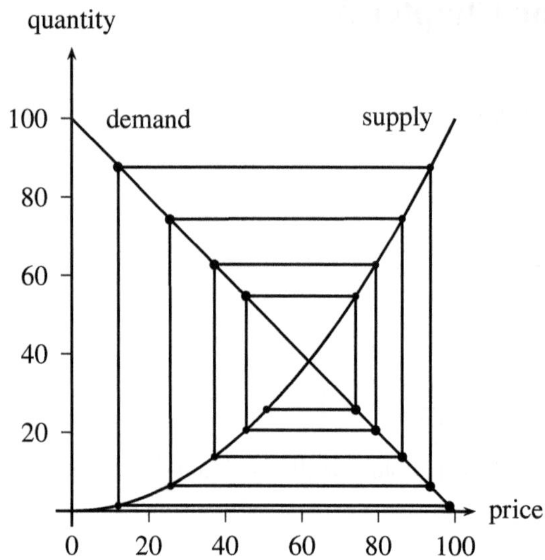

Solution to Exercise 5

f_1 is continuous for $(x, y) \neq (0, 0)$. However, at $(x, y) = (0, 0)$, f_1 is not continuous.

f_2 is continuous for $(x, y) \neq (0, 0)$. At $(x, y) = (0, 0)$, f_2 is also continuous. Thus, f_2 is continuous on all of $\mathbb{R}^2$.

Solutions for Chapter 4

Solution to Exercise 1

a) $f_1'(x) = 8x - 7 - \frac{1}{x^2}$

b) $f_2'(x) = 20x^3 - 3$

c) $f_3'(x) = 4x^3 - 3x^2 - 18x + 9$

d) $f_4'(x) = 700x^{699} + 5000x^{-5001}$

e) $f_5'(x) = 4x^3 - 64x$

Solution to Exercise 2

a) $g_1'(x) = 1$

b) $g_2'(x) = \frac{2}{x+7}$

c) $g_3'(x) = -\frac{4}{(x+5)^5}$

d) $g_4'(x) = \frac{e^x}{e^x+2}$

e) $g_5'(x) = -\frac{2e^{(x-\ln(x))}\left(1-\frac{1}{x}\right)}{(e^{(x-\ln(x))}+2)^3}$

Solution to Exercise 3

a) $h_1'(x) = 2e^{-\frac{x}{2}}(-2+x)$

b) $h_2'(x) = -e^x\left(\ln(x-2) + \frac{1}{x-2}\right)$

c) $h_3'(x) = \frac{e^x\left(\ln(x)-\frac{1}{x}\right)}{\left(\ln(x)\right)^2}$

d) $h_4'(x) = \frac{3x-4}{x^5}$

Solution to Exercise 4

a) $g_1'(x) = f'(x)e^{f(x)}$ and $g_1''(x) = f''(x)e^{f(x)} + (f'(x))^2e^{f(x)}$

b) $g_2'(x) = \frac{1}{3}f'(x)\frac{1}{\sqrt[3]{(f(x)+7)^2}}$, for all x with $f(x) \neq -7$,
$g_2''(x) = \frac{1}{3}f''(x)\frac{1}{\sqrt[3]{(f(x)+7)^2}} - \frac{2}{9}(f'(x))^2\frac{1}{\sqrt[3]{(f(x)+7)^5}}$, for all x with $f(x) \neq -7$.

c) $g_3'(x) = \frac{2f'(x)}{(1-f(x))^2}$, for all x with $f(x) \neq 1$,
$g_2''(x) = \frac{2f''(x)(1-f(x))+4(f'(x))^2}{(1-f(x))^3}$, for all x with $f(x) \neq 1$.

d) $g_4'(x) = 0$ and $g_4''(x) = 0$.

Solution to Exercise 5

The following conditions must hold: $\Pi(1000) = 0$, $\Pi(4000) = 0$, and $\Pi(0) = -1000000$. The profit function is then $\Pi(x) = -\frac{1}{4}x^2 + 1250x - 1000000$. The profit is maximized at a quantity of 2500 (hl).

Solution to Exercise 6

a) f_1 has zeros at $x = -1$, $x = 1$, and $x = 3$, a local maximum at $x = 1 - \frac{2}{\sqrt{3}}$, a local minimum at $x = 1 + \frac{2}{\sqrt{3}}$, and an inflection point at $x = 1$.

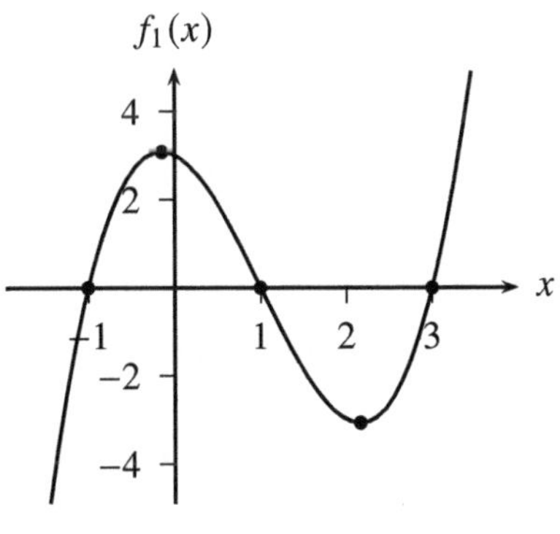

b) f_2 is not differentiable at $x = 0$, so the following considerations are made: The continuous function e^x is strictly monotonically increasing, e^{-x} is strictly monotonically decreasing. Thus,

$$e^{|x|} = \begin{cases} e^{-x} & \text{for } x < 0\,, \\ e^{x} & \text{for } x \geq 0 \end{cases}$$

is strictly monotonically decreasing on $(-\infty, 0]$ and strictly monotonically increasing on $[0, \infty)$ with a global minimum at $x = 0$. Therefore, f_2 has a global maximum at $x = 0$. The zeros of f_2 are at $x = -\ln(2)$ and $x = \ln(2)$. Further, $\lim_{x \longrightarrow -\infty} f_2(x) = -0,5$ and $\lim_{x \longrightarrow \infty} f_2(x) = -0,5$.

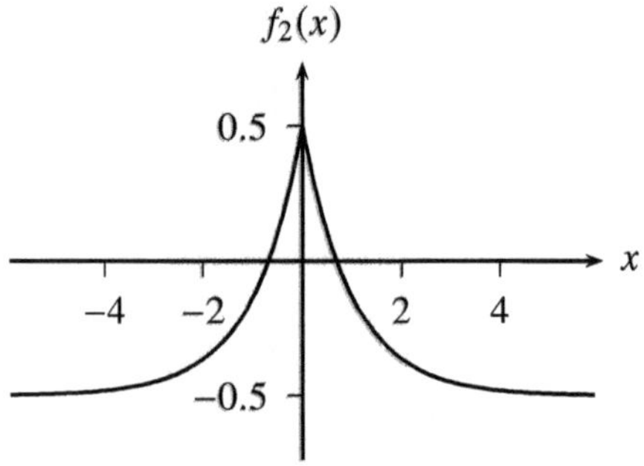

Solution to Exercise 7

a) $x_1 = -2, x_2 = -1, x_3 = 0, x_4 = 1, x_5 = 2$

b) $x_1 = -1\,, x_2 = 0\,, x_3 = 1$

c) no zeros

d) all $x \in \mathbb{R}$

e) $x_1 = -1, x_1 = 0$

Solution to Exercise 8

a) $f'(x_1) = 12 \cdot x_1^2 \cdot b \cdot c + b^2 \cdot c^2$

b) $f'(x_2) = 4 \cdot a^3 \cdot c + 2 \cdot a \cdot x_2 \cdot c^2 + \frac{\ln(c)}{d}$

c) $f'(x_3) = 4 \cdot a^3 \cdot b + 2 \cdot a \cdot b^2 \cdot x_3 + \frac{b}{d \cdot x_3}$

d) $f'(x_4) = -\frac{b \cdot \ln(c)}{x_4^2}$

Remark: In this exercise you have already determined the partial derivatives of the function

$$f(x_1, x_2, x_3, x_4) = 4x_1^3 x_2 x_3 + x_1 x_2^2 x_3^2 + \frac{x_2 \ln(x_3)}{x_4}\,,$$

$x_1, x_2 \in \mathbb{R}$, $x_3 \in \mathbb{R}_+ \setminus \{0\}$, $x_3 \in \mathbb{R} \setminus \{0\}$, see Chapter 5.

Solution to Exercise 9

a) $k'(x) = \frac{h'(x)g'(h(x))f'(g(h(x)))g(f(x))-f(g(h(x)))f'(x)g'(f(x))}{(g(f(x)))^2}$

b) $k(x) = \frac{e^{2\sqrt{x}}-4}{e^{x^2-4}}$

$k'(x) = \frac{\frac{1}{\sqrt{x}}e^{2\sqrt{x}}-2x\left(e^{2\sqrt{x}}-4\right)}{e^{x^2-4}}$

Solution to Exercise 10

a) $f_1'(x) = x^4 + 8x$

b) $f_2'(x) = 4x(x^2-1)^9(2-x^2)^5(13-8x^2)$

c) $f_3'(x) = \frac{2(1-x^2)}{(1+x^2)^2}$

d) $f_4'(x) = -3 \cdot \cos^2\big(\sin(x)\big) \cdot \sin\big(\sin(x)\big)\cos(x)$

e) $f_5'(x) = \frac{2e^{x+1}}{(e^{x+1}+1)^2}$

f) $f_6'(x) = x^x(\ln(x)+1)$

Solution to Exercise 11

a) $\varepsilon_{f_1}(x) = 17$

b) $\varepsilon_{f_2}(x) = 5x$

c) $\varepsilon_{f_3}(x) = 17 + 5x$

d) $\varepsilon_{f_4}(x) = \frac{1}{\ln(x)}$

e) $\varepsilon_{f_5}(x) = -2$

f) $\varepsilon_{f_6}(x) = -\frac{2x^2}{x^2+5}$

Solutions for Chapter 5

Solution to Exercise 1

a) $\operatorname{grad} f(x,y,z) = \left[\, 3x^2yz^2 + 4xy^2z, x^3z^2 + 4x^2yz, 2x^3yz + 2x^2y^2 + 15z^2 \,\right]$

b) $\operatorname{grad} f(x,y,z) = \left[\, -e^{-x}\cos(yz), -e^{-x}\sin(yz)\cdot z, -e^{-x}\sin(yz)\cdot y \,\right]$

c) $\operatorname{grad} f(x,y,z) = \left[\, \ln(y^2+2z^3+1), \frac{2xy}{y^2+2z^3+1}, \frac{6xz^2}{y^2+2z^3+1} \,\right]$

Solution to Exercise 2

a) $\varepsilon_{f,x}(x,y) = \dfrac{x}{xy}\cdot y = 1$ and $\varepsilon_{f,y}(x,y) = \dfrac{y}{xy}\cdot x = 1.$

b) $\varepsilon_{f,x}(x,y) = \dfrac{x}{x^2y^5}\cdot 2xy^5 = 2$ and $\varepsilon_{f,y}(x,y) = \dfrac{y}{x^2y^5}\cdot 5x^2y^4 = 5.$

c) $\varepsilon_{f,x}(x, y) = n + x$ and $\varepsilon_{f,y}(x, y) = n + y$.

Solution to Exercise 3

For the partial derivatives, we obtain:

$$\frac{\partial f}{\partial x}(x, y) = \begin{cases} \dfrac{2xy}{\sqrt{2x^2 + y^2}}, & \text{if } (x, y) \neq (0, 0), \\ 0, & \text{if } (x, y) = (0, 0), \end{cases}$$

and

$$\frac{\partial f}{\partial y}(x, y) = \begin{cases} \dfrac{2x^2 + 2y^2}{\sqrt{2x^2 + y^2}}, & \text{if } (x, y) \neq (0, 0), \\ 0, & \text{if } (x, y) = (0, 0). \end{cases}$$

Solution to Exercise 4

a) $\dfrac{\partial f}{\partial y}(x, y)\, dy = -(8xy + 3y^2)dy$

b) $df_{(x,y)} = (3x^2 - 4y^2 + 2x)dx - (8xy + 3y^2)dy$

c) $df_{(3,1)} = 29\, dx - 27\, dy$

Solution to Exercise 5

a) The total differential at the point $(x, y, z) = (80, 100, 120)$ is

$$df_{(80,100,120)}(dx, dy, dz) = 0.5574\, dx + 0.8065\, dy + 0.7235\, dz\,.$$

For $dx = dy = dz = 2$ the approximate increase in output is 4.1748.

b) f is homogeneous of degree 1, so the output increases by 30%.

Solution to Exercise 6

a) For $c > 0$ the contour lines are given by

$$\left\{ (x, y) \in \mathbb{R}^2 \,:\, y = \pm\frac{1}{2}\sqrt{x^2 - c},\ |x| \geq \sqrt{c} \right\},$$

for $c \leq 0$ by

$$\left\{ (x, y) \in \mathbb{R}^2 \,:\, y = \pm\frac{1}{2}\sqrt{x^2 - c} \right\}.$$

For the levels $c = 0, 1, 4, 9$ (black) and $c = -1, -4, -9$ (gray), respectively, one obtains the following sketch:

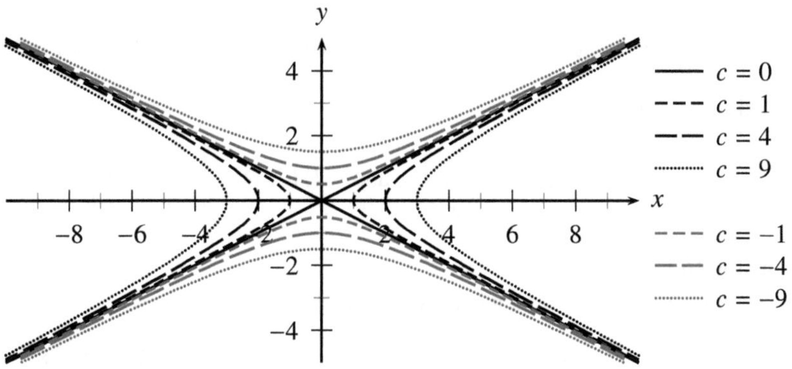

b) The tangent functions are $t_{f,(0,1)}(x,y) = 4 - 8y$ and $t_{f,(2,1)}(x,y) = 4x - 8y$.

Solution to Exercise 7

a) f is homogeneous of degree $r = 2$.

b) g is not homogeneous.

c) h is homogeneous of degree $r = 1.5$.

d) i is homogeneous of degree $r = 1$.

e) j is homogeneous of degree $r = 1$.

Solution to Exercise 8

a) f is homogeneous of degree 3.

b) The output decreases by 27.1%.

Solution to Exercise 9

a) Partial elasticities and elasticity of scale:

$$\varepsilon_{f,x} = \frac{6x^2y^3}{3x^2y^3+1}, \quad \varepsilon_{f,y} = \frac{9x^2y^3}{3x^2y^3+1}, \quad \text{and} \quad \varepsilon_{f,t} = \frac{15x^2y^3}{3x^2y^3+1}.$$

b) f is not homogeneous.

c) The output increases approximately by 9.2308%.

Solution to Exercise 10

a) $\frac{dy}{dx} = -\frac{3x^2-2y}{-2x+2y}$, b) $\frac{dy}{dx} = -\frac{ye^{xy}+2y^2}{xe^{xy}+4xy}$.

Solution to Exercise 11

a) $\frac{dz}{dt} = 4t^3 + 3t^2 + 1$, b) $\frac{dz}{dt} = 4t^3$, c) $\frac{dz}{dt} = 4te^{2t} + 4t^2e^t + 2e^{2t} + 8te^t$.

Solution to Exercise 12

$\frac{dq}{dt}/q = 0.004405$, the demand increases approximately by 0.4405% per unit time.

Solution to Exercise 13

$$J_f(x,y) = \begin{bmatrix} 2x & 1 \\ 1 & -2y \\ -y & -x \end{bmatrix} \quad \text{and} \quad J_f(2,1) = \begin{bmatrix} 4 & 1 \\ 1 & -2 \\ -1 & -2 \end{bmatrix}.$$

Solution to Exercise 14

The compositions $g \circ f$ and $g \circ g$ are not defined,

$$(f \circ f)'(0,1) = \begin{bmatrix} 0 & 4 \\ -4 & 0 \end{bmatrix}, \qquad (f \circ g)'(1,0,-1) = \begin{bmatrix} 2 & 2 & 8 \\ -4 & -4 & 4 \end{bmatrix}.$$

Solution to Exercise 15

a) The directional derivative is $\frac{\partial f}{\partial p} = \frac{3\sqrt{3}}{5}x + \frac{4}{5}$.

b) The ball rolls in the direction $\left[-\frac{\sqrt{3}}{2}, -\frac{1}{2} \right]$.

Solutions for Chapter 6

Solution to Exercise 1

There are no extrema at the stationary points $(1,0)$, $(-1,1)$, and $(-1,-1)$. At $(1,1)$ and $(1,-1)$, there are local minima, while at $(-1,0)$ there is a local maximum. There are no global extrema.

Solution to Exercise 2

The stationary points are $(0,0)$, $(1,0)$, and $(-1,0)$. At $(1,0)$ and $(-1,0)$, there are local minima. The sufficient criterion is inconclusive at $(0,0)$. Therefore, we consider $f(t,0)$ and $f(0,t)$ for t near zero and conclude that there is no extremum at $(0,0)$.

Solution to Exercise 3

The stationary points are $(0,-2)$, $(0,2)$, $(-4,2)$, and $(4,-2)$. At $(-4,2)$, there is a local maximum, while at $(4,-2)$, there is a local minimum. The other two stationary points are not extrema, and there are no global extrema.

Solution to Exercise 4

There are two local extrema: a local minimum at $(3,2)$ and a local maximum at $(0,0)$. The minimum at $(3,2)$ is even a global minimum. There is no global maximum.

Solution to Exercise 5

At all points $\boldsymbol{x}$ where $x_1 = 0$ or $x_2 = 0$ or $x_3 = 0$, f attains a global maximum.

Solution to Exercise 6

The profit is maximized at $p_1 = 49$ and $p_2 = 40$, yielding 4808 monetary units. In this case, 68 units of the first and 48 units of the second good are produced.

Solution to Exercise 7

On the domain $\mathbb{R}^2_+$, f is convex, while on $\mathbb{R}^2$, f is neither convex nor concave.

Solution to Exercise 8

a) convex, b) concave, c) neither convex nor concave.

Solution to Exercise 9

The stationary points are $(-2, -1)$, $(2, 1)$, $(-2, 1)$ and $(2, -1)$.

Solution to Exercise 10

The stationary points are $(0, \sqrt{2}, 1)$ and $(0, -\sqrt{2}, 1)$.

Solution to Exercise 11

At $(6, 8, 6)$, f attains a minimum subject to the given constraints.

Solution to Exercise 12

a) The minimum cost combination is $(x_1, x_2) = (4, 1)$.

b) The costs increase approximately by $\frac{3}{20} = 0.15$.

Solution to Exercise 13

There are four points that satisfy the KKT conditions:

$(0, 0)$ is a candidate for a minimum, $\mu_1 = 0$, $\mu_2 = 2$.
$(0, \sqrt{5})$ is a candidate for a maximum, $\mu_1 = -1/\sqrt{5}$, $\mu_2 = 0$.
$(\pm 2, 1)$ are candidates for maxima, $\mu_1 = -1$, $\mu_2 = 0$.

Checking the sufficient conditions yields:

There is a minimum at $(0, 0)$.
There is no extremum at $(0, \sqrt{5})$.
There are two maxima at $(\pm 2, 1)$.

Solution to Exercise 14

There are four points that satisfy the KKT conditions:

$(3, 2)$ is a candidate for a minimum or maximum, $\mu_1 = \mu_2 = 0$.
$(0, 2)$ is a candidate for a maximum, $\mu_1 = -27$, $\mu_2 = 0$.
$(3, 0)$ is a candidate for a maximum, $\mu_1 = 0$, $\mu_2 = -4$.
$(0, 0)$ is a candidate for a maximum, $\mu_1 = -27$, $\mu_2 = -4$.

Checking the sufficient conditions yields:

There is a local minimum at $(3, 2)$.
There is no extremum at $(0, 2)$.
There is no extremum at $(3, 0)$.
There is a local maximum at $(0, 0)$.

Solutions for Chapter 7

Solution to Exercise 1

$$F_1(x) = \frac{x^4}{4} - x^2 + 5x + c, \quad c \in \mathbb{R},$$

$$F_2(x) = -\frac{1}{x} + \frac{3}{2}\ln(|x|) - \frac{1}{3}(x+1)^3 + c, \quad c \in \mathbb{R},$$

$$F_3(x) = \frac{2}{3}(x+1)^{\frac{3}{2}} + 2x^{\frac{1}{2}} + c = \frac{2}{3}\sqrt{(x+1)^3} + 2\sqrt{x} + c, \quad c \in \mathbb{R},$$

$$F_4(x) = x - \arctan(x) + c, \quad c \in \mathbb{R}, \text{ since } f_4(x) = \frac{1+x^2-1}{1+x^2} = 1 - \frac{1}{1+x^2},$$

$$F_5(x) = e^{x^2} + c, \quad c \in \mathbb{R}.$$

Solution to Exercise 2

a) $1 - 3e^{-2} + c, c \in \mathbb{R}$ b) $\frac{1}{2} + c, c \in \mathbb{R}$ c) $\frac{1}{2}e^x(\sin(x) - \cos(x)) + c, c \in \mathbb{R}$

Solution to Exercise 3

a) 79.2 b) $\frac{1}{3}$ c) $-\frac{1}{6(x^2-6x)^3} + c, \quad c \in \mathbb{R}$

Solution to Exercise 4

a) $\frac{1}{2}$ b) -1 c) $\frac{1}{3}e^3 + e^2 + e$ d) ∞

Solution to Exercise 5

a) 1 b) $\frac{1}{2}$ c) $\frac{4}{5}$ d) 0

Solution to Exercise 6

a) $\frac{1}{2}$ b) $\frac{64}{15}$

Solutions for Chapter 8

Solution to Exercise 1

$$L(\boldsymbol{A}, \boldsymbol{0}) = \left\{ \lambda \begin{bmatrix} 3 \\ 4 \\ -1 \\ 0 \\ 0 \end{bmatrix} + \mu \begin{bmatrix} -1 \\ -1.5 \\ 0 \\ -1 \\ 0 \end{bmatrix} : \lambda, \mu \in \mathbb{R} \right\}.$$

Solution to Exercise 2

a) $L(\boldsymbol{A}, \boldsymbol{b}) = \left\{ \begin{bmatrix} 3 \\ \frac{1}{2} \\ 0 \end{bmatrix} + \lambda \begin{bmatrix} 1 \\ \frac{1}{3} \\ -1 \end{bmatrix} : \lambda \in \mathbb{R} \right\}$, b) $L(\boldsymbol{A}, \boldsymbol{b}) = \emptyset$.

Solution to Exercise 3

a) $L(\boldsymbol{A},\boldsymbol{b}) = \left\{ \begin{bmatrix} 1 \\ 4 \\ 0 \\ 0 \end{bmatrix} + \lambda \begin{bmatrix} -2 \\ 1 \\ 4 \\ -1 \end{bmatrix} : \lambda \in \mathbb{R} \right\}$,

b) $L(\boldsymbol{A},\boldsymbol{b}) = \left\{ \begin{bmatrix} 0 \\ 3 \\ 1 \\ 0 \\ 0 \end{bmatrix} + \lambda \begin{bmatrix} -1 \\ 2 \\ 0 \\ -1 \\ 0 \end{bmatrix} + \mu \begin{bmatrix} 4 \\ 1 \\ 1 \\ 0 \\ -1 \end{bmatrix} : \lambda, \mu \in \mathbb{R} \right\}$,

c) $L(\boldsymbol{A},\boldsymbol{b}) = \left\{ \begin{bmatrix} 8 \\ 0 \\ 2 \\ 0 \\ 4 \end{bmatrix} + \lambda \begin{bmatrix} -1 \\ 0 \\ 3 \\ -1 \\ 0 \end{bmatrix} + \mu \begin{bmatrix} 3 \\ -1 \\ 2 \\ 0 \\ -6 \end{bmatrix} : \lambda, \mu \in \mathbb{R} \right\}$.

Solution to Exercise 4

$\boldsymbol{A}^{-1} = \frac{1}{2} \begin{bmatrix} 1 & -1 & 0 \\ 1 & 1 & -2 \\ -1 & 1 & 2 \end{bmatrix}$, $\quad L(\boldsymbol{A},\boldsymbol{b}) = \left\{ \begin{bmatrix} -1 \\ -1 \\ 4 \end{bmatrix} \right\}$.

Solution to Exercise 5

$\boldsymbol{A}^{-1} = \begin{bmatrix} -3 & 3 & -\frac{1}{2} \\ -2 & 2 & -\frac{1}{2} \\ 3 & -\frac{5}{2} & \frac{1}{2} \end{bmatrix}$, $\quad \boldsymbol{B}$ is not invertible.

Solutions for Chapter 9

Solution to Exercise 1

a) The span is $\mathbb{R}^4$; the vectors are linearly independent.

b) The span is $\left\{ \begin{bmatrix} \lambda, \mu, \nu, 0 \end{bmatrix}^T : \lambda, \mu, \nu \in \mathbb{R} \right\}$; the vectors are linearly independent.

Solution to Exercise 2

a) yes b) no c) yes

Solution to Exercise 3

a) The vectors are linearly independent if $a, b, c \in \mathbb{R} \setminus \{0\}$.

b) The vectors are linearly dependent, regardless of the values of a, b, and c.

Solution to Exercise 4

a) $\dim \operatorname{sp}(\boldsymbol{x}, \boldsymbol{y}, \boldsymbol{z}) = 2$

b) The vectors $\{\boldsymbol{x}, \boldsymbol{y}\}$ form a basis (but also $\{\boldsymbol{x}, \boldsymbol{z}\}$ and $\{\boldsymbol{y}, \boldsymbol{z}\}$).

c) With respect to the basis $\{\boldsymbol{x}, \boldsymbol{y}\}$, it holds $\boldsymbol{z} = (-2) \cdot \boldsymbol{x} + 1 \cdot \boldsymbol{y}$.

Solution to Exercise 5

a) $\text{rk}(\boldsymbol{A}) = 4, \quad \text{rk}(\boldsymbol{B}) = 2.$

b) $\dim L(\boldsymbol{A}, \boldsymbol{0}) = 0, \quad \dim L(\boldsymbol{B}, \boldsymbol{0}) = 3.$

c) $L(\boldsymbol{A}, \boldsymbol{0}) = \{\boldsymbol{0}\}$ and

$$L(\boldsymbol{B}, \boldsymbol{0}) = \left\{ \boldsymbol{x} = \lambda \begin{bmatrix} 3 \\ 4 \\ -1 \\ 0 \\ 0 \end{bmatrix} + \mu \begin{bmatrix} -1 \\ -\frac{3}{2} \\ 0 \\ -1 \\ 0 \end{bmatrix} + \nu \begin{bmatrix} 2 \\ \frac{7}{2} \\ 0 \\ 0 \\ -1 \end{bmatrix} \; : \; \lambda, \mu, \nu \in \mathbb{R} \right\}.$$

Solutions for Chapter 10

Solution to Exercise 1

$\det \boldsymbol{A} = -48, \quad \det \boldsymbol{B} = -30, \quad \det \boldsymbol{C} = 53.$

Solution to Exercise 2

a) $\det(\boldsymbol{ABC}) = \det(\boldsymbol{A}) \det(\boldsymbol{B}) \det(\boldsymbol{C}) = 76320$

b) Since the determinants are non-zero, the matrices are invertible,

$$\det(\boldsymbol{A}^{-1}\boldsymbol{B}\boldsymbol{C}^{-1}) = \frac{\det \boldsymbol{B}}{\det \boldsymbol{A} \cdot \det \boldsymbol{C}} = 0.01179\,.$$

c) $\det\left(-\frac{1}{2}\boldsymbol{A}\right) = \left(-\frac{1}{2}\right)^3 \det \boldsymbol{A}^T = 6$

Solution to Exercise 3

$\det \boldsymbol{A} = -47, \quad \det \boldsymbol{B} = 16.$

Solution to Exercise 4

The matrix $\boldsymbol{A}$ has the triple eigenvalue $\lambda_{1,2,3} = 3$. The corresponding eigenspace is

$$V(3) = \left\{ c_1 \begin{bmatrix} 1 \\ 1 \\ 0 \end{bmatrix} + c_2 \begin{bmatrix} 0 \\ 0 \\ 1 \end{bmatrix} \; : \; c_1, c_2 \in \mathbb{R} \right\}.$$

The matrix $\boldsymbol{B}$ has the double eigenvalue $\lambda_{1,2} = 2$ and the simple eigenvalue $\lambda_3 = 1$. The corresponding eigenspaces are

$$V(2) = \left\{ c_1 \begin{bmatrix} 4 \\ -1 \\ 0 \end{bmatrix} + c_2 \begin{bmatrix} 4 \\ 0 \\ 1 \end{bmatrix} \; : \; c_1, c_2 \in \mathbb{R} \right\} \quad \text{and} \quad V(1) = \left\{ c \begin{bmatrix} 3 \\ 0 \\ 1 \end{bmatrix} \; : \; c \in \mathbb{R} \right\}.$$

The matrix $\boldsymbol{C}$ has the simple eigenvalues $\lambda_1 = 1$, $\lambda_2 = 4$ and $\lambda_3 = -1$. The corresponding eigenspaces are

$$V(1) = \left\{\boldsymbol{x} \in \mathbb{R}^3 \,:\, \boldsymbol{x} = c[0,1,-1]^T,\, c \in \mathbb{R}\right\},$$
$$V(4) = \left\{\boldsymbol{x} \in \mathbb{R}^3 \,:\, \boldsymbol{x} = c[1,0,1]^T,\, c \in \mathbb{R}\right\},$$
$$V(-1) = \left\{\boldsymbol{x} \in \mathbb{R}^3 \,:\, \boldsymbol{x} = c[1,2,0]^T,\, c \in \mathbb{R}\right\}.$$

Solution to Exercise 5

$\boldsymbol{A}$: $\Delta_1 = -9 < 0$, $\Delta_2 = 32 > 0$, and $\Delta_3 = -128 < 0$, so $\boldsymbol{A}$ is negative definite.

$\boldsymbol{B}$: $\Delta_1 = 4 > 0$, $\Delta_2 = 15 > 0$, and $\Delta_3 = 44 > 0$, so $\boldsymbol{B}$ is positive definite.

$\boldsymbol{C}$: $\Delta_1 = -2 < 0$, $\Delta_2 = 5 > 0$, and $\Delta_3 = 45 > 0$, so $\boldsymbol{C}$ is indefinite.

Solution to Exercise 6

The matrix $\boldsymbol{A}$ is positive definite if and only if $c > |d|$.

Solution to Exercise 7

For example, the matrix

$$\boldsymbol{A} = \begin{bmatrix} 28 & -12 \\ 60 & -26 \end{bmatrix}$$

has the required properties.

Solution to Exercise 8

The matrix $\boldsymbol{A}$ is not diagonalizable, as for the eigenvalue $\lambda_{1,2,3} = 3$ algebraic and geometric multiplicity do not coincide.

The matrix $\boldsymbol{B}$ is diagonalizable, as there is a double eigenvalue and a simple eigenvalue and for each eigenvalue algebraic and geometric multiplicity coincide. $\boldsymbol{S}$ can be chosen as

$$\boldsymbol{S} = \begin{bmatrix} 4 & 4 & 3 \\ -1 & 0 & 0 \\ 0 & 1 & 1 \end{bmatrix}$$

Then, $\boldsymbol{S}^{-1}\boldsymbol{B}\boldsymbol{S} = \operatorname{diag}(2,2,1)$ holds.

The matrix $\boldsymbol{C}$ is diagonalizable, as there are three simple eigenvalues and for each eigenvalue algebraic and geometric multiplicity coincide. $\boldsymbol{S}$ can be chosen as

$$\boldsymbol{S} = \begin{bmatrix} 1 & 0 & 1 \\ 2 & 1 & 0 \\ 0 & -1 & 1 \end{bmatrix}$$

Then, $\boldsymbol{S}^{-1}\boldsymbol{A}\boldsymbol{S} = \operatorname{diag}(-1,1,4)$ holds.

Solutions for Chapter 11

Solution to Exercise 1
From the sketch, one observes that $x_1 + x_2$ attains its maximum at the point $(3, 2.5)$ in the feasible region (gray).

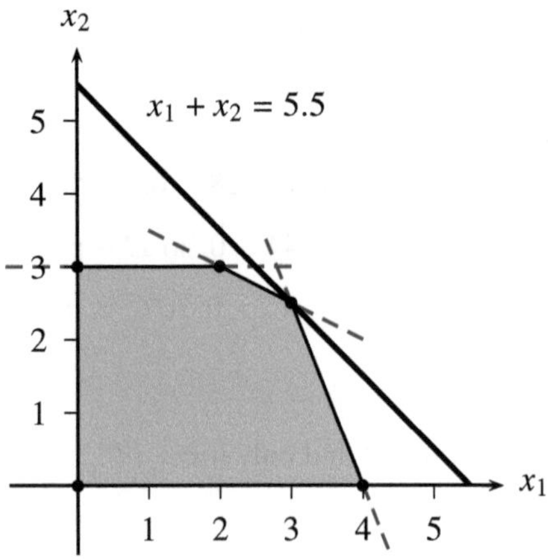

Solution to Exercise 2
The solution is $x_1 = x_2 = x_3 = 1$. The optimal value is 4 .

Solution to Exercise 3
The solution is the vertex $x_1 = 3$, $x_2 = 2.5$, $x_3 = 3$ or the vertex $x_1 = 3$, $x_2 = 4$, $x_3 = 1.5$, each with optimal value 11.5.

Solution to Exercise 4
The solution is $x_1 = x_2 = 2$.

Solution to Exercise 5
One solution is $x_1 = 10$, $x_2 = 20$. Another solution is $x_1 = 0$, $x_2 = 40$. In each case, the optimal value of the objective function is 40.

Solution to Exercise 6

a)

$$\begin{array}{rrl} \min & y_1 + y_2 & \\ \text{s.t.} & -y_1 + y_2 \leq & 1 \\ & y_1 + y_2 \geq & 7 \\ & y_1 \leq 0,\ y_2 \leq 0 & \end{array}$$

b) The feasible region is empty.

Solution to Exercise 7

$$\begin{array}{ll} \min & 12y_1 + y_2 + 5y_3 + 3y_4 \\ \text{s.t.} & 5y_1 + y_2 + y_3 + y_4 \geq 3 \\ & 7y_1 + y_2 + 2y_3 + 2y_4 \geq 4 \\ & 3y_1 + y_2 + 3y_3 + y_4 \geq 5 \\ & 2y_1 + y_2 + y_3 + y_4 = 1 \\ & y_1, y_3 \geq 0\,, y_2 \leq 0\,, y_4 \text{ unbounded} \end{array}$$

Solutions for Chapter 12

Solution to Exercise 1

For the values $p_0 = 2$ and $q_0 = 100$ one obtains

$$q(p) = 100e^{-2p+4}\sqrt{\frac{2}{p}} \qquad \text{and} \qquad q(2,5) = 32.9041\,.$$

Solution to Exercise 2

In general, the demand function is

$$q(p) = q_0\left(\frac{p}{p_0}\right)^a \exp(-b(p - p_0))$$

and for the specific values

$$q(p) = \frac{10}{p^2 2^p}\,.$$

Solution to Exercise 3

$C(q) = 2\sqrt{q}$

Solution to Exercise 4

a) General solution to the homogeneous ODE: $y = \gamma\sqrt{1 + x^2}$ with $\gamma \in \mathbb{R}$.

b) Particular solution to the nonhomogeneous ODE: $y = 2(1 + x^2)$.

c) General solution to the nonhomogeneous ODE: $y = 2(1+x^2)+\gamma\sqrt{1 + x^2}$ with $\gamma \in \mathbb{R}$.

d) Solution satisfying the initial condition: $y = 2(1 + x^2) - \sqrt{1 + x^2}$.

Solution to Exercise 5

General solution of the ODE: $y = \gamma e^{(x^2)} - \frac{1}{2}(x^2 + 1)$, $\gamma \in \mathbb{R}$.

Solution to Exercise 6

General solution of the ODE: $y = \ln\left(\frac{x^3}{27} + C\right)$, $C \in \mathbb{R}$.
Solution satisfying the initial condition: $y = 3\ln\left(\frac{x}{3}\right)$.

Solution to Exercise 7

Solution satisfying the initial condition: $y = \sqrt[3]{\frac{8.25}{x} - \frac{1}{4}x^3}$.

Solution to Exercise 8

a) General solution: $y(x) = \gamma_1 e^{0.8x} + \gamma_2 e^{0.4x}$, $\gamma_1, \gamma_2 \in \mathbb{R}$.
Solution satisfying the initial condition: $y(x) = 2e^{0.8x} - 2e^{0.4x}$.
The differential equation is not stable.

b) General solution: $y(x) = \gamma_1 \cos(2x) + \gamma_2 \sin(2x)$, $\gamma_1, \gamma_2 \in \mathbb{R}$.
Solution satisfying the initial condition: $y(x) = 4\cos(2x) - 2\sin(2x)$.
The differential equation is not stable.

c) General solution: $y(x) = (\gamma_1 + \gamma_2 x)e^{0.2x}$, $\gamma_1, \gamma_2 \in \mathbb{R}$.
Solution satisfying the initial condition: $y(x) = (4 - 2x)e^{0.2x}$.
The differential equation is not stable.

d) General solution: $y(x) = \gamma_1 e^{2x} \cos(x) + \gamma_2 e^{2x} \sin(x)$, $\gamma_1, \gamma_2 \in \mathbb{R}$.
Solution satisfying the initial condition: $y(x) = 2e^{2x} \cos(x) - e^{2x} \sin(x)$.
The differential equation is not stable.

Solution to Exercise 9

a) Particular solution: $y(x) = 25 + 1.5625\, e^{2x}$.
General solution: $y(x) = 25 + 1.5625\, e^{2x} + \gamma_1 e^{0.8x} + \gamma_2 e^{0.4x}$, $\gamma_1, \gamma_2 \in \mathbb{R}$.

b) Particular solution: $y(x) = \frac{1}{4}x^2 - \frac{1}{8}$.
General solution: $y(x) = \frac{1}{4}x^2 - \frac{1}{8} + \gamma_1 \cos(2x) + \gamma_2 \sin(2x)$, $\gamma_1, \gamma_2 \in \mathbb{R}$.

c) Particular solution: $y(x) = 3x^2 e^{0.2x}$.
General solution: $y(x) = 3x^2 e^{0.2x} + (\gamma_1 + \gamma_2 x)e^{0.2x}$, $\gamma_1, \gamma_2 \in \mathbb{R}$.

d) Particular solution: $y(x) = 2x^2 + 3.2x + 1.76$.
General solution: $y(x) = 2x^2 + 3.2x + 1.76 + \gamma_1 e^{2x} \cos(x) + \gamma_2 e^{2x} \sin(x)$, $\gamma_1, \gamma_2 \in \mathbb{R}$.

Solution to Exercise 10

a) According to the Hurwitz criterion, the differential equation is not stable.

b) According to the Hurwitz criterion, the differential equation is stable.

Solution to Exercise 11

The general solution of the homogeneous system is

$$y(x) = \gamma_1 \left(\begin{bmatrix} 1 \\ 0 \end{bmatrix} + \begin{bmatrix} 3 \\ -1 \end{bmatrix} x \right) e^{-2x} + \gamma_2 \begin{bmatrix} 3 \\ -1 \end{bmatrix} e^{-2x}.$$

Solutions for Chapter 13

Solution to Exercise 1

a) $y_t = 4 \cdot 5^t$

b) $y_t = 0$

c) $y_t = \frac{3}{5} + (-4)^t$

d) $y_t = 1 + (-3)^t$

Solution to Exercise 2

a) The equilibrium price is $\overline{p} = 20$.

b) General solution: $p_t = 20 + \gamma \left(-\frac{1}{2}\right)^t$, $\gamma \in \mathbb{R}$.

c) Starting from period 12, the distance to the equilibrium price is less than 0.01.

Solution to Exercise 3

a) Roots of the characteristic polynomial: $z_{1,2} = \pm 2$.
General solution: $y_t = \gamma_1 2^t + \gamma_2(-2)^t$, $\gamma_1, \gamma_2 \in \mathbb{R}$.
Solution satisfying the initial conditions: $y_t = \frac{1}{4}2^t - \frac{1}{4}(-2)^t = 2^{t-2} - (-2)^{t-2}$.

b) Roots of the characteristic polynomial: $z_{1,2} = \pm 2i$.
General solution: $y_t = \gamma_1 2^t \cos\left(\frac{\pi}{2}t\right) + \gamma_2 2^t \sin\left(\frac{\pi}{2}t\right)$, $\gamma_1, \gamma_2 \in \mathbb{R}$.
Solution satisfying the initial conditions: $y_t = 2^t \left[\cos\left(\frac{\pi}{2}t\right) - \sin\left(\frac{\pi}{2}t\right)\right]$.

c) Roots of the characteristic polynomial: $z_{1,2} = 0.2$.
General solution: $y_t = \gamma_1\, 0.2^t + \gamma_2\, t\, 0.2^t$, $\gamma_1, \gamma_2 \in \mathbb{R}$.
Solution satisfying the initial conditions: $y_t = 0.2^t + 9 \cdot t \cdot 0.2^t = 0.2^t(1 + 9t)$.

d) Roots of the characteristic polynomial: $z_1 = 2$, $z_2 = -3$.
General solution: $y_t = \gamma_1 2^t + \gamma_2(-3)^t$, $\gamma_1, \gamma_2 \in \mathbb{R}$.
Solution satisfying the initial conditions: $y_t = \frac{2}{5}2^t - \frac{2}{5}(-3)^t = \frac{2}{5}\left(2^t - (-3)^t\right)$.

e) Roots of the characteristic polynomial: $z_{1,2} = 2 \pm i$.
General solution: $y_t = (\sqrt{5})^t\, [\gamma_1 \cos(0.4636t) + \gamma_2 \sin(0.4636t)]$, $\gamma_1, \gamma_2 \in \mathbb{R}$.
Solution satisfying the initial conditions:
$$y_t = 2(\sqrt{5})^t\, [\cos(0.4636t) - \sin(0.4636t)].$$

f) Roots of the characteristic polynomial: $z_1 = 1$, $z_2 = 2$, $z_3 = -1$.
General solution: $y_t = \gamma_1 1^t + \gamma_2 2^t + \gamma_3(-1)^t$, $\gamma_1, \gamma_2, \gamma_3 \in \mathbb{R}$.
Solution satisfying the initial conditions: $y_t = -\frac{1}{2} + \frac{1}{3}2^t + \frac{1}{6}(-1)^t$.

g) Roots of the characteristic polynomial: $z_{1,2} = \frac{1}{2}$, $z_3 = 2$.
General solution: $y_t = \gamma_1\left(\frac{1}{2}\right)^t + \gamma_2 t\left(\frac{1}{2}\right)^t + \gamma_3 2^t$, $\gamma_1, \gamma_2, \gamma_3 \in \mathbb{R}$.
Solution satisfying the initial conditions: $y_t = -4\left(\frac{1}{2}\right)^t + 2t\left(\frac{1}{2}\right)^t + 2^t$.

Solution to Exercise 4

a) Roots of the characteristic polynomial: $z_1 = 0.8$, $z_2 = 0.4$.
Solution satisfying the initial conditions: $y_t = 25 + 8 \cdot 0.8^t + 4 \cdot 0.4^t$.

b) $\lim_{t\to\infty} y_t = 25$.

c) As in b), $\lim_{t\to\infty} y_t = 25$.

d) Yes.

Solution to Exercise 5

a) $y_t = (\gamma_1 + \gamma_2 \cdot t) \cdot 1.6^t$

b) $y_t = y^* = 5$

c) $y_t = 5 + (\gamma_1 + \gamma_2 \cdot t) \cdot 1.6^t$

d) $y_t = 5 + (8 + 2t) \cdot 1.6^t$

e) The solution does not converge but diverges to infinity.

Solution to Exercise 6

a) The solution of the difference equation is $Y_t = G\,(5 + 2 \cdot 2^t + 5 \cdot 1.2^t)$.

b) The difference equation is not stable.

Solution to Exercise 7

a) Particular solution: $y_t = 25 + 6.25 \cdot 2^t$.
General solution: $y_t = 25 + 6.25 \cdot 2^t + \gamma_1 0.8^t + \gamma_2 0.4^t$, $\gamma_1, \gamma_2 \in \mathbb{R}$.

b) Particular solution: $y_t = \left(\frac{1}{2}t + \frac{1}{4}t^2\right) 2^t$.
General solution: $y_t = \left(\frac{1}{2}t + \frac{1}{4}t^2\right) 2^t + \gamma_1 2^t + \gamma_2(-2)^t$, $\gamma_1, \gamma_2 \in \mathbb{R}$.

c) Particular solution: $y_t = \frac{1}{2}t^2 0.2^t$.
General solution: $y_t = \gamma_1\, 0.2^t + \gamma_2\, t\, 0.2^t + \frac{1}{2}t^2\, 0.2^t$, $\gamma_1, \gamma_2 \in \mathbb{R}$.

d) Particular solution: $y_t = 0.4\, t\, 2^t$.
General solution: $y_t = \gamma_1 2^t + \gamma_2(-3)^t + 0.4\, t\, 2^t$, $\gamma_1, \gamma_2 \in \mathbb{R}$.

e) Particular solution: $y_t = 15 + 6t + t^2$.
General solution:
$$y_t = 15 + 6t + t^2 + (\sqrt{5})^t \left[\gamma_1 \cos(0.4636t) + \gamma_2 \sin(0.4636t)\right],\ \gamma_1, \gamma_2 \in \mathbb{R}.$$

Solution to Exercise 8

a) Not stable. b) Not stable. c) Stable. d) Not stable.
e) Not stable. f) Not stable. g) Not stable.

Solution to Exercise 9

a) Necessary stability criterion (Smithies): not stable.
b) Sufficient stability criterion (Smithies): stable.
c) Sufficient stability criterion (Sato): stable.

Solution to Exercise 10
General solution:

$$\begin{bmatrix} x_t \\ y_t \end{bmatrix} = \gamma_1 3^t \begin{bmatrix} 1 \\ 1 \end{bmatrix} + \gamma_2 (-2)^t \begin{bmatrix} 4 \\ -1 \end{bmatrix} .$$

Solution satisfying the initial conditions:

$$x_t = 1.2 \cdot 3^t + 0.8 \cdot (-2)^t ,$$
$$y_t = 1.2 \cdot 3^t - 0.2 \cdot (-2)^t .$$

Solution to Exercise 11

a) General solution:

$$\mathbf{y}_t = \gamma_1 \cdot 3^t \begin{bmatrix} 1 \\ -1 \\ 1 \end{bmatrix} + \gamma_2 \cdot 2^t \begin{bmatrix} 2 \\ 0 \\ 1 \end{bmatrix} + \gamma_3 (-1)^t \begin{bmatrix} 0 \\ -1 \\ 1 \end{bmatrix} .$$

b) Solution satisfying the initial condition:

$$\mathbf{y}_t = \begin{bmatrix} -3^{t+1} + 2^{t+2} \\ 3^{t+1} - 2(-1)^t \\ -3^{t+1} + 2^{t+1} + 2(-1)^t \end{bmatrix} .$$

Solution to Exercise 12
A particular solution is

$$\mathbf{y}_t = \begin{bmatrix} -4 \\ 7t \\ -1 - 5t \end{bmatrix} .$$

Selected textbooks

- **Elementary undergraduate:**

  Asano, A. (2013). *An Introduction to Mathematics for Economics.* Cambridge University Press, Cambridge.

  Rosser, M. J.,, and Lis, P. (2025). *Basic Mathematics for Economists* (4th ed.). Routledge, London.

- **Intermediate undergraduate:**

  Allen, R. G. D. (1938). *Mathematical Analysis for Economists.* Macmillan, London.

  Jacques, I. (2023). *Mathematics for Economics and Business* (10th ed.). Pearson, Harlow.

  Sydsæter, K., Hammond, P., Strøm, A., and Carvajal, A. (2026). *Essential Mathematics for Economic Analysis* (7th ed.). Pearson, Harlow.

  Werner, F., and Sotskov, Y. N. (2006). *Mathematics of Economics and Business.* Routledge, London.

- **Advanced undergraduate/early graduate:**

  Chiang, A. C., and Wainwright, K. (2005). *Fundamental Methods of Mathematical Economics* (4th ed.). McGraw-Hill, New York.

  Hoy, M., Livernois, J., McKenna, C., Rees, R., and Stengos, T. (2022). *Mathematics for Economics* (4th ed.). MIT Press, Cambridge.

  Pemberton, M., and Rau, N. (2023). *Mathematics for Economists: An Introductory Textbook* (5th ed.). Manchester University Press, Manchester.

  Simon, C. P., and Blume, L. (1994). *Mathematics for Economists.* Norton, New York.

K. Mosler et al., *Mathematical Methods for Business and Economics*, Mathematics Study Resources 20, https://doi.org/10.1007/978-3-662-73831-3

- **Graduate/PhD:**

CARTER, M. (2001). *Foundations of Mathematical Economics*. MIT Press, Cambridge.

DE LA FUENTE, A. (2000). *Mathematical Methods and Models for Economists.* Cambridge University Press, Cambridge.

FLEMING, W. (1977). *Functions of Several Variables* (2nd ed.). Springer, New York.

SEIERSTAD, A., and SYDSÆTER, K. (1987). *Optimal Control Theory with Economic Applications*. North-Holland, Amsterdam.

SYDSÆTER, K., HAMMOND, P., SEIERSTAD, A., and STRØM, A. (2008). *Further Mathematics for Economic Analysis* (2nd ed.). Pearson, Harlow.

VOHRA, R. V. (2005). *Advanced Mathematical Economics*. Routledge, London.

Index

K. Mosler et al., *Mathematical Methods for Business and Economics*, Mathematics Study Resources 20, https://doi.org/10.1007/978-3-662-73831-3

E

F

G

H

T

MIX
Papier aus verantwortungsvollen Quellen
Paper from responsible sources
FSC® C105338

If you have any concerns about our products,
you can contact us on
ProductSafety@springernature.com

In case Publisher is established outside the EU,
the EU authorized representative is:
Springer Nature Customer Service Center GmbH
Europaplatz 3, 69115 Heidelberg, Germany

Printed by Libri Plureos GmbH
in Hamburg, Germany